A TEXTBOOK
OF
ADVANCED PRACTICAL PHYSICS

A TEXTBOOK
OF
ADVANCED PRACTICAL PHYSICS

[Covering Complete Honours Syllabi of All Indian Universities]

[Revised and Enlarged Fourth Edition]

Samir Kumar Ghosh MSc, PhD

Professor, Department of Physics (Retd.)
Visva-Bharati University, Santiniketan, WB

Author of

A Textbook of Practical Physics (General Course)
Numerical Problems in Physics (Hons Course)
Vyabaharic Padarthavidya (General Course)
Paribarti Prabaha (a.c.) (Hons Course)
Padarthavidyar Ruprekha (H.S. Course)
Joint Entrance Physics Questions and Answers
University Physics (Engineering Course)
Physics Quiz (Bengali)

New Central Book Agency (P) Ltd

LONDON

HYDERABAD ERNAKULAM BHUBANESWAR

NEW DELHI KOLKATA PUNE GUWAHATI

NCBA

REGD OFFICE

8/1 Chintamoni Das Lane, Kolkata 700 009, India
email: ncbapvtltd@ncbapvtltd.com

OVERSEAS

NCBA (UK) Ltd, 149 Park Avenue North, Northampton, NN3 2HY, UK
email: ncbauk@yahoo.com

EXPORT

NCBA Exports Pvt. Ltd, 212 Shahpur Jat, New Delhi 110 049
email: ncbaexports@ncbapvtltd.com

BRANCHES

2/27 Ansari Road, 1st Floor
Daryaganj, **New Delhi** 110 002
email: ncbadel@ncbapvtltd.com

House No. 3-1-315, 1st Floor
Nimboliadda, Kachiguda, **Hyderabad** 500 027
email: ncbahydb@ncbapvtltd.com

Shop Nos. 3 & 4, Vinayak Towers
681/B Budhwar Peth, Appa Balwant Chowk
Pune 411 002
email: ncbapune@ncbapvtltd.com

GSS Shopping Complex, 1st Floor
Opposite College Post Office, Convent Road
Ernakulam 682 035
0484-4021054 / 53

Shop No. 15, Bharati Towers, Ground Floor
Forest Park
Bhubaneswar 751 001
email: ncbabub@ncbapvtltd.com

Radhamayee Complex [Second Floor]
Opp. Axis Bank, Panbazar, Jaswanta Road
Guwahati 781 001
email: ncbaguwahati@ncbapvtltd.com

A TEXTBOOK OF ADVANCED PRACTICAL PHYSICS ● Samir Kumar Ghosh

First Published: September 1985
Second Edition: April 1992
Third Edition: January 1997
With Additions: July 2000
Fourth Edition [Thoroughly Revised and Enlarged]: May 2008
Reprinted: April 2013, January 2016, October 2019

PUBLISHER AND TYPESETTER
New Central Book Agency (P) Ltd
8/1 Chintamoni Das Lane, Kolkata 700 009

PRINTER
New Central Book Agency (P) Ltd
Web-Offset Division, Dhulagarh, Sankrail, Howrah

TECHNICAL EDITOR
Dipan Roy

PROJECT TEAM
Sutapa Kar, Prabhat Jas, Joydeep Ghosh, Partha Biswas, Kushaldev Bhattacharya, Bikash Mukherjee and Suman Dutta

ISBN: 978 81 7381 243 9

Price ₹ 485.00

Dedicated
to my revered and esteemed
teacher
Late Dr D. P. Ray Chaudhuri
who taught me first to
appreciate the
charms and to
unravel the mysteries of
Experimental Physics

Contents

Chapter 11 : APPENDIX **526-536**

PREFACE TO THE FOURTH EDITION

The fourth edition of *A Textbook of Advanced Practical Physics* is brought out in a completely revised, enlarged, and updated form in accordance with the newly revised syllabus of Calcutta University effective from this academic year. All the new experiments introduced by Calcutta University have been incorporated in this revised edition but, at the same time, all the experiments contained in the previous edition have been retained in order to cater to the needs of the students of other universities.

The merit of this edition, however, will be best judged by the students concerned for whom it is meant.

I am very much indebted to Prof. Bikash Ch. Pal of Asutosh College, who has helped me a lot in preparing this revised edition. I am also thankful to Dr C. M. Pal of R. K. Mission Vidyapith, Belur; Dr N. R. Chakrabarty of Jamshedpur Cooperative College; Dr Nutan Chandra of Jamshedpur Women's College; and Prof. (Dr) Dipali Bannerjee, Head of the Department of Physics, Bengal Engineering and Science University, Sibpore, for their help and kind cooperation. My sincere thanks and gratitude flow to them.

Finally, I convey my sincere thanks to Messrs New Central Book Agency (P) Ltd, Kolkata, for their keen interest in bringing out this revised edition and in promoting the book in the market to reach the larger physics-aficionado student community.

"Paramanu" (Atom)
Purbapalli (N)
Santiniketan
Poila Boishakh, 1415 Bangabda
14 April 2008

Samir Kumar Ghosh

PREFACE TO THE FIRST EDITION

The methods of experiment used in physical investigations become more varied at the same time as our knowledge of physical facts and laws grows more extensive. But little conscious effort is made to draw up the practical curriculum accordingly. The common purpose of all education is the development of the ability to think, but with our eagerness to help the students in the laboratory, we usually leave them never in need of thinking. In the public estimation of the gigantic achievements of physics ... the queen of all sciences ... nourished on the many admirable popular expositions, there is a natural tendency to view the achievements resulting primarily from the speculative genius of a few great thinkers. It is easy to forget the innumerable achievements of precise experimentations, of subtle instrument design and of supreme manipulative skill, which have furnished the facts on which the great theoretical generalisations are founded. The craft and art of an experimental scientist, in fact, form the basis of modern science and, so, of our knowledge of the world we live in.

The preface of a laboratory handbook of Physics is, of course, no place for a sermon but during my long thirty years' experience as a Physics teacher, and from my close association with the laboratories, it has been felt keenly, the gradual deterioration which a disinclination to think inevitably brings in its wake. The teachers only, I feel firmly, can turn the tide by their concerted efforts and by impressing on the students that the laboratory work may be made highly commendable or entirely worthless by the attitude the students take towards it.

Keeping these ideas in view and to cope with the existing dearth of a good book on Practical Physics for the Honours students, I had decided to undertake the responsibility of bringing out this volume with a view to catering to the needs of the Honours students of different Indian universities. It has been written in accordance with the revised syllabi of different universities, particularly the experiments newly included in the syllabus of the Calcutta University. Emphasis has been made on the systematic and detailed instructions about how to perform the experiment, how to tabulate the data properly and precautions need to be adopted.

An experiment has little value unless the student can judge the accuracy of his result. The importance of recognising and minimising personal and instrument error is emphasised at the end of each experiment together with the estimation and validation of the reliability of the end results. In this edition, an estimate of errors, likely to arise from the circumstances under which the experiment is performed, is given for each experiment and suggestions are made as to the most effective way in which theory and practice can be brought into the closest accord. To further these ends, additional sections on 'Errors' and 'Degree of Accuracy' are included in Chapter 1 for guiding the students to calculate themselves the errors in their performed experiments. A test of the students' knowledge, as well as the details of the intricacies of the experiment, are given at the end of each experiment in the form of 'Questions and Answers'.

I am glad to record my gratitude to my many well-wishers, friends, and students who had encouraged me in writing this book, to some of my ex-colleagues at Asutosh College, Calcutta, on whose knowledge and time I have drawn for advice and criticism, and to my wife and son who have given not only sympathy but very active assistance.

I shall be failing in my duties if I do not acknowledge my indebtedness and express my thanks to Sri J. N. Sen of New Central Book Agency (P) Ltd, Kolkata, but for whose assistance, cooperation, and keen interest this volume could not be marketed in such a short time.

Lastly, I shall be grateful to all my fellow colleagues if they kindly call my attention to mistakes or obscurities, if any, in this very first impression. I believe, however, that all my labour will be amply rewarded if this book proves its worth not only to the students for whose creative minds it is ultimately meant but also to their teachers of this ever exciting subject.

Samir Kumar Ghosh

"Paramanu" (Atom)
Purbapalli (N)
Santiniketan
15 August 1985

Chapter 1
Introduction

Science may be defined as the systematic investigation of the physical world and the first and the fundamental steps to ascertain the facts in connection with a problem under investigation and experimentation. A scientific experiment is a delibrately arranged series of events intended to reveal as clearly as possible with some regularity in the behaviour of objects—a so-called natural law. Foundation of science, particularly of Physics, is therefore experiment-based and when an experiment is performed, we cannot rely too much on our senses of organs to reach the accuracy in our observations. Many scientific discoveries, it is true, have been made in the past with what we now call the crudest of apparatus and equipments, but it is the development of precision instruments and apparatus during the last few decades that has led to discoveries of many scientific phenomena of extreme practical importance and verification of several theoretical postulates. Scientific work, therefore, requires making of measurements and observations with the greatest possible accuracy and reliability. Lies in this the importance of the study of experimental Physics.

1.1 Importance and Object of an Experiment

Laboratory experiments not only help us in verifying certain theoretical statements and laws but also enable us to frame laws that conform to the data obtained as a result of such experiments. In quantitative experiments, however, exact measurements are necessary and in general, experiments in Physics are mainly designed to achieve definite quantitative results; while in qualitative experiments, the scope is restricted only to the observations of sequences and relations of phenomena. Primary objects of an experiment in general are :

(i) Determination of a physical constant, such as Stefan's constant; Mechanical Equivalent of Heat, etc.

(ii) Measurement of a quantity related to a particular body, such as Young's Modulus of a material, focal length of a lens, etc.

(iii) Comparison of two quantities : such as e.m.f.'s of two cells; resistance of two wires, etc.

(iv) Testing and verification of qualitative laws : such as Stokes' Law, Jurin's Law, etc.

In fact, the general laws of Physics have already been verified and physical constants have also been determined much more accurately than what can be expected of a new experimentalist in the laboratory. One might then ask, 'What is the use of doing such experiments in the laboratory again?' The answer will be that a laboratory experiment affords an opportunity to the students to familiarise themselves with various instruments which are taught in their classes and which find applications in industry, engineering and in everyday life. It further develops in them a habit of doing things regularly, systematically and efficiently. In the laboratory, a student gets a training of these physical senses and create in themselves a scientific outlook. The work in the laboratory, in fact, trains a student in the habit of patience, power of observation, calm reasoning and systematisation, which enable him to think and act with precision and accuracy in his future life. Therein lies the success of a good practical training.

1.2 General Instructions

Before performing an experiment, the student must be clear in his mind as what he is going to do. He must never be in a hurry nor should he be casual in his work. He must have a sound and good theoretical knowledge of his experiment in order to achieve the success. This enables him to understand the reason of every step in his experiment. The habit of mental prejudice and his carelessness may spoil the whole experiment. Therefore, while performing an experiment, the following points must be borne in mind.

(a) Background knowledge of the theory of the experiment must be possessed.

(b) The working principle of all the instruments and apparatus used in a particular experiment must be known.

(c) The apparatus must be properly arranged and set, so that readings can be conveniently taken.

(d) A definite plan of performing the experiment must be made before starting the actual experiment.

(e) As many readings as possible, of the same quantity under altered conditions, should be taken in order to reduce the random error in the measurement. (See. **1.6a**)

(f) Repeat the experiment under suitably altered conditions in order to eliminate any systematic error that may creep in. (See. **1.6b**)

(g) All data recorded must be entered in a tabular form under suitable headings and with proper units. These should be done according to the order and sequence in which they are taken.

(h) Physical quantities involved should be entered up to an accuracy permitted by the particular instrument with which they are measured. No advantage is gained by using more numerical digits in expressing the value of the measured quantity. (See **1.7**)

(i) It is very important to have a clear notion of the different sources of error that may creep in the process of recording data. This enables to have an idea as to where to pay the greatest attention in order to get the best result with least labour (See Errors **1.6**)

(j) In expressing the final result, emphasis should never be made to give to it an accuracy higher than that in the quantity which has the least accuracy.

1.3 How to perform an experiment and to record data

All the instruments and apparatus required for the particular experiment should be collected before starting the experiment and those should be arranged on the working table in such a manner that the parts which are to be read or adjusted, lie at a convenient place on the table, so that it becomes easy to handle them or to make observations. Physical instruments and apparatus, particularly the electrical ones, should be handled very carefully, since careless handling or lack of experience, may cause a great damage to them. While making observations, one should keep one's mind free from any prejudice and all the readings taken, correct or incorrect, must be record. Even if the observations seem to be wrong, they should never be set right by manipulation, since all the credit gained is lost if that is detected.

Date and name of the experiment should be entered on the top of the record book. Circuit diagrams for electrical experiments must be drawn before starting the experiment. All observations must be recorded in tabular forms under proper headings and in the sequence in which they are observed. Since all measurements are affected by errors of observation, one should not, in general, rely upon a single measurement for the value of the quantity. Several measurements should be made and their arithmetic mean will give the most probable value of the quantity measured. Only the mean value should never be recorded but all the observations taken should be recorded in the notebook. A wrong entry if made, should be deleted neatly by a line and the correct one entered above or below it. Overwriting should never be made in recording data. In any experiment, divergences from the mean value are not due to errors of observation only, but represent real differences in the quantities measured. For example, diameter of a wire when measured at different places, difference in the values may be due to non-uniformity of the wire. Special care, however, must be taken in mesuring those quantites which have great significances on the result.

Recording of observations and data should be directly entered in the practical notebook in the same order in which they have been taken during the course of the experiment. *Recording of data on loose sheets, should never be made.*

1.4 Calculation of Results

Having recorded all the data and observations, calculations may be made by applying the appropriate formula. Logarithmic tables should always be used in making calculations and all the calculations must be shown legibly on the left side of the margin. Calculations must never be done at the back page or elsewhere, since all of them constitute a part of the experiment and they are subject to inspection and verification. Care should be taken not to carry out calculations beyond what is justified by the least count of the apparatus. All quantities, including the final result, should be provided with proper units, if any. The percentage error in the result must also be calculated and shown.

It should always be borne in mind that in performing laboratory experiments, intelligence and honesty in work is more important than good results.

1.5 Laboratory Notebook

A special laboratory notebook has to be maintained in Physics practical, particularly from the examination point of view. It should contain a neat and a systematic record of each of the experiments with a list of contents at the beginning. A neat and well-maintained note book creates a very good first-hand impression of the student. Accordingly, the record of each experiment should be made, as far as possible, carefully and neatly, preferably in the following scheme :

(i) Date and title of the experiement.

(ii) Necessary theory of the experiment in brief with related formulae.

(iii) List of apparatus required for performing the experiment.

(iv) Record of observations in tabular forms.

(v) Calculations and results with units, if any.

(vi) Graphs drawn, if any.

(vii) Sources of errors and necessary precautions taken to minimise errors.

(viii) Percentage error.

1.6 Errors in Physical Measurements

While performing an experiment, it is important to have a clear idea of the various sources of errors that may creep in. The word 'error', however, does not carry any sense of blame. It is not a mistake. In repeated observations of a quantity, which is in principle a constant, it is often found that slightly different value of that quantity may be obtained. By the term *'error'* is meant, the difference between the true value of a quantity and the best measured value.

The errors may arise from sources which are unknown and uncontrollable. The errors normally involved in the measurements of the quantities in the formula, may be broadly classified as (a) Random errors and (b) Systematic errors.

(a) Random Errors

Errors of the type, which are beyond the control of the observer under the given conditions of a measurement, are known as *random* or *accidental errors*. They are ascribed to unknown factors and might be due to (i) small changes in the condition of the experiment or (ii) the incorrect judgement of the observer in taking different observations. The resulting error is a matter of chance.

Positive and negative errors are, therefore, equally probable but large random errors are less likely to occur than small ones, the errors being due to chance. If we take a large number of observations of the same quantity, then it is likely that the majority of observations may involve small errors. Since these errors have equal chances of being positive or negative, they can be made minimum by taking the arithmetic mean of large number of observations of the same quantity. The arithmetic mean, of course, will be very close to the correct result. If one or two observed readings differ appreciably from the rest ones, then they should be rejected in finding the arithmetic mean.

In order to know how the arithmetic mean is related to the 'true' value, an error known as, *'probable error'* is used. This is expressed by the error of the measurement which is the amount by which the true values may differ. If A is the arithmetic mean of a set of measurements, and is the *probable error*, it implies that the true value is as likely to be within the range $A \pm a$ as outside it.

In calculating the probable error, the following procedures may be adopted :

(i) The arithmetic mean is calculated.

(ii) The difference between each observed value and the arithmetic mean is calculated. It is known as deviation (δ).

(iii) Ignoring the signs, the average value (d) of the deviation is calculated.

(iv) The average deviation (d) is divided by $\sqrt{n-1}$, where n is the number of observations. The quotient so obtained is α, the average deviation of the mean.

(v) The final probable error is 0.8 times α.

(b) Systematic Error

During experimentation, certain sources of errors operate systematically or constantly making the observed values always higher or lower than the true value. These errors are called the *systematic errors*. For example, all instrumental errors, like zero error, come under this category. To avoid this type of error, different methods are adopted in different cases. In some cases, the errors are detected beforehand and the readings are corrected accordingly. Or in some other cases, the error is allowed to occur but finally is eliminated with the help of data taken during experimentation. In some cases also, the errors are eliminated by repeating the experiment under altered condition. In fact, systematic errors may have origins of three types : (i) *personal*, (ii) *external* and (iii) *instrumental*. Of course, the personal errors are due to some bias of the observer who may have the tendency to read too low or too high. These may again be due to parallax, eye-strain or fatigueness of the observer. External errors are due to external conditions like temperature, humidity, vibration, etc. Instrumental errors are within the instruments and are mainly due to manufacturing defects or wear and tear of this instruments.

When the actual source of systematic error is known, necessary actions may be taken to eliminate it but when the source is not known definitely but its existence is apprehended, the best way to minimise the error is to repeat the observations under altered conditions.

In view of all these sources of errors, one should adopt the following methods in performing experiments :

(i) Instruments are to be checked for zero errors, if any.

(ii) Several readings of the same quantity under a given set of conditions are to be taken, in order to reduce the random errors in the instruments.

(iii) Observations are to be repeated under suitably altered conditions to detect systematic errors, if any. Care should be taken of existence of any instrumental errors.

1.7 Degree of Accuracy and Percentage Error

Since no physical measurement is always exact, the accuracy of a result is limited by the precision of the apparatus used and the skill of the observer. In all measurements, the *final* digit in the result is always *estimated* and as such becomes doubtful. This doubtful last digit, however, is in a sense significant, as it gives some informations about the quantity being measured. A *significant figure* is defined as one that is known to be reasonably trustworthy. In expressing a figure in decimal notation, there is no sense in retaining a figure following the one that is estimated. The recorded values should always be expressed with upto that decimal place, which the measuring instrement can record with the maximum degree of accuracy. For example, if a voltmeter can read up to one-hundreth of a volt, the value of e.m.f. of a cell (say, a lead storage) should be recorded as 2.00 volts and not as simply 2 volts. On the contrary, there is no sense in expressing the e.m.f. with three figures at the decimal place. In all kinds of calculations, insignificant figures should be dropped. This simplifies calculation without affecting any accuracy of the result. In calculating a mean value, retention of significant figures more than those contained in the individual values, is of no use. The mean of two individual readings (say 32.5 and 32.8) should never be written as 32.65. The second decimal figure in the calculated mean value is insignificant and hence should be dropped. While dropping the insignificant figure, the last figure is to be retained unchanged if the insignificant figure is less than 5, while if it is 5 or more, the significant figure is increased by unity. It is always to be kept in mind that the accuracy of the final result is determined by the least accuracy of the quantities involved in the computation.

The importance of an error in any recorded value lies not in the absolute value of the error but in its value relative to the observed and recorded value. The relative error is expressed as

$$\text{Proportional or Relative Error} = \frac{\text{Error}}{\text{Observed Value}}$$

$$\text{Percentage Error} = \frac{\text{Error}}{\text{Observed Value}} \times 100$$

$$= \text{Relative Error} \times 100.$$

To calculate the error, the following example may be given. Suppose, a quantity Q is related to three other quantities x, y, z by the relation $Q = k.x^{\alpha}.y^{\beta}.z^{\gamma}$, where k, α, β, γ are constants. If δQ, δx, δy and δz represent the errors in the observed quantities, then the maximum value of $\frac{\delta Q}{Q}$ is obtained by logarithmic differentiation, i.e.,

$$\left(\frac{\delta Q}{Q}\right)_{max} = \alpha\frac{\delta x}{x} + \beta\frac{\delta y}{y} + \gamma\frac{\delta z}{z}.$$

The quantities α, β, γ should be treated as positive even though in the expression for Q, one or more of the indices may be negative.

The expression $\frac{\delta Q}{Q}$, $\frac{\delta x}{x}$, etc. are the proportional or relative errors in the measurement of respective quantities and each of them multiplied by 100, gives the corresponding percentage error. Since the relative error in each quantity is multiplied by corresponding index (c.f. $\alpha.\frac{\delta x}{x}$, etc.), it follows that the quantity having the highest power must be measured with the maximum care and precision than the rest ones.

Example : To illustrate the case, let us take the expression for Young's modulus by bending of beam method. Young's modulus is given by $Y = \frac{gL^3}{4bd^3}\left(\frac{M}{x}\right) = \frac{gL^3}{4bd^3}.\alpha$ (where $\alpha = \frac{M}{x}$). As α is determined from the mean straight line curve, proportional error in α (i.e., $\frac{\delta\alpha}{\alpha}$) is not very large. The proportional error in Y is thus $\frac{\delta Y}{Y} = 3\frac{\delta L}{L} + \frac{\delta b}{b} + 3\frac{\delta d}{d}$ taking the errors in L, b and d as $+\delta L$, $-\delta b$ and $-\delta d$, so that error δY in Y may be maximum.

As L is measured by a metre scale, its maximum error may be $\delta L = 0.2$ cm. If the depth (d) and breadth (b) of the beam are measured by slide callipers, the mximum error in them may be one vernier constant which is 0.01 cm. Suppose the hypothetical experimental data are $L = 100$ cm, $b = 1.6$ cm, $d = 0.8$ cm.

Thus the maximum proportional error in Y would be

$$\frac{\delta Y}{Y} = \frac{3\times 0.2}{100} + \frac{0.01}{1.6} + \frac{3\times 0.01}{0.8}$$

$$= 0.006 + 0.0062 + 0.0375 = 0.0497.$$

This shows that the greatest contribution to the proportional error in Y is due to the proportional error in the depth d. In order to bring this error in the same order as those of other, the depth (d) of the beam should be measured by a screw gauge when the maximum error may be 0.001 cm. In that case,

$$\frac{\delta Y}{Y} = 0.006 + 0.0062 + 0.00375$$

$$= 0.01595.$$

The percentage error is thus reduced from 4.9% to 1.59%. This percentage error in Y may be further reduced by taking the menu of six values of Y—three values for three different lengths (L) and for each length, two values of Y are obtained, one for breadth (b), depth (d); another for breadth (d), depth (b)—the beam being rotated through 90°.

Thus the following principle may be adopted as a guide in recording experimental data.

"The measurement of different quantities in an experiment, should be so arranged and planned that the percentage error of each, contributes nearly the same amount to the total error in the final result."

This can be achieved by (i) repeating the readings, (ii) using measuring instruments having higher accuracy and precision and (iii) increasing, if possible, the value of the quantity to be measured.

In planning an experiment, it is often necessary to have an idea about the amount of error involved in a single measurement of a physical quantity. Generally, the error in a single observation lies within the limits stated below with the most commonly used instruments.

Slide Callipers (v.c. = 0.01 cm)	Error : 0.01 cm
Screw Gauge (l.c. = 0.001 cm)	Error : 0.001 cm
Metre Scale in mm ...	Error : 0.1 cm
Spherometer (l.c. = 0.001 cm)	Error 0.002 cm
Balance (Sensitivity 2 mg/div)	Error 2 mg
Stopwatch (0.2 sec div)	Error 0.4 sec
Thermometer (0.5° div)	Error 0.2° using eye estimation
Thermometer (0.1° div)	Error 0.05° using eye estimation

In general, errors always correspond to one small division of the measuring scale but when difference of two readings are taken or concerned, it may be as twice.

1.8 Mean Error (Deviation) and Standard Deviation

***Mean Deviation* :**

If $a_1, a_2, a_3, \ldots, a_n$ are the values of n different observations of the same quantity Q, say, then the arithmetic mean of the values of the n observations is given by

$$M = \frac{a_1 + a_2 + a_3 + \ldots\ldots + a_n}{n} = \frac{\Sigma a}{n}.$$

The error or deviation of each individual observation from the mean value (M) of the whole series is given by $\delta_1 = (a_1 - M)$; $\delta_2 = (a_2 - M)$, ... etc. If A be the arithmetic mean of the numerical values of these deviations (without considering their signs), then

$$A = \frac{\delta_1 + \delta_2 + \delta_3 + \ldots + \delta_n}{n} = \sum \frac{\delta}{n}.$$

The average or mean error is given by

$$A = \pm \sum \frac{\delta}{n} \text{ for a single resullt}$$

and $$= \pm \sum \frac{\delta}{(n\sqrt{n})} \text{ for the mean of } n \text{ results.}$$

Standard Deviation :

Instead of considering the arithmetic mean of the numerical values of the deviation sometimes the root-mean-square (r.m.s.) value of these deviations are taken. This value is known as the standard deviation or mean deviation (D).

$$\text{Hence, Standard deviation } D = \pm \sqrt{\frac{\delta_1^2 + \delta_2^2 + \ldots + \delta_n^2}{n}}$$

$$= \pm \sqrt{\frac{\Sigma \delta^2}{n}}.$$

1.9 Graphs and how to plot them

A graph which is simply a pictorial representation of the variation of one quantity with respect to another, shows more clearly than a tabulation form of records. In fact, graphs are more useful and helpful than tabulated record of data since, the graphs (i) enable us to determine the value of a quantity not available in the data recorded, and (ii) they can tell us the amount of error made while taking a particular reading. For experiments like calibration of instruments, graph drawings, in fact, are indispensable. In drawing graphs, the following rules should be followed :

(a) Independent variables should be plotted along X-axis and the dependent variable along Y-axis.

(b) The origin need not necessarily be always zero. It should simply be the nearest rounded off number less than the smallest value of the abscissa and ordinate values given.

(c) Proper choice of units of both abscissa and ordinate values should be made and those must be indicated on the graph paper. In choosing the unit, care should be taken that the larger part of the graph paper is being utilised and unit of length on the graph paper and the unit of the quantity to be plotted, should bear to each other a simple ratio.

(d) At large division marks alone each axis, the value of the quantity to which they correspond, must be put down. These values should preferably be integrs, hundreths or tenths but never bad irregularr fractions.

(e) Each pair of the variables (point) is located on the graph paper by a small dot in its proper place and surrounded by a small circle lightly drawn.

(f) The marked points are joined by the best smooth curve. One or two points may not lie on this curve but it should be drawn passing through the majority of the points. Other parts not lying on the curve, may be distributed evenly on two sides of the curve. If the graph is not a straight line, a smooth curve may be drawn passing through most of the points. But these should not be any sudden change in the curvature in the graph drawn. This may be tested by holding the graph at eye-level and looking at it tangentially.

(g) Ink should never be used in drawing graphs.

(h) If any value is determined from the graph, the point should be marked and the corresponding ordinate (or abscissa) should be drawn in broken lines.

(i) It is advisable that in drawing graphs millimetre graph papers should be preferred to inch graph papers.

Chapter 2

Some Basic Instruments and Apparatus— Their Uses and Applications

While performing physical experiments in the laboratory, it is very often required to measure the length or mass of a body or the distance between two points. The instruments which need to be frequently handled for these measurements are described below with their principles of operations and aplications.

2.1 Vernier Scale

When any length is measured by an ordinary scale, it is not possible to measure accurately any length less than the smallest division of the ordinary scale. For this, a device was made by the French mathematician, P. Vernier, by which a length smaller than even a millimetre can be measured accurately. Such a scale is known as **Vernier Scale**.

The Vernier is an auxiliary scale which is used in conjunction with a common scale, called the main scale, for measuring fractions of the smallest division of the main scale. While a linear scale, a straight vernier [Fig. 2.1 (a)] and with a circular scale, a circular vernier [Fig. 2.1(b)] are used. The principle of both is the same. If there are n divisions in a

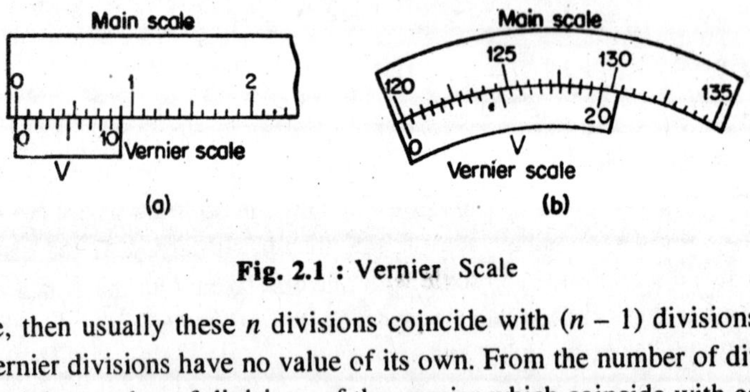

Fig. 2.1 : Vernier Scale

vernier scale, then usually these n divisions coincide with $(n - 1)$ divisions of the main scale. The vernier divisions have no value of its own. From the number of divisions of the main scale and the number of divisions of the vernier which coincide with each other, the value of the gap between one vernier division and one smallest division of the main scale is determined. The value of this gap is known as the *Vernier constant*.

Let n-divisions of a vernier scale coincide with $(n - 1)$ divisions of the main scale.

i.e., n-divisions of the vernier scale = $(n - 1)$ divisions of the main scale.

or, 1 division of the vernier scale $= \dfrac{(n-1)}{n}$ divisions of the main scale.

Thus, **Vernier constant** (v.c.) = 1 m.s.d. – 1 v.s.d.

$$= 1 \text{ m.s.d.} - \frac{n-1}{n} \text{ m.s.d.}$$

$$= \left(1-\frac{n-1}{n}\right)\text{m.s.d.} = \frac{1}{n}\text{m.s.d.}$$

So, **Vernier constant** = $\frac{1}{n}$ × value of 1 small main scale division.

Normally, all vernier scales have not the same value of vernier constants. Sometimes, 10 vernier divisions coincide with 9 main scale divisions or sometimes 20 v.s.d. with 19 m.s.d. or, 50 v.s.d. with 49 m.s.d., etc.

2.2 Diagonal Scale : Its construction and uses

It is a special type of scale which is capable of measuring a length upto 0.01 cm or 0.01 inch. To construct this scale, on the right or left of the zero mark of the main scale, another length (1 cm or 1 inch) is taken (*ABEF*). The two horizontal lines (*AB* and *EF*) of this space are further subdivided into ten equal spaces (Fig. 2.2). The points *B* and *E* are marked as zero. Similarly, the vertical side *AF* of the area *ABEF* is also divided into ten equal divisions with zero mark at *A*. From each points on the line *AF* (0, 1, 2, ...) lines parallel to the base *AB* are drawn. Then the points 0, 1, 2, ..., 9 on *AB* are joined with the points 1, 2, 3, ..., 10 of *EF* respectively giving ten diagonal lines.

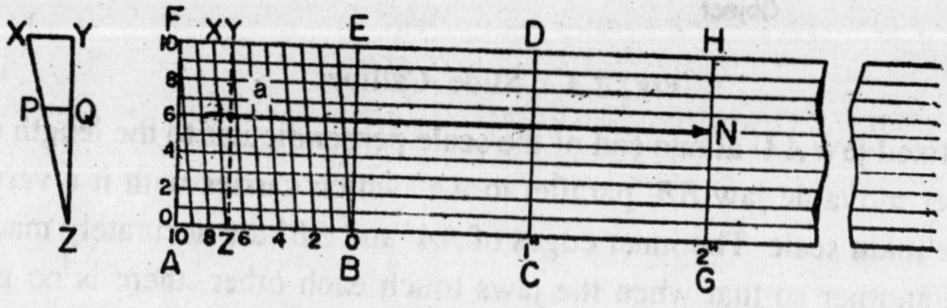

Fig. 2.2 : Diagonal Scale

Now suppose, we are to measure the length of an object which seems to be within 2 inches and 3 inches. If measured with an ordinary scale, this length is obtained up to first decimal place but if measured with a diagonal scale, it is possible to get the length up to two decimal places. At first, a pair of dividers are set to the end of the distance to be measured. Transfer the divider to the scale with one end fixed with the 2 inch mark on the scale (*G*). The other end is placed on the line *AB*. It is obvious from the figure that the measurable length (*NP*) is between 2.7 and 2.8 inches. To find the second decimal place, the divider is moved across the scale upward and parallel to itself, one end always remaining on the vertical line through *G*. The other end will describe the dotted line as drawn (*ZY*) in the figure. This end cuts the intersection point of any diagonal line (*XZ*) and a horizontal line (6 in this case from the bottom). In figure 2.2, the length *NP* is some extra portion to be added to 2.7 inches. The amount of this extra portion is determined in the following way.

From the figure,

$$\frac{PQ}{XY} = \frac{ZQ}{ZY} = \frac{0.6ZY}{ZY} = 0.6$$

∴ $PQ = 0.6\, XY$. But since, $XY = 0.1$ inch,

hence, $PQ = 0.6 \times 0.1 = 0.06$ inch.

So, the length of the distance $PN = 2.7 + 0.06 = 2.76$ inches.

2.3 Slide Calliper and its uses

This is an instrument for accurate of the thickness or small length of a substance in the form of a slab or the diameter of a round object. Measurements are done with the help of a main scale and a vernier scale.

It consists of a thin nickel plated steel scale usually graduated in centimetres and millimeters on one side and inches on the other side. The scale is known as the main scale (Fig. 2.3).

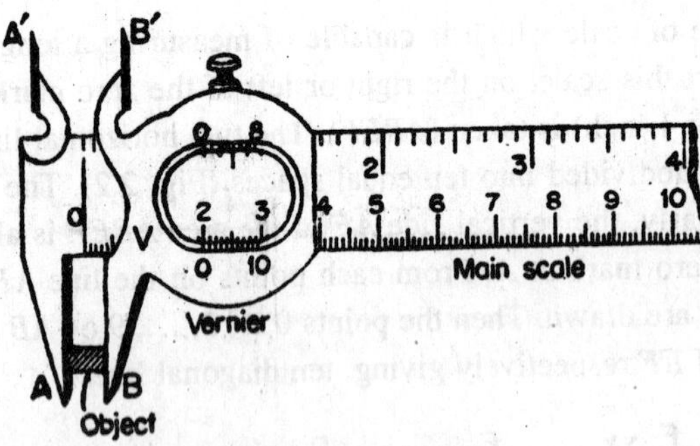

Fig. 2.3 : Slide Calliper

There is a fixed jaw AA' at one end of the scale perpendicular to the length of the scale. There is another movable jaw BB' parallel to AA' which carries with it a vernier scale V sliding over the main scale. The inner edges of AA' and BB' are accurately machined to be parallel to one another so that when the jaws touch each other, there is no gap between them. Under such circumstances, the zero of the vernier scale coincides with the zero of the main scale. If this is not so, then there is a zero error of the instrument which should be determined initially.

The body whose length or diameter or thickness is to be measured, is placed between the two jaws and the sliding jaw moved till the body is held lightly between the jaws. The readings of the main scale and the vernier scale are noted in the usual manner. The zero error of the instrument, if any, has also to be taken into account.

Instrumental Error

In a slide calliper, normally the instrumental error is due to its *zero error*. When the fixed and the movable jaws touch each other, then if the *zero* marks of the main and the vernier scale do not coincide, the instrument is then said to have *zero error*. The zero mark of the vernier may be behind or ahead of the zero mark of the main scale. If x-division of the vernier scale coincides with a certain mark of the main scale, then the *zero error* (y) will be

$$y = \pm x \times \text{(vernier constant)}.$$

This error should be positive (+) or negative (–) according as the vernier zero mark is on the right or left side of the zero mark of the main scale.

To dertermine the length of a body

The body whose length is to be determined, is placed between the jaws of the callipers and the jaws just touch the body. The reading of the main scale up to the left of the vernier zero mark is noted (S). The additional fractional length (M) beyond this reading has to be added to S. Thus, the total length (R) will be

$$R = S + M.$$

To determine the value of M, it is to be seen that how many divisions of the vernier scale have coincided with a mark on the main scale. If this number of division of vernier be n, then $M = n \times$ (vernier const.)

In this way, the length of the body is to be measured at different positions and at right angles at each point and the mean value of the readings is to be determined.

Results :

(A) Vernier constant of the instrument :

1 small main scale division = ... cm. Vernier constant = ... cm.

(B) Instrumental Error :

Error is x-divisions of the vernior, i.e., error $y = \pm x \times$ (v.c.)

No. of Obser-vation	Main Scale reading (S) cm	Vernier Reading (n)	Value of Vernier (M) cm	Total $R = S + M$ cm	Mean R cm	Instrumental Error (y) cm	Corrected length ($R \pm y$) cm
1. (a)							
(b)							
2. (a)							
(b)							
3. (a)							
(b)							
...	...	...	...	...			
...	...	...	...	...			

(a) and (b) denote readings at two mutually perpendicular directions at each place.

Calculations :

(i) Volume of a rectangular parallelopiped,

$V = a \times b \times c$ cc (a, b, c are the length, breadth and height respectively).

(ii) Volume of a sphere,

$V = \dfrac{\pi d^3}{6}$ cc (d is the diameter of the sphere).

(iii) Volume of a solid cylinder,

$V = \dfrac{\pi d^2 l}{4}$ cc (d is the diameter of the cylinder, l is the length of the cylinder).

(iv) Volume of a hollow cylinder,

$V = \frac{\pi L}{4}\left(d_1^2 - d_2^2\right)$ cc (d_1 and d_2 are the outer and inner diameters of the cylinder respectively.)

2.4 Screw-gauge (Micrometer Screw) and its uses

This is an instrument by which small lengths, thicknesses, diameters, etc. can be measured with great accuracy, even up to one-thousandth of an inch or centimetre. (Fig. 2.4).

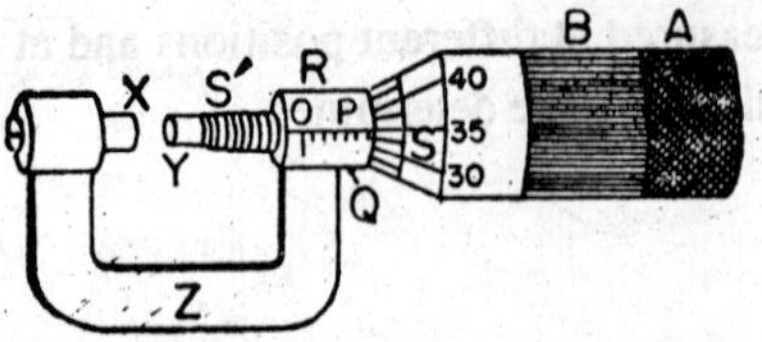

Fig. 2.4 : Screw-gauge

It consists of a U-shaped piece of steel (*Z*) having two arms. A fixed stud (*X*) is attached to one arm and the other arm (*Y*) is provided with a cylindrical tube (*R*) on which a linear scale (*P*) is attached. This is the fixed reference line which is marked with millimetre or half-millimetre scale. An accurately cut screw (*S'*), provided with a collar (*B*), works in the tube and can move axially when it is turned by a milled head (*A*). The end (*S*) of the collar is divided into a number of equal divisions (5 or 100) which behaves as the circular scale. When the ends *X* and *Y* just touch each other, the zero mark of the circular scale must coincide with the zero mark of the linear scale. Otherwise, the instrument is said to have an instrumental error which may be positive or negative.

When the screw is given a complete turn, its end moves forward by the distance between two successive threads of the screw. This distance is called the pitch of the instrument. The linear displacement of the screw for a rotation of one circular division is called the *least count of the instrument*. If for one complete rotation of the screw, the linear displacement of the screw is *P* mm, then the pitch = *P* mm.

If the circular scale has *N* divisions, then the *least count* = $\frac{P}{N}$ mm.

So, if the smallest linear scale division in a screw-gauge is 1 mm and there be 100 divisions in its circular scale, then the least count of the instrument will be $\frac{1}{100}$ mm = 0.01 mm = 0.001 cm. Thus, one-thousandth of a centimetre distance can be measured with the instrument.

Zero Error :

To find the zero or instrumental error of the instrument, the screw-head is so turned that the end *Y* touches the fixed end *X*. If in this position, the zero mark on the circular scale coincides with the reference line (*P*), then there is no zero error. On the other hand, if the zero mark crosses the line, then there is zero error. If *y* circular scale divisions are above or below the reference line, then the value of zero error will be ± *y* × (least count). If the zero mark of the circular scale is above the reference, then the error is negative and the error is to be added while, if the zero mark is below the reference, the error is positive and its value is to be substracted from the observed reading. Thus, the corrected reading

= Observed reading – (± zero error).

To determine the length or diameter of a wire

Determination of the least count :

Pitch of the screw-gauge = ... mm (P)

Number of circular scale divisions = N

Least count of the instrument = $\frac{P}{N}$

Instrumental (zero) error (e) = ... mm

= ± y divisions of circular scale.

= ± y × (least count) mm

No. of observation	Linear scale reading (N) (mm)	Circular scale reading (r)	Value of circular scale reading (r × l.c) mm	Total reading ($N + r$ × l.c) mm	Mean length or diameter (d) mm	Instrumental Error (± e) mm	Corrected length or diameter $d_1 = (d - e)$ mm
1. (a)							
(b)							
2. (a)							
(b)							
3. (a)							
(b)							
.....							
.....							
.....							
.....							

Note : Observations (a) and (b) denote readings at perpendicular directions at the same place.

2.5 Spherometer and its uses

Spherometer is an instrument which is specially designed for measuring the radius of curvature of a spherical surface or to measure the thickness of a very thin glass plate. It consists of a micrometer screw (D) moving through a nut (N) supported on three legs A, B, C (Fig. 2.5). The points A, B, C of the legs form an equilateral triangle. The tip of the screw D passes through the centre of the triangle ABC. The pitch of the screw is normally either 1 mm or 0.5 mm. The circumference of the metallic circular disc (P) is divided into 100 equal divisions. A linear vertical scale (S_1) attached to the edge of the circular scale gives the vertical displacement of the circular

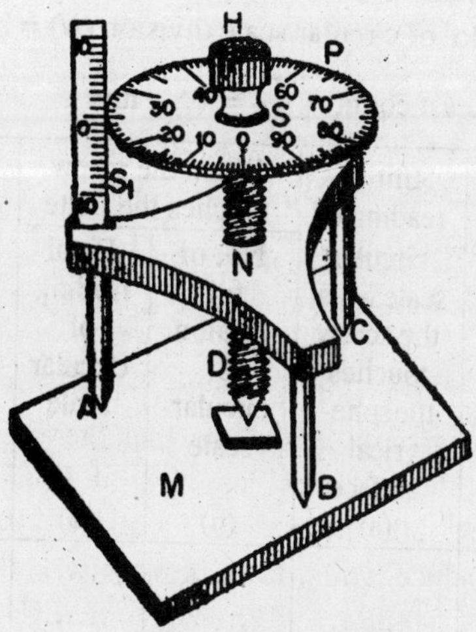

Fig. 2.5 : Spherometer

scale. Normally, when the three legs and the tip of the screw, are at the same plane, then the zero mark of the circular scale coincides with the zero of the linear scale.

To use the instrument, at first the least count of the instrurment is determined. This is done in the same manner as is done in case of a screw-gauge. The circular scale is given a complete rotation and for this, the shift of the position of the circular scale across the linear scale is noted. This gives the *pitch* of the instrument. This pitch divided by the number of the total circular scale divisions, gives the least count of the spherometer.

To measure the radius of curvature of a given spherical surface, the spherometer is placed on the spherical surface and the screw is raised or lowered, so that the tip of the screw just touches the curved surface. The reading of the circular scale is noted which is the initial reading. Then the spherical surface is removed and the instrument is placed on the plain glass plate. The screw is gradually lowered till the tip of the screw just again touches the glass plate. The exact point of touching the plate is determined by looking tangentially along the glass and finding when the gap between the tip of the screw and its image just vanishes. During lowering down the screw, the number of complete revolution which the circular scale makes, is noted. This number multiplied by the pitch, gives the value of the shift of the screw for its complete rotation. When the tip just touches the glass plate, the reading of the circular scale is again noted. This is the *final* reading of the circular scale. The initial reading of the circular scale substracted from this final reading, gives the value of the additional circular scale readings. This additional reading multiplied by the least count, gives the value of the additional circular scale readings. This value when added to the value of the complete rotation, gives the total value of the *elevation* of the spherical surface.

Now, if h be this elevation and if d be the average distance between any two legs of the spherometer, then the radius of curvature of the spherical surface is given by

$$R = \frac{d^2}{6h} + \frac{h}{2}. \qquad \text{... (2.1)}$$

Results :

Determination of the least count :

Value of one small division of the linear scale = ... mm

Pitch of the screw (p) = ... mm

No. of circular scale division (N) = ...

Least count = $\frac{p}{N}$ = ... mm

No. of Obs.	Initial reading of circular scale when the screw touches the spherical surface (a)	When the screw touches the plate: No. of full rotaton of circular scale (n)	When the screw touches the plate: Final reading of circular scale (b)	Value of complete rotation, $M = n \times$ pitch (nm)	Additional circular scale reading, $C = (b - a)$	Value of additional circular scale reading, $N = C \times$ l.c. (mm)	Value of h $(M + N)$ (mm)	Mean value of h (mm)
1.	......	......	......	......	......	......	......	
2.	......	......	......	......	......	......	......	......
3.	......	......	......	......	......	......	......	
......	......	......	......	......	......	......	......	

Distance between the outer legs of the Spherometer

Distance between A and B = ... (x) cm

Distance between B and C = ... (y) cm

Distance between C and A = ... (z) cm

Average distance between any two legs,

$$d = \frac{x+y+z}{3} = \text{......} \text{ cm}$$

Thus, the radius of curvature of the spherical surface

$$R = \frac{d^2}{6h} + \frac{h}{2} = \text{......} \text{ cm}$$

Note : While taking readings, the screw must be rotated always in the same direction, otherwise back-lash error may creep in. Readings must be taken at least at three different positions of the spherical surface.

2.6 The Balance : Its adjustments and use

The measurement of mass by means of an ordinary balance consists in balancing the weights of two masses suspended from the beam against each other so that their turning moments on a lever are equal and opposite. Balances used in the laboratory can measure weights even up to 1 milligram or less and such balances are known as physical balance.

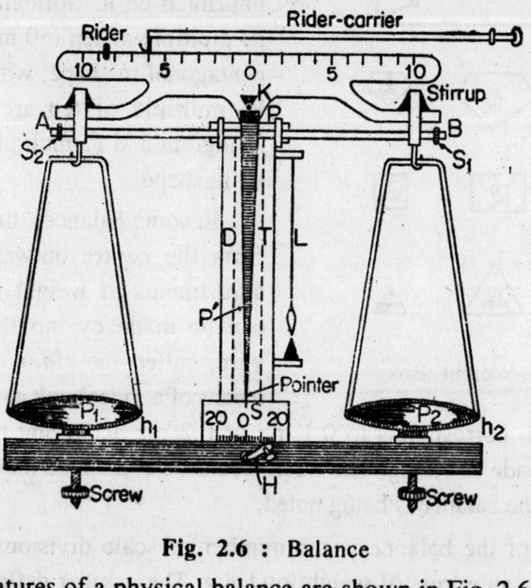

Fig. 2.6 : Balance

The essential features of a physical balance are shown in Fig. 2.6. It has a brass beam (AB) which remains horizontal and has two equal halves. At the middle of this beam, there is an agate knife-edge (K) placed on an agate plate (P). The balance can oscillate with the knife-edge as its pivot. The agate plate is fixed on a solid rod (D) placed inside a hollow pillar (T). By rotating a handle (H) at the base, the solid pillar can be raised or lowered. Two aluminium pans P_1 and P_2 are placed on two sides of the rod AB. Two triangular agate knife-edges are placed towards the end of the rod AB and at equal distances from K. Two stirrups are placed on these knife-edges. The two pans are suspended from these stirrups with the aid of hooks. A fine pointd-needle P' is suspended vertically from the mid-point of AB. Its narrow tip at the lower end can move freely on a graduated scale (S) at the base. When the balance remains horizontal, the pointer rests just at the mid-point of the scale. There are two screws S_1 and S_2 at the ends of the arm, by means of which the beam can be made horizontal. A plumbline L remains suspended across the pillar T. This plumbline helps to ascertain whether the balance pillar is perfectly vertical. The balance may be levelled with the help of levelling screws fixed at the base. When the balance is not in use, it rests on two supports (h_1, h_2). The entire balance is caged inside a glass box so that during its use, it remains unaffected by air-current. A movable lid in front of the glass case can be raised for handling the balance.

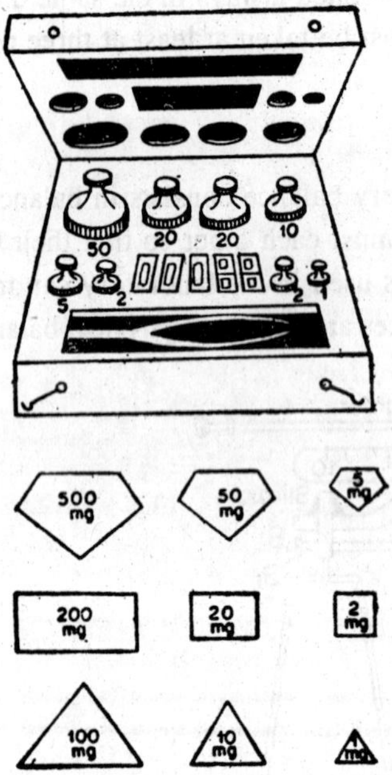

Fig. 2.7 : Weight Box

The mass of a body is determined by comparison with bodies of known masses. These known masses are contained in a box, called the weight box. It generally contains masses from 50 gm to 1 milligram (Fig. 2.7). Masses weighing 1 gm or more are made of brass while the lighter masses are of aluminium. These weights should be handled with a pair of forceps and should never be touched with hands. The mass of each is imprinted on it. Normally, 5 milligrams and its multiple of ten (50 mg, 500 mg, etc.) are pentagonal in shape; while 2 milligrams and its multiple of ten are rectangular and 1 milligram and its multiple of ten are triangular in shape.

In some balances, the beam is graduated from the centre outwards. In such cases, adjustments of weight smaller than 10 mg can be made by moving a small piece of wire, called the *rider*, along the beam by means of a rider hook and rider carriage. Each division of the beam corresponds to 0.1 mg or 0.2 mg according to mass of the rider, which is generally made of 10 mg. The rider should be at the zero mark of the beam, when the zero position of the balance is being noted.

The sensitivity of the balance is the number of scale divisions through which the pointer moves due to an excess of weight of 1 mg. The angular deflection α of a balance beam due to an excess weight w is given by

$$\tan \alpha = \frac{wd}{Wh},$$

where, h— distance between C.G. and point of support of the beam.

d—length of the balance arm.

W—weight of the beam.

Thus, higher the *sensitivity* , the lighter is the beam, the smaller is the value of h and the longer is the arm.

2.7 To weigh a body by the method of oscillations

In the process of weighing, when the beam of the balance is lowered, the pointer swings back and forth across the scale with a slowly decreasing amplitude. To wait for it to come to rest would make weighing a slow process and even then the true position of the rest-point would be affected by errors due to friction. The resting point can be, however, determined by the *method of oscillation* without waiting for the pointer for a number of complete oscillations, so that an odd number of readings are obtained, the first and the last turning points being observed on the same side. The mean of the readings, say three, corresponding to one side and mean of the readings, say two, corresponding to the other side, are taken. The mean of these two means, give the *zero-point* or the *rest point*. Let it be denoted by P. The body to be weighed is placed on the left pan and weights are added on the right pan in a systematic order. Always the beam is arrested before any weight is placed on the pan.

By trial, enough weight (W) on the right pan is put, so that the new position of rest of the pointer appears to lie within 5 divisions to the right of the zero-point. The beam is allowed to oscillate and five successive turning points are taken as before. From these, the rest position of the pointer is obtained. Let it be Q. Next a small weight (say, 10 mg) is added to the right pan and the corresponding restposition is again obtained. Let it be called R. Then the weight of the given body W' is

$$W' = \left(W + \frac{Q-P}{Q-P} \times \frac{10}{100}\right) \text{gm.}$$

Results :

Load on left pan	Load on right pan	Turning point Left	Turning point Right	Mean Left	Mean Right	Zero-point	Rest position	Mass of the body W' gm
Nil	Nil	4	18	5.7	17.5	11.6 (P)		
		6	17					
		7						
Body	W gm	3	13	4	12.5		8.25 (Q)	
		4	12					
		5						
Body	W gm + 10 mg	1	10	2	9.5		5.75 (R)	
		2	9					
		3						

2.8 Fortin's Barometer : Its construction and use

Of the instruments which are generally used for the measurement of atmospheric pressure, Fortin's barometer (Fig. 2.8) is the most common one. It acts on the principle of Torricelli's experiment. The apparatus consists of a mercury cistern *GDEF* in communication with the atmosphere, over which a glass tube of about one metre long and one square centimetre crosssection, fllled with pure, dry and air-free mercury, is inverted. It is enclosed in a metal tube on which a scale is etched near the top of the mercury column. A vernier V can slide in a slot made in the metal tube and it can be operated by a screw S_2.

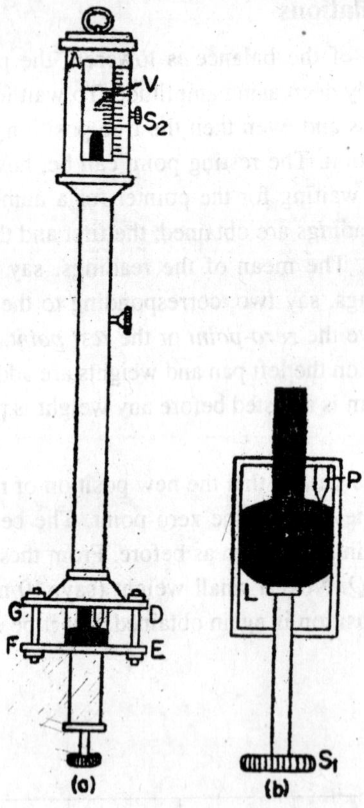

Fig. 2.8 : Fortin's Barometer

The slot enables to see the mercury column near the top. The zero of the scale is at the tip of an ivory pointer P which projects downwards from the ceiling of the mercury cistern. The surface of the mercury in the cistern can be adjusted by a screw S_1, so that the tip of the pointer just touches it. The bottom of the cistern (L) is made of samoy leather on which the screw S_1 presses.

To determine the air-pressure, the instrument is hung vertically on the wall. By operating the screw S_1, the ivory pointer is made to just touch the mercury surface in the cistern. In this condition, the surface of mercury in the cistern coincides with the zero mark of the main scale. By operating the screw S_2, the lower edge of the vernier is made tangential to the curved mercury surface. Readings of both main scale and vernier scale are taken at this stage. The temperature is noted from the thermometer attached to the instrument. The observations are repeated several times, by resetting the mercury level.

When the tip of the ivory pointer touches the mercury surface in the cistern and the bottom portion of the vernier scale is just tangential to the upper curved surface of the mercury column, then if the main scale reading is M cm and the vernier reading is x cm, the atmospheric pressure will be equivalent to a height (H) of the mercury column, given by

$$H = M + x \times \text{(vernier const.)}$$

Thus, the atmospheric pressure (P) in absolute unit will be

$$P = H\rho g \text{ dynes/cm}^2,$$

when, ρ = density of mercury (gm/cc)

g = acceleration due to gravity (cm/sec^2).

Results :

Vernier constant = ... cm

= ...... inch

Air temperature :

Before experiment ($t_1°$) = ... °C

After experiment ($t_2°$) = ... °C

$\therefore$ Mean temperature ($t°$) = $\dfrac{t_1° + t_2°}{2}$ = ... °C.

Scale	No. of Obs.	Main scale reading (M)	Vernier reading (x)	Value of vernier	Total (H) ($M + x \times$ v.c)	Mean (H)
cm	1.	... cm	...	... cm	... cm	... cm
	2.	... cm	...	... cm	... cm	
	3.	... cm	...	... cm	... cm	
	4.	... cm	...	... cm	... cm	
inch	1.	... inch	...	... inch	... inch	... inch
	2.	... inch	...	... inch	... inch	
	3.	... inch	...	... inch	... inch	
	4.	... inch	...	... inch	... inch	

Discussion :

The value of the atmospheric pressure so obtained at the temperature t °C, is not accurate, since the material by which the scale is constructed, also expands with rise of temperature. If the coefficients of expansion of scale material and of mercury be respectively α and γ, then the correct height (H_0) of the mercury column will be

$$H_0 = H\{1 - (\gamma - \alpha)t\}$$

and thus the correct air-pressure will be

$$P = H_0 \rho g.$$

Normally, the scale is made of brass whose coefficient of linear expansion (α) is 18.9×10^{-6}/°C and the volume coefficient of mercury (γ) is 18.2×10^{-5}/°C.

2.9 Travelling (Vernier) Microscope

In some experiments, it becomes often necessary to measure short vertical or horizontal distances with great accuracy. The travelling or the vernier microscope is an instrument well suited for the purpose. It is essentially a compound microscope (Fig. 2.9) capable of independent horizontal and vertical movements. Its displacement in either direction can be had with the help of a scale and a vernier.

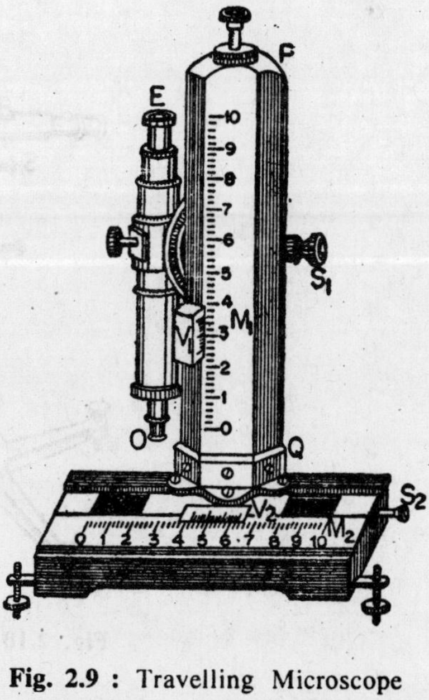

Fig. 2.9 : Travelling Microscope

The displacement in the vertical direction can be measured with the main scale M_1 and vernier V_1, while that in the horizontal direction, can be read off from the scale M_2 and vernier V_2. The microscope can be raised or lowered along a vertical pillar PQ and its axis can be kept fixed horizontally, vertically or at any angle to them by means of a screw S_1. The microscope is viewed through the eyepiece *E* while the objective lens *O* faces the object to be viewed. The microscope is focussed with the help of a screw attached to the body of the microscope. The eyepiece is provided with a cross-wire which is brought into focus by sliding the eyepiece in or out. The levelling screws fitted at the base enables the stage to be horizontal and the pillar vertical. Normally, the focal length of the objective is generally 3 or 4 cm.

To measure the distance (horizontal or vertical) between two points, the cross-wire is first focussed with the image of one point. Then by rotating the screw (S_1 or S_2), the microscope is displaced vertically or horizontally and again the cross-wire is focussed with the image of the other point. The readings of the microscope (main scale and vernier) are taken for the images of both the points. The difference (horizontal or vertical) between these two readings will be the distance between those two points.

Before using the microscope for taking readings, the vernier constants of both the verniers (V_1 and V_2) should be determined. Normally, in a microscope, both the verniers have the same vernier constants. With the aid of the base screws, the microscope must be levelled before taking readings.

2.10 Cathetometer : its construction and uses

Sometimes it may be necessary to measure a vertical distance accurately but the range of a travelling microscope may not be adequate for the purpose. In such a case, an instrument known as *Cathetometer* is used. It consists of a graduated vertical rod *PQ* attached to a heavy metallic base *R* attached with three levelling screws (Fig. 2.10). The rod carries a slide to which is fixed a telescope (*T*) with its axis horizontal. The rod carries a main scale (*S*) and a movable frame (*F*). When this frame moves across the rod, then a vernier

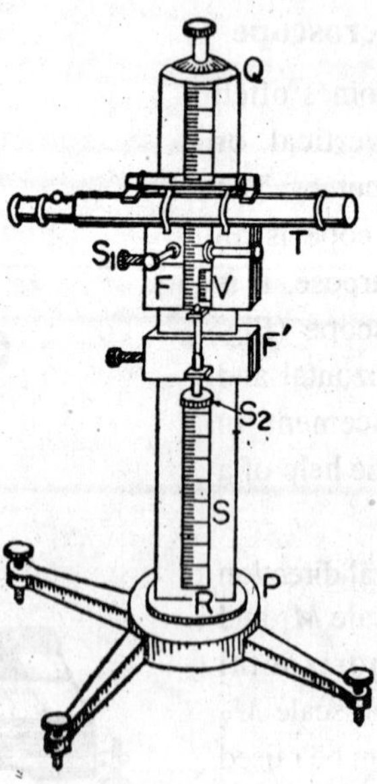

Fig. 2.10 : Cathetometer

(V) attached to the frame, touches the main scale. The position of the telescope can be known from the readings of the main scale and the vernier. A spirit level attached to the telescope enables to keep it horizontal. The axis of the telescope can be tilted with the help of another screw S_1. There is another frame F' below the frame F, to which is also attached another micrometer screw S_2. With the help of this screw, the frame F alongwith the telescope can be raised or lowered. The telescope eyepiece contains a cross-wire which enables the object to be focused.

To take readings with the aid of a cathetometer, at first the instrument is levelled with the help of the levelling screws. The rod is made vertical and the axis of the telescope made horizontal. The cross-wire is then accurately focused and made to coincide with the image of the object point. The difference of the readings of the scale attached to the cathetometer, gives the amount by which the telescope is moved vertically and the difference of the readings of the scale for two positions of the telescope, is actually the distance between the two points which needs to be measured. The principle of taking these readings is the same as in case of a travelling microscope.

Chapter 3

Properties of Matter

3.1 Determination of the value of the acceleration due to gravity (*g*) with the help of a compound (bar) pendulum

Theory :

The time-period (T) of a compound pendulum is given by

$$T = 2\pi\sqrt{\frac{K^2 + l^2}{gl}} \qquad \text{... (3.1a)}$$

where l—the distance of the point of suspension from the centre of gravity,

K—radius of gyration of the pendulum about an axis passing through its C.G.

Since the periodic time of a simple pendulum is given by $T = 2\pi\sqrt{\frac{L}{g}}$, the period of the rigid body (compound pendulum) is the same as that of a simple pendulum of length,

$$L = \frac{l^2 + K^2}{l} = l + \frac{K^2}{l} \qquad \text{... (3.1b)}$$

This length (L) is known as the length of the *Simple Equivalent Pendulum*. The expression for L can be written as a quadratic in l.

Thus from Eq. (3.1b), $l^2 - lL + K^2 = 0$. ... (3.1c)

This gives two values of l (l_1 and l_2) for which the body has equal times of vibration. From the theory of quadratic equations,

$$l_1 + l_2 = L$$

and $$l_1 l_2 = K_2$$

As the sum and the products of two roots are positive, the two roots are both positive. This means that there are two positions of the centre of suspension on the same side of C.G. about which the periods (T) would be the same. On the other side of the C.G., similarly there will be two more points of suspension, about which the time periods (T) will again be the same. Thus, there are altogether four points, two on either side of the C.G., about which the time periods of the pendulum are the same (T). The distance between two such points, asymetrically situated on either side of the C.G., will be the length (L) of the simple equivalent pendulum. If the period of oscillation about these points are T, then from the expression, $T = 2\pi\sqrt{\frac{L}{g}}$, we get

$$g = 4\pi^2 \frac{L}{T^2}. \quad \text{... (3.1d)}$$

By finding L graphically, and determining the value of the period T, the acceleration due to gravity (g) at the place of the experiment, can be found out.

Apparatus :

Bar pendulum, knife-edge, stopwatch, telescope, metre scale.

A simple form of the compound pendulum in the form of a bar, was designed by D. Owen in 1939 and is as shown in Fig. 3.1. It consists of a metallic bar about one metre long, in which a series of circular holes of neraly 5 mm in diameter, are bored at equal distances (about 5 cm) along its length. The bar can be suspended from a knife-edge with the help of these holes and made to oscillate. The knife-edge is fixed in a platform supported on three screws. The rear screw can be adjusted to make the platform horizontal. Two pointed marks at the two ends of the bar are used to find periods in conjunction with a fiduciary mark which may be the cross-wire of a telescope.

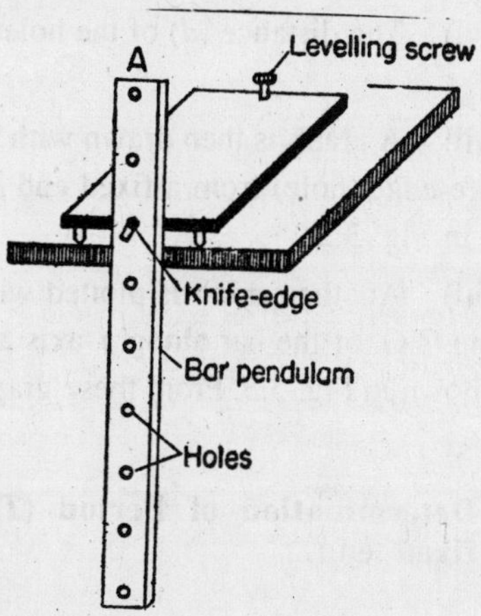

Fig. 3.1 : Bar Pendulum

Methods :

(i) With the help of a levelling screw, the knife-edge is first made horizontal and it is tested with a spirit-level. The bar is then suspended on the knife-edge by introducing it through the hole nearest to one end.

(ii) The lower mark at the end of the bar, is focused through a telescope avoiding parallax.

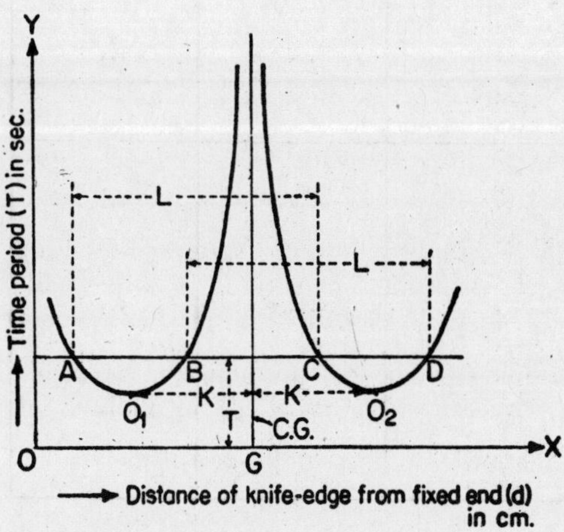

Fig. 3.2 : T versus d graph

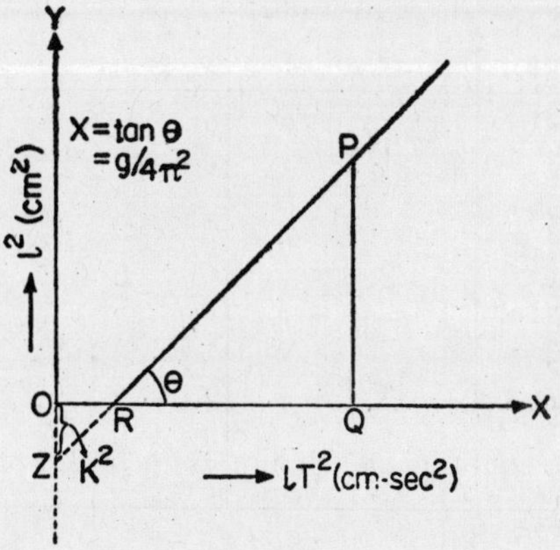

Fig. 3.3 : lT^2-l^2 graph

(iii) The bar is made to oscillate with a very small amplitude and care is taken to see that the plane of oscillation is always vertical. Time taken for 30 oscillations is noted with the help of a stopwatch and therefrom the *period* of oscillation (T) is determined.

(iv) Similar operation is repeated for other successive holes on one side and period for each hole is determined. For each hole, time taken for 30 oscillations is noted independently thrice, from which the mean time period (T) is determined.

(v) The bar is now inverted and by introducing through the holes on the other side, similar operations are repeated to determine the period corresponding to each hole.

(vi) The distance (d) of the holes from a certain fixed end (say A) is measured by a scale.

(vii) A graph is then drawn with then period (T) along Y-axis and the distance (d) of the knife-edge (hole) from a fixed end A, along X-axis. The nature of the graph will be as shown in Fig. 3.2.

(viii) Another graph is plotted with the square of the distance (l^2) of the knife-edge from the C.G. of the bar along Y-axis and lT^2 along X-axis. The nature of this graph will be as shown in Fig. 3.3. From these graphs, the value of g can be obtained as shown later.

Results :

(A) Determination of Period (T) and Distance (d) of knife-edge from one fixed end.

Table 1

No. of holes from one fixed end (A)		Distance of kinfe-edge (d) from the fixed end (cm)	Time for 30 oscillations (min & sec)	Mean time taken (min & sec)	Time period (T) (sec)
On one side of C.G.	1.	...	... ...	...	...
	2.	...	... ...	...	...
	3.	...	... ...	...	...
On other side of C.G.	1.	...	... ...	...	...
	2.	...	... ...	...	...
	3.	...	... ...	...	...

(B) Determination of the value of *g* from graph :

Table 2

From Graph 1 (Fig. 3.2)

No. of obs.	Length AC (cm)	Length BD (cm)	Length of Eq. Simple Pendulum $L = \frac{AC+BD}{2}$ cm	Corresponding value of time period (T) from graph (sec)	$g = 4\pi^2 \frac{L}{T^2}$ (cm/sec^2)	Mean value of 'g' (cm/sec^2)
1.	...	...	...	...	...	
2.	...	...	...	...	...	
3.	...	...	...	...	...	

Table 3

From Graph 2 (Fig. 3.3)

No. of Obs.	Distance of knife-edges from C.G. (l) cm	Correspon-ding Time period (T) from Table 1 (sec)	Value of l^2 (sec^2)	Value of lT^2 (cm-sec^2)	Length PQ from graph (cm)	Length QR from graph (cm)	Slope of graph $X = \frac{PQ}{QR}$	Mean value of X
1.	...	...	...	...	...	...	...	
2.	...	...	...	...	...	...	...	
3.	...	...	...	...	...	...	...	
...	...	...	...	...	...	...	...	
...	...	...	...	...	...	...	...	

Value of $g = 4\pi^2 X$ cm/sec^2

For determination of the value of g from *Graph* 1 (Fig. 3.2), horizontal line is drawn parallel of X-axis cutting the curves in four points A, B, C, D. The distance between the asymmetric points on either side (AC and BD) will evidently be equal to the length of the equivalent simple pendulum (L). The mean value of L is obtained as $\frac{1}{2}(AC + BD)$. The periods are the same (T) for these four points. The value of the ordinate corresponding to the horizontal line drawn, gives the value of this common period (T) for the equivalent length (L). Putting the values of L and T in equation $T = 2\pi\sqrt{\frac{L}{g}}$, the value of g can be determined. Several parallel lines (horizontal) are drawn to cut the curves in four points and the mean value of g is obtained from the values of g obtained for different horizontal lines.

For determination of the value of g from *Graph* 2 (Fig. 3.3), the position of the C.G., of the bar is located by pivoting it and balancing on a sharp point. The distance of the knife-edges from the C.G., are measured. From equations 3.1(a), it is evident that the nature of the curve $l^2 - lT^2$ will be a straight line (Fig. 3.3). The slope of the curve is obviously

$\frac{g}{4\pi^2}$ (since from Eq. (3.1a), $l^2 = \frac{g}{4\pi^2} \, lT^2 - K^2$). That is the slope of the curve $X = \frac{g}{4\pi^2}$. The straight line cuts the X-axis at the point R and an ordinate dropped from any arbitrary point P on the curve, meets the X-axis at Q.

Thus $\tan \theta = X = \frac{PQ}{QR}$ = slope of the curve. Hence, by measuring PQ and QR, the value of g is obtained from the relation,

$$g = 4\pi^2 X = 4\pi^2 \frac{PQ}{QR}.$$

Taking three different points (P) on the curve, the mean value of X (slope) is determined from three sets of calculations and hence the value of g is determined.

[Calculation of Percentage Error to be done as shown in Article 1·7, Ch. 1]

Precautions :

(i) The oscillation must be of very small amplitude and the period must be determined very accurately.

(ii) The knife-edge must be kept horizontal and the oscillation must take place in a vertical plane. Care must be taken to check rotational motion of the pendulum during oscillation.

(iii) It is better to use a telescope to note the period of oscillation, by concentrating attention on a fixed mark on the pendulum, which passes through the cross-wire of the telescope when viewed through.

(iv) During experimentation, care should be taken to avoid existence of air current in the vicinity of the pendulum.

[Question and Answers on pendulums are given after the next experiment.]

3.2 Determination of the acceleration due to gravity by means of a Kater's Pendulum

Theory :

It is seen that in case of a compound pendulum, if it is possible to obtain two parallel axes of suspension on either side of the C.G., which have exactly the same time of swing, then the distance between them is the length (L) of the simple equivalent pendulum. If the equal periodic time about these two knife-edges are T, then

$$g = \frac{4\pi^2}{T^2} . L. \qquad \text{... (3.2a)}$$

Bessel showed that the accurate determination of g did not require the difficult process of getting the two periods exactly alike. It was sufficient if they were nearly equal.

Thus, if T_1 and T_2 are the periods about the two positions of the knife-edges, which are very nearly equal and if l_1 and l_2 are the distances of the two knife-edges from the C.G., of the pendulum, then

$$T_1 = 2\pi \sqrt{\frac{l_1^2 + K^2}{l_1 g}} \text{ and } T_2 = 2\pi \sqrt{\frac{l_2^2 + K^2}{l_2 g}} \qquad \text{... (3.2b)}$$

$$\therefore \quad l_1 g T_1^2 = 4\pi^2\left(l_1^2 + K^2\right)$$

$$\text{and} \quad l_2 g T_2^2 = 4\pi^2\left(l_2^2 + K^2\right).$$

Subtracting, $g\left(l_1T_1^2 - l_2T_2^2\right) = 4\pi^2\left(l_1^2 - l_2^2\right)$

$$\text{or,} \quad \frac{4\pi^2}{g} = \frac{l_1T_1^2 - l_2T_2^2}{l_1^2 - l_2^2} = \frac{1}{2}\left[\frac{T_1^2 + T_2^2}{l_1 + l_2} + \frac{T_1^2 - T_2^2}{l_1 - l_2}\right]. \qquad \text{... (3.2c)}$$

Hence, measuring l_1 and l_2 and determining the periods T_1 and T_2 (which are very nearly equal), the value of g can be obtained from Eq. (3.2c).

Apparatus :

Kater's pendulum, stopwatch, telescope.

Captain Kater of England constructed a pendulum which enabled to determine the value of g on the principle stated above. It consists of a brass rod about a meter long (Fig. 3.4) which is provided with two knife-edges K and K' turned inwards to face each other. Beyond the knife-edges, there are two equal cylinders, one of boxwood (A) and the other of brass (B). Two other equal cylinders, smaller in size and one of boxwood (C) and the other of brass (D), are placed between the knife-edges which are movable along the rod. By shifting these, the C.G. of the rod can be altered. The knife-edges are generally movable and can be fixed at any desired positions. The distance between the knife-edges has to be determined very accurately. The pendulum may be made to oscillate about any of the knife-edges by placing them on metallic plates rigidly fixed on any permanent support.

Fig. 3.4 : Kater's Pendulum

Method :

(i) The pendulum is suspended from one of the knife-edges (say K), so that the cylinder B is in the downward position. A sharp mark is placed at the lower tip of the pendulum at this position and a low power telescope is fitted at a distance, wherefrom the sharp mark at the lower tip can be focused through the telescope.

(ii) The pendulum is given oscillation with a very small amplitude, so that it can freely oscillate about the knife-edge K.

(iii) The time taken for a small number of oscillations (say 5) about this knife-edge is determined with the help of a stopwatch reading at least up to one-tenth of a second.

(iv) The pendulum is then suspended about the other knife-edge (K') and again the time for the same number of oscillations as in (iii) is noted. These two times usually may differ much.

(v) The smaller cylinder (D) is now shifted in one direction slightly and again time for the same number of oscillations as before is noted for both the knife-edges K and K'. If by this shifting of the weight D, the difference in the times for oscillation about the two knife-edges increases, then the weight (D) should be moved in the opposite direction. Otherwise, it should be moved in the same direction in subsequent adjustments.

(vi) This process (shifting of the weight D) should be continued in the same direction as before until the difference between the times for a given number of oscillations (say 5) about the two knife-edges, gradually decreases and ultimately, the two times are nearly equal.

(vii) The process is now repeated with more number of oscillations being taken into account (say 10, 15, 20 ..., etc.) and finally when the two times about the two knife-edges are very nearly equal for 50 oscillations, then the two times for 50 oscillations about the two knife-edges are noted.

(viii) While observing times for oscillations more than 20 in number, finer adjustments for equalising the times, may be made with the finer adjustments attached with the other smaller cylinder (C) and shifting it very precisely.

The difference in the times for 50 oscillations about the two knife-edges should never exceed more than 2 seconds.

(ix) The time for 50 oscillations about each knife-edge, is determined thrice and the mean time for 50 oscillations about each of the knife-edges is calculated. From this, the time periods about each knife-edge (T_1 and T_2) are obtained.

(x) The pendulum is then removed and balanced on a sharp wedge placed on a horizontal table. The balancing point marks the C.G. of the pendulum. From this point, the distances of the two knife-edges (l_1 and l_2) are determined.

Thus determining T_1 and T_2 and mesuring l_1 and l_2, the value of g is calculated from Eq. (3.2c).

Results :

Least count of the stopwatch = ... sec.

(A) Determination of Periodic Time (Preliminary observations) :

Table 1

No. of Obs.	Adjustment done by shifting	No. of oscillations observed	Total times for oscillation about the knife-edges			
			K		K'	
			min	sec	min	sec
1.	Cylinder D	5	...	...	...	...
2.		5	...	...	...	...
3.		10	...	...	...	...
4.		10	...	...	...	...
...		...				
...		...				
1.	Cylinder C	20	...	...	...	...
2.		20	...	...	...	...
3.		25	...	...	...	...
4.		...	...	...	...	...
...		...				
...						

(B) Determination of final Periodic Time :

Table 2

No. of Obs.	Oscillations about the knife-edge	Time for 50 oscillations		Mean time for 50 oscillations		Time Period (sec) $(\frac{t_1}{50} \text{ or } \frac{t_2}{50})$
		min	sec	min	sec	
1.		......	......			
2.	K	......	......	......	......	......
3.		......	......	(t_1)		(T_1)
4.		......	......			
1.		......	......			
2.	K'	......	......	......	......	......
3.		......	......	(t_2)		(T_2)
4.		......	......			

(C) Determination of the distances of the knife-edges from the C.G. :

Table 3

No. of obs.	Position in cm of					Distance of knife-edge from C.G.	
	C.G. (x_0)	Knife-edge K (x_1)	Mean of x_1	Knife-edge K' (x_2)	Mean of x_2	K (l_1 cm)	K' (l_2 cm)
1.		...	...	...			
2.		...	...	...			
3.	...	...	...	...	...	...	...
						$(x_1 \sim x_0)$	$(x_2 \sim x_0)$
4.	...	...	...				

Calculations :

$T_1 = \ldots\ldots$ sec $\qquad l_1 = \ldots\ldots$ cm

$T_2 = \ldots\ldots$ sec $\qquad l_2 = \ldots\ldots$ cm

From eq. (3.2c),

$$\frac{4\pi^2}{g} = \frac{1}{2}\left[\frac{T_1^2 + T_2^2}{l_1 + l_2} + \frac{T_1^2 - T_1^2}{l_1 - l_2}\right]$$

or, $\quad g = \ldots\ldots\ldots$ cm/sec^2.

Precautions :

(i) The amplitude of oscillations must be very small.

(ii) The knife-edges should be adjusted parallel to each other.

(iii) Final observations for time period to be noted for a large number of oscillations, while preliminary observations can be made for small number of oscillations.

(iv) Correction for finite arc of swing may be made by observing half angles of swing in radians (α_1, α_2) at the start and finish respectively and from the formula, $T_0 = T\left(1 - \frac{\alpha_1\alpha_2}{16}\right)$, where T and T_0 are the observed and corrected periods respectively.

(v) For greater accuracy of the result, time periods can be determined by coincidence methods.

Error Calculation :

[To be made as per method shown in Art. 1.7, Chapter 1.]

Questions and Answers

Q.1. What is a compound pendulum and what are its advantages over a simple pendulum?

A. : An weighted rigid body of any shape swinging under gravity in a vertical plane about any horizontal axis passing through it, is called a compound pendulum. A simple pendulum in practices is not according to its definition while a compound pendulum is realised in the same from as it is defined. (Other points of difference to be seen from a text-book).

Q.2. What are 'centre of suspension' and 'centre of oscillation'?

A. : Centre of suspension is the point when the horizontal axis of rotation meets the vertical section of the pendulum taken through its C.G. while the centre of oscillation is another point on the other side of the C.G. at a different distance than that of the centre of suspension, such that the period of oscillation is the same as that about the centre of suspension. These two points are interchangeable.

Q.3. Are there only two points about which the periods are the same?

A. : No, altogether there are four points, two on each side and collinear with the C.G., about which the periods are the same.

Q.4. What is simple equivalent pendulum?

A. : It is a simple pendulum whose length is such that its period is the same as that of the compound pendulum.

Q.5. When is the period of a compound pendulum minimum?

A. : When the distance of the centre of suspension from the C.G. is equal to the radius of gyration of the pendulum about an axis parallel to the axis of rotation and passing through the C.G.

Q.6. What happens when the centre of suspension coincides with C.G.?

A. : The period of oscillation becomes infinite.

Q.7. Can you find the value of radius of gyration from Graphs 1 & 2?

A. : Yes, they can be determined (See Figs. 3.2 & 3.3).

Q.8. Why is it necessary to make the knife-edges horizontal?

A. : Otherwise, the bar may slip off and may not oscillate in a vertical plane.

Q.9. Why are there two pairs of weights in a Kater's pendulum?

A. : Rough adjustments are done by the larger weights and fine adjustments are done by the smaller weights.

Q.10. Is it necessary that the two time periods must be exactly equal?

A. : No, if they are nearly equal, formula in Eq. (3.2c) may be applied to find the value of g.

Q.11. Does the formula for Kater's pendulum experiment give accurate result?

A. : The formula in Eq. (3.2c) contains two terms. The first term contains $(l_1 + l_2)$—the distance between the two knife-edges, which can be measured accurately. But the second term contains $(l_1 - l_2)$ which means that the position of the C.G. has to be determined accurately to get $(l_1 - l_2)$. This determination involves certain inaccuracy. But since, this second term contains $\left(T_1^2 - T_2^2\right)$ which is very very small $\left(T_1 \simeq T_2\right)$, this term does not effectively affect the result.

Q.12. What are the main sources of error in those experiments?

A. : The error may be due to (a) buoyancy and viscosity of air, (b) yielding of the support, (c) curvature of the knife-edges, (d) finite amplitude of oscillation and (e) flow of air alongwith the pendulum.

3.3 Determination of Young's Modulus of the material of a beam by the method of Flexure (bending)

Theory :

If a light beam of uniform cross-section rests symmetrically on two knife-edges separated by a distance L and if on applying a load of mass M at its mid-point, a depression x is produced, then the Young's modulus (Y) of the material of the beam is given by

$$Y = \frac{MgL^3}{4bd^3x}, \qquad \text{... (3.3a)}$$

where b and d are the breadth and depth of the beam respectively and g is the acceleration due to gravity.

Measuring depression of the mid-point (x) for a given load M and knowing L, b, d and g, the value of Y of the material of the beam can be obtained.

Apparatus :

Rectangular uniform beam, two heavy iron stands (with levelling screws); two knife-edges; scale pan or hanger with a vertical pointer; weights, travelling microscope.

The given material in the form of a rectangular bar PQ about a metre long is placed horizontally (Fig. 3.5) on two knife-edges K_1 and K_2. The knife-edges are fixed on two vertical stands having levelling screws at the base. The bar carries at its mid-point (M) a frame (F) which has a pan or hanger (H) on which weights can be placed. A microscope sliding along a vertical scale is placed to focus the tip of the pointer.

Method :

(i) After measuring the length of the bar, a well-defined mark is given at the mid-point of the bar. A pair of marks A and B are given at equal distances from the mid-point (M) on either side of it. The two knife-edges (K_1, K_2) are placed at these points A and B. The distance between these two knife-edges (L) is the effective length of the bar.

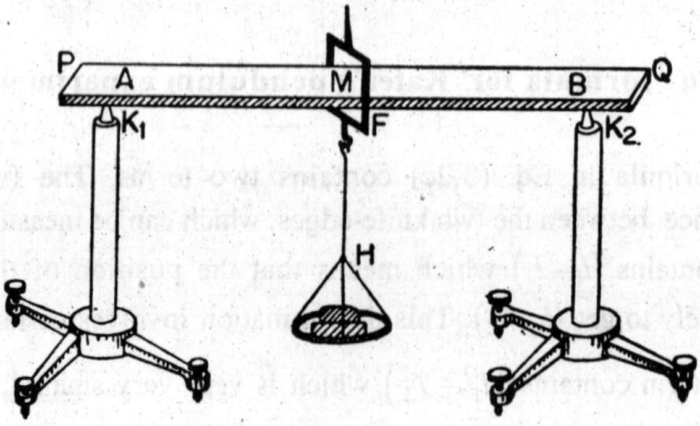

Fig. 3.5 : Bending of beam

(ii) The frame with the pointer is placed just at the mid-point (*M*) of the bar and the scale pan is hung from the hook of the hanger (*H*).

(iii) The distance between the knife edges is measured which gives the value of *L*. (Generally this distance is kept at 80 cm or 90 cm or 100 cm).

(iv) When there is no load on the pan, the tip of the pointer is focused through the travelling microscope and its readings (in main scale and vernier) are noted. This gives the initial reading.

(v) Now, gradually weights of 0.5 kg are placed on the hanger and for each case, the depression of the pointer is focused with the help of the microscope, and the microscope readings are taken. From each of these readings, if the initial reading is subtracted, then the depression corresponding to that respective weight is obtained.

(vi) Then the loads are gradually decreased stepwise as before and the pointer readings are again noted. So, for a given load, two readings are obtained, one while the load is increasing and the other while it is decreasing. Mean of these two values gives the actual reading for the corresponding load.

(vii) Then changing the position of the knife-edges, two more sets of readings may be taken for two different values of the length (*L*) of the bar.

(viii) For each length of the bar, after taking a set of readings for both loads increasing and decreasing, the bar is rotated through 90°, when the former breadth now becomes the depth of the bar. The operations (v) and (vi) are again repeated for this setting.

(If the bar supplied is of very small depth, then, it should be rotated through 180° and operations (v) and (vi) should be repeated as before.)

(ix) Therefore, three sets of readings are taken for loads increasing and decreasing for three different lengths of the bar (say, 80 cm, 90 cm and 100 cm) and in each case, readings are taken by rotating the beam either through 90° or 180°.

(x) Finally, the breadth (*b*) and the depth (*d*) of the bar are measured by a screw-gauge or a slide calliper at different position of the bar and their mean values are obtained.

(xi) Since, in the theory, the expression for *Y* contains depth to the third power, it is always better to measure depth by a screw-gauge which will minimise the error in the experiment.

(xii) A graph is plotted with loads on the *X*-axis and corresponding depression (*x*) on the *Y*-axis. The graph will be a straight line, from which taking any arbitrary point, the value of *Y* can be found out. [See 'Drawing of Graph' in (E) later.]

(A) Load-depression data :

Table 1

Distance between the knife-edges (L) (cm)	No. of Obs.	Load (M) kg	Microscope readings for						Mean Reading $\frac{P+Q}{2}$ cm	Depression x in cm
			Load increasing			Load decreasing				
			Main scale (cm)	Vernier Scale (cm)	Total (P) (cm)	Main Scale (cm)	Vernier Scale (cm)	Total (Q) (cm)		
	1.	0	...	...	...	...	...	...	...... (a)	0
	2.	0.5	...	...	...	...	...	...	...... (b)	$(b-a)$
	3.	1.0	...	...	...	...	...	...	...... (c)	$(c-a)$
L_1	4.	1.5	...	...	...	...	...	...	...... (d)	$(d-a)$
	5.	2.0	...	...	...	...	...	...	...... (e)	$(e-a)$
	6.	2.5	...	...	...	...	...	...	...... (f)	$(f-a)$
	7.	3.0	...	...	...	...	...	...	...... (g)	$(g-a)$

[Similar Tables may be made for recording depressions for two other different distances (L_2 and L_3) between the knife-edges.]

Results :

Vernier constant of the microscope :

1 small main scale division = ... cm

No. of vernier divisions = ...

Vernier constant = 1 m.s.d. – 1 v.s.d.

= .... cm

(A) Load-depression data :

Table 1 (on previous page)

(B) Load-depression record when the beam is turned through 90° (or 180°) :

Table 2

[Similar Table like Table 1 is made.]

(C) Measurement of depth of the beam :

Table 3

Pitch of the screw-gauge = ... mm

No. of circular scale divisions = ...

Least count = ... mm

No. of Obs.	Linear Scale Reading (mm)	Circular Scale Reading (mm)	Total Reading (mm)	Depth (d) (cm)	Mean depth (d) (cm)
1.	...	...	...	...	
2.	...	...	...	...	
3.	...	...	...	...	
4.	...	...	...	...	
5.	...	...	...	...	

(D) Measurement of breadth of the beam :

Table 4

Vernier Constant of Slide Calliper :

1 m.s.d = ... cm

...... vernier div. = ... m.s.d

∴ vernier const. = ... cm.

No. of Obs.	Main Scale Reading (cm)	Vernier Reading (cm)	Total Reading (cm)	Mean breadth (b) (cm)
1.	...	...	...	
2.	...	...	...	
3.	...	...	...	
4.	...	...	...	
5.	...	...	...	

(E) Drawing of graphs :

The value of Y can be determined from graphs. For this, a graph is drawn with loads (M) plotted along X-axis and corresponding depressions (x) along Y-axis. The graph will be a straight line passing through the origin. On the graph, the origin must be taken as (0, 0). The nature of the graph will be as shown in Fig. 3.6. On this straight line graph, any arbitrary point P is taken and substituting the value of ordinates and abscissae of this point in Eq. (3.7a) the value of Y can be obtained. For three different lengths of the beam being taken (L_1, L_2, L_3), three different straight lines are obtained and all of them can be plotted on the same graph paper. Depressions for the same load for all the three lengths of the beam can be obtained, as shown in the graph. Similar graphs are to be plotted when the beam is rotated through 90° or 180°.

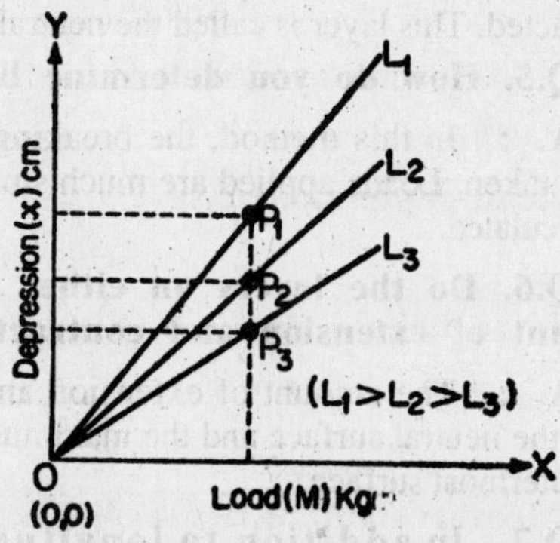

Fig. 3.6 : Load-Depression graph

Calculations :

From the graph, when load (M) = ... gm

Depression (x) = ... cm

$\therefore$ From Eq. (3.3a); $Y = \dfrac{MgL^3}{4bd^3x}$ = ... dynes/cm^2

Precautions :

(i) The beam must be kept horizontal and the pointer alongwith the frame must be suspended at its mid-point.

(ii) Since the value of depth (d) is small and it occurs to the third power in the expression for Y, it must be measured with a screw-gauge.

(iii) While taking readings, the microscope must be rotated in the same direction, so as to avoid the back-lash error.

(iv) After loading or removing weights on and from the pan respectively, some time must be allowed before taking the readings.

Error Calculation :

[See Example given in Art. 1.7]

Questions and Answers

Q.1. Define Young's modulus, Bending moment, Flexural rigidity, Moment of inertia and Neutral axis.

A. : [See any textbook.]

Q.2. On what factors does the value of Y depend?

A. : It depends on the material of the wire.

Q.3. The weights (force) are applied at right angles to the length of the beam, then how does longitudinal stress and strain come into play?

A. : When the beam is bent, its filaments on the convex side are extended while those on the concave side are contracted. As a result, longitudinal strain will occur. As a reaction, forces of elongation and contraction along the length of the beam will set up longitudinal stress.

Q.4. What is neutral surface of the beam?

A. : There is a layer along the length of the beam which is neither extended nor contracted. This layer is called the neutral surface.

Q.5. How do you determine breaking load in this case?

A. : In this method, the breaking load is very high for the particular shape of the beam taken. Loads applied are much smaller than the breaking load. So, its value need not be calculated.

Q.6. Do the layers on either side of the neutral surface undergo same amount of extension and contraction?

A. : The amount of extension and contraction increases as we go away and away from the neutral surface and the maximum extension/contraction occurs for the filaments in the outermost surface.

Q.7. In addition to longitudinal stress is any other kind of stress produced?

A. : Shearing stress is also produced in the process of bending which acts across the cross-section of the beam. The amount of this stress is very small which is not taken into account since the beam is long and thin.

Q.8. What is a cantilever?

A. : It is a thin uniform beam clamped horizontally at one end and bent under the action of a load placed at the other end.

Q.9. Does the weight of the beam produce any depression and is it taken into account?

A. : The weight causes some depression but since the changes in depression are calculated for different loads, hence the depression of the beam weight cancels out.

Q.10. Why is the beam placed symmetrically on the knife-edges?

A. : To keep the reactions at the knife-edges equal, as assumed in the theory.

Q.11. What type of beam is preferred in this experiment?

A. : A moderately long and fairly thin beam is preferred, whence the depressions are large but depression due to shearing stress is small.

Q.12. Why the length of the beam is measured in between the two knife-edges?

A. : As per theory, the beam is considered equivalent to two inverted cantilevers whose loaded ends are taken at the knife-edges.

Q.13. Which quantity is to be measured very accurately?

A. : The depression and the depth, because both are very small and depth occurs to the third power.

3.4 Determination of Young's modulus (Y), Rigidity modulus (n) and Poisosn's ratio (σ) of a wire by Searle's method

Theory :

Two exactly identical bars *AB* and *CD* are joined at their centres by a wire *GG* of the material whose elastic constants are to be determined (Fig. 3.7). The system is suspended by two parallel torsionless threads so that in the equilibrium position, the bars are parallel to each other and lie in the same horizontal plane [Fig. 3.7(a)]. If the two bars be slightly rotated through small equal angles in opposite directions and then be released [Fig. 3.7(b)], the bars will begin to oscillate in the horizontal plane executing S.H.M. and each vibrating with the same period (T_1) given by

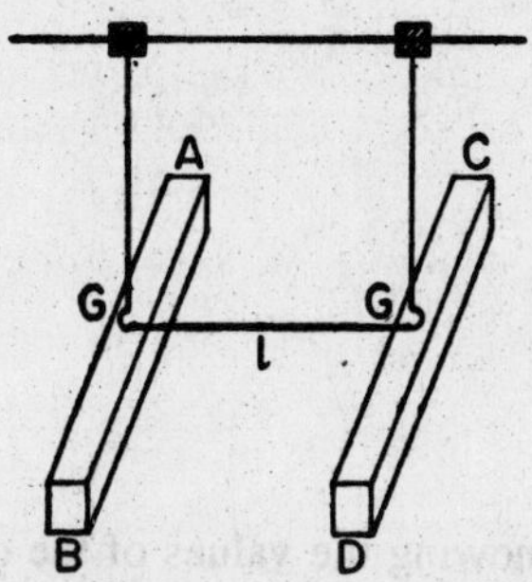

(a)

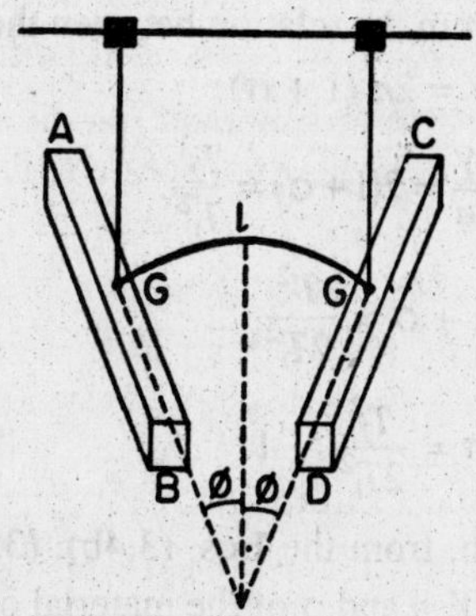

(b)

Fig. 3.7 : Determination of Y

$$T_1 = 2\pi\sqrt{\frac{2lI}{\pi r^4 Y}}, \qquad \text{... (3.4a)}$$

where I = moment of inertia of each bar about a vertical axis throughts its C.G.

l = length of the wire GG.

r = radius of the wire GG.

From Eq. (3.4a), Young's modulus of the wire,

$$Y = \frac{8\pi l I}{r^4 T_1^2}. \qquad \text{... (3.4b)}$$

Now, if the threads are removed and one of the bars is fixed horizontally on a support while the other is suspended below from the wire (Fig. 3.8), and if the lower bar is slightly turned in its plane and then released, it will execute torsional oscillations with the time-period T_2 given by

$$T_2 = 2\pi\sqrt{\frac{2l'I}{\pi n r'^4}}, \qquad \text{... (3.4c)}$$

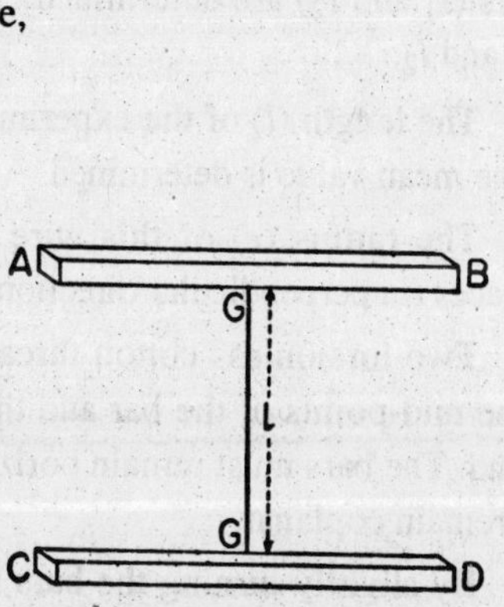

Fig. 3.8 : Determination of n

where I = moment of inertia of the bar CD about a vertical axis through its C.G.

l' = length of the wire between the bars AB and CD.

r' = radius of the wire.

From Eq. (3.4c), the rigidity modulus (n) of the wire will be,

$$n = \frac{8\pi l' I}{r'^4 T_2^2}. \qquad \text{... (3.4d)}$$

Now, if the same wire is used for both the parts of the experiment, then $l = l'$ and $r = r'$. Thus, from Eqs. (3.4b) and (3.4d),

$$\frac{Y}{n} = \frac{T_2^2}{T_1^2}. \qquad \text{... (3.4e)}$$

But from the relation between the elastic constants, we have

$$Y = 2n\,(1 + \sigma). \qquad \text{... (3.4f)}$$

$$\therefore \quad \frac{Y}{n} = 2(1+\sigma) = \frac{T_2^2}{T_1^2}.$$

$$\text{or,} \quad 1 + \sigma = \frac{T_2^2}{2T_1^2}$$

$$\text{or,} \quad \sigma = \frac{T_2^2}{2T_1^2} - 1. \qquad \text{... (3.4g)}$$

Hence, from the Eqs. (3.4b), (3.4d) and (3.4g), knowing the values of the quantities, values of Y, n and σ of the material of the wire can be determined.

Apparatus :

Apparatus required in Searle's method : Stopwatch, thread, balance and weight box, metre scale; screw-gauge; slide calliper and experimental wire.

Method :

(i) The bars AB and CD are weighed separately and their masses are determined. Their lengths are also measured by a metre scale. Their diameters (if circular) or breadths (if rectangular) are determined by a slide calliper. From these data, the moments of inertia of the two bars (I_1 and I_2) are determined. The value of I is thus obtained by taking the mean value of I_1 and I_2.

(ii) The length (l) of the experimental wire is measured thrice by means of a metre scale and the mean value is determined.

(iii) The radius (r) of this wire is determined by a screw-gauge at least at five different places (in perpendicular directions at each place) and the mean value is determined.

(iv) Two torsionless cotton threads, each of about 50 cm in length, are tied to the hooks at the mid-points of the bar and they are suspended from a fixed rigid support from the other end. The bars must remain horizontal and parallel to each other. These bars and the wire must remain coplanar.

(v) By slightly turning the bars in opposite directions in their own plane and then releasing them, the beams are set to oscillations. This can be achieved by drawing the two ends of the bar (B and D) slightly towards each other with the help of a loop of a thin thread slipped over them and then by burning the thread.

(vi) When set to oscillations, the ends B and D of the bars will alternatively approach towards and recede from each other.

(vii) With the help of a stopwatch, the time for 30 oscillations of the bars is noted and at least five such observations are taken. From these the mean time taken for 30 oscillations are determined and the time period (T_1) is thus obtained.

(viii) Knowing thus the values of I, l, r and T_1, the value of Y of the material of the wire is determined from Eq. (3.4a).

(ix) Next to determine the value of n, the suspension threads are removed and the bar CD is kept suspended below the bar AB by the same wire (GG). The bar AB is clamped to a rigid support.

(x) The time period (T_2) of torsional oscillation of the bar CD is determined as before from several observations and their mean value.

(xi) The moment of inertia of the bar CD about the central axis of suspension, the radius (r') and the length (l') of the wire (GG) are obtained as before. If the same wire be used for determination of Y, then $l = l'$ and $r = r'$, I being already determined previously.

(xii) Thus, knowing the values of T_2, l', r' and I, the value of n can be obtained from Eq. (3.4d).

(xiii) Finally, using Eq. (3.4g), the Poisson's ratio (σ) of the material of the wire can be obtained from the knowledge of T_1 and T_2.

Results :

Mass of the bar AB = ... gm

Mass of the bar CD = ... gm

Length of the bar AB = ... cm

Length of the bar CD = ... cm

(A) Determination of radius/breadth of the bars :

Vernier constant of the slide calliper = ... cm

Table 1

[A table for Slide Calliper's readings to be made as shown in Article 2.3]

(B) Moments of Inertia of the bars :

Table 2

Bar	Mass of the bar (gm)	Length of the bar (cm)	Breadth (b) or Radius (R) of the bar (cm)	Moment of Inertia of the bar (gm-cm^2)	Mean Moment of Inertia (I) (gm-cm^2)
AB	... (M_1)	... (L_1)	... (b_1 or R_1)	$I_1 = \frac{M_1}{12}(L_1^2 + b_1^2)$ or $= M_1\left(\frac{L_1^2}{12} + \frac{R_1^2}{4}\right)$	$I = \frac{1}{2}(I_1 + I_2)$
CD	... (M_2)	... (L_2)	... (b_2 or R_2)	$I_2 = \frac{M_2}{12}(L_2^2 + b_2^2)$ or $= M_2\left(\frac{L_2^2}{12} + \frac{R_2^2}{4}\right)$	

(C) Length (l) and radius (r) of the wire GG :

Table 3

No. of Obs.	Length (cm)	Mean Length (l) (cm)
1.	....	....
2.	....	....
3.	....	....

Table 4

[For measurement of diameter (radius) of the wire, a table for screw-gauge reading is to be made as shown in Article 2.4]

(D) Determination of Time Period of Oscillation (T_1) :

Table 5

No. of Obs.	No. of Oscillations	Time taken	Mean Time taken	Time Period (T_1) (sec)
1.	30	...... min ...... sec		
2.	30	...... min ...... sec		
3.	30	...... min ...... sec	...... min ...... sec.	...... sec
4.	30	...... min ...... sec		
5.	30	...... min ...... sec		
6.	30	...... min ...... sec		

(E) Determination of Time Period (T_2) when torsional oscillation taking place :

Table 6

[Same Table as of Table 5 is to be made.]

Calculations :

From Eq. (3.4b), $Y = \dfrac{8\pi l I}{r^4 T_1^2} =$ .... dynes/cm^2

From Eq. (3.4d), $n = \dfrac{8\pi l' I}{r'^4 T_2^2} =$ .... dynes/cm^2

From Eq. (3.4g), $\sigma = \dfrac{T_2^2}{2T_1^2} - 1 =$ ....

Precautions :

(i) Care should be taken to see that the oscillations of the bars take place in a horizontal plane. For this, the distance between the points wherefrom the upper ends of the suspension wires hang, must be equal to the distance between the hooks attached at the mid-points of the bars.

(ii) The amplitude of oscillations of the bar should be small so that the suspension threads remain vertical and the wire is not strained beyond the elastic limit.

(iii) In order that the bending of the wire may be symmetrical, the cotton thread loop, used to start the oscillations should touch the bars at equal distance from the respective ends. It is further preferable to burn the upper half of the loop so that the thread falls off rapidly.

(iv) The radius of the wire and both the time periods (T_1 and T_2) of oscillations are small and since these quantities occour in the expressions with higher powers, these quantities must be measured very accurately to minimise the errors.

(v) The length of the wire should be made fairly large so as to make the time-period of oscillation fairly large and appreciable making the possible error in them minimum.

(vi) In this experiment, the chief source of error lies in the measurement of the time-period of oscillations and also in the measurement of radius of the wire.

Error Calculation :

From the expressions of Y and n, the proportional errors are,

$$\frac{\delta Y}{Y} = \frac{\delta l}{l} + \frac{\delta I}{I} + \frac{4\delta r}{r} + 2\frac{\delta T_1}{T_1}$$

and $$\frac{\delta n}{n} = \frac{\delta l'}{l'} + \frac{\delta I}{I} + \frac{4\delta r'}{r'} + 2\frac{\delta T_2}{T_2}.$$

Since the quantities T_1, T_2 and r_1, r_2 are very small, hence they contribute maximum error to the results obtained. Thus, the periods (T_1 and T_2) should be measured taking large number of oscillations a number of times with a stopwatch reading fraction of a second and the radius of the wire should be measured with a screw-gauge at different positions of the wire and at right angles to each other at each position. The quantities l, l' and I are fairly large and they can be measured quite accurately. Hence, their contributions towards the proportional error are not very significant.

Questions and Answers

Q.1. For a material, how many elastic constants are there?

A. : There are four elastic constants for a material, namely, Young's modulus (Y), Rigidity modulus (n), Bulk modulus (K) and Poisson's ratio (σ).

Q.2. Can all of them be determined by this method?

A. : Yes, the values of Y, n and σ are determined as described and from the knowledge of the relation existing between the elastic constants, the value of K can also be determined.

Q.3. What are the relations?

A. : (See any textbook).

Q.4. Is there any restrictions on the angular displacements of the bar during its oscillations?

A. : In the first case of oscillation, the amplitude should not preferably exceed 5°, because then the wire is slightly bent and the supporting threads remain vertical. In the second oscillation, there is no restriction in the value of the amplitude but the wire should not be twisted beyond its elastic limit.

Q.5. How do Young's modulus and rigidity modulus come into play in this experiment?

A. : Young's modulus is involved in the first part of the experiment, while rigidity modulus comes in the second part. When the bar ends come closer to each other, the wire bends into a circular arc. On bending, some of its filaments extend and some contract, while some remain unchanged. As a result of these extensions and contractions of the filaments, longitudinal stresses are called into play, bringing Young's modulus in the picture. Again, when one bar is clamped and the other undergoes rotation, the wire is twisted. Due to this twisting, shearing stress is called into play bringing rigidity modulus into the picture.

Q.6. Can you use metallic wires instead of threads?

A. : For threads, torsional reaction is small and the only couple acting on the bars is one to the bending of the wire. Metallic wires, if used, will exert torsional reaction sufficiently.

Q.7. Which factors control the values of T_1 and T_2?

A. : Bending moment of the wire controls T_1 while torsional couple of the wire for unit twist controls T_2.

Q.8. Is there any difference in the bending of wire in this experiment and that in the bending of beam?

A. : In this experimental, the wire is bent in the form of a circular arc while in bending of beam experiment, the beam is bent such that the curvature is minimum in the middle and increases towards the ends.

Q.9. Will you prefer heavy or light bars?

A. : Heavier bars are preferred because their moments of inertia and hence the time-periods will be large. But they should not be too heavy to break the suspension system.

Q.10. Which quantities are to be measured very accurately in this experiment?

A. : See Precautions (iv).

Q.11. What are the advantages of this method over the other methods?

A. : In this method a small length of the experimental wire is required for determination of Y, while in the extension method, a long wire is required. Here Y is measured by measuring time-period of oscillation and not by observing any deflections or depressions. This method enables to get all the four elastic constrants simply by observing only two time-periods. Poisson's ratio is obtained by this method without measuring lateral and longitudinal strains. It involves measurements of only two time-periods without knowing the values of Y and n.

3.4(A) Determination of the Moment of Inertia of a body about an axis passing through its C.G. and perpendicular to its length

Theory :

The time period T of the torsional oscillation of a body of moment of inertia I is given by $T = 2\pi\sqrt{\frac{I}{C}}$, where C is the torque per unit twist in the suspension wire.

If T_0 is the time period of the cradle alone and I_0 be its moment of intertia about the given axis, then

$$T_0 = 2\pi\sqrt{\frac{I_0}{C}} . \quad \text{... 3.4(A)a}$$

Again, if I_1 is the known moment of inertia of a body about an axis through its C.G., then when the body is placed on the table with the suspension axis coinciding with the above axis, the time period T_1 of torsional oscillation is given by

$$T_1 = 2\pi\sqrt{\frac{I_0 + I_1}{C}} \qquad \text{... 3.4(A)b}$$

Finally, if T_2 be the time-period of oscillation when the body (whose moment of inertia is to be determined) is placed on the cradle and if I_2 be its unknown moment of inertia, then

$$T_2 = 2\pi\sqrt{\frac{I_0 + I_2}{C}}. \qquad \text{... 3.4(A)c}$$

From these equations, $I_0 = \frac{CT_0^2}{4\pi^2}$; $I_1 + I_0 = \frac{CT_1^2}{4\pi^2}$; $I_2 + I_0 = \frac{CT_2^2}{4\pi^2}$

$$\text{Hence, } I_2 = I_1\frac{T_2^2 - T_0^2}{T_1^2 - T_0^2}. \qquad \text{... 3.4(A)d}$$

Thus, if I_1 is known for a given body, then the value of I_2 can be found out from the measurements of time-periods.

Usually, the body of known moment of inertia is a right circular cylinder of mass M and length and radius being l and r respectively. In that case,

$$I_1 = M\left(\frac{l^2}{12} + \frac{r^2}{4}\right)$$

For a rectangular body,

$$I^2 = M\left(\frac{l^2 + b^2}{12}\right), \qquad \text{...3.4(A)e}$$

where l and b are the length and breadth of the body.

Apparatus :

The apparatus consists of a circular, flat metal disc X having a circular groove (H) in it. The disc is rigidly fixed to a frame F (Fig. 3.9) suspended by a wire T. The groove contains several masses (w, w, w), by moving which the disc may be made horizontal with the help of a spirit level. A mirror m is fixed to the wire T, by which the period of oscillation of the disc can be noted by telescope and scale arrangment. The whole apparatus is kept enclosed in a glass case to protect it from disturbances due to wind.

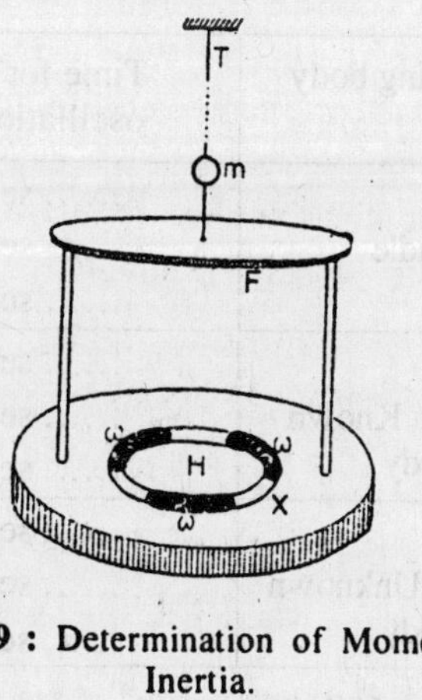

Fig. 3.9 : Determination of Moment of Inertia.

Procedure :

(i) The mass M of the known cylinder is determined by a balance. Its length (l) is measured by a scale and its radius or breadth (r or b) is measured by a slide calliper. Substituting these values in Eq. 3.4(A)e, its moment of inertia (I_1) is calculated.

(ii) The cradle alone is then made to oscillate with a small amplitude by giving it a small twist. The time taken for its 20 oscillations is noted and from that, its time-period (T_0) is calculated.

(iii) Next the body of known moment of inertia is placed horizontally on the cradle in such a way that its C.G. is along the suspension wire. The cradle with the body is again made to oscillate as before and the period of oscillation (T_1) is calculated.

(iv) Finally, in the similar way, the period of oscillation (T_2) is determined with the unknown body being placed on the cradle.

(v) Substituting the values of T_0, T_1, T_2 and I_1 in Eq. (3.1d), the value of I_2 can be determined.

Results :

(A) The length (l), the radius (r) or the breadth (b) of the known body are determined by a scale and a slide calliper.

[Make table as shown in ∮2.3]

(i) Length of the cylinder (l) = $\frac{\ldots+\ldots+\ldots}{3}$ = ... cm

(ii) Diameter/Breadh of the cylinder/rectangular body = ... cm

(iii) Mass of the cylinder = (... + ... + ...) gm + (... + ... + ...) mg = ... gm

(B) Determination of moment of inertia (I_1) of the known body :

Table 2

Mass of the body (M) (gm)	Length of the body (l) (cm)	Radius (r) or the breadth (b) of the body (cm)	Moment of Inertia (I_1) or the body (gm-cm^2)
...	...	...	...
...	...	...	...
...	...	...	...

(C) Determination of the Time-Periods :

Table 3

Oscillating body	Time for 20 oscillations	Mean time (t)	Time period $\left(\frac{t}{20}\right)$
Cradle	...... sec ...... sec ...... sec	...... sec	...... sec (T_0)
Cradle + Known body	...... sec ...... sec ...... sec	...... sec	...... sec (T_1)
Cradle + Unknown body	...... sec ...... sec ...... sec	...... sec	...... sec (T_2)

(D) Calculations :

Moment of inertia of the known body,

I_1 = ... gm-cm^2.

Moment of inertia of the unknown body,

$$I_2 = I_1 \times \frac{T_2^2 - T_0^2}{T_1^2 - T_0^2} = \ldots\ldots\ldots \text{ gm-cm}^2.$$

Precautions :

(i) The amplitude of oscillations should not be very large.

(ii) The oscillations should be due to twisting only and no other oscillaion should affect the free oscillation of the system.

(iii) The axis of oscillation must always pass through the C.G. of the body.

Questions and Answers

Q.1. What are meant by 'moment of inertia' and 'radius of gyration'?

A. : See definitions from any textbook.

Q.2. Will the moment of inertia be different if the axis of rotation changes?

A. : Yes, if the axis of rotation changes, the value of K (radius of gyration) changes and hence M.I. also changes, although mass remains the same.

Q.3. What type of oscillation is this?

A. : This is of the simple harmonic oscillation type.

Q.4. On what factors does the time-period depend?

A. : It depends upon the M.I. of the body, the dimension and the material of the suspension wire.

Q.5. What type of suspension wire is preferred?

A. : The wire should be long and thin when the period will be large and can be measured precisely.

Q.6. How would you ascertain that the C.G. is on the axis of the suspension wire?

A. : The M.I. of the body is determined by making the axis of oscillation passing through the C.G. of the body (by guess work). Then the M.I. of the same body is determined when the axis of oscillation passes through two other points situated on either side of suspected C.G. The lowest of these three values of M.I. is the desired M.I. of the body about the axis passing through the C.G. of the body.

3.4(B) Determination of the Rigidity modulus of the material of a wire by Statical method

Theory

If a homogeneous wire of length l and of uniform radius r be fixed at one end and if the other end be twisted by application of a couple of moment C applied at right angles to its length, then the twist θ produced at the free end is given by the equation

$$C = \frac{\pi n r^4 \theta}{2l}, \qquad \ldots 3.4(B)a$$

where n—rigidity modulus of the material of the wire.

If the couple applied is due to a load of M gm acting with an arm of diameter d cm, then

$$C = Mgd. \qquad \ldots 3.4(B)b$$

If the observed angle of twist be ϕ degrees, then

$$\theta = \frac{\pi}{180}\phi \text{ radians.}$$

Substituting the values of C and ϕ, we get

$$Mgd = \frac{\pi n r^4 . \pi\phi}{2l \times 180}$$

$$\text{or,} \quad n = \frac{360 lgdM}{\pi^2 r^4 \phi} \quad \text{... 3.4(B)c}$$

$$= K\left(\frac{M}{\phi}\right), \quad \text{... 3.4(B)d}$$

where $K = \dfrac{360lgd}{\pi^2 r^4}$.

If l, r and C are in c.g.s. units and θ in radians, then n will be in dynes per sq. cm. Knowing the different values of angle of twist (ϕ) for different values of M, and knowning the value of K from the geometrical dimensions of the wire, the value of n can be obtained from Eq. 3.4(B)d.

Apparatus :

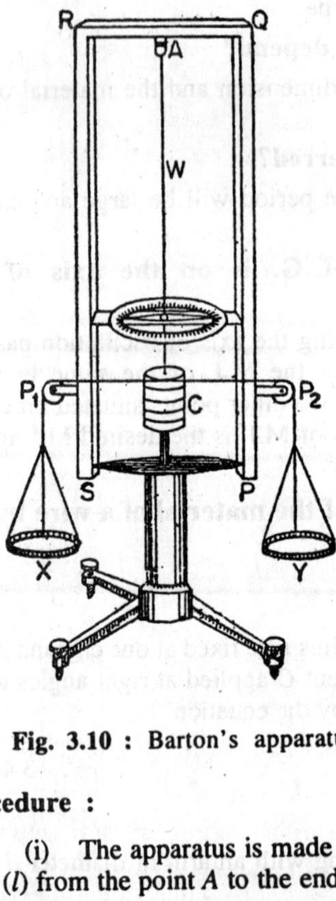

Fig. 3.10 : Barton's apparatus

Barton's apparatus : Standard weights, screw-gauge, slide calliper, metre scale.

Barton's apparatus (Fig. 3.10) is a special type of apparatus used for determining the rigidity modulus by statical method. It consists of a heavy iron frame having a heavy base with levelling screws. A vertical frame $PQRS$ is attached to the base. The experimental wire (W) is rigidly fixed at the top of the frame at A. The lower end of the wire is attached to a solid cylinder C, known as the flywheel, on which two threads with one end of each attached to the cylinder and wound in the same direction, pass round tangentially in opposite directions and carry two scale pans at their other ends (X, Y) after passing over two pulleys (P_1, P_2) mounted on ball bearings. By placing weights on the pans (X, Y), a couple can be applied on the fly wheel. For the couple applied, the lower end of the wire is twisted and the amount of twist can be measured on a circular scale by means of a pointer moving across the scale.

Procedure :

(i) The apparatus is made vertical by the levelling screws. Next, the length of the wire (l) from the point A to the end of the flywheel where the pointer is fixed, is measured by a scale.

(ii) The diameter of the wire is measured by a screw-gauge at least at five different places of the wire along the entire length. At each place, two readings at right angles are taken. The mean diamter and hence the mean radius (r) is obtained.

(iii) The diameter of the flywheel is measured with the help of a slide calliper in different directions at the place where the two strings leave the wheel. The mean of these readings will give the value of d.

(iv) The values of l, d and r being known, the value of K in Eq. 3.4(B)d can be obtained.

(v) Different equal loads are then placed on the two pans and at each time, the reading of the pointer on the circular scale, is noted. For both increasing and decreasing loads, readings are noted and from the mean of two readings, the twist (ϕ) in the wire for each load is determined.

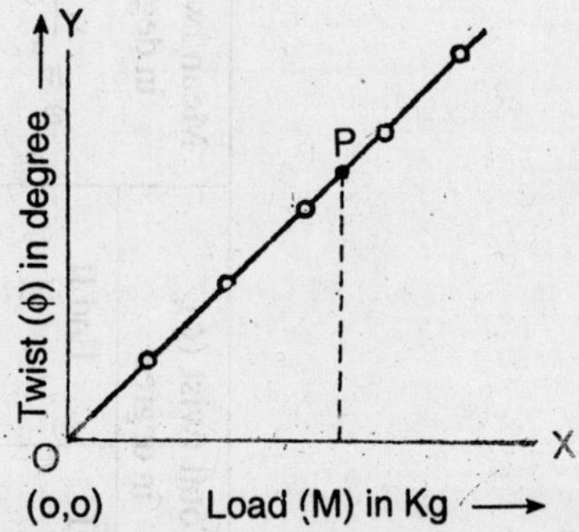

Fig. 3.11 : (M-ϕ) graph

(vi) Finally, a graph is drawn with the applied loads (W) being plotted along X-axis and the twists (ϕ) along Y-axis. The graph will be a straight line (Fig. 3.11) passing through the origin.

(vii) Taking an arbitrary point P on the graph, the value of $\frac{M}{\phi}$ will be obtained from the absciassa and ordinate of the point, and substituting the values of $\frac{M}{\phi}$ in Eq. 3.4(B)d) the value of n can be obtained.

Results :

(A) Determination of diameter of the wire :

Table 1

[The diameter is measured by a screw-gauge at five different places of the wire and the results are tabulated as shown in Table 1 of $\oint$3.5]

(B) Determination of the length (l) of the wire :

Table 2

[The length of the wire is determined by a metre scale and then tabulated as shown in Table 2 of $\oint$3.5]

(C) Determination of diameter (d) of the flywheel :

Table 3

[The diameter of the flywheel is determined at different places by a slide calliper and the readings are tabulated as shown in Table in $\oint$2.3]

(D) Determination of twists (ϕ) for different loads (M) :

Table 4 (vide next page)

Drawing of Graph :

As mentioned earlier, a graph is drawn with M plotted along X-axis and ϕ along Y-axis. The graph is a straight line passing through the origin. For any arbitrary point P on the graph (Fig. 3.11), the value of $\frac{M}{\phi}$ is obtained and substituting this value in Eq. 3.4(B)d, the value of n is obtained.

(D) Determination of twists (ϕ) for different loads (M) :

Table 4

No. of Obs.	Load of each pan (gm)	Pointer reading in degrees				Mean of End I readings (a) in degree	Mean of End II readings (b) in degree	Total twist (ϕ) in degree		Mean twist (ϕ) in degree
		Load increasing		Load decreasing				End I	End II	$\phi = \frac{x_1 + x_2}{2}$
		End I	End II	End I	End II			(x_1)	(x_2)	
1.						...... (a_0)	...... (b_0)	0	0	0
2.						...... (a_1)	...... (b_1)	$(a_1 - a_0)$	$(b_1 - b_0)$	...
3.						......	......	$(a_2 - a_0)$	$(b_2 - b_0)$	...
4.										
5.										
6.										

Precautions :

(i) Since the radius (r) of the wire occurs to the fourth power in the theory, it should be measured very carefully at different places by means of a screw-gauge.

(ii) The experimental wire must pass through the centre of the flywheel and the threads must leave the wheel tangentially and horizontally.

(iii) After placing the loads on the pans, some time must be allowed before taking the readings of the pointer.

(iv) By removing the scale and the pointer (if such arrangement exists) another set of readings may be taken. The shifting of the pointer makes a change in the length of the experimental wire.

Questions and Answers

[Questions related to this particular experiment are given at the end of the next experiment (Dynamical method) alongwith questions of that experiment.]

3.4(C) Determination of the Rigidity modulus of the material of a wire by dynamical method

Theory :

When a heavy body is suspended by a wire, the weight of the wire being negligible compared to that of the body, and is made to execute torsional oscillations about the axis of suspension, the time period of oscillations is given by

$$T = 2\pi\sqrt{\frac{I}{C}} \qquad \text{... [3.4(C)a]}$$

where I is the moment of inertia of the body about the axis of suspension and C is the torque per unit twist. If the cross-section of the wire be circular (πr^2) where r is the radius of the wire and l be the length of the wire, then

$$C = \frac{\pi n r^4}{2l} \qquad \text{... [3.4(C)b]}$$

where n is rigidity modulus of the material of the wire.

From Eq. 3.4(C)a and 3.4(C)b

$$C = \frac{4\pi^2 I}{T^2} = \frac{\pi n r^4}{2l} \quad \text{or, } n = \frac{8\pi l I}{r^4 T^2} \qquad \text{... [3.4(C)c]}$$

knowing the values of l, I, T and r, the value of n can be determined.

If the heavy body used be a solid metallic cylinder of mass m and Radius R and is allowed to oscillate about its axis of symmetry, then its moment of inertia about the axis of suspension is given by

$$I = \frac{1}{2}MR^2 \qquad \text{... [3.4(C)d]}$$

Apparatus :

Dynamical method apparatus; Stop watch; Slide calliper; Screw-gauge; Balance; Metre scale.

The dynamical method apparatus consists of a long wire PQ suspended from the point P and carries a heavy solid cylinder (X) at its lower end (Fig. 3.12). The suspension wire passes through the axis of the cylinder. A cross mark (×) is given on the surface of the cylinder by which the time period of oscillation of the cylinder can be determined. The whole instrument is kept under a glass cover to protect it from air current.

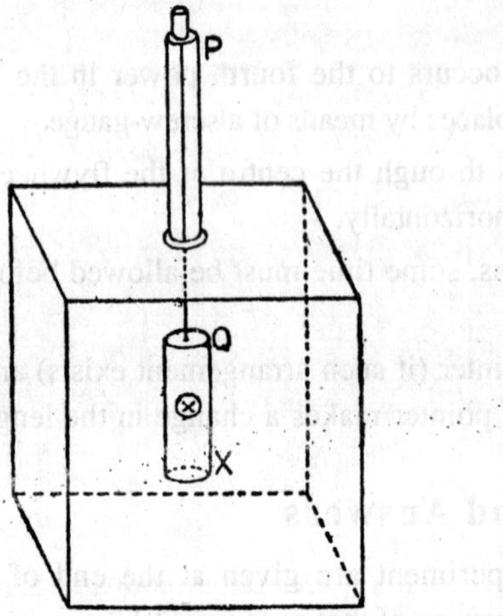

Fig. 3.12 : Dynamical Method Apparatus

Procedure :

(i) The cylinder is removed from the wire and its mass (M) is determined by a balance. Its diameter is measured by a slide calliper from which its radius (R) to obtained. Using Eq. 3.4(C)d, its moment of inertia is thus determined.

(ii) After suspending the cylinder from the wire, the length (l) of the wire is measured from the point of suspension to the point at which the wire is attached to the cylinder.

(iii) The radius of the wire is measured by a screw-gauge and this measurement is done at least at five different places of the wire taking two readings at right angles at each place. The mean value of the radius (tr) of the wire is thus obtained.

(iv) The cylinder is then given a small twist and allowed to oscillate about the axis of suspension. Keeping attention on the (×) mark on its surface, the time taken for 30 oscillations is recorded. Thrice such observations are taken and from these the time period (T) is determined.

(v) Finally, using Eq. 3.4(C)c the value of n is obtained, substituting the value of l, I, r and T.

Results :

Mass of cylinder (M) = ... gm

(A) Determination of radius (R) of the cylinder :

Table 1

Vernier constant of the Slide Calliper = ... cm

[A table similar to Table 2.3 is to be made for recording data for radius determination.]

(B) Determination of Moment of Inertia of the cylinder :

Table 2

No. of Obs.	Mass. of the cylinder (M)	Radius of the cylinder (R)	Moment of Inertia of the cylinder $I = \frac{1}{2} MR^2$	Mean Value of I
1.	...... gm	...... cm	...... gm-cm^2	...... gm-cm^2
2.	...... gm	...... cm	...... gm-cm^2	

(C) Determination of the length (l) of the wire :

(i) ...... cm; (ii) ...... cm; (iii) ...... cm; (iv) ...... cm.

Mean length (l) = ... cm

(D) Determination of the radius (r) of the wire :

Table 3

[Reading at five different places are to be taken and a table similar to Table 1 of Expt. No. 3.5 is to be made.]

(E) Determination of Time Period (*T*) :

Table 4

No. of Obs.	Time for 30 Oscillations (sec)	Mean Time (sec)	Time Period (*T*) (sec)
1.	......		
2.	......	......	......
3.	......		
4.	......		

Calculations :

From Eq. 3.4(C)c, $n = \dfrac{8\pi l I}{r^4 T^2}$

Substituting the values of *l*, *I*, *r* and *T* in this equation, the value of *n* can be obtained.

$\therefore n = \ldots\ldots\ldots$ dynes/cm^2

Precautions :

(i) Since the value of *r* occurs to be fourth power and that of *T* to the second power, their values should be determined very precisely and carefully.

(ii) The suspension wire must pass through the axis of the cylinder.

(iii) During torsional oscillation of the cylinder, care should be taken to avoid pendulum type of oscillation of the cylinder.

3.5 Determination of Young's modulus of a bar by Koenig's method

Theory :

Young's modulus of the material of a beam can be found out by a method devised by Koening, in which the depression of the beam due to a load suspended from the beam, is observed and measured with the help of two reflecting mirrors and a telescope. In the arrangement as shown in Fig. 3.13, if *M* is the load applied in gm, L is the distance between the knife-edges K_1 and K_2, D is the distance between the scale and the remote mirror (M_2), *l* is the distance between the mirrors, and *b* and *d* being the breadth and depth of the beam, then Young's modulus (*Y*) of the material of the beam will be given by

$$Y = \frac{3MgL^2(t+2D)}{2bd^3x} \qquad \ldots(3.5a)$$

where *x* is the difference in the scale readings as observed through the telescope after reflections from the mirrors and caused due to the application of the load *M*. Noting this difference of scale readings for a given load and measuring the different constants of the apparatus and dimensions of the beam, the value of *Y* can be obtained from Eq. (3.5a).

Plotting a graph with load (*M*) versus scale reading difference (*x*), which will be a straight line graph, the value of *Y* can also be obtained from this graph.

Apparatus :

Koenig's apparatus (Fig. 3.13), Telescope, scale, loads (1 kg, 2 kg ..., etc.); experimental bar, etc.

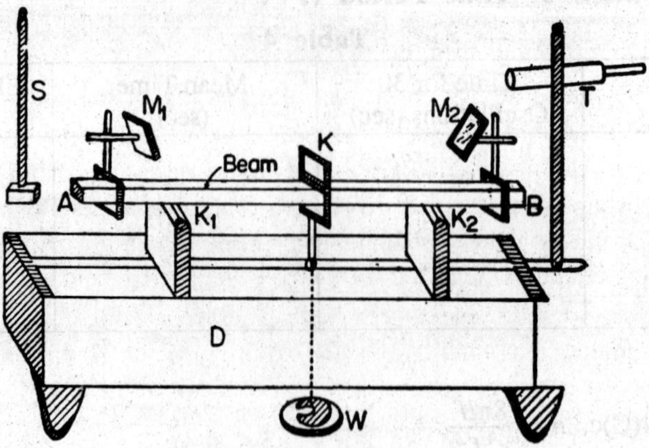

Fig. 3.13 : Koenig's method to find Y

The apparatus holds the experimental bar (AB) which carries a knife-edge (K) supporting the loads on a pan (W). At the ends of the bar are two mirrors (M_1 and M_2), almost normal to the bar but displaced slightly to enable a scale S to be seen through a telescope T, the light from S having suffered two reflections at the mirrors. The telescope carries a sharply focused cross-wire which is made to coincide with the scale divisions of the attached scale (S). When the beam is loaded, the mirrors will be turned towards each other (inward) as a result of the depression produced in the beam and the scale divisions seen through the telescope, will thus be altered. The difference in the scale readings due to a given load is denoted by x. The beam is supported on two more knife-edges (K_1 and K_2) placed on a rigid base (D) and the hanger carrying the load attached to a hook suspended from the knife-edge (K), passes through a slot cut through the baseboard.

The schematic diagram for the shift of the scale readings is shown in Fig. 3.14.

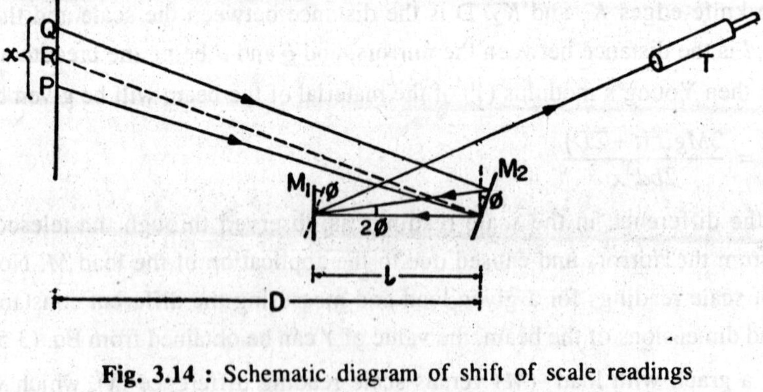

Fig. 3.14 : Schematic diagram of shift of scale readings

Method :

(i) The experimental beam (AB) is placed horizontally on two knife-edges (K_1, K_2) symmetrically. The knife-edge K is placed properly at the mid-point of the beam and the scale pan W is suspended from the hook attached to the knife-edge.

(ii) The mirrors (M_1 and M_2) are properly set (both in same inclined position), so that the illuminated scale (S) readings are visible through the properly focused telescope (T) after the successive reflections from the two mirrors.

(iii) With no load initially at the pan, the reading of the scale is noted as visible through the telescope. This is the initial reading (x_0).

(iv) Now, gradually weights of 1 kg (or 0·5 kg) are hung from the hanger and for each case, as the beam is depressed, the readings of the scale as visible through the telescope are recorded. From each of these readings, if the initial reading is subtracted, then the shift of the scale reading corresponding to tht respective weight will be obtained.

(v) The loads are then gradually decreased stepwise as before and scale readings are again noted. So, for a given load, two scale readings are obtained, one while the load is increasing and the other while it is decreasing. The mean of these two scale readings gives the actual scale reading for the corresponding load.

(vi) By changing the position of the knife-edges (K_1 and K_2), two more sets of readings are taken for two different values of the distance (L) between the knife-edges. Altogether thus three sets of readings are obtained for three values of L, each time readings being taken for loads increasing and decreasing.

(vii) The bar AB is now rotated through 90°, when the former breadth now becomes the depth of the bar. The operations (iii) to (vi) are repeated for this setting. (If the bar supplied is of very small depth, it may not be possible to make it stable by rotating it through 90°. Then operation (vii) should be done by rotating the bar through 180°).

(viii) Finally, the breadth (b) and depth (d) of the bar are measured by a screw-gauge/slide calliper at different positions of the bar and their mean values are determined.

(ix) Since, in the theory, the expression for Y contains depth to the third power, so it is always advisable that the depth should be measured very accurately with the help of a screw-gauge to avoid any possible error in the experiment.

(x) Finally, a graph may be plotted with loads (M) on the X-axis and corresponding shift (x) of scale readings on the Y-axis. The graph will be a straight line passing through the origin. From this graph, taking an arbitrary point, the value of Y can also be determined (See 'Drawing of Graphs')

Results :

(A) Load-Scale reading data :

Table 1

Distance between the knife-edges (K_1 and K_2) (cm)	No. of Obs.	Load (M) (kg)	Scale reading for		Mean scale reading $\frac{P+Q}{2}$ (cm)	Shift of scale readings (x) (cm)
			Load increasing (P) (cm)	Load decreasing (Q) (cm)		
	1.	0	......	......	...... (a)	0 (x_0)
L_1	2.	0·5	......	......	...... (b)	($b-a$) ... x_1
	3.	1·0	......	......	...... (c)	($c-a$) ... x_2
	4.	1·5	......	......	...... (d)	($d-a$) ... x_3
	5.	2·0	......	......	...... (c)	($e-a$) ... x_4
	6.	2·5	......	......	...... (f)	($f-a$) ... x_5
	7.	3·0	......	......	...... (g)	($g-a$) ... x_6

[Similar tables are to be made for two more values of L_2 and L_3 by changing the distance between the knife-edges K_1 and K_2].

(B) Load-Scale reading data when the beam is turned through 90° (or 180°) :

Table 2

[Similar Table like Table 1 is to be made].

(C) Distance between the scale and the remote mirror M_2 :

$$D = \frac{\dots+\dots+\dots}{3} = \dots\dots \text{ cm}$$

(D) Distance between the mirrors M_1 and M_2 :

$$l = \frac{\dots+\dots+\dots}{3} = \dots\dots\dots \text{ cm}$$

(E) Measurement of depth of the bar :

[A table similar to Table 3 of Experiment No. 3.3 is to be made.]

(F) Mesurement of breadth of the bar :

[A table similar to Table 4 of Experiment No. 3.3. is to be made].

(G) Drawing of graphs :

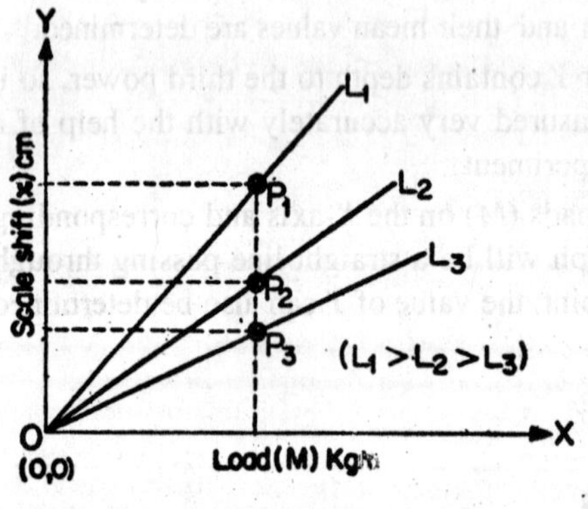

Fig. 3.15 : Load-scale shift curve

A graph is plotted with the load (M) on the X-axis and the shift of scale readings (x) along the Y-axis. The graph will be a straight line passing through the origin (Fig. 3.15). The origin should be taken as (0,0). On this graph, from and arbitrary point (P) taken, the values of M and x can be had from the values of abscissa and ordinate of the point and substituting these values in Eq. (3.5a), the value of Y can be obtained. For three different values of L (distance between the knife-edges K_1 and K_2), three straight line graphs will be obtained, which can be plotted on the same graph paper with the same scale. Scale shifts for the same load but for different values of L can be had and values of Y for different values of L can thus be determined. When the bar is rotated through 90° (or 180°), similar graphs are to be plotted again.

Calculations :

From the graph,

When load (M) = ... gm

Scale shift (x) = ... cm

From Eq. (3.5a), $Y = \dfrac{3MgL^2(l+2D)}{2bd^3x}$ = ... dynes/cm²

Three values of Y are determined for three different values of L (from the graph) and hence the mean value of Y is obtained.

Precautions :

(i) The beam must be kept horizontal and the hanged suspension system must not touch the baseboard.

(ii) The scale and the mirrors must not be disturbed by any means during the experiment.

(iii) Scale reading must be focused with the cross-wire avoiding parallax.

(iv) Depth must be measured by a screw-gauge to minimise error.

(v) The scale must be kept vertical so that it always remains within the field of view of the telescope.

[See other Precautions of Experiment No. 3.3].

Error Calculations :

From the expression,

$$Y = \frac{3MgL^2(l+2D)}{2bd^3x}$$

we get

$$\frac{\delta Y}{Y} = \frac{2\delta L}{L} + \frac{\delta(l+2D)}{(l+2D)} + \frac{\delta b}{b} + 3\frac{\delta d}{d}.$$

Since $\left(\frac{M}{x}\right)$ is determined from the mean straight line curve, proportional error in $\left(\frac{M}{x}\right)$ is not very large. By calculating proportional error for each term on the R.H.S., the final percentage error can thus be calculated. (See Art. 1.7).

Questions and Answers

Q.1. In determining the value of Y of a beam, which method is more advantageous—the flexure method or this method?

A. : This method is more accurate, since for any small depression of the beam, the depression being measured from the shift of scale readings reflected from two mirrors, we get accurate value of the depression caused.

Q.2. Does the position or the inclination of the mirrors affect the scale reading?

A. : From the theory and expression of Y, it is obvious that the scale reading (x) depends on the value of the distance between the mirrors (l) but the inclination of the mirrors for a particular set of readings must not change during the experiment although they can be set at any angle at the start of the experiment.

Q.3. How can the accuracy of the experiment be increased?

A. : The greater the value of the scale-shift (x) for a given load, greater will be the accuracy of the experiment. Now for a given load (M), the value of x can be made to be greater by increasing the value of L or l or D. Thus it is advised to take a long length of beam (distance between the knife-edges K_1 and K_2), to set the mirrors long distance (l) apart and to keep the scale at a distance (D) from the remote mirror (M_2).

[For other questions on depression of beam under load, see Questions and Answers under Experiment No. 3.3].

3.6 Determination of Surface Tension of a liquid by capillary tube method and verification of Jurin's law

Theory :

If a capillary glass tube of uniform circular corss-section is partly immersed in a liquid which wets glass, the liquid rises the tube to a certain height above the surface of the liquid. If h be the height of the base of the concave meniscus of the liquid column and r be the internal radius of the rube at which the menisucs stands; then the surface tension T of the liquid is given by

$$T = \frac{1}{2} r \rho g \left(h + \frac{1}{3} r\right) = \frac{1}{2} r \rho g h' \quad \text{... (3.6a)}$$

where $h' = \left(h + \frac{1}{3} r\right)$.

where, ρ is the density of the liquid and g the acceleration due to gravity.

Eq. (3.6a) is applicable in case of pure water and glass free from any contamination. In case of other liquids like turpentine, paraffin, etc., which have an angle of contact θ with the glass, Eq. (3.6a) will be modified as

$$T = \frac{r \rho g}{2 \cos \theta} \left(h + \frac{1}{3} r\right). \quad \text{... (3.6b)}$$

In fact, for pure water and clean glass, $\theta = 0$.

Experimental Arrangement :

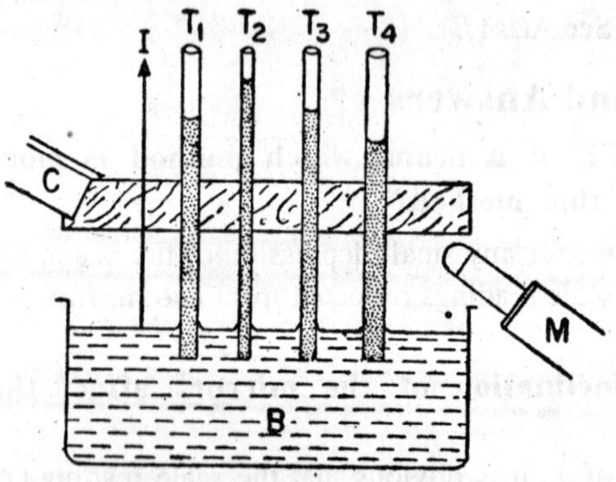

Fig. 3.16 : Surface tension by Capillary tube method

A long narrow glass tube (diameter 4 to 5 mm) is taken which is cleansed and washed inside carefully. From this tube, narrow capillary tubes of different diameters are drawn. The tube drawn should be of uniform diameter each. These tubes are attached on a glass plate (Fig. 3.16), each being equidistant from each other and parallel to one another. They may be fixed by wax. a needle pointer (I) about 10 or 15 cm long is also attached on the same glass plate by the side of the tubes and also parallel to them. All the tubes and the needle must remain vertical and parallel, which can be ensured by taking help of a piece of graph paper while fixing them.

Method :

(i) The glass plate with the tubes and the pointer, is held by a clip (C) on stands. It is checked that the tubes remain vertical in this position.

(ii) The experimental liquid is taken in a clean glass trough (B) and its temperature (t°C) is noted. Next the glass plate along with the tubes and clip is lowered and the tubes are made to immerse partially (their lower ends) in the liquid as shown in the figure. While doing this, care must be taken to see that the lower tip of the needle **just** touches the liquid surface.

(iii) A travelling microscope (M) is set with its axis horizontal and bed levelled. Its horizontal traverse must be parallel to the plane of the glass plate. Its eyepiece is focused on the cross-wires.

(iv) The axis of the microscope is brought near the tubes and the liquid levels inside the tubes are focused. The levels of liquid inside the tubes are concave but through the microscope, they appear to be convex. Due to capillary action, the liquid rises inside the tubes and the heights upon which they rise, depend on the radius of the tube.

(v) The tangential surfaces of the concave liquid meniscus in each tube are focused successively by the microscope and for each case, the readings of the main scale (vertical) and the vernier are noted.

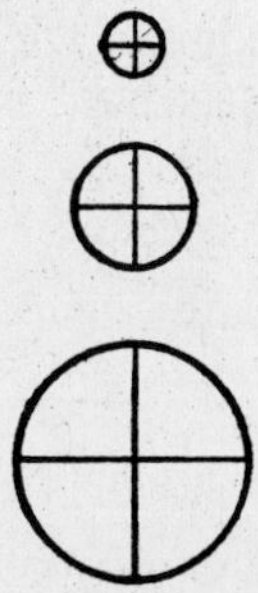

Fig. 3.17 : Radius of capillary tube

(vi) Next the microscope is moved horizontally and the upper tip of the needle is focused and readings from the scale are noted. From these two readings, the distance between the needle tip and the liquid meniscus in each tube (above or below the needle tip) can be obtained.

(vii) The position of the concave meniscus of each tube is marked by an ink dot on the outer surface of the tube and all the tubes are cut at the ink marks with a sharp glass cutting file.

(viii) The tubes are then held with the glass strip in a horizontal plane such that the cut ends of the tubes face the microscope.

(ix) The microscope is focused on the cut ends of the tubes successively when more or less circular cross-sections will be visible. The horizontal and the vertical diameters of the inner circle of the section, i.e., of the bore of the capillary (Fig. 3.17) are measured. Two readings at each place may be taken preferably.

(x) The needle is removed and its length is determined by means of a metre scale.

(xi) Finally, the density of the liquid has to be determined by a specific gravity bottle. In case, the liquid taken is water, its density is obtained from the standard value, noting its temperature.

Results :

Vernier constant of the microscope :

One small division of the main scale = ... cm

.... vern. divn. = ... main scale div.

One vern. divn. = ... main scale div.

∴ vernier constant = ... cm

Temperature of the liquid (t°C) = ... °C

(A) Determination of height (h) of the liquid :

Table 1

Tube No.	Microscpe readings for						Difference of two readings	Height of the liquid column
	Concave liquid meniscus			Top end of the pointer				
	Main Scale (M) (cm)	Vernier (V) (cm)	Total ($M + V = X$) (cm)	Main Scale (M) (cm)	Vernier (V) (cm)	Total ($M + V = Y$) (cm)	($X \sim Y$) (cm)	$h = l + (X \sim Y)$ (cm)
1.	...	...	...	...	...	...	...	...
2.	...	...	...	...	...	...	...	...
3.								
4.								
5.								
6.								

Table 2

Tube No.	No. of Obs.	Readings for						Horizontal/ Vertical diameter (d_1/d_2) (cm)	Meandiameter $d = \frac{d_1 + d_2 + d_3 + d_4}{4}$ (cm)	Mean radius (r) of the tube, $r = \frac{d}{2}$ (cm)
		Left/Lower end			Right/Upper end					
		Main Scale (M) (cm)	Vernier (V) (cm)	Total (M + V) (cm)	Main Scale (M) (cm)	Vernier (V) (cm)	Total (M + V) (cm)			
I.	1. (a)	...	...	... (x_1)	...	...	... (y_1)	$d_1 = x_1 \sim y_1$		...
	(b)	...	...	... (x_2)	...	...	... (y_2)	$d_2 = x_2 \sim y_2$		
	2. (a)	...	...	... (x_3)	...	...	... (y_3)	$d_3 = x_3 \sim y_3$		
	(b)	...	...	... (x_4)	...	...	... (y_4)	$d_4 = x_4 \sim x_4$		
II.	1. (a)									
	(b)									
	2. (a)									
	(b)									
III.	1. (a)									
	(b)									
	2. (a)									
	(b)									
...	...									
	...									

[(a) and (b) are readings for horizontal and vertical diameters at the same place]

(C) Determination of the length (l) of the needle :

Table 3

No of Obs.	Metre scale readings for		Length of the needle $l=(l_2-l_1)$ (cm)	Mean value of the length of the needle (l) (cm)
	Left end of the needle (l_1) (cm)	Right end of the needle (l_2) (cm)		
1.	......	......	......	
2.	......	......	......	......
3.	......	......	......	

(D) Determination of the density (ρ) of the liquid :

Table 4

Mass of empty bottle (W_1 gm)	Mass of bottle + liquid (W_2 gm)	Mass of bottle + water at t°C (W_3 gm)	Density of the liquid $\rho = \frac{W_2 - W_1}{W_3 - W_1} \times \rho_t$
	...	...	...... gm/cc

[If the experimental liquid is water, then the temperature of the water (t°C) is noted and the density of water at t°C is determined from the standard table.]

(E) Calculation of Surface Tension (T) :

Table 5

Tube No.	Height of liquid column (h) (cm)	Effective height of liquid column $\left(h' = h + \frac{1}{3}r\right)$ (cm)	Radius of the tube (r) (cm)	Value of ($h' \times \rho$) (cm²)	Mean value of ($h' \times r$) (cm²)	Surface tension $T = \frac{1}{2} r\rho g h'$ (dynes/cm)	Men value of surface tension T (dynes/cm)
1.	...	...	...	...	...	(i) from numerical value of ($h' \times r$)	
2.	...	...	...	...	...	=	...
3.	...	...	...	...	...	(ii) from graph	at t°C
4.	...	...	...	...	...	=	...
...	...	...	...	...	...	...	...
...	...	...	...	...	...	...	...

Verification of Jurin's Law and the Graph

If the radius of a capillary tube be r and the height to which a liquid rises in this tube be h, then according to Jurin's law, the product of h' and r is constant, where h' is the effective height $\left(= h + \frac{1}{3}r\right)$ of the liquid column.

From Eq. (3.6a) for a given liquid at a given temperature, the value of ($h' \times r$) should be constant since the other terms of the equation remain constant. From Table 5, it will be observed that the values of $(h' \times r)$ obtained in column V, appear to be remaining constant which verifies Jurin's law numerically.

Again, the law can also be verified graphically. If a graph be drawn with the radii of the tubes (r) plotted along X-axis and the corresponding effective heights (h') of the liquid column along the Y-axis, then the graph will be a *rectangular hyperbola* (Fig. 3.18a). If $(h' \times r)$ is constant, then a graph of h' versus r will be a rectangular hyperbola which proves Jurin's law. Thus if such a graph is obtained, then Jurin's law is verified. Further, another graph may also be plotted with h versus $\frac{1}{r}$ (Fig. 3.18b) in which case, it will be a straight line curve as shown.

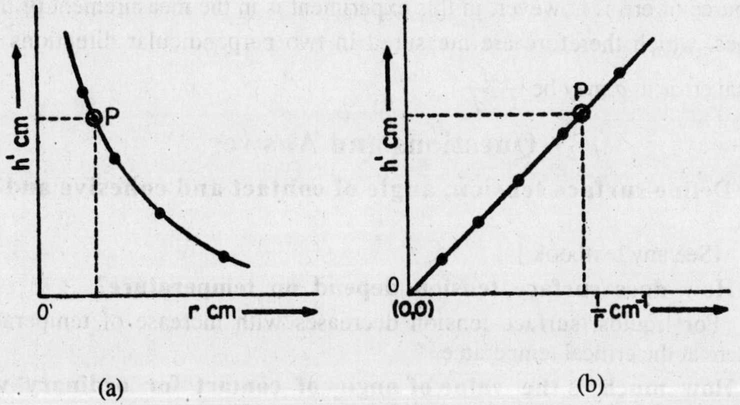

Fig. 3.18 : Graphs of $(h'\text{-}r)$ and $(h'\text{-}\frac{1}{r})$

Calculation :

From Table 5, the mean value of $(h' \times r)$ when substituted in Eq. (3.6a), the value of T is obtained. Again, from the graph of $(h' - r)$, if any arbitrary point P is taken (Fig. 3.18a), then from the value of the abscissa and ordinate of this point, the value of T can also be obtained. The mean value of T at the noted temperature is obtained from these two values.

Precautions :

(i) Since the value of surface tension of a liquid is very much dependent on the temperature of the liquid, hence the temperatue of the liquid must be noted before and after the experiment and the mean temperature is calculated.

(ii) The glass tube from which the capillaries are drawn, must be cleansed before use so as to make it free from grease and other contaminations, since the presence of any contamination will affect the value of the surface tension.

(iii) The radius of the tube is supposed to be measured at the position where the liquid meniscus rises in the tube. Further, since the amount of rise of liquid depends on the radius of the tube, the tubes should be drawn with uniform bore as far as practicable.

(iv) The tubes must be set vertical and parallel to each other and to the needle.

(v) Since the height of the liquid column and the radius of the tubes are of very small magnitudes, they must be measured very carefully.

(vi) While taking readings with the microscope, care must be taken to avoid any back-lash error.

Error Calculations :

$$T = \frac{1}{2}\rho grh' = \frac{1}{4}\rho gdh' \text{ where } d \text{ is the diameter of the tube.}$$

Hence, maximum proportional error in T will be

$$\frac{\delta T}{T} = \frac{\delta d}{d} + \frac{\delta h'}{h'} \text{ (since } \rho \text{ and } g \text{ are constants).}$$

The diameter is measured by taking the difference of two readings and thus the maximum error in this measurement will be 2 vernier divisions of the microscope, i.e., $2 \times v.c.$ Again, h' $[= l + (X - Y)]$ is obtained by taking the difference of two readings. Hence, total maximum error in measuring h' may be 4 vernier divisions, i.e. $4 \times v.c.$ The greatest source of error, however, in this experiment is in the measurement of the diameter of the tubes, which therefore are measured in two perpendicular directions so that the proportional error in d may be $\frac{\delta d}{\sqrt{2}d}$.

Questions and Answers

Q.1. Define surface tension, angle of contact and cohesive and adhesive forces.

A. : [See any textbook.]

Q.2. How does surface tension depend on temperature?

A. : For liquids, surface tension decreases with increase of temperature and it becomes zero at the critical temperatue.

Q.3. How much is the value of angle of contact for ordinary water and glass interface and for mercury-glass interface?

A. : For ordinary water and glass $\theta = 18°$ for mercury-glass $\theta = 138°$.

Q.4. Do all liquids rise in a capillary tube?

A. : No, only those rise which wet glass, while others which do not wet glass (mercury-glass), the liquid depresses downward.

Q.5. Is this method applicable for all types of liquids?

A. : No, this method is not applicable for volatile liquids.

Q.6. Why does a liquid rise in a capillary tube?

A. : [See the theory of 'capillarity' from any textbook.]

Q.7. On what factors does the ascent of liquid depend?

A. : It depends upon the radius of the tube, the angle of contact and the S. T. of the liquid.

Q.8. No rise is observed when an ordinary tube is immersed in water, why?

A. : The radius of an ordinary tube being very high, the rise is too small and becomes inappreciable.

Q.9. Will a given liquid rise up to the same height in different tubes having same radii but of different materials?

A. : The rise depends upon the angle of contact and this angle is different for different materials.

Q.10. What will happen if the length of the capillary tube is less than the height to which water is expected to rise?

A. : The height to which water rises, depends upon the radius of the curvature of the meniscus. If the tube is of insufficient length, this water rises to the top of the tube and forms there a surface with a new radius of curvature (r'), such that the product of r' and h (length of the tube) remains constant and equilibrium is maintained.

Q.11. Wherefrom does the energy come when water rises in a capillary tube against gravity?

A. : As water rises in the tube, the water-glass interface increases while air-glass interface decreases by the same amount. The water-glass interface thus gains energy due to S.T. and the latter loses energy. But the S.T., which is equal to the surface energy, of air-glass surface is greater than that of water-glass surface. Thus, loss in energy of air-glass surface is greater than the gain by water-glass surface. The balance of lost energy is used up in making the water rise against gravity.

Q.12. Can you give some practical examples of capillarity?

A. : Rise of kerosene oil in the wick of a lamp, absorption of ink by blotting paper and sucking water from ground by root-hairs of plants are the examples of capillarity.

Q.13. Why radius is measured at a point in perpendicular directions?

A. : This reduces the error due to ellipticity, if any, of the bore to a minimum. It Further reduces the proportional error as well.

Q.14. Why should the tubes be vertical?

A. : If the tube is inclined, the meniscus will become elliptical and thus the liquid rises to a small height and the formula does not hold good.

Q.15. Which is preferred—distilled water or tap water?

A. : Distilled water is likely to have traces of grease which makes the angle of contact of water considerable and thus the rise will be small. This makes the obtained value of S.T. lower.

Q.16. What are the disadvantages of this method?

A. : (a) Availability of preferably clean and uniformly-bored capillary tube is difficult.

(b) Experiment is static and thus chances of contamination are liekly to take place.

(c) It is difficult to measure the height of liquid and radii of tubes very precisely.

(d) Assumption that the angle of contact is zero and the meniscus is spherical, is not fully justified.

(e) Variation of S.T. with temperature, cannot be studied by this method.

3.7 Determination of the Surface Tension of a liquid by the drop-weight method

Theory :

When a liquid drop is allowd to fall from a vertical tube, then when the drop is about to break away from the tube, it is assumed that the drop has the cylindrical form as shown in Fig. 3.19. If r be the radius of the orifice, the excess pressure inside the drop at this level is equal to $\frac{T}{r}$ due to the cylindrical curvature of the liquid, where T is the surface tension of the liquid forming the drop. This produces a downward thrust on the drop equivalent to

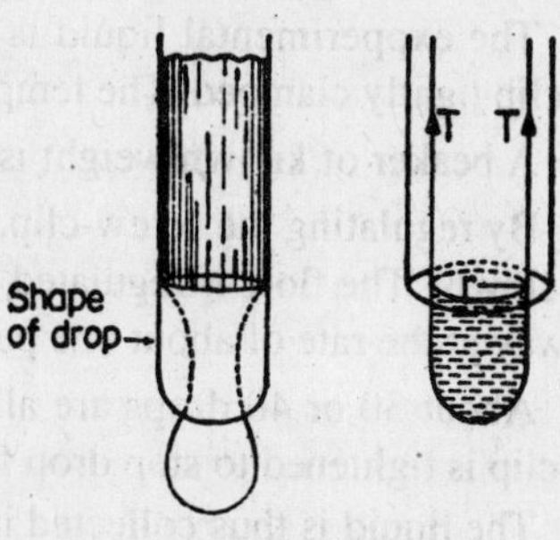

Fig. 3.19 : Shape of the drop

$$\frac{T}{r} \times \pi r^2 = \pi T r. \quad \text{... (3.7a)}$$

If m is the mass of the drop, the total downward force is

$$\pi T r + mg. \quad \text{... (3.7b)}$$

This force is balanced by the upward forces of surface tension round the circle of contact. This force is equal to $2\pi r T$.

Hence, $2\pi r T = \pi T r + mg$

$$\text{or, } T = \frac{mg}{\pi r} \quad \text{... (3.7c)}$$

This expression is obtained on the assumption that the drops break away under ideal statical conditions. To a closer approximation, Lord Rayleigh showed that

$$T = \frac{mg}{3 \cdot 8 r} \quad \text{... (3.7d)}$$

From a counted number of drops collected in a beaker of known mass, the average mass of a drop is found. Knowing the radius of the orifice from measurement by a microscope, the value of T can be found out from Eq. (3.7d).

Apparatus :

A funnel or a burette, rubber tubing, clamp, narrow tube of about 3 or 4 mm diameter, screw-clips, beaker and travelling microscope, experimental and liquid.

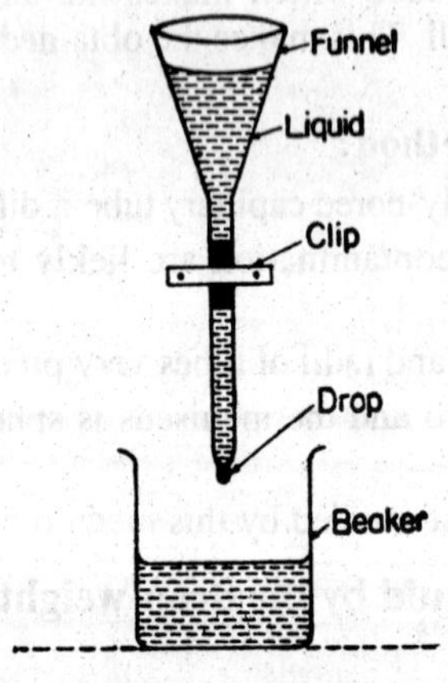

Fig. 3.20 : Surface tension by drop-weight method

A tube is drawn out to a fine narrow shape of about 3 or 4 mm in diameter at the end. It is connected to a burette or a funnel by means of a rubber tubing (Fig. 3.20). The funnel or the burette and the tube are clean and the end of the tube is coated with paraffin wax. A beaker is placed below the tube to collect the drops.

Method :

(i) The funnel/burette and the tube are thoroughly cleansed. The tube is fitted to the rubber tubing and the latter is attached to the end of the funnel/burette. A screw clip is fixed on the rubber tubing which controls the flow of drops.

(ii) The expperimental liquid is poured inside the funnel/burette from above, with the screw-clip tightly clamped. The temperature of the liquid is noted.

(iii) A beaker of known weight is placed first below the tube to collect the drops.

(iv) By regulating the screw-clip, flow of drop is started and the drops are allowed to form very slowly. The flow is regulated in such a manner that drops are formed and allowed to break away at the rate of about one per minute or even at a slower rate.

(v) About 30 or 40 drops are allowed to be collected in the beaker at a time when the screw-clip is tightened to stop drop formation.

(vi) The liquid is thus collected in the beaker and its weight is again taken. From the knowledge of the weight of the empty beaker, the weight of the liquid drops collected (for a known number of drops) is obtained.

(vii) From the known number of drops collected, the average mass of a single drop is calculated.

(viii) Finally, the radius of the orifice is determined by a microscope taking both horizontal and vertical diameters (as in 'capillary tube' experiment).

(ix) From Eq. (3.7d), the value of T is obtained, knowing the values of the other quantities.

N.B. If the radius of the tube is about 3 or 4 mm, the number of drops collected at a time may be 30 or 40. But, if the tube is much narrow (1 mm or less in diameter), then at least 80 or 100 drops are to be collected at a time to get some appreciable weight of the liquid.

Results :

Mass of the empty glass beaker (W) = …… gm

Temperature of the liquid (t) = …… °C.

(A) Determination of mass of liquid drops and their numbers :

Table 1

Mass of empty beaker (W) gm	No. of Obs.	No. of drops counted (N)	Mass of beaker + liquid (W') (gm)	Mass of liquid (M) ($W' - W$) (gm)	Mass (m) of each liquid drop (M/N) (gm)	Average mass of each liquid drop (m) (gm)
	1.	…	…	…	…	
	2.	…	…	…		
	3.	…	…	…		
…						
	4.	…	…	…	…	
	5.	…	…	…	…	
	6.	…	…	…	…	
	…	…	…	…	…	

(B) Determination of radius (r) of the orifice :

Table 2

No. of obs.	Readings for						Hori-zontal/ Vertical diameter (d_1/d_2) (cm)	Mean Diameter (d) (cm)	Mean radius of the orifice ($r=d/2$) (cm)
	Left/lower end			Right/Upper end					
	Main Scale (M) (cm)	Vernier (V) (cm)	Total (M + V) (cm)	Main scale (M) (cm)	Vernier scale (V) (cm)	Total (M + V) (cm)			
1. (a)		……	……	……	……	……	(d_1)		
(b)	……	……	……	……	……	……	… (d_2)		
2. (a)	……	……	……	……	……	……	… (d_1)		
(b)	……	……	……	……	……	……	… (d_2)	…	…
3. (a)	……	……	……	……	……	……	… (d_1)		
(b)	……	……	……	……	……	……	… (d_2)		
……									
……									

[(a) and (b) correspond to horizontal and vertical diameter readings.]

Calculations :

From Eq. (3.7d), the value of surface tension,

$$T = \frac{mg}{3.8r}$$

Substituting the values of mass of each drop (m) and radius of the orifice (r) from Table 1 and Table 2, we get the value of surface tension of the experimental liquid T = ... dynes/cm at t = ...°C.

Discussion :

There remains an uncertainty as to whether the correct expression of T is $\frac{mg}{\pi r}$ or $\frac{mg}{3 \cdot 8r}$.

This uncertainty may be avoided by taking the expression at $T = \frac{mg}{K}$ ... (3.7e)

and finding the value of K for the particular tube used. The experiment is first performed with a liquid of known surface tension (say, water) and the value of K of the tube is determined from Eq. (3.7e). The same tube is then used for determination of surface tension of the unknown experimental liquid after removing all traces of water from the funnel/burette and the tube. From this experiment, since K is known, the value of T of the unknown liquid can be determined using the same equation. Surface tension of alcohol, benzene, paraffin oil, etc. can be determined by this method.

Further, this method is very suitable for comparison of surface tensions of two different liquids. If m_1 and m_2 be the masses of drops of two liquids issuing from the same orifice and if T_1 and T_2 are their surface tensions, then one gets

$$\frac{T_1}{T_2} = \frac{m_1}{m_2}. \quad \text{... (3.7f)}$$

Precautions :

(i) The radius of the orifice must be measured very carefully as it is a very small quantity.

(ii) Large number of drops are to be collected to get the average mass of each drop and to get an appreciable difference of mass of the beaker before and after the drops are collected.

(iii) Drops should be formed at a very slow rate (one or two drops per minute) so that they can break away individually.

(iv) Weights should be taken with precision.

Error Calculations :

From the expression of surface tension

$T = \frac{mg}{3.8r}$, the proportional errors are obtained as,

$$\frac{\delta T}{T} = \frac{\delta m}{m} + \frac{\delta r}{r}.$$

If sufficient amount of liquid drop is collected then $\frac{\delta m}{m}$ is comparatively smaller than $\frac{\delta r}{r}$. Hence, the greatest error creeps in from the measurement of radius of the orifice, which thus should be measured very precisely. The percentage error in the measurement of T can obviously be calculated from the above expression.

Questions and Answers

Q.1. What are the advantages of this method over other methods of determination of surface tension of a liquid?

A.: In this method, the experimental procedure is very simple. Further, this method enables determination of S.T. of a volatile liquid and to compare the S. T. of two liquids. The method can be extended to find the interfacial surface tension of two non-miscible liquids (e.g., oil and water) by allowing drops of denser liquid to fall into the less dense liquid.

Q.2. On what principle is the theory of the experiment based?

A. : On the expression of the excess pressure over a curved liquid surface.

Q.3. If instead of the cylindrical curvature of the liquid surface at the orifice mouth, the curvature is spherical, how would the expression be modified?

A. : The expression for excess pressure in that case would have been $\frac{2T}{r}$ instead of $\frac{T}{r}$.

Q.4. Does the value of T depend on the material of the tube?

A. : No, the value of T depends only on the radius of the orifice and not on the material of the tube.

Q.5. On what factors does the accuracy of the experiment depend?

A. : The accuracy depends on the accuracy of the measurement of the radius of the orifice and of the mass of the drop.

3.8 To study the variation of surface tension of a liquid with temperature by Jaeger's method

Theory :

When an air bubble of radius r is formed at a depth d inside a liquid of density ρ_1, then the excess pressure (p) inside the bubble at this stage is given by

$$p = \frac{2T}{r}, \qquad \text{... (3.8a)}$$

where T is the surface tension of the liquid at the given temperature of the liquid.

The radius of curvature of the bubble is minimum when it is hemispherical, i.e. when its radius is equal to the radius of the orifice where the bubble is blown.

In the arrangement shown in Fig. 3.21, the pressure inside the bubble is equal to the pressure inside the capillary tube C. This is equal to the atmospheric pressure P plus the difference of pressure inside the manometer M. If h is the difference of the heights of the manometric liquid of density ρ_2, then the pressure inside the bubble is

$$p_1 = P + h\rho_2 g. \qquad \text{... (3.8b)}$$

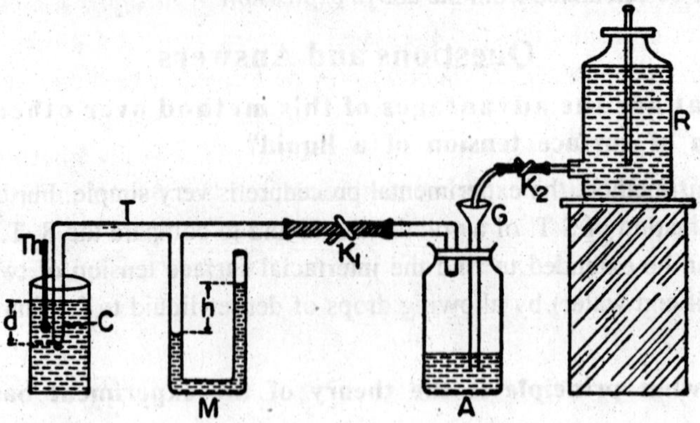

Fig. 3.21 : Jaeger's method to determine surface tension

If the bubble is blown at the end of a narrow capillary tube C whose orifice is immersed up to a depth of d inside the experimental liquid, then the pressure on the outer surface of the bubble formed is given by

$$p_2 = P + d\rho_1 g. \quad \text{... (3.8c)}$$

Thus the excess pressure (p) inside the bubble over the outer pressure will be given by

$$p = p_1 - p_2 = (P + h\rho_2 g) - (P + d\rho_1 g) = g(h\rho_2 - d\rho_1). \quad \text{... (3.8d)}$$

This is obviously equal to $\frac{2T}{r}$.

Hence, $\frac{2T}{r} = g(h\rho_2 - d\rho_1)$ or, $T = \frac{rg}{2}(h\rho_2 - d\rho_1)$. ... (3.8e)

In deriving Eq. (3.8e), it has been assumed that the bubble breaks away when its radius becomes equal to that of the orifice. This assumption is not quite correct and hence the above expression for T is inaccurate. The correct expression for T is given by

$$T = \frac{1}{2} g . f(r) . (h\rho_2 - d\rho_1), \quad \text{... (3.8f)}$$

where $f(r)$ is an unknown quantity and a definite function of the radius r having the same dimension as that of r. This is Jaegar's exact formula and the value of $f(r)$ is first determined from the given value of the surface tension of the liquid at the room temperature applying Eq. (3.8f). Thus knowing the value of $f(r)$, the value of T can be obtained from this equation at different temperatures, when the other quantities are measured.

In the experimental arrangement specially made, the experimental liquid can be heated gradually and its temperature raised and at each temperature, the measurements of the aforesaid quantities being made, the values of T at different temperatures can be obtained and thus the variation of surface tension of the liquid with temperature can be studied. The surface tension decreases with the rise of temperature of the liquid.

Apparatus :

Jaeger's apparatus; Thermometer, travelling microscope.

A thin glass tube C drawn in the form of a narrow capillary of diameter about 1 or 2 mm, is fixed vertically in the liquid whose surface tension is to be determined placed in a beaker. This tube is connected rigidly to a glass tube T which terminates in another rubber tubing to which is attached a pinch-cock K_1. The end of this tube is connected to a glass tube fixed with an aspirator system (A and R) from which by dropping water and by compressing the air confined, air bubbles may be formed at the other extreme end of the entire tube system and escaping through the orifice of the capillary tube C. When the bubble formed, is of the same radius as that of the orifice at C, it becomes unstable and breaks away. During the formation of the bubble, the liquid level in the open limb of the manometer rises and the maximum difference in heights (h) between the two columns is observed. The manometric liquid normally taken is kerosene but preferably, it should be of lower density like xylol, etc. This would give to a greater difference of liquid column (h) in the manometer for a given set-up.

Method :

(i) The capillary tube C after being cleaned, is set vertically and then immersed well inside the experimental liquid taken in a beaker. The manometer is partially filled with the manometric liquid whose density is either known or has to be determined by a sp. gravity bottle, if not known.

(ii) The opposite end of the experimental tube, which remains inside the aspirator A, is taken out of the aspirator and remains in open air. Under these circumstances, the readings for the liquid levels in the two limbs of the manometer are taken by a microscope The manometer will then show the atmospheric pressure (P).

(iii) Setting back the end of the glass tube inside A, the pinch-cock K_2 is slowly opened and the pressure of the air inside the aspirator A is gradually increased. A bubble of air is formed in the liquid in the beaker (at the end of the capillary tube C), and as the air pressure increases, it grows in size, finally breaking away. The pressure indicated by the manometer rises until the bubble becomes unstable, and suddenly falls as the bubble is detached from the tube. Successive bubbles are formed at the orifice of the tube and the pressure inside is so adjusted that bubbles are formed approximately one in every five or ten seconds.

(iv) The maximum difference of the liquid levels in the manometer is noted by a focused microscope. If the two limbs of the manometer are not visible and are not within the field of view of the microscope simultaneously, then the maximum rise or depression of liquid level in any one of the limbs, may be recorded. If the manometric tube is uniform, then the difference of the reading of any one limb from the reading taken under step (ii) will be the half of the total difference of the readings of the two limibs. This recording is reported a number of times from which the value of h is obtained.

(v) The liquid in the beaker is then slowly heated. Until its temperature rises by about 8 °C or 10 °C. Keeping the temperature constant at this new value, the experiment is repeated [(iii) and (iv)] and reading for h is noted at this temperature.

(vi) In this way, the liquid is gradually heated and about 5 or 6 sets of readings are taken for different temperatures.

(vii) Now the microscope is brought near the beaker and its readings for the water level in the beaker and the mouth of the orifice are taken. From the difference of these two readings, the value of d can be obtained.

(viii) Lastly, with the help of a sp. gravity bottle, the densities of the experimental liquid (ρ_1) and of the manometric liquid (ρ_2) are determined. If the experimental liquid be water, then its densities at different temperatures are obtained from the standard table.

(ix) Having known the values of $h, d, f(r), \rho_1$ and ρ_2 the value of T at different temperatures can be obtained from Eq. (3.8e). Thereafter, a graph is drawn with temperatures (θ°C) along the X-axis and surface tension (T) along the Y-axis. The graph (Fig. 3.21) shows the variation of surface tension of a liquid with temperature and it is apparent that surface tension decreases with temperature. From the graph, surface tension of the liquid at any arbitrary desired temperature can also be found.

Results :

(A) Determination of density of the experimental and the manometric liquids :

Table 1

Liquid	Mass of empty bottle (W_1) (gm)	Mass of bottle + liquid (W_2) (gm)	Mass of bottle + water (W_3) (gm)	Density of the liquid $\frac{W_2 - W_1}{W_3 - W_1}$ (gm/cc)
Experimental	...	...	...	ρ_1 = ... gm/cc
Manomertic	...	...	...	ρ_2 = ... gm/cc

(B) Record of temperature and manometric liquid level differences :

Table 2

V.C. of the microscope = ... cm

No. of Obs.	Temp. of bath (θ °C)	No. of bubbles observed	Manometric liquid readings for: Upper limb (cm)	Upper limb Mean (cm)	Lower limb (cm)	Difference of liquid levels in two limbs (h cm)
1.		0	..... (h_0)	..... (h_0)	..... (h_0)	0
2.	θ_1 (room temperature)	1	...			
		2	...			
		3	...	..... (h_1)		... $2(h_1 - h_0)$
		4	...			
		5	...			
3.	θ_2	1	...			
		2	...			
		3	...	..... (h_2)		... $2(h_2 - h_0)$
		4	...			
		5				
4.	θ_3	1				
		2				
		3	...	..... (h_3)		... $2(h_3 - h_0)$
		4	...			
		5				
		......				
		......				
		......				

(C) Reading for determination of the depth of the orifice below the liquid level in the bath :

Table 3

No. of Obs.	Microscope readings for Liquid level in bath (d_1 cm)		Microscope readings for Tip of the orifice inside the liquid (d_2 cm)		Depth of orifice (d cm) $= (d_2 \sim d_1)$
		Mean		Mean	
1.	......		......		
2.	......		......		
3.	......	......	......	......	......
4.	......		......		
5.	......		......		

(D) Determination of f(r) :

Table 4

Room temperature (θ_1) = ... °C; Density of experimental liquid (ρ_1) at θ_1°C = ... gm/cc; Density of the manometric liquid (ρ_2) = ... gm/cc; g = 980 cm/sec^2; Surface Tension (T) of the given liquid at θ_1°C (given) = ... dyne/cm.

Depth of the orifice (d cm)	Difference of liquid levels in two limbs at room temperature [$h = 2(h_1 - h_0)$ cm]	Value of $f(r)$ in cm $\dfrac{2T}{g(h\rho_2 - d\rho_1)}$
...	...	...

(E) Temperature-Surface tension reacords :

Table 5

No. of Obs.	Temperature (θ °C) of the liquid	Density of the liquid: Manometric (ρ_2 gm/cc)	Density of the liquid: Experimental (ρ_1 gm/cc)	Depth of orifice (d cm)	Diff. of manometric liquid levels (h cm)	Surface Tension $T = \dfrac{f(r)g}{2} \times (h\rho_2 - d\rho_1)$ (dynes/cm)
1.	θ_2°	...	...		...	...
2.	θ_3°	...	...	......	...	...
3.	θ_4°	...	...		...	...
4.	θ_5°	...	...		...	...
...	...	...	...		...	...

(F) Drawing of graph :

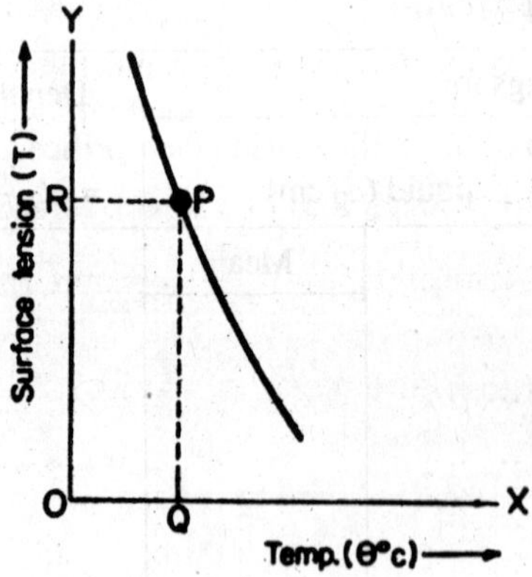

Fig. 3.22 : Temp-S.T. variation graph

On a graph paper, temperature (θ °C) is plotted along X-axis and the corresponding value of the surface tension (T) is plotted along Y-axis. The nature of the curve will be as shown in Fig. 3.22. From the graph, it is evident that the surface tension decreases with the rise of temperature. If it is required to determine the surface tension at any arbitrary temperature, the same can be done by taking the temperature at the point (P) on the graph and finding the corresponding value of surface tension.

The experiment can be repeated by taking different depths of the orifice.

Calculations :

From Eq. (3.8 f), surface tension

$$T = \frac{f(r)g}{2}(h\rho_2 - d\rho_1) = \text{........ dynes/cm.}$$

From the graph at any unknown temperature (OQ), the surface tension (T) = PQ = ... dynes/cm.

Precautions :

(i) The capillary employed should be free from any contamination like grease, etc.

(ii) The apparatus must be perfectly air-tight. It is preferred to have the apparatus in one single-piece avoiding the use of joints.

(iii) Maximum pressure recorded by the manometer is independent of the formation of the bubbles. However, bubbles should not be formed very rapidly.

(iv) The manometer should preferably contain a liquid of low density (xylol is very much preferred).

(v) Temperature of the bath should be kept steady during taking readings of the manometer.

(vi) Back-lash errors in taking microscope readings should be avoided.

Error Calculation :

From Eq. (3.8e),

$$T = \frac{rg}{2}(h\rho_2 - h\rho_1).$$

$$\text{Hence, } \frac{\delta T}{T} = \frac{\delta r}{r} + \frac{\rho_2 \delta h + \rho_1 \delta d}{(h\rho_2 - d\rho_1)} = \text{......}$$

$$\therefore \text{ Percentage error} = \frac{\delta T}{T} \times 100 = \text{...... \%}$$

Questions and Answers

Q.1. How does the S.T. vary with temperature?

A. : S. T. decreases with rise of temperature. Surface tension at any temperature t°C is given by $T_t = T_0(1 - \alpha t)$ where, T_0 is S.T. at 0°C and α is known as the temp. coeff. of S.T. Eotvos formula is $S\left(\frac{M}{\rho}\right)^{\frac{2}{3}} = K(T_c - T)$, where S is the S.T. of a liquid, M and ρ are its molecular weight and density respectively, T_c—critical temp., T—temperature and K is a constant equal to 2.2.

Q.2. If S.T. decreases with temperature, where does it vanish?

A. : S. T. vanishes at the critical temperature of the liquid.

Q.3. How is the bubble formed and why is the pressure inside the bubble greater than that outside it?

A. : [See any textbook]

Q.4. Why does the liquid in the manometer oscillate during formation and breaking away of the bubble?

A. : During bubble formation, the pressure inside gradually increases and the difference in the manometric liquid level gradually increases. Just when the bubble breaks away, the pressure inside it suddenly drops and so does the liquid level. This causes an oscillation of the liquid level.

Q.5. How can the oscillation be damped?

A. : By drawing the open end of the manometer into a capillary.

Q.6. Can water or mercury be used in the manometer?

A. : No, the manometric liquid should be of very low density to give an appreciable difference in liquid levels.

Q.7. Does the manometer reading depend on the depth of the capillary tube inside the bath?

A. : Manometric liquid level difference increases with the increase of the depth of the capillary tube.

Q.8. What are the advantages of this method over the capillary tube method?

A. : By this method, temperature variation of S.T. can be studied but not by the capillary tube method. Also, S. T. of a liquid at different concentrations can be compared by this method.

Q.9. On what factors does the accuracy of this experiment depend?

A. : The accuracy depends on the narrowness of the capillary tube and on the precise measurement of the manometer reading just at the time of breaking away of the bubble.

3.9 Determination of Surface Tension of a liquid by Ripples method. (Rayleigh's method)

Theory :

The method depends on the measurement of the wavelength of ripple formed on the surface of a liquid whose surface tension is to be determined. The velocity v of a harmonic disturbance on the surface of any liquid is given by

$$v = \sqrt{\frac{\lambda}{2\pi} g + \frac{2\pi T}{\lambda \rho}}, \quad \text{... (3.9a)}$$

where λ is the wavelength of the ripples, T the surface tension, ρ the density of the liquid.

$$\text{Thus, } v = \sqrt{\frac{\lambda}{2\pi}\left(g + \frac{4\pi^2}{\lambda^2 \rho}.T\right)}. \quad \text{... (3.9b)}$$

The surface tension, therefore, seems to increase the effective value of g. For large waves the second term inside the bracket can be neglected.

Now, if the harmonic disturbance on the liquid surface is made by an electrically maintained tuning fork of frequency n, then the frequency of the periodic disturbance on the liquid surface will also be n.

$$\text{Hence, } v = n\lambda = \sqrt{\frac{\lambda}{2\pi}\left(g + \frac{4\pi^2}{\lambda^2 \rho}.T\right)}$$

$$\therefore n^2\lambda^2 = \frac{\lambda}{2\pi}\left(g + \frac{4\pi^2}{\lambda^2 \rho}.T\right)$$

$$\text{or, } T = \frac{\lambda^2 \rho}{4\pi^2}\left(2n^2\lambda\pi - g\right). \quad \text{... (3.9c)}$$

Measuring the value of λ of the ripples formed and by knowing the frequency (n) of the periodic disturbance (equal to that of the tuning fork), the value of surface tension (T) of the liquid can be determined from Eq. (3.9c).

Apparatus :

Ripple tank, electrically maintained tuning fork with a dipper, millimetre scale etched on a plate with a wire attached, short focused telescope and an illuminating arrangement.

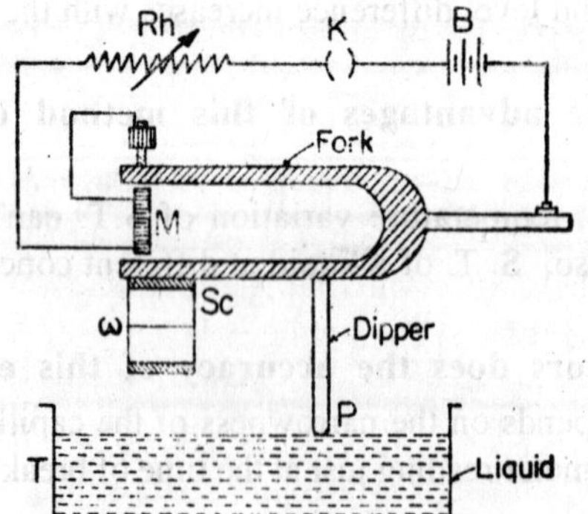

Fig. 3.23 : Surface Tension by Ripple Method.

The arrangement of the apparatus is as shown in Fig. 3.23. A tuning fork of frequency about 80 to 100 cps is made to vibrate and its vibration is maintained electrically. The plane of vibration of the fork is vertical and its arms are horizontal, one over the other. A dipper D is attached to one of its prongs and as the fork vibrates vertically, the dipper, which initially just touches the surface of the experimental liquid taken in a container (tank T), vibrates up and down and gets in and out of the liquid surface (P) periodically. This movement of the dipper produces ripples on the liquid surface. These ripples produced, travel towards two ends of the tank and on reflection from the ends, superpose on the direct ripples producing thereby stationary ripples. To determine the wavelength of these ripples, a small millimetre scale (Sc) etched on a transparent plate, is attached to the same prong of the fork. The scale has a fine wire (w) attached to it horizontally and the plane of the scale is vertical and parallel to the plane of vibration of the prongs.

A beam of strong parallel light is reflected from the ripple surface of the liquid and is made to pass obliquely through the transparent scale and the wire. The transmitted light is viewed through a low power telescope. The magnifications of the images of the scale and the wire are the same. When viewed through the telescope, an observer will see the image of the wire not straight but a wave-like form of the ripples, since the light proceeds in planes which are parallel to the crests and troughs of the real stationary ripples.

Method :

(i) By closing the key K, the electrical circuit is made operative and the tuning fork is made to vibrate. The value of the rheostat is so adjusted that the vibration is just maintained.

(ii) The dipper is set in such a way that it just touches the surface (P) of the experimental liquid taken in the tank T.

(iii) Strong light from a source is rendered parallel by a lens and then made incident obliquely on the ripple surface of the liquid. The reflected light is made to pass through the wire (w) and the transparent scale (Sc) and finally received by a short-focus telescope, whose cross-wire can be moved by a micrometer screw. The wavy picture of the wire is made very sharp by adjusting the source of light.

(iv) Next, the circular scale of the micrometer is to be calibrated. To do this, the total number of the circular scale divisions (N) is noted. By rotating the micrometer screw, the cross-wire is made to coincide with a certain mark of the linear scale on the plate. The reading of the micrometer (say, x_1) is noted. The micrometer is then rotated until the cross-wire shifts on the linear scale by an integral millimeter (p, say), where p may be 1 or 2 or 3, ... mm. To do this, the number of complete rotation (say, R_1) of the micrometer and its final reading (say, x_2) are noted. Then the total number of micrometer circular divisions rotated in $N.R_1 + (x_2 - x_1)$ and this corresponds to a linear shift on the scale equal to p.

Thus the least count of the micrometer is $\dfrac{p}{[N.R_1 + (x_2 - x_1)]}$. The value of this least count of the micrometer may be taken by taking different values of p on the linear scale and the mean value of the least count (l.c.) is thus obtained.

(v) After obtaining the least count, the wavelength (λ) of the ripples has to be determined. It is assumed that the wavelength of the shadow of the wire w is equal to the wavelength of the ripples. At the extreme right of the field of view of the telescope, the cross-wire is made to coincide with a crest of a ripple and the corresponding reading of the micrometer (x_3) is noted. It is now gradually rotated (avoiding back-lash error) to shift the cross-wire to coincide with the next crest of the ripple towards left. The number of full rotation (R_2) of the screw and the final reading (x_4) is noted. From this, the wavelength (λ) of the ripples is found out as $\lambda = [N.R_2 + (x_4 - x_3)] \times l.c.$ This mode of determination of λ is repeated by succesively focussing the cross-wire on successive crests of the ripples gradually towards left until the last crest is visible. Determining the value of λ for each successive crest readings, the mean value of λ can be determined.

(vi) The process can be repeated by moving the cross-wire again from left to right and focusing this time on successive troughs of the ripples and finding thereby the mean value of λ, as is done in operation (v). The mean value of λ is finally obtained from the readings of operations (v) and (vi).

(vii) The frequency (n) of the fork is normally etched on the fork, or if not, the same has to be determined by the stroboscopic method or by the use of a phonic wheel.

(viii) The density of the experimental liquid has to be determined with the help of a sp. gravity bottle at the temperature during the time of the experiment. If the liquid be water, the density (ρ) can be had from the table at the desired temperature.

(ix) Finally, substituting the values of l, n, ρ and g in Eq. (3.9c), the value of surface tension (T) of the liquid can be determined.

Results :

Frequency of the tuning fork (n) = ... cps. (Hertz)

(A) Calibration of the micrometer scale :

Total number of circular scale division of the micrometer (N) = ...

Table 1

No. of Obs.	Initial C.S.R. when cross-wire coincides with one mark of linear scale (x_1)	When cross-wire coincides with another mark of the linear scale				Shift of crosswise on linear scale (p) (mm)	Least count of micro-meter (mm) $p/[N.R_1+(x_2-x_1)]$	Mean value of least count (mm)
		No. of complete rotation of C.S. (R_1)	Final C.S. reading (x_2)	Additional no. of C.S.R. (x_2-x_1)	Total No. of C.S. readings $[N.R_1 + (x_2-x_1)]$			
1.	...	...	...	...	...	...	...	...
2.	...	...	...	...	...	...	...	...
3.	...	...	...	...	...	...	...	...
...	...	...	...	...	...	...	...	...

Table 2

Mode of Obs.	No. of Obs.	Order of Observation	Initial C.S.R. when cross wire coincides with a crest/trough (x_3)	No. of complete rotation of C.S. (R_2)	Final C.S.R. when cross wire coincides with next crest/trough (x_4)	Additional No. of C.S.R. $(x_4 - x_3)\}$	Total No. of C.S.R. $\{N.R_2 + (x_4 - x_3)\}$	Wave length $\lambda = \{N.R_2 + (x_4 - x_3)\} \times l.c.$ (cm)	Mean Wave-length (λ) (cm)
Right to Left	1.	1st to 2nd	…	…	…	…	…	…	…
	2.	2nd to 3rd	…	…	…	…	…	…	
	3.	3rd to 4th	…	…	…	…	…	…	
(Crest to Crest)	…	…	…	…	…	…	…	…	
	…	…	…	…	…	…	…	…	
	…	…	…	…	…	…	…	…	
Left to Right	1.	1st to 2nd	…	…	…	…	…	…	
	2.	2nd to 3rd	…	…	…	…	…	…	
	3.	3rd to 4th	…	…	…	…	…	…	
(Trough to Trough)	…	…	…	…	…	…	…	…	
	…	…	…	…	…	…	…	…	
	…	…	…	…	…	…	…	…	

(C) Density of the experimental liquid (ρ) :

Table 3

Temperature of the liquid .... = ... °C

Mass of empty bottle (W_1) (gm)	Mass of bottle + liquid (W_2) (gm)	Mass of bottle + water (W_3) (gm)	Density of the liquid $\left(\rho = \frac{W_2 - W_1}{W_3 - W_1}\right)$ (gm/cc)
......	......	......	......

Value of g = ... cm/sec^2.

Calculations :

From Eq. (3.9c),

$$T = \frac{\lambda^2 \rho}{4\pi^2}\left(2n^2\lambda\pi - g\right)$$

$$= \text{...... dynes/cm at ...... °C}$$

Precautions :

(i) Since the expression for T contains λ and n in higher powers, these two quantities should be measured very carefully to give accuracy to the result.

(ii) The tank containing the liquid must be free from any grease or other contamination, since presence of them will change the value of the surface tension of the liquid.

(iii) Back-lash error of the rotation of the micrometer screw should be avoided.

(iv) Tuning fork should be of low frequency, otherwise ripples ($\lambda < 1{\cdot}7$ cm) will not be produced.

(v) Care must be taken to see that the electrically maintained tuning fork vibrates uniformly and steadily during the entire experiment, and the vibration is properly maintained.

Error Calculation :

For waves of less than 1.7 cm wavelength, the value of the ripple velocity (v) on the surface of the liquid mainly depends on the surface tension (T) of the liquid. Hence in Eq. (3.9b), the second term in the bracket is more important than the first term. Hence, from Eq. (3.9b),

$$v = n\lambda = \sqrt{\frac{2\pi T}{\lambda\rho}}.$$

$$\therefore \quad T = \frac{n^3\lambda^3\rho}{2\pi}.$$

Thus, $\frac{\delta T}{T} = 2\frac{\delta n}{n} + \frac{3\delta\lambda}{\lambda}$ [ρ and π being constant]

If the frequency (n) of the fork is fixed, then

$$\frac{\delta T}{T} = \frac{3\delta\lambda}{\lambda}.$$

The percentage error can hence be calculated.

Questions and Answers

Q.1. Distinguish between waves and ripples.

A. : Gravity normally controls a wave while surface tension mainly affects the ripples. Critical velocity of a wave is 23 cm/sec and the waves of length less than 1.7 cm are generally called ripples.

Q.2. Of the two methods of determination of surface tension of a liquid (capillary tube method and ripples method), which one is more convenient and accurate?

A. : Capillary tube method involves measurement of height of ascent of liquid and radius of capillary tube at the point of maximum ascent of liquid. Both the measurements are to be done very carefully to get accuracy of the result, while in ripple method, only determination of wavelength (λ) is to be done carefully if the fork frequency is supplied. Further, in the latter method variation of surface tension of a liquid with temperature and concentration can be easily studied in the same set-up, while capillary tube method is not so convenient for the purpose. Of course, the latter method requires difficult setting for experiment arrangement.

Q.3. Is there any practical example of ripples?

A. : Yes, ripples are observed on surface of liquid in a pond or tank due to light breeze.

3.10 Determination of coefficient of viscosity of a liquid by its flow through a capillary tube (Poiseuille's method)

Theory :

If a liquid be allowed to flow in a horizontal capillary tube of internal radius r and length l, under the influence of a steady difference of pressure P across its ends, and if the liquid is assumed to move without appreciable kinetic energy and the streamline motion, then the volume V of the liquid that flows out in one second, is given by

$$V = \frac{\pi P r^4}{8 l \eta},$$ where η is the coefficient of viscosity of the liquid.

Hence, $$\eta = \frac{\pi P r^4}{8 l V}. \qquad \text{... (3.10a)}$$

This is known as Poiseuille's equation and by measuring the values of P, r, l and V, the relation may be utilised to determine the coefficient of viscosity of a liquid.

Calculation of critical velocity (v_c) and maximum height of liquid level (h_{max}) :

It is possible to determine the critical height of the liquid level that creates the pressure difference (P) across the ends of the capillary tube. At this critical height of the liquid level, transition from the streamline to the turbulent motion of the liquid occurs in the capillary tube.

If v_c be the critical velocity, ρ be the liquid density, and r be the radius of the tube, then

$$v_c = \frac{K.n}{\rho r} \qquad \text{... (3.10b)}$$

where K is known as Reynold's number which is approximately 1000. Taking the approximate value of η of the liquid at the room temperature (which is about 0.01 for water) and its density (ρ), the value of v_c may be ascertained. The volume of liquid coming out per second at the critical velocity is $V_c = v_c \times \pi r^2$, which is thus obtained. From the (*h-V*) graph drawn (Fig. 3.22), thus the height of liquid column corresponding to V_c is obtained. This height will obviously be h_{max}, which is the permissible height up to which the liquid level may be raised to have streamline motion.

Apparatus :

Poiseuille's apparatus : Cathetometer, thermometer, scale, mercury, beaker and balance.

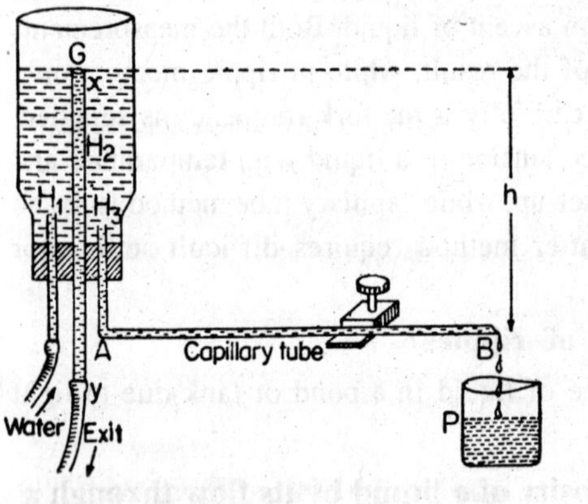

Fig. 3.24 : Poiseuille's method

Poiseuille's apparatus consists of a big glass vessel (*G*) whose upper end is open but at the base, a rubber cork is fitted in which there are three holes (H_1, H_2 and H_3). Through H_1, water enters into the vessel (Fig. 3.24) and in the second (H_2 is entered a glass tube (*xy*) open at both ends. Thus tube acts as a sink and water can emerge out of the vessel through this opening when it reaches a particular height. In the third (H_3), the experimental capillary tube is fitted (*AB*) as shown in the figure. The experimental liquid flows through this tube (*AB*). The open end of *AB* inside the vessel *G* is at a lower height than the level of liquid maintained in the vessel. As a result, the pressure of liquid at the end *A* is much greater than that at *B* and thus the flow of liquid is maintained in the capillary tube. On emergence from the end *B*, the liquid is collected in a beaker (*P*) or in a measuring cylinder.

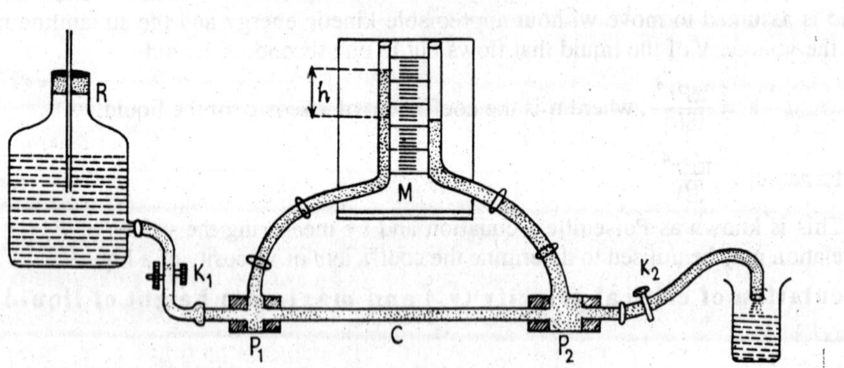

Fig. 3.24(a)

Method :

(i) The length of the capillary tube (*AB*) is measured by a metre scale and the mean length (*l*) of the tube is determined from three observations. Then it is cleansed by nitric acid and caustic soda and finally washed and dried.

(ii) A long pellet of mercury is now introduced in the tube by suction at one end. The length of the pellet inside the capillary tube is measured with a meter scale. The pellet is moved at different positions of the tube, and length of the pellet at each position is determined. From this, the mean length of the pellet (L) is obtained.

(iii) Then this pellet of mercury is poured in a previously weighed empty crucible, and the mass (m) of the pellet of mercury is determined. If r be the internal radius of the capillary tube, then the volume of the mercury pellet is $\pi r^2 L$ and its mass will be $m = \pi r^2 L\rho$, where ρ is the density of mercury.

Hence, $m = \pi r^2 L\rho$. ... (3.10c)

(iv) Now, a cathetometer is focused from a distance and its cross-wire is focused on the upper surface of the liquid in the vessel G. The reading of the cathetometer is taken from its scale. Then the cathetometer is rotated in the same set-up and its cross-wire is again focused at the open end (B) of the tube AB. Reading from the scale is again taken. From the difference of these two readings, the height (h) of the liquid level in the vessel (G) from the axis of the tube (AB) is obtained. Thus the pressure (P) difference between the two ends of the tube which causes the liquid to flow will be given by

$$p = h\rho' g, \qquad \text{... (3.10d)}$$

where ρ' is the density of the experimental liquid at room temperature.

(v) To determine the rate of flow of liquid, a measuring cylinder is held below the open end (B) of the tube and the quantity of liquid collected in it in a given time (t) is noted. The time is recorded by means of a stopwatch. From this, the rate of flow (V) of the liquid is obtained.

(vi) By changing the upper level of the liquid in the vessel, the value of h is changed and the corresponding change in the value of P is effected. The rate of flow through the tube for at least five different values of P is taken and for each case, the rate of flow (V) of the liquid is determined.

(vii) A graph is now plotted with h along X-axis and rate of flow (V) along Y-axis, from which the value of η can be obtained. (See 'Drawing of Graph')

Alternative Method

An alternative arrangement is shown in Fig. 3.24(a). The capillary tube C is horizontal and is connected to two chambers P_1 and P_2, each of which has two more openings. R is a reservoir bottle containing water and is connected to P_1 as shown, through a pinch-cock K_1. Two limbs of a manometer (M) are connected to P_1 and P_2. The exit tube of the water from P_2 is also provided with another pinch-cock K_2 which is used if the water level is not visible in the right limb of the manometer. Air bubble enters through a hollow tube immersed in water at R as water flows out of R. The lower end of that tube in R remains at atmospheric pressure and hence the manometer maintains a uniform difference in level until the water in R falls below the lower end of the tube in R.

The difference of pressure (P) between the two ends of the capillary tube (C) is obtained from the difference (h) in water level in the two limbs of the manometer. The rest of the procedure is the same as before, except that the value of h may be measured from the scale attached to the manometer. The value of h should be noted from time to time to make sure that it remains steady. The volume of water flowing out may be collected by a measuring cylinder as shown in the diagram.

Results :

(A) Determination of length (l) of the capillary tube :

Table 1

No. of Obs.	Length of the tube (cm)	Average length (l) of the tube (cm)
1.	...	...
2.	...	...
3.	...	...

(B) Determination of the radius (r) of the capillary tube :

Table 2

No. of Obs.	Length of mercury column in the tube		Mass of the beaker		Mass of mercury	$r^2 =$	(Radius)^4
	Length (cm)	Mean length (cm)	When empty (m_1) (gm)	With mercury (m_2) (gm)	$(m_2 - m_1)$ = (m) (gm)	$\frac{m}{\pi L \rho}$ (sq. cm)	r^4 (cm^4)
1.	...	...	...				
2.	...	...	...				
3.	...	...	...	...	...	...	...
4.	...	...	...	...	...	...	...
5.	...	...	...	...	...	...	...

(C) Determination of the value of pressure (P) by the cathetometer :

Table 3

Vernier constant of the cathetometer = ... cm

Temperature of the experimental liquid = ... °C

Density of the liquid (ρ') = ... gm/cc

No. of Obs.	For liquid level at G			For the end B of the capillary tube			Height of the liquid level $h = X_1 \sim X_2$ (cm)	Pressure difference (P) $= h\rho' g$ (dynes/cm²)
	Main Scale Reading (M)(cm)	Vernier Reading (V) (cm)	Total $X_1 =$ ($M + V$) (cm)	Main Scale Reading (M)(cm)	Vernier Reading (V) (cm)	Total $X_2 =$ ($M + V$) (cm)		
1.	...	...	...	...	...	...	...	...
2.	...	...	...	...	...	...	...	...
3.	...	...	...	...	...	...	...	...
4.	...	...	...	...	...	...	...	...
5.	...	...	...	...	...	...	...	...
6.	...	...	...	...	...	...	...	...

(D) Determination of the rate of flow (V) of the liquid :

No. of Obs.	Height of the liquid level (h) (cm)	Volume of liquid collected (V') (cc)	Time of collection (t) (sec)	Mean value of V' (cc)	Volume of liquid collected per sec (Rate of flow) $V = V'/t$ cc/sec
1.	...	...	...	...	...
2.	...	...	...	...	...
3.	...	...	...	...	...
4.	...	...	...	...	...
5.	...	...	...	...	...
6.	...	...	...	...	...

Calculation :

From Eq. (3.11a),

$$\text{coefficient of viscosity } \eta = \frac{\pi P r^4}{8lV}.$$

Substituting the values of the measured quantities in the R.H.S. of this equation, the value of η can be obtained. Thus

η = ……… gm/cm/sec or poise at ………°C.

Calculation of critical velocity (v_c) and h_{max} :

From Eq. (3.10b), the value of v_c is determined and hence the value of h_{max} is obtained from the graph (Fig. 3.25) as discussed in the Theory.

Drawing of Graph :

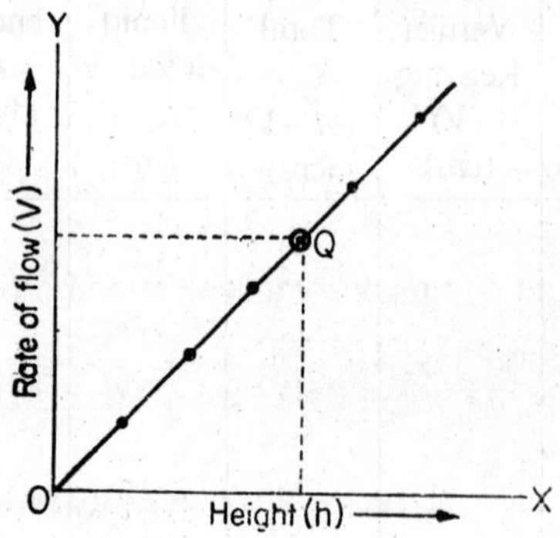

Fig. 3.25 : (h-V) graph

From Eq. (3.10a), since the value of η is constant for a given liquid at a given temperature, $P \propto V$ or $h\rho g \propto V$, or $h \propto V$. Thus if a graph be plotted with h versus V, the curve will be a straight line passing through the origin (0, 0), as shown in Fig. 3.25. From this graph, any arbitrary point Q is taken whose ordinate and abscissa will give the values of corres-ponding V (rate of flow) and height of liquid (h). Substituting these values in Eq. (3.10a), the value of η can be determined.

Precautions :

(i) The main source of error in this experiment may be in the measurement of the radius of the tube. Since in the expression for h, radius occurs to the fourth power, any error in the determination of radius of the tube will seriously affect the result. The tube selected must, therefore, be uniform and its radius be measured very carefully.

(ii) Little vaseline or grease should be attached to the capillary tube near its open end so that liquid coming out of it may not run back.

(iii) Pressure (liquid-height) should be so adjusted that liquid comes out of the capillary tube in a slow trickle. If the flowing liquid possesses appreciable kinetic energy, then the effective pressue should be taken as

$$P = \rho g\left(h - \frac{2V^2}{g\pi^2 r^4}\right). \quad \text{... (3.10e)}$$

(iv) The motion of liquid must be streamline and care must be taken to see that the flow of liquid does not become turbulent.

(v) Sufficient quantity of liquid must be collected if collection is done by a measuring cylinder.

Error Calculation :

Since, $\eta = \dfrac{\pi \mathrm{P} r^4}{8lV} = \dfrac{\pi h\rho g r^4 t}{8lV'}$

when V′ is the volume of liquid collected in t seconds, the maximum proportional error in η is given by

$$\frac{\delta\eta}{\eta} = \frac{\delta h}{h} + \frac{4\delta r}{r} + \frac{\delta t}{t} + \frac{\delta l}{l} + \frac{\delta V'}{V'} \quad \text{... (3.10f)}$$

From Eq. (3.10c), $m = \pi r^2 L\rho$.

Hence, $\dfrac{2\delta r}{r} = \dfrac{\delta m}{m} + \dfrac{\delta L}{L}$

gives the maximum values of possible errors in the measurement in r. Substituting this expression in the value of $\dfrac{\delta\eta}{\eta}$, Eq. (3.10f) becomes

$$\frac{\delta\eta}{\eta} = \frac{\delta h}{h} + \frac{2\delta m}{m} + \frac{2\delta L}{L} + \frac{\delta t}{t} + \frac{\delta l}{l} + \frac{\delta V'}{V'}.$$

Noting the maximum values of possible errors in the measurement of different quantities by different devices, the maximum proportional error in the measurement of η can be obtained and hence the percentage error in the measurement is calculated.

Questions and Answers

Q.1. What is coefficent of viscosity? What are its unit and dimension?

A. : For definition, see any textbook. Its unit is poise or gm/sec/cm and it has the dimension $ML^{-1}T^{-1}$.

Q.2. What is the difference between 'streamline' and 'turbulent' motions?

A. : In streamline motion, the particles of the liquid move in straight paths and parallel to each other, while in turbulent motion, the particles travel in zig-zag paths.

Q.3. What will happen if the diameter of the tube is large?

A. : In that case, the motion will not be streamline.

Q.4. Does the coefficient of viscosity depend on temperature?

A. : Yes, for liquids, the coefficient of viscosity decreases but for gases, it increases with temperature.

Q.5. Should the pressue difference across the ends of the tube remain constant during a particular set?

A. : Yes, otherwise the rate of flow will change during the particular set of reading.

Q.6. Should the tube remain horizontal?

A. : Yes, otherwise effect of gravity will come into play.

Q.7. Is the velocity of liquid same everywhere inside the tube?

A. : No. It is maximum along the axis of the tube and minimum at the surface of contact with the glass tube.

Q.8. Are there any assumptions in the method?

A. : Yes, firstly, it is assumed that there is no acceleration of the liquid along the axis of the tube, which is not true. Secondly, it is assumed that the energy due to difference of pressure between the two ends of the tube is entirely spent to overcome the viscous drag of the liquid—which is also not true.

Q.9. If the value of *h* be high, what will be the nature of the curve of *h*-*V*?

A. : So long as the value of *h* is small, (*h*-*V*) curve is a straight line but if *h* is very high, then the motion becomes turbulent and the curve is no longer a straight line.

Q.10. Is this method suitable for all types of liquids?

A. : This method is suitable for liquids having low viscosity. For high viscous liquids, Stokes' method is suitable.

Q.11. What is the difference between 'friction' and 'viscosity'?

A. : Though friction and viscosity having some similarities, yet there is a lot of difference between them. For liquids at rest, viscosity does not arise although friction still works then. Again, viscosity depends on the relative velocity while friction has nothing to do with the same. Finally, effects of viscosity depend on the area and shape of the liquid surface while friction has nothing to do with them.

Q.12. Are there any examples of viscosity in nature?

A. : (i) Flow of water in a river is much more in the middle than towards the shore.

(ii) Raindrops acquire velocity up to a certain value due to the viscosity of air and reach the earth surface with that velocity.

[For 'origin of viscosity' and 'Reynold's number' see any Textbook]

3.11 To verify Stokes' Law and hence to determine the coefficient of viscosity of a liquid

Theory :

From hydrodynamical considerations, it was shown by Stokes that when a sphere falls under gravity through a fluid of infinite extent, ultimately it moves with a constant velocity, called the terminal velocity, which is given by

$$6\pi\eta r v = \frac{4}{3}\pi r^3 g(\rho - \sigma), \quad \text{... (3.11a)}$$

Where r = radius of the falling sphere

v = terminal velocity acquired by the sphere

g = acceleration due to gravity

ρ = density of the material of the sphere

σ = density of the fluid

η = coefficient of viscosity of the liquid.

The relation is known as Stokes' law. This is valid for streamline motion of the fluid and if vr is small compared to η. The method is suitable for highly viscous liquid.

In practice, the experiment is performed in a liquid of total depth h taken in a cylinder of inner radius R. This finite depth and radius of the liquid seem to call for a correction in the formula. For finite radius R, it was shown by Faxen and Bason that the observed terminal velocity (v) should be multiplied by a factor $\left\{1+\frac{2.4r}{R}\right\}$ to take into account the finite lateral extension of the liquid. Again, for the finite depth H, the corresponding multiplying factor is $\left\{1+\frac{3.3r}{H}\right\}$. These corrections are known as Ladenburg corrections. Thus the corrected value of the terminal velocity (v) will be

$$v\left\{1+\frac{2.4r}{R}\right\}\left\{1+\frac{3.3r}{H}\right\} \quad \text{...(3.11b)}$$

To verify the law, for different values of r, corresponding values of v are determined at the same temperature. The value of v may be determined, from the time t required by a sphere in moving a known distance S in the liquid with the terminal velocity. Thus from Eq. (3.11a), with proper correction, the coefficient of viscosity will be

$$\eta = \frac{2}{9}\,\frac{r^2 g(\rho-\sigma)}{v} = \frac{2}{9}\,\frac{r^2 g(\rho-\sigma)}{v\left(1+\frac{2.4r}{R}\right)\left(1+\frac{3.3r}{H}\right)}. \quad \text{... (3.11c)}$$

So, if Stokes' law is true, then $\frac{r^2}{v}$ should remain constant for a given liquid at a given temperature. So, if a graph be drawn with r^2 against v, then the graph would be a straight line. Hence, if the value of $\frac{r^2}{v}$ is shown to be constant and if (r^2–v) graph is a straight line, then Stokes' law will be verified. From this graph, taking the value of $\frac{r^2}{v}$ and substituting the same in Eq. (3.11c), the value of coefficient of viscosity (η) of the liquid can be determined.

Apparatus :

A glass tube of large diameter (greater than 5 cm) and of length about 50 cm or more with three horizontal marks, small spherical balls of different diameters between 1 mm and 3 mm (used in the cycle as bearing balls, may be used), thermometer, stopwatch, experimental liquid (castor oil or glycerine), screw-gauge, slide calliper, metre scale.

The glass cylinder is about 50 cm long with diameter about 5 cm or more. On its surface, there are three horizontal marks (X_1, X_2, X_3), two of them (X_1 and X_3) about 10 cm from the top and the bottom and the other mark (X_2) just midway between them. The thermometer (*Th*) records the temperature of the liquid (Fig. 3.26). Sometimes only two marks (X_1 and X_3) are provided in the tube.

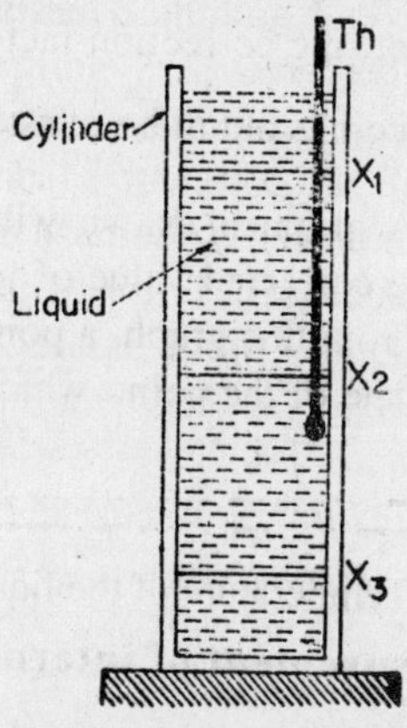

Fig. 3.26 : Stokes' method to find coeff. of viscosity.

Method :

(i) The inner diameter of the cylinder is measured by a slide calliper at different places, and the mean diameter and hence the radius (R) is determined.

(ii) Diameters ($2r$) of the balls are measured by means of a screw-gauge and if the balls are very small, those should be measured by a microscope.

(iii) The cylinder is set vertically and the marks on its outer surface are put as stated. It is filled with the experimental liquid and some time is allowed so that it may be in temperature equilibrium with the surroundings. The thermometer (*Th*) is placed in the liquid and after stirring the liquid for a while, its temperature is recorded.

(iv) Balls of three different diameters are selected and at least five balls of each diameter are taken. The diameters of each category of balls are measured by a screw-gauge as stated before. Density of the material of the balls and also the density of the experimental liquid, are determined by a sp. gravity bottle. While determining the density of the balls, assorted balls of different diameters should be taken to take the average value of the density.

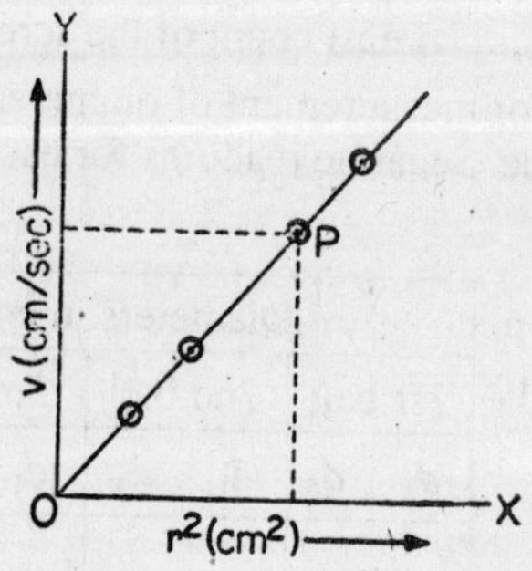

Fig. 3.27 : (r^2-v) graph

(v) The balls of a particular diameter are then wetted by the experimental liquid and then with the help of a spatula, one ball is gently dropped on the liquid surface and the time is takes in crossing the equal distances (X_1X_2 and X_2X_3) between the marks, are noted. These times should be equal if terminal velocity is attained by the ball while it reaches the mark X_1. If times are not equal, either smaller balls may be used or the mark X_1 is lowered and accordingly the marks X_2 and X_3 are adjusted.

(vi) Starting from the smallest size of the balls, gradually the balls are then gently dropped on the liquid surface and for each case, the time taken by the ball in crossing the marks X_1 and X_3 are noted by a stopwatch. From these timings, the terminal velocity (v) is obtained, if the distance X_1X_3 is known by a scale. The procedure is repeated at least for three different sizes of the balls and for at least five balls of each size.

(vii) Having measured the total height (H) of the liquid in the cylinder and radius (R) of the cylinder, the correction factors are obtained for balls of each size. From data, if the value of $\frac{r^2}{v}$ comes out to be constant, then Stokes' law is said to be verified.

(viii) A graph is drawn with r^2 along X-axis and v along Y-axis. This value of v should be the corrected value of v. The graph will be a straight line (Fig. 3.27) if Stokes' law is true. From this graph, a point P, the values of r^2 and v can be had from the abscissa and the ordinate of the point, which when substituted in Eq. (3.12C), will give the value of η.

Results :

Temperature of the liquid = ... °C

(A) Measurement of internal diameter (2R) of the cylinder :

Table 1

V.C. of the slide calliper = ... cm

No. of Obs.	Diameters in cm at several places									Mean diameter (D) in cm	Mean radius (R) in cm
	1	2	3	4	5	6	7	8	9		
	...	...	...	...	...	...	...	...	...		

(B) Measurement of diameters (2r) of the balls :

Least count of the screw-gauge = ... cm.

[For measurement of diameters by a screw-gauge, a Table as shown in Expt. 2.4 has to be made. Separate readings for each group of balls, have to be taken also.]

Table 2

Group of balls	Diameters in cm at several places										Mean diameter (d) (cm)	Mean radius (r) (cm)
	1st ball		2nd ball		3rd ball		4th ball		5th ball			
	d_1	d_2	d_1	d_2	d_1	d_2	d_1	d_2	d_1	d_2		
I.	...	...	...	...	...	...	...	...	...	...	...	...
II.	...	...	...	...	...	...	...	...	...	...	...	...
III.	...	...	...	...	...	...	...	...	...	...	...	...

(d_1 and d_2 correspond to measurements at right angle.)

(C) Determination of density of the ball and the liquid :

Density of water at room temperature (ρ_t) .... gm/cc.

Table 3

Wt. of empty bottle (W) (gm)	Wt. of bottle + solid (W_1) (gm)	Wt. of bottle + solid + water (W_2)(gm)	Wt. of bottle + water (W_3) (gm)	Wt. of bottle + liquid (W_4) (gm)	Density of the solid $\rho = \frac{(W_1 - W)}{(W_3 - W) - (W_2 - W_1)} \times \rho_t$	Density of the liquid $\sigma = \frac{W_4 - W}{W_3 - W} \times \rho_t$
...	...	...	...	...	...... gm/cc	...gm/cc

(D) Determination of terminal velocity (v) of the balls :

Test Run :

Time of fall of the largest ball across (X_1X_2 and X_2X_3) (in seconds)

X_1X_2 =	...	...	...	...		...		...
X_2X_3 =	...	...	...	...	...	...	...	...

[If there are only two marks in the tube (X_1 and X_3), then time of fall across X_1 and X_3 is to be noted.]

(Equality of two times of fall assures attainment of terminal velocity.)

Distance $X_1X_3 = L = \frac{...+...+...}{3}$ = ... cm

Toal height of liquid in the jar = $H = \frac{h_1 + h_2 + h_3}{3}$ = ...cm,

where h_1, h_2 and h_3 are three measurements of the same liquid depth in the tube.

Table 4

Group of balls	Radius of balls (r cm)	Ball No. of particular group	Time of fall (t) (sec)	Mean time (t) in sec	Velocity of fall $v' = L/t$ (cm/sec)	Correction factors $(1+2.4\ r/R) \times (1+3.3r/H)$	Corrected velocity ($v = v' \times$ correction) (cm/sec)
I.		1	...				
		2		...	...		
	...	3	...				
	(r_1)	4					
		5					
II.		1					
		2					
	.....	3	...	...	...		
	(r_2)	4	...				
		5	...				
III.		1	...	...			
		2	...	...			
	...	3	...	...	...		
	(r_3)	4	...	...			
		5	...	...			

(E) Verification of Stokes' law :

Group of balls	Mean radius of balls (r cm)	Square of mean radius (r^2) cm^2	Corrected terminal velocity (n cm/sec)	Value of r^2/v (cm-sec)	Mean value of r^2/v (cm-sec)	Value of Coeff. of viscosity (η) in poise
I.	...	...	...	...	...	
II.	...	...	...	...	...	at ...... °C
III.	...	...	...	...	...	

(F) Evaluation of η from the graph :

From the graph of r^2 versus v (Fig. 3.26), any arbitrary point P is taken on the graph. The values of ordinate and abscissa of this point are determined. These give the values of v and r^2, and when these values are substituted in Eq. (3.11c), the value of the coefficient of viscosity (η) of the liquid can be determined at the given temperature.

Since the graph obtained is a straight line, hence Stokes' law is verified.

Calculation :

From Eq. (3.11c), substituting the values of all the quantities measured,

$$\eta = \frac{2}{9}\cdot\frac{r^2 g(\rho-\sigma)}{v} = \text{........ poise (gm-cm}^{-1}\text{. sec}^{-1}\text{) at .... °C.}$$

Precautions :

(i) The radii of the balls being very small, they should be measured very carefully.

(ii) The balls should be dropped centrally in the cylinder, so that they are quite away from the walls.

(iii) The balls should not be dropped by hand and they should be wetted before dropping.

(iv) As viscosity changes rapidly with temperature, the temperature of the liquid should remain constant during the experiment and this should be checked.

(v) To avoid parallax error, paper collars may be used as the marks on the cylinder surface. The stopwatch should be started at the instant when the ball just touches the upper edge of the paper collar and should be stopped at the moment it touches the upper edge of the lower collar.

(vi) The chief source of error in this experiment is the lack of arrangement to ensure perfect constancy of temperature of the liquid throughout the experiment.

(vii) Care should be taken to start and stop the watch just in time; as some error in measurement of time may easily creep in, while the same is recorded.

Error Calculation :

From Eq. (3.11c),

$$\eta = \frac{2}{9}\frac{r^2 g(\rho-\sigma)}{v'\left(1+\frac{2.4r}{R}\right)\left(1+\frac{3.4r}{H}\right)}$$

$$= \frac{2}{9}\frac{r^2 g\rho' . t}{L.K.K'},$$

where, $\rho' = (\rho - \sigma)$; $v' = \frac{L}{t}$; the correction factors being K and K'.

Hence, the maximum proportional error in η is

$$\frac{\delta\eta}{\eta} = 2\frac{\delta r}{r} + \frac{\delta\rho'}{\rho'} + \frac{\delta t}{t} + \frac{\delta L}{L} + \frac{\delta K}{K} + \frac{\delta K'}{K'}.$$

Since the values of r and t are very small, hence these two measurements contribute the greatest to the proportional error, while the values of L, R and H being large, and fairly accurate weighing can be done while determining the values of ρ and σ, they contribute very little to the error. Hence, maximum care should be taken to determine the radii of the balls and to note the time of their fall.

$\therefore$ Percentage error in η is $\frac{\delta\eta}{\eta} \times 100\%$.

Question and Answers

Q.1. State Stokes' law and under what conditions does this formula hold?

A. : See Theory; the law holds good when a sphere assumes a terminal velocity in an infinite ocean of liquid.

Q.2. How are these conditions satisfied in this experiment?

A. : The method for ascertaining the attainment of terminal velocity is described in method, while for lack of infinite ocean of liquids necessary corrections are made for finite size of the vessel.

Q.3. What are the chief sources of error in this experiment?

A. : See "Precautions" in the previous page.

Q.4. Can you measure the coefficient of viscosity of water by this method?

A. : No, only the viscosity of very viscous liquids can be determined by this method (cf. glycerine, castor oil, etc.). If water is taken, then for its low viscosity the time of fall would be extremely small giving high inaccuracy in the measurements.

Q.5. Of the two methods—capillary flow method and Stokes' method, which one is preferable for measurement of viscosity of a liquid?

A. : The two methods have their distinctive advantages. While high viscous liquids can only be used in Stokes' method, capillary flow method is suitable only for liquids having very low viscosity (cf. water, etc.).

Q.6. If high viscous liquid is used in capillary flow method, what will happen?

A. : In that case, to maintain the flow of liquid, a very high pressure-difference will be needed at the end of the capillary tube and even then the rate of flow will be extremely small and the flow may not be streamline.

[For other questions on viscosity, see at the end of the previous Experiment No. 3.11.]

Chapter 4
Sound and Heat

4.1 Determination of the velocity of sound in rods by Kundt's tube method

Theory :

When a rod of length l_r, clamped at the middle, is set into longitudinal vibrations to emit its fundamental note, the middle point (M) of the rod will act as a node while its two ends P and Q will be antinodes. The fundamental frequency of the rod n will be given by

$$n = \frac{V_r}{\lambda_r} = \frac{V_r}{2l_r}, \qquad \text{... (4.1a)}$$

since, $\lambda_r = 2l_r$ and where V_r is the velocity of the waves in the rod. The vibrations produced are stationary. The vibrations of the rod produce resonant vibrations in the air of the wave tube and stationary waves are also produced in it. If l_a is the distance between successive nodes in air (Fig. 4.2) and V_a the velocity of sound in air, then the frequency of vibration of air is also

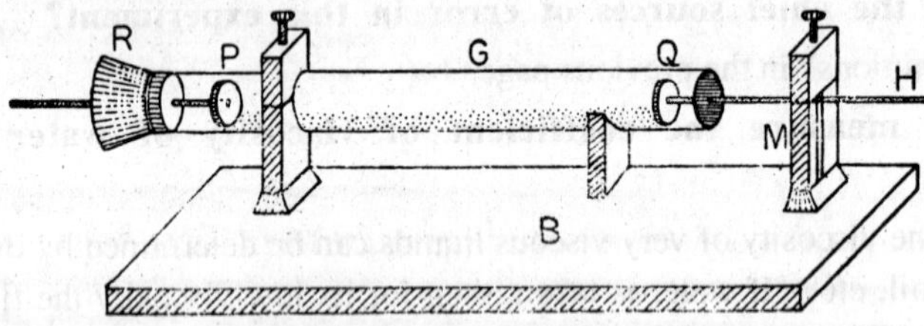

Fig. 4.1 : Kundt's tube apparatus

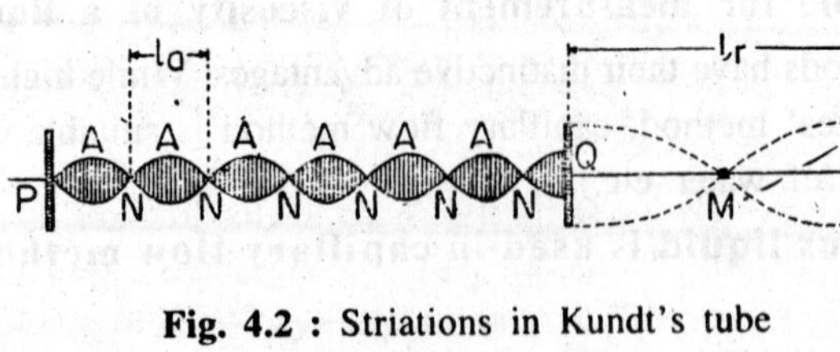

Fig. 4.2 : Striations in Kundt's tube

$$n = \frac{V_a}{2l_a}. \qquad \text{... (4.1b)}$$

Thus, $n = \dfrac{V_r}{2l_r} = \dfrac{V_a}{2l_a}$.

$$\therefore \quad V_r = V_a \cdot \frac{l_r}{l_a}. \qquad \text{... (4.1c)}$$

If V_0 be the velocity of sound in dry air at 0°C, t°C is the temperature of air in the tube, P the atmospheric pressure and f the aqueous vapour pressure of the air at t°C, then

$$V_a = V_0\left(\frac{273+t}{273}\right)^{1/2} \Big/ \left(1-\frac{3}{8}\frac{f}{P}\right)^{1/2}$$

$$= V_0\left(1+\frac{1}{2}\cdot\frac{t}{273}\right)\left(1+\frac{3}{16}\frac{f}{P}\right) \qquad \text{... (4.1d)}$$

Apparatus :

Kundt's tube apparatus : Metre scale, resined leather or cloth, lycopodium powder or cork dust.

Kundt's tube in its simplest from consists of a glass tube G (Fig. 4.1) about one metre long and 4 cm in diameter having a movable piston P at one end, while the other end is open. The experimental rod H, about 1 metre long and 1 cm in diameter, is clamped just at its mid-point M and has at its one end, which is inside the tube, a thin disc Q whose diameter is slghtly smaller than that of the tube, so that the disc just clears the wall of the tube. This is the sounding rod. The tube remains thoroughly dried and a thin layer of lycopodium powder is spread inside it. The tube and the rod remain in the horizontal straight line. When the other end of the rod is stroked longitudinally by means of a resined leather or cloth, longitudinall vibrations are set up in it which is communicated by the disc to the air inside the tube. Air is then forced to vibrate. If the position of P is so adjusted that resonance occurs between the vibrations of H and of the air, stationary waves will be formed in the air of the tube. Nodes will occur at P and a little away from Q since Q is in vibration (Fig. 4.2). The powder is set into vibration and finally it sets itself in small heaps at or near the nodes where the air is in least disturbance. At the antinodes (A), the molecules are in violent motion, where the powder forms prominent *striations* (Fig. 4.2) at right angles to the length of the tube. The striations have maximum length at the antinodes. They are due to circulation and vertex formation of the air inside the tube. The distance between neighbouring striations depends on the intensity of sound wave, density and size of the particles and on the pressure of the gas. Distance between any two consecutive nodes or antinodes is half of the wavelength.

Method :

(i) The length of the rod is measured carefully at least thrice and their mean taken. It is then clamped exactly at the mid-point. The light disc Q is then attached to its one end and it is set inside the tube carefully so that the disc nearly fills the tube but not touches it.

(ii) The glass tube is thoroughly dried and dry lycopodium powder or cork dust is thinly and uniformly spread over a metre scale, and after introducing the scale with the powder on it in the tube, the scale is tilted upside down when the powder collects along the whole length of the tube. It is absolutely necessary that both the tube and the powder must be dry.

(iii) The portion MH of the rod is now held with a resined leather in hand and pulled longitudinally. This excites longitudinal vibrations in the rod. This piston P is gradually moved either inwards or outwards until strong vibrations are set up in the air inside the tube. This will be evident when the powder will be strongly agitated. When resonance occurs, powder gathers at the nodes (*N*) and form antinodes (*A*) in the form of striations in between the nodes (Fig. 4.2). The maximum length of the striation is at the antinodes.

(iv) When the striations are well developed, two extreme distinct antinodes are selected and the distance *x* is measured between them. By counting the number of complete loops (*n*) in between them, the antinodal separation in air is obtained as $l_a = \frac{x}{n}$. If the striations are not well developed, then the distance *y* between two extreme nodes is measured and dividing it by the number (*n*) of complete loops in between them, again the nodal separation in air is obtained as $l_n = \frac{y}{n}$. Since antinodal or nodal separations are the same, hence $l_a = l_n$.

(v) The piston *P* is then displaced, readjusted and observations are repeated several times.

(vi) The temperature of air inside the tube is obtained by introducing a thermometer in it. The atmospheric pressure is obtained from a barometer. The pressure *f* mm of mercury of aqueous vapour in the air of the tube at the observed temperature is obtained from a standard table.

(vii) Taking the velocity of sound in air at 0°C as V_0 = 331.5 metres/sec, the values of *t*, *f* and *P* are substituted in Eq. (4.1 d), from which the value of V_a is obtained.

(viii) Substituting this value of V_a and the measured lengths l_a and l_r in Eq. (4.1c), the value of velocity of sound in rod (V_r) can be obtained.

Results :

(A) Length of the sounding rod (l_r) :

$$l_r = \frac{l_1 + l_2 + l_3}{3} = \ldots\ldots \text{ cm.}$$

(B) Determination of nodal (antinodal) separation in air :

Table 1

No. of Obs.	Temp. of air in the tube (*t* °C)	Distance between two extreme nodal/antinodal separation (*y*/*x* in cm)	No. of nodes or antinodes between them (*n*)	Nodal (l_n) or l_n antinodal (l_a) distance $\left(l_a = l_n = \frac{x \text{ or } y}{n}\right)$ (cm)	Mean value of (cm)	Pressure of water vapour (*f*) in mm at *t* °C	Barometric pressure (*P*) in cm
1.		...	...	...			
2.		...	...	...			
3.	...	...	...	...	...	...	...
4.		...	...	...			
5.		...	...	...			

(C) Determination of velocity in rod (V_r) :

Table 2

Atmospheric Pressure (P) in mm of Hg	Value of f in mm of Hg	Temp. of air (t°C)	Value of l_a (cm)	Value of l_r (cm)	Value of V_a as calculated from Eq. (4.1d) in cm/sec	Value of V_r (cm/sec) as calculated from Eq. (4.1 c) at t°C
...	...	...	...	...	...	...

Calculations :

From Eq. (4.1d) : $V_a = V_0\left(1+\frac{1}{2}\frac{t}{273}\right)\left(1+\frac{3}{16}\frac{f}{P}\right)$ = ... cm/sec.

From Eq. (4.1c) : $V_r = V_a.\frac{l_r}{l_a}$ = ... cm/sec at t°C.

Precautions :

(i) The tube and the powder must be prefectly dry.

(ii) The powder should be spread evenly in the tube.

(iii) The tube should be rotated slightly about its axis so that the powder is on the point of just slipping down.

(iv) The rod should be stoked lightly and the hand should not slip off the end, for in that case, alongwith the fundamental, higher harmonics will be excited.

(v) If too much heating is developed due to vigorous stroking, the rod should be allowed to cool down before next stroking.

Error Calculation :

From Eq. (4.1c), $V_r = V_a.\frac{l_r}{l_a}$, the maximum proportional error in measurement of V_r will be

$$\frac{\delta V_r}{V_r} = \frac{\delta V_a}{V_a}+\frac{\delta l_r}{l_r}+\frac{\delta l_a}{l_a}.$$

Since the measurement of V_a involves readings from a barometer, a thermometer and other data from standard tables, there is least possibility of any errors and hence $\frac{\delta V_a}{V_a}$ is negligibly small. Again, since the rod is about one metre long, the value of $\frac{\delta l_r}{l_r}$ is also very small. The main source of error is in the measurement of l_a since its value is small. To minimise this error, a large number of nodal/antinodal distances are measured from which the distance between two consecutive nodes or antinodes is determined.

Questions and Answers

Q.1. Why is the rod clamped in the middle? Can it be clamped at any other place?

A. : The rod is clamped in the middle so that it can vibrate in its fundamental mode with node at mid-point and antinodes at the ends. If it is clamped at one-fourth points of its length, then the clamped points would be nodes and the mid-points and the two free ends will be antinodes. The frequency of the rod will then increase.

Q.2. Why the tube and the powder should be dry?

A. : The presence of any moisture affects the velocity of sound and also that will affect the free movement of the powder.

Q.3. Why is the disc Q loosely fitted?

A. : The disc is loosely fitted so that it may vibrate freely inside the tube.

Q.4. Why temperature recording is essential?

A. : Because the velocity of sound depends very much on temperature of air.

Q.5. If the rod is replaced by a similar rod of some other material, will then the distance between the nodes/antinodes change?

A. : The velocity of sound and hence the frequency will be different for different materials. Hence for a different rod, the air column will vibrate with a different frequency resulting a change in the nodal/antinodal separation.

Q.6. What types of wave are produced in the air in the tube?

A. : Longitudinal stationary waves.

Q.7. Is it possible to determine the value of Y of the material of the rod?

A. : Velocity of sound in the rod is given by $V_r = \sqrt{\frac{Y}{\rho}}$, where Y and ρ are the Young's modulus and density of the material of the rod respectively. Hence knowing ρ, the value of Y can be determined.

Q.8. Can you determine the velocity of sound in liquids by this method?

A. : Using iron filings instead of powder and filling the tube with air-free liquids, the velocity of sound in liquids can be obtained in the same way.

Q.9. Will the value of V_a be the same if a bigger or a narrower tube is taken?

A. : No, the velocity V_a obtained depends on the value of the diameter of the tube. Velocity V' in a tube of radius r is given by $V' = V\left(1 - \frac{1}{r}\sqrt{\frac{\gamma}{4\pi n}}\right)$, where V is the velocity in a large volume of gas, n is the frequency of sound and γ is a constant known as kinematic viscosity of the gas—depending upon the density and static coefficient of viscosity of the gas.

Q.10. For accurate determination, whether nodal or antinodal distances should be measured?

A. : For accuracy, of antinodal distance of is preferred, since the striation effects increase towards the antinode and are marked by well-defined disc of dust particles.

Q.11. Will the dust figure alter for a change of (i) pressure, (ii) temperature and (iii) moisture inside the tube?

A. : See 'effects of temperature, pressure and moisture on velocity of sound' from any textbook.

Q.12. At which time of the year, the experiment is supposed to give good results?

A. : The experiment is best performed in winter when the atmosphere remains fairly dry. In rainy season, the experiment cannot be successfully performed for presence of excessive moisture in air which causes a lot of trouble in free formation of striations of the powder.

4.2 Determination of the ratio of the two specific heats of air by Kundt's tube

Theory :

If V_a is the velocity of sound in air at a certain temperature, then V_a is given by

$$V_a = \sqrt{\frac{\gamma P}{\rho}}, \qquad \text{... (4.2a)}$$

where P is pressure and ρ is the density of air under given condition and γ is the ratio of the specific heat of air at constant pressure (C_P) to that at constant volume (C_V). If the values of P, V_a and ρ are known, then the value of γ can be determined from Eq. (4.2a).

When Kundt's tube apparatus is used for this experiment, then P is the atmospheric pressure, obtainable from a barometer and ρ is the density of air at room temperature. The air contains some amount of moisture always and the partial pressure of existing moisture is also contained in the value of P.

If P is the atmospheric pressure in mm of Hg,

θ is the temperature of air (room) in °C,

f is the partial pressure of existing moisture in mm of Hg

and ρ_0 is the density of dry air at S.T.P. (= 0.001293 gm/cc),

then the density (ρ) of the moist air at pressure P and temperature θ is given by

$$\rho = \rho_0 \frac{273}{273+\theta} \times \frac{P - 0.375f}{760}. \qquad \text{... (4.2b)}$$

From the theory of the previous experiment (Kundt's tube experiment), the expression for velocity of sound in air for longitudinal wave is given as

$$V_a = V_r \cdot \frac{l_a}{l_b} \text{ (from Eq. 4.1c).}$$

Hence, if V_r is obtained, then performing the previous experiment (Expt. 4.1), the value of l_a and l_r can be had and V_a can be determined. Now, the velocity of sound in a solid (V_r) is given by

$$V_r = \sqrt{\frac{Y}{d}}, \qquad \text{... (4.2c)}$$

where Y and d are the Young's modulus and density of the material of the rod. Since the values of Y and d are known, by measuring l_a (the nodal/antinodal separation) inside the Kundt's tube and l_r, the length of the rod, the value of V_r and hence of V_a are determined. Obtaining ρ from Eq. (4.2b) and substituting the values of ρ and V_a in Eq. (4.2a), the value of γ can be easily obtained.

Apparatus :

The same apparatus as in Expt. No. 4.1.

Method :

The same procedure as in Expt. 4.1 has to be followed to measure the value of the nodal/antinodal distances (l_a) in the tube. The value of Y and d are supposed to be supplied and the values of P and f are to be obtained from a barometer and a standard table respectively. The length of the rod (l_r) is also obtained beforehand as in Expt. No. 4.1.

Results :

[Same tabulations as in Expt. No. 4.1 are to be done.]

Calculation :

From the knowledge of Y and d, the value of V_r is obtained from the equation,

$$V_r = \sqrt{\frac{Y}{d}}.$$

Then from Eq. (4.1c), $V_a = V_r \cdot \frac{l_a}{l_r}$, the value of V_a is obtained. Then in Eq. (4.2a), putting the value of V_a and that of P obtained from the barometer, the value of γ is obtained as

$$\gamma = V_a^2 \cdot \frac{\rho}{P} \quad \text{(value of } \rho \text{ is obtained from Eq. 4.2 b).}$$

Precautions :

[The same precautions are to be adopted as in Expt. No. 4.1.]

Error Calculation :

[The same calculation is to be made as is done in Expt. No. 4.1.]

Question and Answers

Q.1. What is γ?

A. : It is the ratio of two sp. heats of a gas (constant pressure to constant volume).

Q.2. Is it the same for all gases?

A. : No, it is maximum for monoatomic gases (cf. He, Ne, Argon, etc.) but its value decreases with the number of atoms in a molecule.

Q.3. What is the importance of the knowledge of γ?

A. : From its value, the number of atoms in a molecule can be ascertained.

Q.4. Is its value constant at all temperatures?

A. : No, since the value of sp. heats changes at very low temperatures.

Q.5. Is its value changed by the presence of any moisture in air?

A. : Yes, if p_1 and p_2 are the partial pressures of dry air and moist air and if γ_1 and γ_2 be their respective ratio of sp. heats, then

$$\frac{P}{\gamma-1}=\frac{p_1}{\gamma_1-1}+\frac{p_2}{\gamma_2-1},$$

where P is the total pressure and γ is the ratio of sp. heats of the mixture.

4.3 Determination of the coefficient of apparent expansion of a liquid by Weight Thermometer method

Theory :

If a weight thermometer filled with a liquid, contains, m_1 gm of liquid at t_1°C and when heated t_2°C (t_2°C > t_1°C) contains m_2 gm of liquid, then the apparent coefficient of expansion (γ') of the liquid for the mean temperature between t_1°C ant t_2°C is given by

$$\gamma'=\frac{m_1-m_2}{m_2(t_2-t_1)} \text{ per °C.} \qquad \ldots (4.3a)$$

[For proof of this expression, see any textbook on Heat]

In fact, $(m_1 - m_2)$ denotes the mass of the liquid expelled from the weight thermometer for its rise of temperature from t_1°C to t_2°C.

Apparatus :

Weight thermometer, beaker, balance, thermometer, water bath for heating.

Weight thermometer is an ordinary glass bulb B whose mouth is very narrow and bent. The bulb along with its neck is filled completely with the experimental liquid (Fig. 4.3). The bulb remains immersed in water, taken in a glass beaker (G) in such a way that its neck remains outside the beaker. When the beaker is heated, water in the beaker becomes heated and the bulb alongwith the experimental liquid is also heated. A thermometer (T) inserted in the beaker, records the temperature of the water bath. The bent narrow mouth of the glass bulb remains immersed in another beaker (G') also containing the experimental liquid.

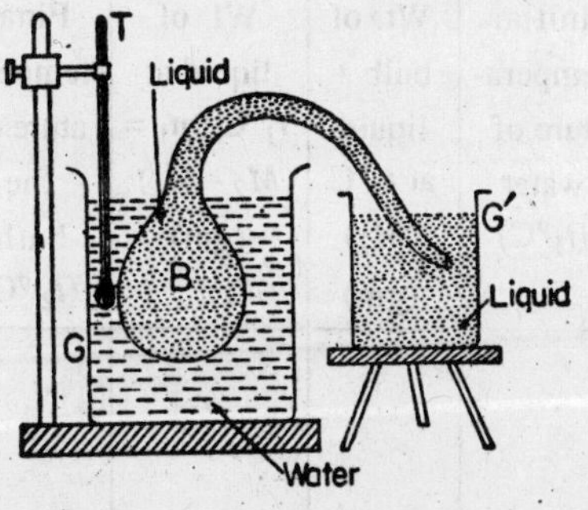

Fig. 4.3 : Weight Thermometer

Method :

(i) From a clean glass tube, a bulb B is blown with a narrow neck, which serves as the weight thermometer. This bulb is dried and its empty weight (M_1) is taken by a balance.

(ii) This bulb has to be filled with the experimental liquid. To do this, the neck of the bulb is partially immersed in the experimental liquid taken in a beaker G' and the bulb is slightly heated in the suspended position. When it is cooled down, air inside it contracts and some liquid is sucked inside the bulb. By repeating this process of alternate heating and cooling several times, the bulb alongwith the neck is completely filled with the liquid.

(iii) The bulb with the liquid is kept at the room temperature for at least ten minutes and the weight of the bulb alongwith the liquid is again determined. Let this weight be M_2. Thus the quantity of the liquid at room temperature t_1°C is ($M_2 - M_1 = m_1$ say).

(iv) The bulb is then placed inside a water bath (*G*) in such a way that its neck remains outside and is inside the liquid taken in another beaker (G′). The temperature of the water in the beaker (*G*) is noted. This should be almost equal to the room temperature (t_1°C). If not, then this temperature should be taken as the initial temperature (t_1°C).

(v) The water is now slowly heated and brought to its boiling point (t_2°C). The temperature is noted and this condition should be maintained at least for ten minutes. Due to heating, the liquid inside the bulb will expand and some liquid will come out of the neck of the bulb which is kept immersed in the liquid in G′ all through. When it is observed that no more liquid is coming out of the neck, then the bulb is carefully withdrawn from the bath.

(vi) The bulb is now dried in air, so that no water sticks to its surface and it is again brought back to the room temperature (t_1°C). Its weight is again taken. Let this weight be M_3. Hence, the mass of liquid remaining in the blub is ($M_3 - M_1 = m_2$) at temperature (t_2°C).

(vii) Substituting the values of m_1, m_2, t_1 and t_2 in Eq. (4.3a), the value of γ' can be obtained.

Results :

Table 1

Wt. of the empty bulb (M_1) (gm)	Initial temperature of water (t_1°C)	Wt. of bulb + liquid at t_1°C (M_2) (gm)	Wt. of liquid at t_1°C ($m_1 = M_2 - M_1$) (gm)	Final temperature of the bath (t_2 °C)	Wt. of bulb + liquid at t_2 °C (M_3) (gm)	Wt. of liquid at t_2 °C ($m_2 = M_3 - M_1$) (gm)	Mass expelled due to heating ($m_1 - m_2$) (gm)	$\gamma' = \dfrac{m_1 - m_2}{m_2(t_2 - t_1)}$ per °C
...	...	...	...	...	...	...	...	...

[The experiment can be repeated by taking another bulb.]

Calculations :

From Eq. (4.3a),

$$\gamma' = \frac{m_1 - m_2}{m_2(t_2 - t_1)} \text{ per °C} = \ldots\ldots \text{ per °C}.$$

Precautions :

(i) The weights must be taken carefully with the help of a sensitive balance.

(ii) Before taking the weight, the bulb must be cooled down to room temperature.

(iii) During the process of alternate heating and cooling while filling the bulb, care should be taken to see that the end of the neck must always remain immersed inside the liquid at G'. Similar is the case while heating the bulb to temperature t_2°C.

(iv) Since the neck with some liquid remains outside the water bath, this portion of the liquid does not attain the higher temperature (t_2 °C), for which some error may creep in. As a precaution, the portion of the neck of the bulb should be as short and narrow as possible.

(v) Before taking the weights, the bulb should be dried by wiping out the water on its outer surface.

Error Calculation :

From Eq. (4.3a),

$$\gamma' = \frac{m_1 - m_2}{m_2(t_2 - t_1)} = \frac{z}{m_2(t_2 - t_1)}, \text{ where, say, } z = (m_1 - m_2) \text{ and } t = (t_2 - t_1)$$

$$\therefore \quad \gamma' = \frac{z}{m_2(t_2 - t_1)} = \frac{z}{m_2 \times t}.$$

Proportional error (maximum) in γ' is

$$\frac{\delta\gamma'}{\gamma'} = \frac{\delta z}{z} + \frac{\delta m_2}{m_2} + \frac{\delta t}{t}.$$

Since m_2 is fairly large, the second term contributes very little to the percentage error, while the weights and the temperature should be measured very carefully with a sensitive balance and a sensitive thermometer respectively.

$$\therefore \quad \text{percentage error} = \frac{\delta\gamma'}{\gamma'} \times 100 = \ldots\ldots\ldots \%.$$

Questions and Answers

Q.1. What is meant by the coefficient of apparent expansion of a liquid and how is it related to the coefficient of real expansion?

A. : [See any textbook].

Q.2. Why is the apparatus called a 'weight thermometer'?

A. : Since the experiment involves in measuring several 'weights' and since from the relation, if γ' is known for a liquid, then by taking weights of the liquid (m_1 and m_2) at temperatures t_1°C and t_2°C, the unknown temperature t_2°C of the liquid can be obtained, when t_1°C is known. Since an unknown temperature can be obtained by this method, the apparatus is called a 'thermometer'.

Q.3. At what temperatue γ' is obtained here?

A. : The mean coefficient of apparent expansion between two given temperatures (t_1°C and t_2°C) is obtained.

Q.4. Why is it called the 'mean coefficient of expansion'?

A. : Since the expansion of the liquid and the vessel are not the same at all temperatures.

Q.5. Can the coefficient of linear expansion of glass be found out from this experiment?

A. : Yes, from the relation $\gamma = \gamma' + \gamma_g$, knowing γ—the coefficient of real expansion of the liquid, and finding out γ' from this experiment, γ_g (the coefficient of cubical expansion of the material of the vessel) is obtained, from which α_g (coefficient of linear expansion) is found out from the relation, $\gamma_g = 3\alpha_g$.

Q.6. If the experimental liquid be not pure, what will happen?

A. : If any impurities are present, then the value of γ' will increase.

Q.7. Is there any assumption in this method?

A. : Yes, it is assumed that during experiment, the density of the liquid does not change with temperature, which is not true.

4.4 Determination of the vapour density of a liquid by Victor Meyer's method

Theory :

Under steady condition of temperature and pressure, when a volatile liquid, whose vapour is heavier than air, is vaporised in a Victor Meyer tube, it displaces its own volume of air having the same temperature and pressure. If this displaced air, when collected over water in a eudiometer tube at pressure (atmospheric) P and temperature t°C, occupies a volume V cc, and if f is the saturation vapour pressure of water at t°C, then the actual pressure of the air at t°C will be $(P-f)$ cm of mercury. The volume (v) of this displaced air when reduced to N.T.P. will be

$$v = \frac{V(P-f)}{76} \times \frac{273}{t+273} \text{ cc.} \qquad \text{... (4.4a)}$$

If m is the mass of the volatilised liquid expressed in gm, then the absolute density of its vapour at N.T.P. would be

$$\rho = \frac{m}{v} = \frac{m}{V} \times \frac{76}{(P-f)} \times \frac{t+273}{273} \text{ gm/cc.} \qquad \text{... (4.4b)}$$

Since the absolute density of hydrogen at N.T.P. is 0.0000898 gm/cc, the density of the vapour relative to hydrogen at N.T.P. would be

$$\rho' = \frac{\rho}{0.0000898} = \frac{m}{V} \times \frac{76}{(P-f)} \times \frac{t+273}{273 \times 0.0000898}. \qquad \text{... (4.4c)}$$

From Eqs. (4.4b) and (4.4c), the absolute and relative densities of the vapour can be determined. If M be the molecular weight of the vapour, then obviously

$$M = 2\rho'. \qquad \text{... (4.4d)}$$

Apparatus :

Victor Meyer's tube, Hoffmann bottle, Eudiometer tube, Burner, etc.

The apparatus Victor Meyer tube (Fig. 4.4) consists of a long glass tube G provided with a long stem and a cylindrical bulb at one end (B). The other end of the tube may be closed by a soft rubber cork (S_1). Near the top of the stem is a side tube (T) serving as a delivery tube, the open end of which dips into water taken in a trough (A). Air coming out of the delivery tube through this open end may be collected in a graduated eudiometer tube

(*C*) held vertically and completely filled with water and set in an inverted position over the open end of the delivery tube. Victor Meyer tube can be heated by introducing its major portion of the stem and bulb inside a metallic jacket (*J*) containing a liquid whose boiling point is at least 15°C higher than that of experimental liquid. If the experimental liquid has b.p. less than 80°C, then water may safely be used inside the jacket. The jacket has a rubber cork (*S*) at its mouth through which the stem of the tube passes and contains an exhaust for steam (*St*). The Victor Meyer tube is held vertically and can be heated by a burner. The bulb (*B*) should be kept a few centimeter above the liquid level in the jacket. A small glass bottle (*H*), known as Hoffmann bottle, fitted with a glass stopper is used as a container of the experimental liquid. It is avdisable to use a small quantity of burnt asbestos or glass-wool inside the Victor Meyer tube so as to guard it against breakage when the bottle filled with liquid is dropped inside it.

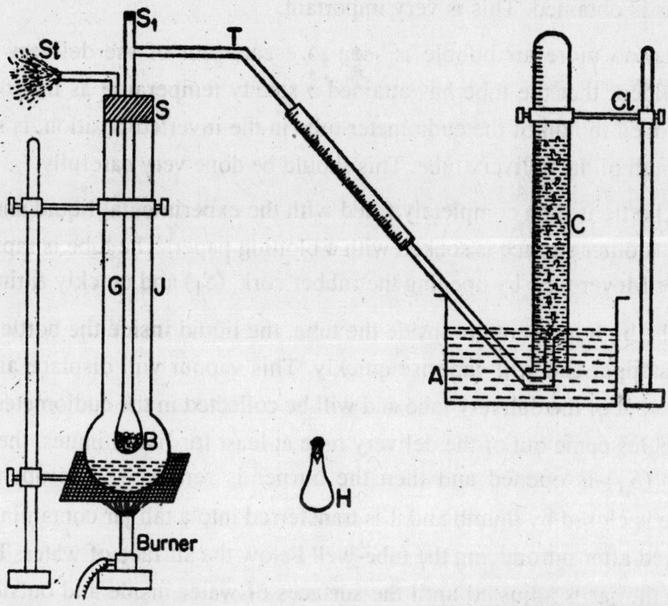

Fig. 4.4 : Victor Meyer's apparatus

Method :

(i) The Victor Meyer tube and the Hoffmann bottle are thoroughly cleaned with caustic soda, nitric acid and water. They are then completely dried by blowing hot air into them.

(ii) The water jacket is filled with water and arrangement for heating the water inside the jacket is made. The Victor Meyer tube is taken and some quantity of burnt asbestos or glass-wool are introduced in it, so that when the liquid-filled bottle would be dropped inside it, the tube may not break. The mounth of the tube is closed and set inside the jacket in such a way that the bulb (*B*) remains a few centimeter above the water surface in the jacket. The mouth of the jacket is closed with the rubber cork (*S*) fitted with the steam exhaust (*St*). The delivery tube (*T*) is kept under water level in the trough (*A*).

(iii) The eudiometer tube (C) is completely filled with water and is kept inverted in a vertical position and fixed by a clamp (Cl). It is kept immersed in water.

(iv) The jacket is now gradually heated and due to heating, air inside the Victor Meyer tube expands and come out of the delivery tube as evidenced by bubbles coming out of the delivery tube immersed under water in the trough.

(v) During this process of heating and attainment of steady temperature of the jacket and the tube, weighing operation of the empty bottle and the bottle filled with the experimental liquid, is done. At first the weight of the empty bottle (M_1) is taken (using a rider) and then the weight of the bottle filled with the liquid, is taken (M_2). The difference of those two weights will give the mass (m) of the liquid taken. Since the experimental liquid in this experiment is normally volatile, the liquid evaporates to some extent while its weight is determined. Hence, the process of taking the weight of the bottle filled with the liquid, is repeated several times, until at least for three or four successive weighings, a constant weights is obtained. This is very important.

(vi) When no more air bubble is seen to escape out of the delivery tube during heating, it is evident that the tube has attained a steady temperaure as that of the jacket. Right then, the open mouth of the eudiometer tube in the inverted position, is set vertically just over the mouth of the delivery tube. This should be done very carefully.

(vii) The bottle is then completely filled with the experimental liquid and any excess liquid lying on its outer surface is soaked with a blotting paper. The tube is rapidly dropped inside the Victor Meyer tube by opening the rubber cork (S_1) and quickly fitting it again.

(viii) At the high temperature inside the tube, the liquid inside the bottle will expand opening up its stopper and will vaporise quickly. This vapour will displace air in the tube which will come out of the delivery tube and will be collected in the eudiometer tube. When no more air bubbles come out of the delivery tube at least for five minutes, then at first the rubber stopper (S_1) is opened and then the burner is removed. The open end of the eudiometer tube is closed by thumb and it is transferred into a tall jar containing water. The thumb is removed after introducing the tube-well below the surface of water. The height of the tube within the jar is adjusted until the surfaces of water inside and outside are in the same horizontal line. The volume of the displaced air (V) inside the eudiometer tube is noted from its graduations. The temperature (t°C) of the water in the jar is also noted. The atmospheric pressure (P) is noted from the barometer and the S.V.P. of water at t°C is obtained from the table.

(ix) Thus knowing P, V, f and t, the vapour density (ρ') of the given liquid can be obtained from Eq. (4.4c), whence m is previously determined.

(x) With the same Hoffmann bottle, two more collections can be done, each time drying the tube and the bottle and taking the mass of the liquid freshly determined.

(xi) Taking a different bottle, the whole experiment may be repeated with a new set of readings and from all these, the mean value of ρ' (vapour density) can be determined.

Results :

(A) Determination of mass (m) of the liquid vaporised :

Table 1

Bottle No.	Mass of empty bottle (M_1) in gm	Mass of the bottle + liquid (M_2) in gm	Value of M_2 when found constant (gm)	Mass of the liquid in gm ($m = M_2 - M_1$)
1.	...	(a) ...... (b) ...... (c) ...... (d) ...... (e) ...... } constant	......	......
2.	...	....	......	......

(B) Readings for Barometric Pressure (P) :

Table 2

[A table as shown in Art. 2.8 is to be made.]

(C) Determination of Vapour Density (ρ) of the liquid :

Table 3

Bottle No.	No. of Obs.	Mass of liquid taken (m gm)	Temp. of water bath (t°C)	Aqueous tension at t°C in cm. of mercury (f)	Barometric Pressure (P cm)	Volume of displaced air at t°C (V cc)	Pressure of displaced dry air ($P - f$) cm	Vapour density (ρ′) in gm/cc	Mean value of vapour density (ρ′) gm/cc
1.	1.	...	...	...	...	...	...	...	
	2.	...	...	...	...	...	...	...	
	3.	...	...	...	...	...	...	...	
2.	1.	...	...	...	...	...	...	...	...
	2.	...	...	...	...	...	...	...	
	3.	...	...	...	...	...	...	...	

Calculations :

From Eq. (4.4c), vapour density of the liquid (relative)

$$\rho' = \frac{m}{V} \times \frac{76}{(P-f)} \times \frac{t+273}{273 \times 0.0000898} = \ldots\ldots$$

Again, the molecular weight (M) of the liquid

$$M = 2\rho' = \ldots\ldots\ldots$$

Precautions :

(i) Both the Victor Meyer tube and the Hoffmann bottle must be kept dry and also the burnt asbestos introduced must be dry.

(ii) The weighings should be done very carefully, particularly in case of bottle filled with the liquid. Since the liquid is volatile, successive weighings must be done until constancy in weight is obtained. The greatest source of error in this experiment lies in this respect.

(iii) Before removing the burner from the jacket, the rubber stopper (S_1) must be opened first, otherwise due to fall of temperature inside the tube, water may suck inside the tube causing a crack therein.

(iv) Cotton should never be used inside the tube, which may absorb some liquid from the bottle when it opens out inside the tube.

(v) After final constant weight (M_2) being obtained, the bottle should be very quickly dropped inside the tube, so that any liquid may not evaporate out during transit from the balance to the tube.

(vi) Care should be taken to check that no part or portion of the tube or jacket or the delivery tube has any leakage anywhere, which would make the quantity of displaced air collected less than what it should be.

(vii) The lower end of the bulb of the tube (B) should not touch the water surface in the jacket.

Error Calculations :

From Eq. (4.4c), it is evident that the maximum proportional error in ρ' is given by

$$\frac{\delta\rho'}{\rho'} = \frac{\delta m}{m} + \frac{\delta V}{V} + \frac{\delta(P-f)}{(P-f)} + \frac{\delta t}{t+273}.$$

Since the value of $(P-f)$ is fairly large, hence its contribution to the error is negligibly small. The value of V and t are also fairly large and also they can be measured accurately. But since the value of m is very small, the greatest source of error is this term $\frac{\delta m}{m}$. Hence, the mass of the liquid must be measured very accurately, or in other words, the weighings must be done accurately using a rider.

Questions and Answers

Q.1. What is the difference between 'vapour density' and 'absolute density'?

A. : Vapour density of a substance is the weight of a certain volume of the vapour at N.T.P. to the weight of the same volume of Hydrogen at N.T.P., whereas the absolute density is the mass of a certain quantity of vapour at a given temperature per unit volume of it. The latter has the unit gm/cc while the former is a pure number.

Q.2. Is the vapour whose density is measured, saturated or unsaturated?

A. : It is unsaturated vapour approximately obeying the gas laws.

Q.3. In the eudiometer tube, is the vapour collected?

A. : No, the displaced air from the tube is collected in it.

Q.4. Can all types of liquid be used in this method?

A. : No, only volatile liquids having b.p. less than approximately 80 °C may be used here.

Q.5. Why a larger Hoffman's bottle is not used to improve the accuracy of the result?

A. : In that case, though the accuracy may increase to some extent but if the mass of the liquid is great, then it may occupy a volume which is greater than the internal volume of the Victor Meyer tube. Then some vapour may come out of the delivery tube and may condense while coming in contact of the cold water in the trough.

Q.6. What happens if the burner is suddenly removed before opening the stopper of the Victor Meyer tube.

A. : [See Precautions (iii)].

Q.7. Why burnt asbestos or glass-wool are placed inside the Victor Meyer tube?

A. : That prevents breaking up of the glass tube (Victor Meyer) when the bottle is dropped inside it.

Q.8. Why the bottom of the V.M. tube is kept slightly above the water surface inside the jacket?

A. : Ordinary water taken inside the jacket contains impurities whose b.p. is not the same as the steam temperature. Then different parts of the V.M. tube will have different temperatures, but if the entire tube remains surrounded by steam, then the temperature of all its parts and portions are the same.

Q.9. Which measurement controls the greatest accuracy in this experiment?

A. : [See 'Error Calculations'.]

4.5 Determination of the thermal conductivity of a good conductor by Searle's method

Theory :

If a steady temperature gradient be maintained along a bar having cross-sectional are A and if the temperatures of two layers of the bar separated by a distance d be θ_1°C and θ_2°C respectively, then the quantity of heat (Q) flowing through them in a time t sec is given by

$$Q = \frac{KA(\theta_1 - \theta_2)t}{d}, \qquad \text{... (4.5a)}$$

where K is the thermal conductivity of the material of the bar. This means that if the cross-section is unity and if the temperature difference is 1°C between any two points, then K calories of heat will flow per unit time between those two points.

Now if this amount of heat (Q) raises the temperature of M gm of water flowing during that time (t) from θ_4 to θ_3 ($\theta_4 < \theta_3$), then

$$Q = M(\theta_3 - \theta_4). \qquad \text{... (4.5b)}$$

Thus, $$Q = \frac{KA(\theta_1 - \theta_2)t}{d} = M(\theta_3 - \theta_4)$$

or, $$K = \frac{M.d}{tA} \times \frac{\theta_3 - \theta_4}{\theta_1 - \theta_2}. \qquad \text{... (4.5c)}$$

Hence, measuring the quantities on the right hand side of this equation, the value of K can be determined.

Apparatus :

Searle's apparatus: Thermometers, water circulation system and steam production arrangement.

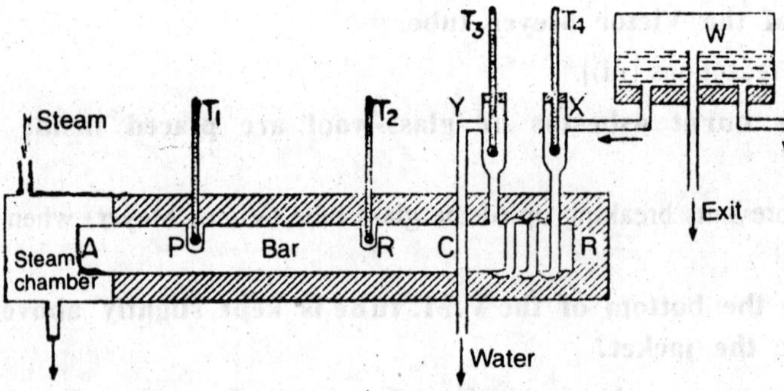

Fig. 4.5 : Searle's apparatus

Searle's apparatus consists of a thick polished metallic bar whose conductivity is to be determined. The bar is about 30 cm in length and 5 cm in diameter (Fig. 4.5). The bar AB is surrounded by dry felt and enclosed in a wooden box. One end of the bar (A) is placed inside a steam chamber through which by passing steam, the end A of the bar can always be maintained at the temperature of the steam. The other end of the bar (B) is surrounded by a copper tube (C) through which water can be passed. This water is passed at a constant pressure and rate, controlled by a water reservoir (W). The tube C has two ends. Through the end X, cold water enters and through the end Y water flows out after absorbing heat from the bar. The thermometers T_3 and T_4 record the temperatures of the outgoing and incoming waters respectively (θ_3 and θ_4). The outcoming water is collected by a measuring cylinder for a given time (t). Near the middle of the bar two holes are drilled (P and R) at a known distance (d) apart and thermometers T_1 and T_2 are placed in these fixed holes containing mercury to ensure good contacts. These thermometers record the temperatures θ_1 and θ_2 respectively.

Method :

(i) Arrangements for boiling water in a vessel is being made and during the process, the following measurements are made.

(ii) The bar is taken out of the box and its diameter is measured at different places with the help of a slide calliper. Similarly, the distance (d) between the pair of holes (P, R)

in the bar is measured. While measuring this, the distances between the centres of the holes are to be taken.

(iii) The four thermometers are then set in their respective positions, and before setting, they are to be checked whether all of them give the same room temperature inside the laboratory. Otherwise, necessary corrections are to be made.

(iv) The end *A* of the bar is suitably placed inside the steam chamber through which steam is continuously passed. When this end attains the temperature of steam, then the thermometers T_1 and T_2 would show steady temperatures. During attainment of this steady state, readings of these two thermometers are to be taken at intervals of 4 to 5 minutes. When the steady temperatures are recorded at least for 10 minutes, they are then noted down.

(v) At this stage, water is made to flow through the copper tubes fitted at the other end of the bar. Water is collected in a measuring cylinder for a fixed time (t sec). During this collection, the temperatures recorded by the thermometers T_4 and T_3 are noted which give the temperatures of the inflowing and outgoing water.

(vi) By changing the rate of flow of water, the readings of T_4 and T_3 can be altered for two more sets and the experiment is repeated. Again, by changing the time of collection (t) of water, the value of M can also be changed and a new set of reading can be taken.

(vii) After all the quantities are measured, substituting them in Eq. (4.5c), the value of K can be obtained.

Results :

(A) Measurement of diameter and distance between the holes of the bar :

Table 1

Vernier constant of the slide calliper = ... cm.

No. of Obs.	Diameter of the bar					Distance (d) betweeen the holes at P and R (cm)	Mean distance (d) between the holes at P and R (cm)
	Main scale reading (cm)	Vernier scale reading (cm)	Total reading (cm)	Mean diameter (r) (cm)	Mean cross-section (A) (sq. cm)		
1.	...	...	...			...	
2.	...	...	...			...	
				...	...		...
3.	...	...	...			...	
4.	...	...	...			...	

(B) Reading of the thermometers :

Table 2

	Time in minutes	No. of Obs.	Thermometer readings				Difference of temperature between T_1 and T_2 $(\theta_1 - \theta_2)$ (°C)	Difference of temperature between T_3 and T_4 $(\theta_3 - \theta_4)$ (°C)
			T_1 (θ_1) (°C)	T_2 (θ_2) (°C)	T_3 (θ_3) (°C)	T_4 (θ_4) (°C)		
Before passing steam	0	1.	...	...	...	...	...	...
	2	2.	...	...	...	...	...	...
	4	3.	...	...	...	...	...	...
	6	4.	...	...	...	...	...	...
	8	5.	...	...	...	...	...	...
After passing seam	0	1.	...	...	...	...	...	...
	2	2.	...	...	...	...	...	...
	4	3.	...	...	...	...	...	...
	6	4.	...	...	...	...	...	...
	8	5.	...	...	...	...	...	...
	...	...	...	...	...	...	...	...
	...	...	...	...	...	...	...	...

[Similar tables are to be made for different sets when the rate of flow (*M*) of water, time of flow (*t*) and values of $(\theta_3 + \theta_4)$ are changed]

(C) Determination of mass of water collected :

Table 3

No. of Obs.	Time of collection (*t* sec)	Volume of water collected (cc)	Mean volume (*V*) (cc)	Density (ρ) of water at mean temp. of θ_4 and θ_3 (gm/cc)	Mass of water (M) collected (gm) $(M = V.\rho)$
I.	... ...	... ...	... ...	...	...
II.	... ...	... ...	... ...	...	...
III.	... ...	... ...	... ...	...	...

(D) Determination of the value of K :

Table 4

No. of Obs.	Mass of collected water (M) (gm)	Time of collection (t) (sec)	$(\theta_1 - \theta_2)$ (°C)	$(\theta_3 - \theta_4)$ (°C)	Value of K in c.g.s. unit	Mean Value of K in c.g.s. unit
I.	...	...	...	...	...	...
II.	...	...	...	...	...	...
III.	...	...	...	...	...	...

Calculations :

From Eq. (4.5c),

$$K = \frac{M.d(\theta_3 - \theta_4)}{t.A(\theta_1 - \theta_2)} = \text{........ cal/sec/cm/°C (c.g.s. unit)}$$

Precautions :

(i) Before performing the experiment, all the four thermometers are to be calibrated among themselves.

(ii) All readings and collection of water are to be made when steady state is reached and the thermometers show constant readings at least for 10 minutes.

(iii) Water should be collected very carefully, so that no water spills out of the measuring cylinder. Care should be taken to see that there is no leakage of water at the corks through which the thermometers pass.

(iv) Difference of temperatures between thermometers T_3 and T_4 should preferably be maintained at about 8 to 10 degrees celsius.

(v) Sufficient quantity (about 500 cc) of water should be collected by the measuring cylinder so as to minimise the error in the experiment.

Error Calculations :

From Eq. (4.5c),

$$K = \frac{M.d(\theta_3 - \theta_4)}{t.A(\theta_1 - \theta_2)}$$, the proportional error in the measurement of K is

$$\frac{\delta K}{K} = \frac{\delta M}{M} + \frac{\delta d}{d} + \frac{\delta A}{A} + \frac{\delta(\theta_3 - \theta_4)}{(\theta_3 - \theta_4)} + \frac{\delta(\theta_1 - \theta_2)}{(\theta_1 - \theta_2)} + \frac{\delta t}{t}.$$

To minimise the error in the experiment, sufficient quantity (M) of water is to be collected for which time of collection (t) will be fairly large. Hence, error due to M and t are minimised. Since the bar is of sufficient length, the distance between the holes (d) is fairly large and the radius of the bar is also not very small. Hence, the errors due to d and A are also very small. However, as a precaution, diameter should be measured by a slide calliper and length by a scale reading up to milimetre. To get accuracy of the result, the difference of temperatures, particularly $(\theta_3 - \theta_4)$, must be made at least 8 to 10 degrees celsius and the most probable error lies in its measurement. Thermometers reading up to one-tenth of a °C are to be used. The percentage error in K is thus

$$\frac{\delta K}{K} \times 100 = \text{.........} \%.$$

Questions and Answers

Q.1. What is meant by thermal conductivity of a substance?

A. : See 'Theory'.

Q.2. What is its difference with the 'thermometric conductivity?

A. : If the substance has the density ρ and sp. heat S, then its thermometric conductivity will be $K/\rho S$, where K is its thermal conductivity.

Q.3. On what factors its value depends?

A. : The value of thermal conductivity depends on the nature of the material but not on its shape and size.

Q.4. Why the bar is surrounded by a jacket?

A. : To minimise radiation loss radially and thus enabling to attain steady state.

Q.5. Is this method applicable for bad conductors?

A. : No, in that case the conduction of heat being poor, the value of $(\theta_1 - \theta_2)$ will be large causing large error due to radiation loss.

Q.6. How can the temperatures be more accurately measured?

A. : By using thermocouples instead of thermometers.

Q.7. Would the bar be thin or thick?

A. : It should preferably be thick to minimise errors.

Q.8. What is temperature gradient in this experiment?

A. : Its magnitude in this experiment is $\dfrac{(\theta_1 - \theta_2)}{d}$.

Q.9. Is it essential to pass water coming out of a constant pressure head?

A. : Yes, otherwise during the experiment, the rate of flow will change causing an error in the mass of water collected.

Q.10. What improvements can be suggested for accuracy of the result?

A. : First, the steam-chest may be replaced by an electrically heated coil, so the amount of heat supplied to the bar may be calculated. Secondly, the temperature of the rod at various points may be recorded by thermocouples from which a graph (temp. distance) may be drawn. From this graph, the temperature gradient may be calculated.

Q.11. Has the temperature of the bar any effect on the value of its conductivity?

A. : Conductivity of pure metals generally decreases with rise of temperature while that of alloys tends to increase. Conductivity of water rises with temperature but for oils it decreases with temperature.

4.6 Determination of thermal conductivity of a bad conductor in the form of a disc by Lees' and Chorlton's method

Theory :

In the arrangement shown in Fig. 4.6, when steam is passed through A, the base B or A is warmed up and heat is conducted through the bad conductor S and finally is transmitted to the brass disc C which is also warmed up. After some time, a steady state is reached when the rate of flow of heat through S is equal to the heat lost from C by convection and radiation.

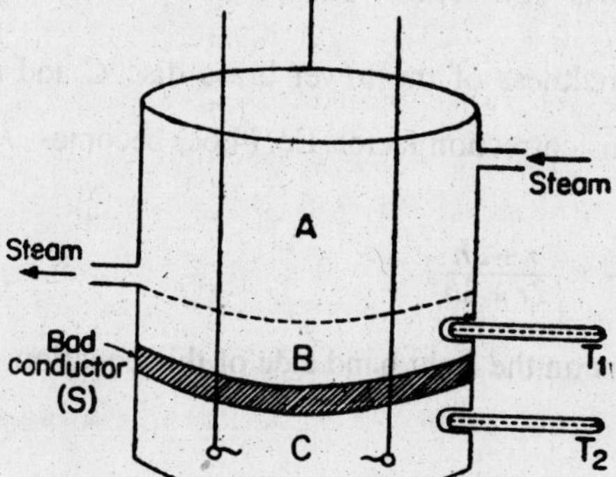

Fig. 4.6 : Lees' apparatus

If θ_1 = temperature of B in the steady state

θ_2 = temperature of C in the steady state

A = area of cross-section of the badly conducting specimen

d = thickness of the conducting slab S

K = thermal conductivity of the material of the specimen (S),

then the quantity of heat (Q) conducted per second through the badly conducting sheet (S) is given by

$$Q = \frac{KA(\theta_1 - \theta_2)}{d}. \quad \text{... (4.6a)}$$

In the steady state, from the lower disc C, this amount of heat is radiated away per second. If M and s be the mass and specific heat of the material of the lower disc C, and $\left(\frac{d\theta}{dt}\right)$ is its rate of cooling at temperature θ_2, then the heat radiated away per second from C is

$$Q = Ms\left(\frac{d\theta}{dt}\right)_{\theta_2} \quad \text{... (4.6b)}$$

Equating the expressions of (4.6a) and (4.6b),

$$\frac{KA(\theta_1 - \theta_2)}{d} = Ms\left(\frac{d\theta}{dt}\right)_{\theta_2}$$

or, $$K = \frac{M.d.s\left(\frac{d\theta}{dt}\right)_{\theta_2}}{A(\theta_1 - \theta_2)}. \quad \text{... (4.6c)}$$

In the present method, the rate of cooling of the disc (C) is determined without the experimental slab S' on it. To obtain the correct value of $\left(\frac{d\theta}{dt}\right)$ at θ_2 under the prevailing condition of the experiment, the quantity $\left(\frac{d\theta}{dt}\right)$ at θ_2 is multiplied by a correction factor F, which is known as *Bedford's correction*. The correction factor is $F = \frac{\pi r^2 + 2\pi rh}{2\pi r^2 + 2\pi rh} = \frac{r+2h}{2r+2h}$, where h is the thickness of the lower brass disc C and r is the radius of the specimen (S). Introducing this correction factor, Eq. (4.6c) becomes

$$K = \frac{M.d.s\left(\frac{d\theta}{dt}\right)_{\theta_2}}{A(\theta_1 - \theta_2)} \times \frac{r+2h}{2r+2h}. \quad \ldots (4.6d)$$

Measuring the quantities on the right hand side of this equation, the value of K can be determined.

Apparatus :

Lees' apparatus : Thermometers $\left(\frac{1}{10}\text{°C}\right)$, steam chest, boiler, experimental bad conductor in the form of a circular disc.

The apparatus consists of a brass disc C supported by means of strings from a ring fixed on a retort stand (Fig. 4.6). The experimental material (bad conductor) is taken in the form of a circular slab (S) of the same diameter as that of C. It is of uniform thickness and placed on the disc C. A steam chamber A is placed on the slab (S) and the bottom of the steam chamber (B) is a thick circular metal plate. Arrangements for entry and exit of steam are there is A. Diameters of B, C and S are all the same. Through two holes drilled in B and C, two thermometers T_1 and T_2 are inserted to record the temperatures of B and C.

Method :

(i) The mass (M) of the lower disc (C) is determined with the help of a spring balance. The diameter of the slab (S) is also determined by measuring its circumference and from it the radius and hence the area A. The specific heat (s) of the material of the slab is obtained from the table of constants.

(ii) To find the thickness (d) of the slab, the measurement has to be done *in situ* at several places. This means that the thickness of the slab is to be measured by keeping it in its position with the disc placed on it. This can be done by making several cross-marks at different places on a paper attached to the upper slab B. Keeping the slab S in its position between B and C, a travelling microscope is focused on one of the cross-marks and the reading (X_1) is taken. The sheet is again focused and the corresponding reading of the microscope (X_2) is noted. Hence, the thickness of the slab will be $d = X_1 \sim X_2$. Similar readings are taken for other cross-marks and for each case, the value of d is obtained. From all these values, the mean value of the thickness is obtained.

(iii) The two thermometers are then compared for any initial errors and then are introduced in the respective holes. Steam is passed into the chamber A from a boiler placed at a distance. Care should be taken to screen off the apparatus from direct heating from the boiler. Temperatures of T_1 and T_2 are recorded at intervals of 5 minutes until steady temperratures are attained.

(iv) Next the supply of steam is cut off and the upper disc *B*, the chamber *A* and the sheet *S* are all removed. The lower slab *C* is then heated directly by a burner until its temperature is raised about ten degrees higher than the steady value recorded by T_2. Then removing the burner, the disc (*C*) is allowed to cool down in the same conditions as during the first part of the experiment. The temperatures during cooling as indicated by T_2 are noted at intervals of half a minute.

(v) From these data, the rate of cooling $\left(\frac{d\theta}{dt}\right)$ and the average temperature of the disc during the interval can be found out. For example, if θ′ and θ″ be the temperatures of the disc at the beginning and at the end of an interval of 30 sec, then the rate of cooling during this interval will be $\frac{(\theta' - \theta'')}{30}$, while interval would be $\frac{(\theta' - \theta'')}{2}$. A graph then can be plotted with the rate of cooling $\left(\frac{d\theta}{dt}\right)$ along *X*-axis and the corresponding average temperature (θ) along *Y*-axis. (Fig. 4.7), which would be a straight line. From this graph, the rate of cooling $\left(\frac{d\theta}{dt}\right)$ at θ_2, the steady temperature of *C* during the first part of the experiment, can be found out.

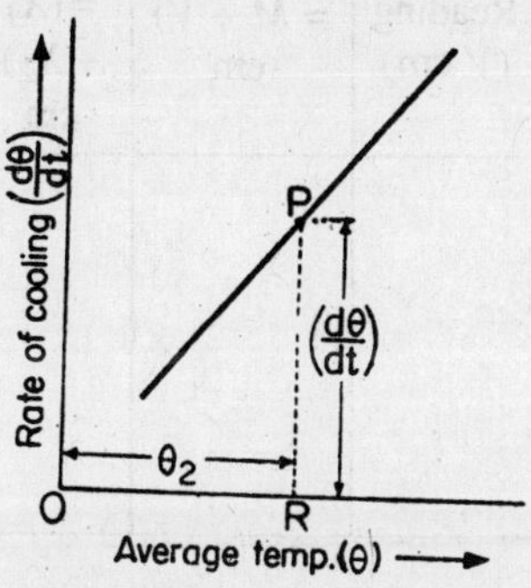

Fig. 4.7 : Cooling rate-temperature

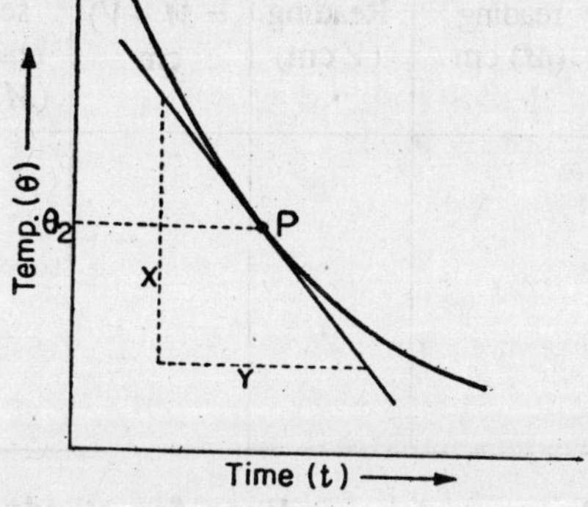

Fig. 4.8 : Cooling curve

(vi) Another graph may also be plotted with the temperature (θ) of the disc (*C*), recorded during cooling, and time (*t*). From this graph (Fig. 4.8) a tangent is drawn at the point on the curve corresponding to θ_2, the steady temperature attained by the lower disc during the static part of the experiment. From this tangent, the slope gives the value of $\left(\frac{d\theta}{dt}\right)$ at θ_2.

(vii) Knowing thus the values of M, s, $\left(\frac{d\theta}{dt}\right)_{\theta_2}$, A, d and $(\theta_1 - \theta_2)$ and also the values of r and h the value of the thermal conductivity K of the material of the slab can be obtained from Eq. (4.6d).

Results :

Mass (M) of the lower disc C = ... gm

Specific heat (s) of the material of the disc = ... cals/gm/°C

(A) Determination of radius and hence area (A) of the sheet

Table 1

No. of Obs.	Circumference of the slab (S) (L cm)	Mean value of the circum-ference (L cm)	Radius of the slab in cm ($r = L/2\pi$)	Area of the slab (S) $A = \pi r^2$ sq. cm
1.	......			
2.	......			
		...	...	...
3.	......			
4.	......			

(B) Determination of thickness (d) of the slab by a microscope

Table 2

No. of Obs.	Reading of a cross-mark on B with the slab S placed in between B and C			Reading of the cross-mark on B without the slab S in between B and C			Thick-ness of the slab d $=(X_1 \sim X_2)$ cm	Mean value of the thick-ness (d) cm
	Main scale reading (M) cm	Vernier Reading (V cm)	Total ($X_1 = M + V$) cm	Main scale reading (M) cm	Vernier Reading (V cm)	Total ($X_2 = M + V$) cm		
1.								
2.								
3.								
4.								
5.								

(C) Thermometer readings for steady-state temperatures :

Table 3

Initial corrections, if any, between the thermometers = ... °C

Time in minutes	0	5	10	15	...	...	...
Temperature (θ_1°C) of T_1	...	...	...	...	...	...	...
Temperature (θ_2°C) of T_2	...	...	...	...	...	...	...

(D) Time-Temperature record during cooling of the disc C :

Table 4

No. of Obs.	Time in min. (t)	Temperature of the disc C in °C	Average temperature during the interval in °C	Rate of cooling of the disc C during the interval $\left(\frac{d\theta}{dt}\right)$	Rate of cooling $\left(\frac{d\theta}{dt}\right)$ at θ_2°C from Graph I	Rate of cooling $\left(\frac{d\theta}{dt}\right)$ at θ_2°C from Graph II
1.	0	...				
2.	$\frac{1}{2}$	...	...	...		
3.	1	...	...	...		
4.	$1\frac{1}{2}$	...	...	...	...	
5.	2	...	...	...	°C per sec	°C per sec
...	...	...	...	...		
...	...	...	...	...		
...	...	...	...	...		

(E) Bedford's correction term (F) :

Table 5

Radius of the specimen (*r*) from Table 1 in cm	Thickness (*h*) in cm of the lower disc *C*		Correction factor $F = \frac{r+2h}{2r+2h}$
		Mean	
....	....		....
	....		
	....	....	

(F) Drawing of graphs :

The methods of drawing two graphs (I and II) have been discussed in operations (v) and (vi) under Methods and from these graphs, the value of $\left(\frac{d\theta}{dt}\right)$ at temperature θ_2 can be obtained.

Calculations :

From Eq. (4.6d), $K = \dfrac{M.d.s.\left(\frac{d\theta}{dt}\right)_{\theta_2}}{A(\theta_1 - \theta_2)} \times \dfrac{r+2h}{2r+2h}$.

Substituting the values of the quantities measured and obtaining $\left(\frac{d\theta}{dt}\right)_{\theta_2}$ from the graph, the value of *K* can be obtained.

Precautions :

(i) Steady-state temperatures shown by thermometers T_1 and T_2 should be recorded at least for ten minutes.

(ii) During cooling of the lower disc, readings of temperatures should be taken at an interval of half-a-minute or even more rapidly.

(iii) The lower disc should be heated by the burner to a temperature at least 10 °C above the steady temperature (θ_2°C) shown by it during the static state.

(iv) Cooling of the lower disc must be allowed with the slab (*S*) placed on it.

(v) The diameter of the experimental slab (*S*) must be large compared to its thickness (*d*) to minimise heat loss by radiation.

(vi) Diameter of the slab (*S*) should be made equal to those of the upper and the lower discs.

Error Calculations :

From Eq. (4.6c), $K = \dfrac{M.d.s.\left(\dfrac{d\theta}{dt}\right)_{\theta_2}}{A(\theta_1 - \theta_2)}$ can be written as

$$K = \frac{M.s.d\dfrac{(\theta' - \theta'')}{t}}{\left(\dfrac{L^2}{4\pi}\right)\times(\theta_1 - \theta_2)},$$

where $\left(\dfrac{d\theta}{dt}\right) = \dfrac{(\theta' - \theta'')}{t}$ = rate of cooling and $A = \pi r^2 = \dfrac{L^2}{4\pi}$.

Hence, maximum proportional error in *K* is given by

$$\frac{\delta K}{K} = \frac{\delta M}{M} + \frac{\delta d}{d} + \frac{\delta(\theta' - \theta'')}{(\theta' - \theta'')} + \frac{\delta t}{t} + \frac{2\delta L}{L} + \frac{\delta(\theta_1 - \theta_2)}{(\theta_1 - \theta_2)}$$

Since the thickness (*d*) is measured by a microscope and the difference of temperatures recorded by T_1 and T_2 (θ_1, θ_2) is sufficient, the maximum source of error is in the measurement of the rate of cooling, The value of $(\theta' - \theta'')$, the fall of temperature in a given time *t*, is fairly small and this term contributes the maximum error to the result. This term thus should be calculated with the greatest care.

Questions and Answers

Q.1. What is 'thermal conductivity' and in what respect does it differ from 'thermometric conductivity'?

A. : For 'thermal conductivity', see any textbook on Heat. Thermometric conductivity is defined as $\dfrac{K}{\rho.s}$ where, *K* is the thermal conductivity, ρ and s are the density and sp. heat of the material respectively.

Q.2. What are the unit and dimension of *K*?

A. : Unit of *K* is cal per sec per cm per °C. Dimension of *K* is $MLT^{-3}\theta^{-1}$.

Q.3. Does the value of *K* depend on the dimension of the specimen?

A. : No, it depends only on the material of the specimen.

Q.4. Is there any reason to take the specimen in the form of a disc?

A. : Yes, in that case for a small thickness with large area of cross-section of the specimen, the amount of heat conducted increases.

Q.5. Can this method be used for good conductors?

A. : No, in that case, due to large conduction of heat, the temperatures recorded by T_1 and T_2 will be very nearly the same.

Q.6. Can this method be applied for liquids also?

A. : Yes, then the liquid is to be placed in a very thin-walled container and set in place of the slab.

Q.7. Is it essential that the steam chamber be placed at the top?

A. : No, it can be placed also below thc bad conductor and heated from beneath.

Q.8. Why the thickness of the slab is measured *in situ*?

A. : Because when the slab is placed in between the discs, its effective thickness is slightly less than its actual thickness due to pressure and the thickness (*d*) is to be measured for that condition under which the flow of heat takes place through it.

Q.9. In case of liquid as specimen, should the steam chamber be placed at the top or below and why?

A. : Since liquids are better heated by convection current apart from conduction, the steam chamber should preferably be placed below the liquid specimen.

Q.10. How do you ensure that θ_1 and θ_2 give the temperatures of the upper and lower surfaces of the slab (specimen)?

A. : Since the discs are made of very good conductors, the temperatures recorded by them are practically the same as those of the surfaces of the specimen slab.

4.7 Determination of thermal conductivity of glass in the form of a tube

Theory :

The thermal conductivity of glass in the form of a tube, can be determined by measuring the heat conducted through its surface from its outer face to the inner face when the outer face is maintained at a constant higher temperature.

If θ_1 = temperature of the outer surface of the tube,

θ_2 = temperature (average) of the inner surface of the tube,

l = length of the tube exposed to the steam which maintains constant outer temperature,

t = time of flow of heat through the surface,

K = thermal conductivity of glass,

r_1 and r_2 = outer and inner radii respectively of the glass tube,

then the heat conducted through the walls of the glass tube in t seconds is given by

$$Q = \frac{K.2\pi l(\theta_1 - \theta_2)t}{\log_e \frac{r_1}{r_2}} \text{ cal.} \quad \ldots (4.7a)$$

Again, if this quantity of heat is used up in heating M gm of water flowing through the glass tube in the same time (t), thereby raising the temperature of water from θ_3 to θ_4, then

$$Q = M(\theta_4 - \theta_3) \text{ cal.} \quad \ldots (4.7b)$$

In the steady state, equating Eqs. (4.7a) and (4.7b)

$$\frac{K.2\pi l(\theta_1 - \theta_2)t}{\log_e \frac{r_1}{r_2}} = M(\theta_4 - \theta_3)$$

$$\text{or,} \quad K = \frac{M(\theta_4 - \theta_3)}{2\pi lt(\theta_1 - \theta_3)} \times \log_{10} \frac{r_1}{r_2} \times 2.303 \quad \ldots (4.7c)$$

Measuring the quantities of the right-hand side of Eq. (4.7c); the value of K can be determined.

Apparatus :

Glass tube with jacket, thermometers, stopwatch, boiler, water circulation arrangement, etc.

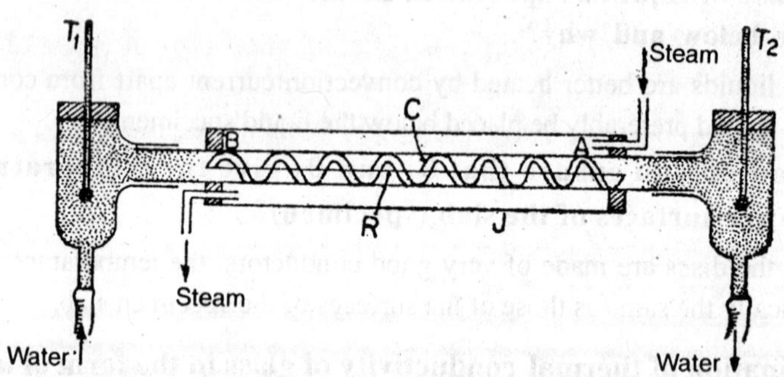

Fig. 4.9 : Conductivity of glass

AB is the experimental glass tube. It is surrounded by a jacket J into which steam enters at the end A and leaves at the end B. Arrangements for passing water through the tube is made from a reservoir which maintains a constant head. Water enters at the end B and leaves at the other end. The end A is kept at a higher level than B which enables bubbles of air, liberated from heated water to escape. Thermometers T_1 and T_2 record the temperatures of ingoing and outcoming water. It is assumed that the temperature of the water at any section of the tube is the same as that of the inner wall at that section. To realise this, the streamline flow of water has to be broken up, which is usually done by introducing into the glass tube a thin copper rod (R) around which a rubber tube (C) is wound in a spiral form. This causes the spiral flow of water along the tube AB making the temperature of water remaining the same all over the cross-section of AB.

Method :

(i) Arrangements for heating water in a boiler is made and the apparatus is screened off from the heat of the boiler.

(ii) The apparatus is arranged as shown in (Fig. 4.9). Arrangements for flow of water from the constant head reservoir to the sink through the glass tube, is made. Flow is regulated by a pinch-cock until a steady stream of water comes out of the tube through the outlet. At this stage, readings of the two thermometers are taken at intervals of 2 minutes until the readings are steady. The steady temperatures are recorded.

(iii) The boiler is then connected to the end A of the jacket and waited till steam comes out freely through the end B. The temperature difference between T_1 and T_2 are noted and this difference adjusted to about 8 °C to 10 °C by controlling the rate of water flow through the tube with the help of the pinch-cock. Temperatures of T_1 and T_2 are recorded at intervals of 2 minutes till they are steady.

(iv) With the help of a large beaker or a measuring cylinder, about a litre of water flowing out of the tube, is collected, noting with a stopwatch the time required for this collection. The temperature of the water collected is noted, and from its volume and density at the temperature, the mass of water (M) collected is determined.

(v) The operation (iv) is repeated twice more for other two values of difference of temperatures between T_1 and T_2 (say, 10 °C and 12 °C) by reducing the rate of water flow through the tube.

(vi) Removing the burner and disconnecting the supply of steam the apparatus is cooled down.

(vii) If the tube (experimental) is permissible for handling by taking it out of the jacket, then its radius (inner and outer) and effective length are to be measured before performing the actual experiment. Otherwise a sample of the tube cut off from the piece which provided AB should be taken.

(viii) To measure the outer and inner radii of the tube, either a travelling microscope may be used directly, or the following method may be employed. The outer radius (r_1) may be measured by a slide calliper at several places at right angles. To find the inner radius (r_2), one end of the tube is closed by a cork and water is poured into it to fill it completely. The tube is held vertically and the length of the water column is measured by a scale. Water is then poured into a measuring cylinder from the tube and its volume is noted. If the length of the water column be L, its volume be V and the inner radius of the tube be r_2, then

$$V = \pi r_2^2 . L, \text{ or, } r_2 = \sqrt{\frac{V}{\pi L}} . \qquad \text{... (4.7d)}$$

(ix) The effective length (l) of the tube AB has to be measured after it has been placed inside the jacket. The effective length is the distance between the inner ends of the corks closing the ends of the jacket tube and can be measured by means of a metere scale while the tube is in position.

(x) The inner radius (r_2) of the tube has to be measured at least thrice, taking different heights (L) of the water column inside the tube and measuring the volume of water (V) in each case. From this, the mean value of r_2 is to be calculated.

Results :

(A) Comparison of readings of the two thermometers :

Table 1

Before passing steam in the jacket	Time (min)	Readings of the thermometers in °C		Value of $(T_2 - T_1)$ in °C	Mean value of $(T_2 - T_1)$°C = θ°C
		T_1	T_2		
	0	...	...	...	...
	2	...	...	...	...
	4	...	...	...	...
	6	...	...	...	...
	8	...	...	...	...
	...	...	...	...	...

[The mean value of $(T_2 - T_1)$ is to be subtracted from the final difference of temperatures $(\theta_4 - \theta_3)$ recorded by the thermometers].

(B) Determination of steam temperature :

Table 2

[Make a Table as shown in **Art. 2.8** for determination of barometric height.]

Barometric reading = ... cm

Steam temperature corresponding to the barometric reading = ... °C (= θ_1°C))

(C) Time-temperature record during flow of water :

Table 3 (see in the next page)

(D) Determination of length (*l*) of the tube :

The effective length (*l*) of the tube *AB* exposed to the steam is measured by a metre scale thrice and the mean value is noted.

Effective length $l = \dfrac{l_1 + l_2 + l_3}{3}$ = ... cm

(E) Determination of outer and inner radii of the tube :

Table 4

(a) Outer radius (r_1) :

Vernier constant of the slide calliper = ... cm

[Make a table of Slide Calliper reading for radius measurement as given in Article 2.3]

Table 5

(b) Inner radius (r_2) :

No. of Obs.	Length (*L*) of water column inside the tube (cm)	Volume of water (*V*) filling the tube (cc)	Value of inner radius (r_2) in cm. $r_2 = \sqrt{\dfrac{V}{\pi L}}$	Mean value of the inner radius (r_2) in cm
1.	...	...	...	...
2.	...	...	...	...
3.	...	...	...	...
4.	...	...	...	...

(C) Time-temperature record during flow of water :

Table 3

	No. of Obs.	Time (min)	Readings of thermometers in °C		Difference of temperatures $= (\theta_4 - \theta_3) - \theta'$ in °C	Average temperature of inner wall of the tube in °C $\theta_2 = (\theta_3 + \theta_4)/2$	Volume of water (V) collected (cc)	Density of water at θ_2°C (gm/cc)	Time of collection (sec)	Mass of water (M) collected (gm)
			T_1 (θ_3°C)	T_2 (θ_4°C)						
While steam flows through the jacket	I.	0	...	...	......	......	...	...	...	...
		2	...	...						
		4	...	...						
		6	...	...						
		...	...	...						
		...	...	...						
	II.	0	...	...	......	......	...	...	...	...
		2	...	...						
		4	...	...						
		6	...	...						
		...	...	...						
		...	...	...						
	III.	0	...	...	......	......	...	...	...	...
		2	...	...						
		4	...	...						
		6	...	...						
		...	...	...						
		...	...	...						

(F) Determination of the value of K :

Table 6

Sets	Mass of water (M) collected (gm)	Time of collection (t) (sec)	Inner radius (r_2) (cm)	Outer radius (r_1) (cm)	Value of $(\theta_1 - \theta_2)$ (°C)	Corrected value of $(\theta_4 - \theta_3)$ (°C)	Value of K (cal/sec/cm/°C)	Mean value of K (cal/sec/cm/°C)
I.	...	...	...	...	...	...	...	
II.	...	...	...	...	...	...	...	...
III.	...	...	...	...	...	...	...	

Calculations :

From Eq. (4.7c),

$$K = \frac{M(\theta_4 - \theta_3)}{2\pi \ell t(\theta_1 - \theta_2)} \times \log_{10}\frac{r_1}{r_2} \times 2.303.$$

Substituting the values of the quantities on the R.H.S. of this equation, the value of K is obtained.

$\therefore \quad K = \ldots\ldots\ldots$ cal/sec/cm/°C.

Precautions :

(i) The experimental tube should preferably be thick as the error in the value of $(r_1 - r_2)$ is fairly large.

(ii) To ensure that steady state is reached, the readings of the thermometers should remain steady at least for 6-8 minutes.

(iii) During flow of water, the out-flowing rubber tube should not be disturbed under any circumstances.

(iv) The greatest source of error lies in the measurement of $(\theta_4 - \theta_3)$. Hence steady state must be attained and thermometers used should be of $\frac{1}{5}$th or $\frac{1}{10}$th of a degree ranges.

Error calculation :

From Eq. (4.7c), it is evident that the maximum proportional error in the value of K is given by

$$\frac{\delta K}{K} = \frac{\delta M}{M} + \frac{\delta t}{t} + \frac{\delta l}{l} + \frac{\delta(\theta_4 - \theta_3)}{(\theta_4 - \theta_3)} + \frac{\delta(\theta_1 - \theta_2)}{(\theta_1 - \theta_2)} + \frac{\delta \log\left(\frac{r_1}{r_2}\right)}{\log\left(\frac{r_1}{r_2}\right)}$$

Now, as the mass (M) of water collected and its time (t) of collection are both large, their contributions towards the proportional error in K are very small. Again, the length (l) of the tube is large and error due to this is also very small. Since the outer radius (r_1) is measured by a slide calliper and the inner radius (r_2) by a method which contributes little error, measurements of radii do not offer significant errors towards the value of K. The greatest source of error lies in the measurement of $(\theta_4 - \theta_3)$, since its value is small and less accuracy is obtained by using ordinary thermometers. Hence, to measure $(\theta_4 - \theta_3)$ carefully, sensitive thermometers ($\frac{1}{5}$°C or $\frac{1}{10}$°C) should be used and the initial errors lying, if any, between them should positively be determined. As the value of $(\theta_1 - \theta_2)$ is fairly large, error due to this is not significant.

The percentage error in the measurement of K is thus

$$\frac{\delta K}{K} \times 100 = \ldots\ldots \%.$$

Questions and Answers

[Q. No. 1 and 2 of the previous experiment (4.6) are to be seen]

Q.1. What is the use of using a spiral inside the tube?

A. : This is done just to keep the temperature of water constant over any cross-section of the tube.

Q.2. Why is the glass tube kept inclined, not horizontal?

A. : The tube does not become empty at any time if it is kept inclined, and secondly, any air bubble produced due to heating of water, is easily swept out.

Q.3. Why is the constant water head used?

A. : If the pressure under which the water flows in the tube changes, the rate of flow will also change causing difficulty in attaining a steady difference of temperature during flow.

Q.4. Why are the two ends of the tube covered with felt?

A. : It is so done to avoid any gain or loss of heat by radiation.

Q.5. Why is the entire length of the tube not measured?

A. : That portion of the tube length is considered which is exposed to steam only.

Q.6. Why is the temperature difference not made very high as 25°C or 30°C?

A. : In that case, the rate of flow should be very small giving inaccuracy to the result, and further for high difference of temperature, radiation loss will be more.

Q.7. Where is the greatest source of error in the experiment?

A. : (See 'Error calculations'.)

Chapter 5
Light

5.1 General precautions for optical experiments

While performing experiments on light, instruments like spectrometers, optical benches, microscopes, polarimeter, etc. are to be used. In using these precision instruments, some general precautions are to be taken while handling them and also to get accurate results, minimising the possible sources of errors. These precautions to be adopted are briefly discussed below.

(A) Parallax and its elimination

If an observer moves his eyes in a direction perpendicular to the line of sight, the relative positions of objects lying at a distance appear to change. The apparent change of position of distant objects due to actual change of position of the observer, is called *parallax*. In all experiments on light, error due to parallax has to be avoided. Normally, the image of a particular object has to be formed on a screen or plane; and the positions of both the screen and the image are adjustable. To avoid parallax, the position of the image or the screen or the both are to be adjusted so that on moving the eye laterally, no relative motion between the image and the screen can be detected.

In Fig. 5.1, if an eye E is placed in one straight line with P_1 (screen) and P_2 (the image of an object) both P_1 and P_2 standing perpendicular to the plane of the paper, the eye sees both of them coincident. But if the eye is moved to positions E_1 or E_2, a relative motion between P_1 and P_2 appears. This is because the image P_2 is not formed in the plane of the screen (P_1). This means, there is parallax in between them. The same thing happens even when P_2 is nearer to eye than P_1. This relative motion will disappear when P_2 and P_1 coincide and parallax is then said to be avoided.

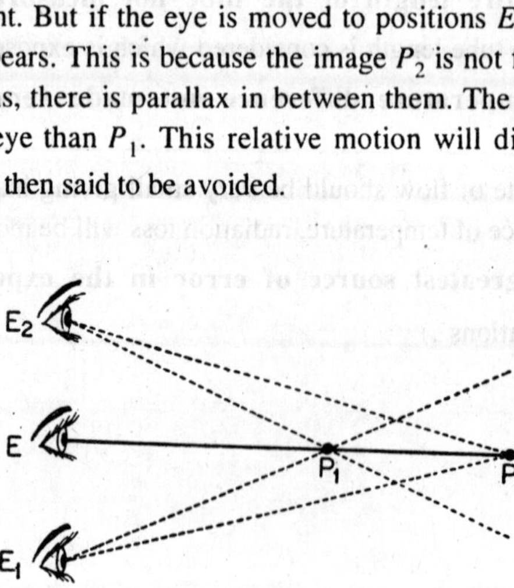

Fig. 5.1 : Parallax error

In order to ascertain the direction of movement of the screen or object to be made, the eye across the line joining P_1 and P_2 is moved. If the *image moves* in the *same direction as the eye*, then the image is *farther* away from the eye than the screen, while if the image moves in the *opposite direction*, then it is *nearer* to the eye than the screen. Accordingly, the screen or the image has to be moved until with the movement of the eye, both of them appear to move *simultaneously* in the *same direction*. Parallax error is then said to be eliminated.

(B) Index error and its correction

While working with experiments on light, one has to often use an optical bench, which has several uprights mounted on it to carry the lens, object and the screen. Their positions on the bench are obtained from a scale attached or etched on the bench. While taking readings for the positions of the screen or the object or the lens, the actual distances between the different parts are needed. But the distances between the index marks of the uprights, as read from the bench-scale, may not be equal to the actual distance between them.

In order to obtain the actual distance, an index rod of known length, say l, is placed horizontally in between a pair of uprights (say, of lens and object) and the positions of the uprights are adjusted (Fig. 5.2) until the ends of the rod just touch the surfaces of the lens and the object. The reading of the upright positions are obtained from the bench-scale, which

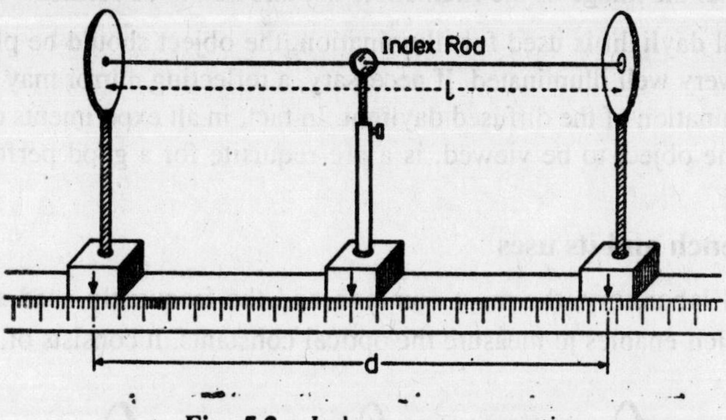

Fig. 5.2 : Index error correction

is (d, say) the *apparent* length of the index rod. Hence, while the *actual* distance between the lens and the object is l, the apparent distance from the bench reading is d. Thus to get the true distance, the value of $(l - d)$ is to be algebraically added to the apparent distance. The magnitude of $(l - d)$ is called the index error. In a similar way, the error for the distance between the lens and the image has to be calculated and necessary corrections have to be made.

(C) Focusing of microscopes and telescopes and proper illumination of the source

In performing experiments on light, telescopes and microscopes are often used and whenever they are employed to view a distance or a near object, they must be properly focused for correct viewing. The cross-wire which generally lies in the focal plane of the eyepiece of the instrument, must be properly focused first. To focus it sharply, the telescope or the microscope should be directed towards a bright white wall and by drawing the focusing lens in or out, the cross-wire should be sharply focused.

To focus a near object by a microscope, it should be placed towards the object, and the microscope tube should be moved as a whole either towards or away from the object until the image of the object appears to be very sharp. Then the microscope tube should be moved very slowly until there is no relative movement between the image of the object and the cross-wire as the eye is moved *laterally*. Under these circumstances, there is no parallax error in between the image of the object and that of the cross-wire.

To focus a distant object by a telescope, it is directed towards the object, and the eye-piece is drawn *in* or *out* by hand or rack-and-pinion arrangement till the image of the object appears to be very sharp. Then the eyepiece tube is to be drawn in or out very *slowly* until there is no *relative movement* between the images of the object and that of the cross-wire, as the eye is moved *laterally*. Under these circumstances, there is no parallax error in between the image of the object and that of the cross-wire.

When an object is illuminated by a source of light, it is to be seen that the object is illuminated as strongly as possible. Normally the sources for illumination used are either sodium light or electric lamps or ordinary daylight. In any case, the source used whatsoever, must be placed in such a way that the object gets the maximum illumination. For a sodium flame or light, it should be placed at such a distance from the object that the latter is illuminated maximum. While using an electric light, always bulbs made of frosted glass should be taken. In case of an ordinary glass electric bulb, it should be wrapped by a white tissue paper, so that the image of the filament is not allowed to be formed.

When normal daylight is used for illumination, the object should be placed in such a manner that it is very well illuminated. If necessary, a reflecting mirror may also be used to increase the illumination of the diffused daylight. In fact, in all experiments on light, proper illumination of the object to be viewed, is a pre-requisite for a good performance of the experiment.

5.2 Optical Bench and its uses

In an optical laboratory, the most common and the frequently used apparatus is an optical bench which enables to measure the optical constants. It consists of, in its simplest

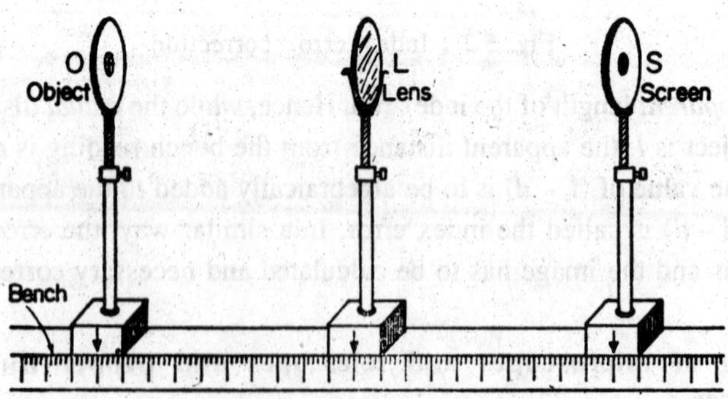

Fig. 5.3 : Optical Bench

form, a long horizontal bench having several uprights mounted on it (Fig. 5.3) to carry the object, lens and the screen. The uprights bear an index mark on each and they can freely slide across the bench. The index marks touch a scale, graduated in millimetre, etched on the bench. From the position of the mark on the scale, the position of the upright carrying the object or the lens or the screen, can be ascertained. The heights of the uprights can be adjusted and placed at any desired position. The object may be an illuminated slit, scratch or a cross-wire. The screen is generally a frosted glass with lines drawn on it to judge parallax. The uprights can be removed from the bench and can be placed at any desired position. An index correction is always necessary as discussed in the previous article (**5.1B**).

For Biprism and Lloyd's mirror experiments, special types of uprights are used in an optical bench, which will be discussed in details in the corresponding experiments.

5.3 Determination of the focal length and the power of a convex lens by displacement method using an optical bench

Theory :

If the distance (D) between the object and the screen is greater than four times the focal length (f) of the given convex lens, then for a given distance between the object and the screen, sharp images can be cast on the screen for two different positions of the lens. If x be the distance between the two positions of the lens, which means the displacement of the lens (Fig. 5.4), then the focal length (f) of the given lens will be given by

$$f = -\frac{D^2 - x^2}{4D}. \qquad \text{... (5.3a)}$$

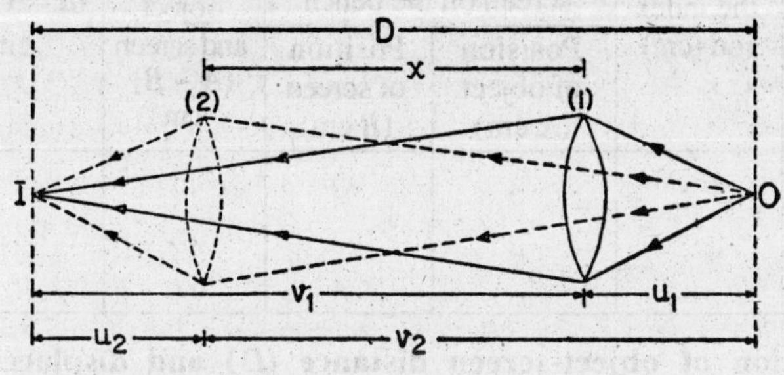

Fig. 5.4 : Displacement method

The power (P) of a lens is thus calculated from the relation

$$P = +\frac{100}{f} \text{ dioptres,} \qquad \text{... (5.3b)}$$

where f is expressed in cm.

Apparatus :

Optical bench, source and screen with stands, lens with holder, index rod.

Method :

(i) The lens, the object and the screen are fixed on the three uprights, having index marks, on the optical bench (See Fig. 5.3). The object may be a cross-mark (×) set at one end of the bench. This mark is illuminated by a source.

(ii) The heights of the uprights are adjusted so that the centres of the cross-mark, the lens and the screen lie in the same horizontal line. The index error (δ) between the object and the screen is also determined by the method discussed in 5.1(B).

(iii) The object and the screen are placed at the two extreme ends of the bench and for two positions of the lens (once nearer the object and once nearer the screen), sharp images of the objects are obtained on the screen. Let the distance between the lens for these two positions be x. The distance between the object and the screen is measured from the bench-scale. Let this distance after index correction be D. This observation is repeated thrice and the mean value of D and x are obtained. It may be noted that when the lens is nearer the object, the real image obtained on the screen is magnified, while when it is near the screen, the image is diminished.

(iv) Next the position of the screen is changed keeping the object fixed. This means that the value of D is changed. Again, the experiment is repeated for this value of D.

(v) Similarly, for three different values of D (in any case the value of D must not be less than $4f$), the experiment is performed.

(vi) For each value of D, the value of f is obtained applying Eq. (5.3a) and from three values of f so obtained, the mean value of f is obtained from which the power of the lens (P) is obtained by using Eq. (5.3b)

Results :

(A) Index correction for object and screen positions :

Table 1

No. of Obs.	Length of the index rod (cm)	Mean length of the index rod (cm)	When the rod touches the object and the screen on the bench		Distance between object and screen $(A \sim B)$ cm	Mean distance $(d = A \sim B)$ cm	Index error $\delta = (l - d)$ cm
			Posision of object (A cm)	Position of screen (B cm)			
1.	...	...	...	...	...		
2.	...	...	...	...	...	...	...
3.	...	...	...	...	...	...	...

(B) Determination of object-screen distance (D) and displacement of lens (x) :

Table 2 (see next page)

Calculations :

From Eq. (5.3a),

$$f = -\frac{D^2 - x^2}{4D} = \ldots\ldots \text{ cm}$$

and from Eq. (5.3b),

$$P = +\frac{100}{f} \text{ dioptres.}$$

Precautions :

(i) Smaller the value of x, minimum will be the error in the value of f. Hence, the value of D ($D > 4f$) should be adjusted such that x becomes small.

(ii) The value of x changes by a large amount for a small change in the value of D. Hence, while taking different sets of reading, the value of D should be changed by small amounts.

(B) Determination of object-screen distance (D) and displacement of lens (x) :

Table 2

No. of Obs.	Position of		Apparent distance between object and screen (D cm)	Corrected value of D (= D + δ) cm	Bench reading for the lens in				Displacement of the lens (x cm)	Focal length $f = \frac{D^2 - x^2}{4D}$	Mean focal length (f) cm	Power of the lens $P = + \frac{100}{f(\text{cm})}$
	Object (X cm)	Screen (Y cm)			First position (cm)	Mean (M cm)	Second position (cm)	Mean (N cm)				
1.	...	...	...	...	...	...	...	...	...	...		
					...		...					
					...		...					
2.	...	...	...	...	...	...	...	...	...	...	...	...
					...		...					Dioptre
					...		...					
3.	...	...	...	...	...	...	...	...	...	...	...	
					...		...					

Note : Eq. (5.3a) can be written as

$$4fD = -D^2 + x^2; \text{ or, } \frac{x^2}{D} = D + 4f.$$

Hence, if a graph is drawn with D along X-axis and $\frac{x^2}{D}$ along Y-axis, then it will be a straight line graph. The point where the graph cuts the X-axis, has the co-ordinates $(-4f, 0)$, since $y = 0 = \frac{x^2}{D}$. Thus $D + 4f = 0$; or, $D = -4f$.

So, the point where the straight line cuts the X-axis has the value numerically equal to $4f$.

Error :

To have the minimum proportional error in the value of f, the displacement (x) of the lens should be small and to have it, the value of D should be adjusted accordingly.

Questions and Answers

Q.1. Of the two methods, this one and the (u-v) method, which one is convenient and why?

A. : This method involves measurement of x and D which can be done very accurately, and moreover, here only one index correction is to be made unlike in (u-v) method.

Q.2. How many focal length a lens has and which one is measured here? Are these two focal lengths equal?

A. : A lens has two focal lengths. Here the second focal length is measured. The two focal lengths will be equal if the lens is surrounded by the same media on both sides.

Q.3. What will happen if D is less than $4f$?

A. : Then only one image of the object will be obtained.

Q.4. Can the size of the object be determined from this method?

A. : Yes, if I_1 and I_2 be the sizes of the image for the two positions of the lens, then the object size $I = \sqrt{I_1 I_2}$.

Q.5. What will happen if D is very large?

A. : Then the value of x will also be very large, and once a very diminished image and once a highly magnified image will be obtained which will cause difficulty in focusing sharply.

Q.6. Why index correction for x is not done here?

A. : Because the displacement of the lens is equal to the displacement of the index mark of the stand on which the lens is kept.

[Besides these, definition of focal length of a lens and the factors on which it depends, are to be known from a textbook.]

5.3(A) Determination of the focal length of a concave lens by combination method

Theory :

If two lenses of focal lengths f_1 and f_2 are put in contact, the equivalent focal length of the combination F is given by

$$\frac{1}{F} = \frac{1}{f_1} + \frac{1}{f_2}, \qquad \text{... [5.3(A)a]}$$

where f_1 and f_2 are the focal length of the individual lenses.

If one of them is a convex lens of focal length f_1 and the other is a concave lens of focal length f_2 and further if f_1 is shorter and f_2 is larger, then the combination will behave as a convex or converging lens of focal length F. The combination will thus form a real image of an object placed in front of it. Correcting for signs, then Eq. [5.3(A)a] becomes

$$-\frac{1}{F} = \frac{1}{f_1} + \frac{1}{f_2} \qquad \text{... [5.3(A)b)}$$

$$\text{or,} \quad f_2 = \frac{Ff_1}{F - f_1}. \qquad \text{... [5.3(A)c]}$$

Thus finding the values of f_1 and F and putting their numerical values in Eq. [5.3(A)c], the value of the focal length (f_2) of the concave lens can be found out.

Apparatus :

Optical bench, convex lens, concave lens, stands, index rod.

Procedure :

(i) At first, with the help of the index rod, the index-correction for the distance between the object and the screen on the optical bench, is determined.

(ii) Next the focal length of the convex lens only is determined by 'displacement' method as in Expt. 5.9. Similarly by combining the concave lens with the convex lens (Fig. 5.5), the equivalent focal length (F) of the combination is also determined by the 'displacement' method.

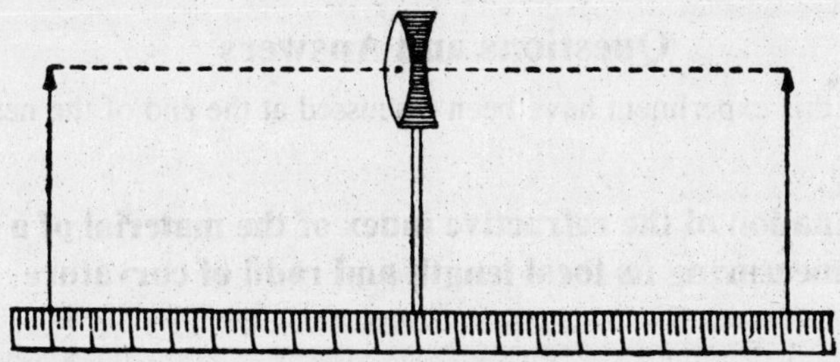

Fig. 5.5 : Combination method

(iii) Substituting the numerical values only of F and f_1 in Eq. [5.3(A)c], the value of f_2 is obtained with sign.

Results :

(A) Determination of the Index correction :

Table 1

[Make same Table as Table 1 of Expt. No. 5.3.]

(B) Determination of the focal length (f_1) of the convex lens :

Table 2

[Make Table as Table 2 of Expt. No. 5.3.]

(C) Determination of the focal length (*F*) of the combination lens : Tables 3

[Make Table as Table 2 of expt. No. 5.3.]

Calculations :

(a) Focal length of the convex lens

$$f_1 = -\frac{D_1^2 - x_1^2}{4D_1} = \ldots\ldots\ldots \text{ cm}$$

(b) Focal length of the combination lens

$$F = -\frac{D_2^2 - x_2^2}{4D_2} = - \ldots\ldots\ldots \text{ cm}$$

(c) Focal length of the concave lens

$$f_2 = \frac{Ff_1}{F - f_2} = + \ldots\ldots\ldots \text{ cm}$$

Precautions :

(i) All the precautions mentioned under Expt. No. 5.3 are also to be followed here.

(ii) The power of the convex lens must be greater than that of the concave lens, otherwise the combination will not behave as a converging lens system. This means $f_1 < f_2$ condition must be fulfilled.

(iii) Index correction will be the same for the focal length determinations of both the convex and the combination lens.

(iv) It is desirable that the difference of focal length of the convex and the concave lens should not be very large.

Questions and Answers

[Questions on this experiment have been discussed at the end of the next Experiment No. 5.3(B).]

5.3(B) Determination of the refractive index of the material of a convex lens by measuring its focal length and radii of curvature

Theory :

The focal length *f* of a convex lens is given by

$$\frac{1}{f} = (\mu - 1)\left(\frac{1}{r_1} - \frac{1}{r_2}\right), \qquad \ldots [5.3(B)a]$$

where r_1 and r_2 are its radii of curvature and μ is the refractive index of its material.

For a convex lens, both *f* and r_1 are negative and r_2 is positive. Thus Eq. [5.3(B)a] becomes

$$\frac{1}{f} = (\mu - 1)\left(\frac{1}{r_1} + \frac{1}{r_2}\right), \qquad \ldots [5.3(B)b]$$

or, $$\mu = 1 + \frac{r_1 r_2}{f(r_1 + r_2)}. \qquad \ldots [5.3(B)c]$$

Hence, by measuring the focal length of the lens and the radii of curvatures of its two surfaces, the value of the refractive index of its material can be determined.

Apparatus :

Convex lens, stand with pin, plane mirror, metre scale and Spherometer.

Procedure :

(i) The convex lens C is placed on a plane mirror (M), and a pin (P), which can be raised or lowered, is set along the axis of the lens (Fig. 5.6). The pin is set in such a position that its tip and its real image exactly coincide without any parallax error. If the distance of the upper surface of the lens from this position of the pin-tip is d_1 and that of the plane mirror from the pin-tip be d_2, then the focal length will be $f = \dfrac{d_1 + d_2}{2}$.

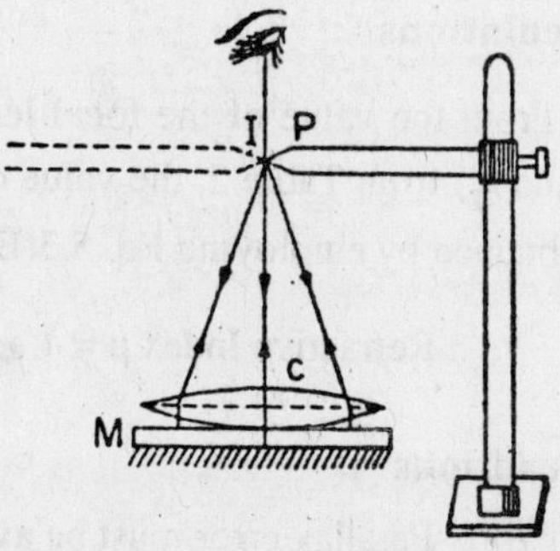

Fig. 5.6 : Focal length of a lens

(ii) In this manner, at least five set of readings are to be taken, displacing the pin position, and from these readings, the mean value of the focal length has to be determined.

(iii) To determine the radii of curvatures of the two curved surfaces of the lens, the use of a spherometer has to be made (Vide Art. 2.5). If d be the mean distance between two adjacent legs of the spherometer and h be the elevation of the lens surface, then the radius of curvature (r) of that surface will be

$$r = \frac{d^2}{6h} + \frac{h}{2}.$$

(iv) Knowing the values of f, r_1 and r_2, the value of μ is obtained by employing Eq. 5.3(B)c.

(v) Applying the formula for the radius of curvature as given in (iii) above, the radii (r_1 and r_2) of both the surface of the lens have to be measured.

Results :

(A) Determination of focal length of the lens :

Table 1

No. of Obs.	Distance of the pin (P) from upper surface of the lens (d_1 cm)	Distance of the pin (P) from plane mirror (d_2 cm)	Focal length of the lens $\left(f = \dfrac{d_1 + d_2}{2} \text{ cm}\right)$	Mean focal length (f cm)
1.	...	...	...	
2.	...	...	...	
3.	...	...	...	...
4.	...	...	...	
5.	...	...	...	

(B) Determination of radii of curvatures of the two surfaces of the lens :

Table 2

[Separate tables for measurements of two radii of curvature (r_1 and r_2) of the lens surfaces are to be made here as per table of spherometer readings shown in Art. 2.5].

Calculations :

From the value of the focal length (f) obtained from Table 1 and radii of curvatures (r_1 and r_2) from Table 2, the value of μ, the refractive index of the material of the lens, can be obtained by employing Eq. 5.3(B)c.

$$\therefore \quad \text{Refractive Index } \mu = 1 + \frac{r_1 r_2}{f(r_1 + r_2)} = \ldots\ldots\ldots$$

Precautions :

(i) Parallax error must be avoided while determining the focal length.

(ii) While viewing the reflected image of the tip of the pin, care should be taken to see that the tip of the pin lies and moves along the axis of the lens. This removes the errors due to spherical and chromatic aberrations.

(iii) Care should be taken to see that real image of the pin is formed by the lens and the mirror (inverted image). If the pin lies within the distance of focal length of the lens, then virtual image will be formed, which should never be considered.

Questions and Answers

Q.1. Define focal length, radius of curvature and refractive index.

A. : [See any textbook]

Q.2. Does the R.I. of a material depend on the velocity of light?

A. : If the velocity of light be c in vacuum and v in the medium, then the R.I. of he medium $\mu = \frac{c}{v}$.

Q.3. Can this experiment be performed with a concave lens?

A. : No, then real image of the pin will not be formed.

Q.4. What is parallax error and how is it removed?

A. : See Art, 5.1 (A).

5.3(C) Determination of the refractive index of a liquid with the help of a convex lens and a plane mirror

Theory :

If a convex lens is placed over a few drops of liquid placed on the surface of a plane mirror, then a plane-concave liquid lens is formed between the lower surface of the convex lens and the plane mirror. The combination behaves as a converging lens since the focal length of the plano-concave lens is greater then the focal length of the convex lens. If f_1, f_2 and F be the focal lengths of the convex, plano-concave lens and the equivalent combination lens, then

$$\frac{1}{F} = \frac{1}{f_1} + \frac{1}{f_2}.$$

Since, the signs for both F and f_1 are negative, we have,

$$\frac{1}{F} = -\frac{1}{f_1} + \frac{1}{f_2} \qquad \text{... 5.3(C)a}$$

or, $$\frac{1}{f_2} = \frac{1}{f_1} - \frac{1}{F} = \frac{F - f_1}{f_1 F} \qquad \text{... 5.3(C)b}$$

But, $$\frac{1}{f_2} = (\mu - 1)\left(\frac{1}{r_1} - \frac{1}{r_2}\right), \qquad \text{... 5.3(C)c}$$

where μ— ref. index of the plano-concave lens (here, that of the liquid).

r_1 and r_2— radii of curvature of the two surfaces of the plano-concave lens.

Here $r_2 = \infty$, for the plano-concave lens and putting $r_1 = r$, we have from Eq. 5.3(C)c,

$$\frac{1}{f_2} = (\mu - 1)\left(\frac{1}{r} - \frac{1}{\infty}\right) = (\mu - 1)\frac{1}{r}$$

or, $$\mu = 1 + \frac{r}{f_2}. \qquad \text{... 5.3(C)d}$$

Hence, by finding the focal length (f_2) of the plano-concave liquid lens by coincidence method and radius of curvature (r) of its upper surface (which is the same as that of the lower surface of the convex lens), the value of μ of the liquid can be found out from Eq. 5.3(C)d.

Apparatus :

Convex lens, plane mirror, stand with a pin, metre scale, spherometer, slide calliper and a few drops of the experimental liquid.

Procedure :

(i) The thickness of the lens at its centre is measured by a slide calliper. Half of this thickness (t) is to be added to the obtained focal length of the lens by coincidence method, to get the correct focal length from the optical centre.

(ii) The surface of the lens which would remain in contact with the liquid, is first ascertained and the radius of curvature of that surface is measured with the help of a spherometer (as described in Art. 2.5).

(iii) Next, only the lens is placed on the plane mirror and its focal length (f_1) is determined by the coincidence method using a pin, by the same way as is described in the previous experiment [Expt. 5.3(B)].

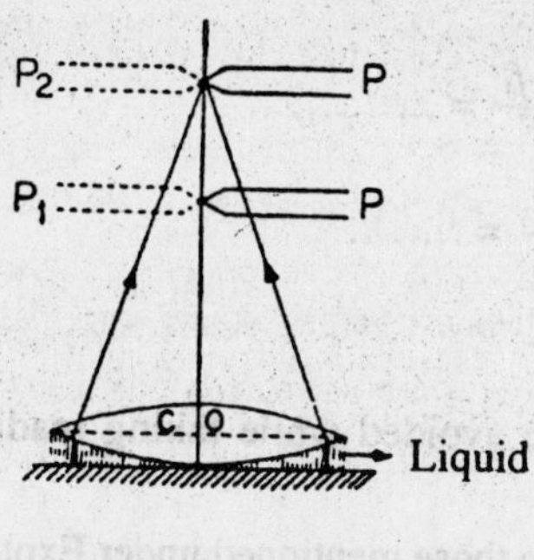

Fig. 5.7 : Determination of refractive index of a liquid

(iv) Some drops of the experimental liquid are then placed on the plane mirror and the lens is placed on the liquid surface gently. This forms a plano-concave lens of liquid (Fig. 5.7). Again, by the help of the pin, the focal length of this combination lens (F) is determined by coincidence method.

(v) Each time, the value of $\frac{t}{2}$ (half the thickness of the lens) is to be added to the value of f and F so obtained, to get their correct values.

(vi) While determining the focal lengths, at least three sets of readings (by displacing the pin position) are to be taken each time.

(vii) Finally, substituting the value of f_1, F and r in Eqs. 5.3(C)b and 5.3(C)d, the value of μ is obtained.

Results :

(A) Determination of the thickness of the lens :

Table 1

V.C. of the slide calliper = ... cm

No. of Obs.	Main scale reading (M cm)	Vernier scale reading V (cm)	Total reading ($M + V$) cm	Mean reading (thickness t) (cm)	Half the thickness ($\frac{t}{2}$ cm)
1.	...	...	...		
2.	...	...	...	...	...
3.	...	...	...		
4.	...	...	...		

(B) Determination of radius of curvature of the lower surface of the convex lens :

Table 2

[Table for radius of curvature measurement by a spherometer is to be made here as shown in Art. 2.5]

(C) Determination of the focal length (F) of the combination lens :

Table 3 (See next page)

Calculation :

From Eq. [5.3(C)], $\frac{1}{f_2} = \frac{F - f_1}{f_1 F} = ...$

and from Eq. [5.3(C)d], $\mu = 1 + \frac{r}{f_2} = ...$

Precautions :

(i) Parallax error must be avoided while taking readings for focal length determination.

[For other precautions, refer to those mentioned under Expt. 5.3(B)]

(C) Determination of the focal length (F) of the combination lens :

Table 3

No. of Bs.	When the pin and its real image coincide						Focal length of the plano-concave lens
	For convex lens			For combined lens			
	Distance of pin from upper surface of the lens (d_1 cm)	Mean distance (d_1 cm)	Focal length $\left(f_1 = d_1 + \frac{t}{2}\right)$ cm	Distance of pin form upper surface of the lens (d_2 cm)	Mean distance (d_2 cm)	Focal length $\left(F = d_2 + \frac{t}{2}\right)$cm	$f_2 = \frac{Ff_1}{F - f_1} = \ldots$ cm
1.	...			...			
2.	...	...	...	...	...	...	...
3.	...			...			

Questions and Answers

[Questions of the last experiment are to be seen first.]

Q.1. What is the nature of the liquid lens?

A. : It is a plano-concave lens.

Q.2. Which lens has the greater focal length?

A. : The liquid lens has the greater focal length than that of the convex lens.

Q.3. Can this method be applied to determine any value of μ?

A. : If the material of the lens has R.I. 1.5, then liquids having R.I. less than 2.0, can be used here.

Q.4. Can the value of μ be determined without measuring the value of *r* by a spherometer?

A. : In that case, another liquid of known R.I. (μ') has to be taken. With that interposed between the lens and the mirror, the value of its focal length (f_2') is to be determined. Then $\mu' = 1 + \frac{r}{f_2'}$, or, $r = f_2' (\mu' - 1)$. Substituting this value of r in Eq. 5.3(C)d, the value of μ can be obtained.

5.4 The spectrometer : Its construction and adjustments

Spectrometer is an indispensable instrument in an optical laboratory. It is usually used in studying the optical spectrum and for determination of refractive indices. The common type of spectrometer normally used in the laboratory has quite a number of parts and requires several careful adjustments before working with it. The construction and the adjustments to be made, are discussed below.

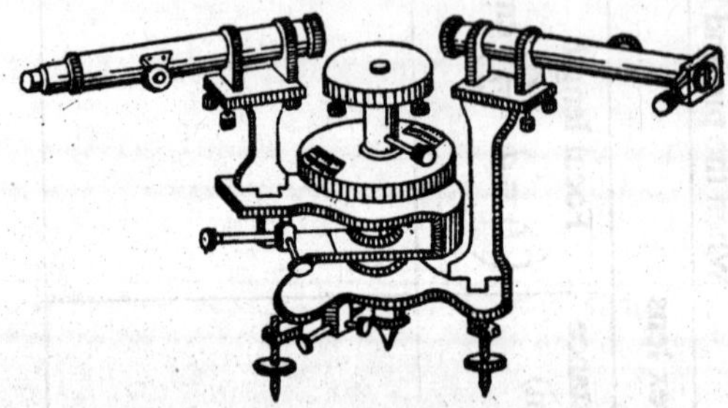

Fig. 5.8 : Spectrometer

The table spectrometer commonly employed in the laboratory has the following parts :

(i) *Collimator*, (ii) *Prism Table*, (iii) *Telescope*.

Collimator :

The main function of the collimator is to produce a parallel beam of light. It consists of a hollow tube provided with an adjustable vertical slit (*S*) at its one end (Fig.5.9) and an achromatic converging lens at the other end. The slit is attached to a smaller tube which is set inside the collimator tube. The distance between the slit and the lens can be altered by a rack and pinion arrangement. Another screw attached to the slit can vary the width of the slit.

The slit consists of a pair of metal jaws with perfectly parallel edges. The collimator tube is kept horizontal by means of screws attached to its ends. Light from a suitable source illuminates the slit facing it. When properly focused, the slit lies in the focal plane of the collimator lens, thereby rendering a parallel beam of light emerging out of the collimator. By altering the width of the slit, the quantity of light entering the collimator can be controlled.

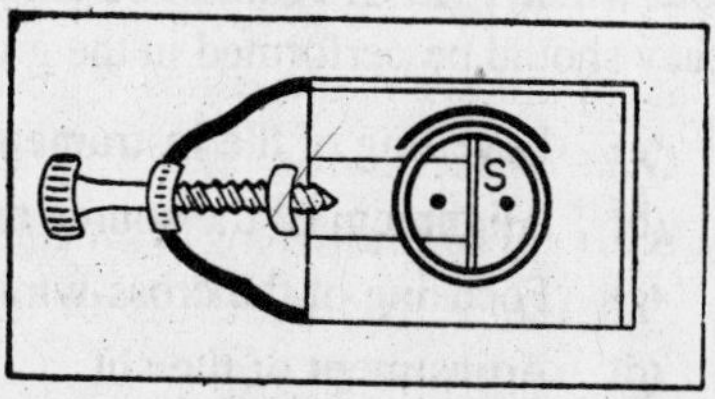

Fig. 5.9 : Slit

Prism-Table :

It is a small circular table with a flat top and mounted on a vertical stand. It is capable of rotation about the vertical axis and can be raised or lowered and can be fixed at any desired position. It is provided with three levelling screws (P, Q, R) by which the table can be made horizontal (Fig. 5.10). A set of parallel, equidistant lines are drawn on the top of the prism-table which are parallel to the line joining two of the levelling screws. These are meant for adjustment of the prism on the table. A set of circular lines, concentric with the axis of rotation of the prism are also drawn on the top. These help in correctly placing the prism. The prism-table can be rotated about the central axis and its amount of rotation can be obtained from the two vernier scales attached to it. The table can be fixed by a screw and there is also a tangential screw attached to it which enables to give the table very small rotation.

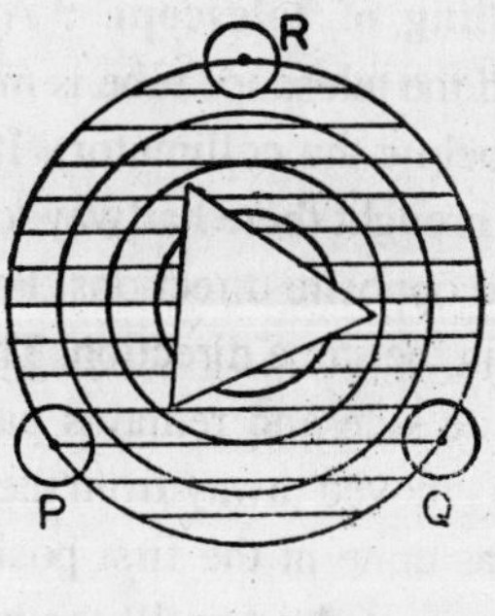

Fig. 5.10 : Prism-table

Telescope :

It is a small astronomical telescope having an objective and a compound eyepiece of Ramsden's type, having a cross-wire attached to it. The telescope axis must be made horizontal by means of a pair of screws attached to it and it is kept perpendicular to the vertical axis of rotation of the prism-table. It can be rotated about the vertical axis of the instrument and the amount of its rotation can be determined from a pair of verniers and circular scales attached to it. It can be kept fixed at any position by means of a screw and fine rotation can be given to it by a tangent screw. During work, the cross-wire remains in the focal plane of the eyepiece and its distance from the objective can be altered by a rack and pinion arrangement. While focused, the image of the cross-wire is at infinity. The whole apparatus stands on three levelling screws, by which the apparatus can be levelled and the main axis may be made vertical. It should be noted that the tangent screws attached to the telescope and the prism-table, do not work unless the corresponding main screws are rigidly tightened.

Adjustments of the spectrometer :

Before starting work with the spectrometer, it requires some preliminary adjustments, without which readings cannot be taken with it. The necessary adjustments are given below and they should be performed in the given order :

(a) Levelling of the instrument (collimator, prism-table and the telescope)

(b) Alignment of the source and the slit

(c) Focusing of the cross-wire

(d) Adjustment of the slit

(e) Focusing for parallel rays.

(a) Levelling :

To work with a spectrometer, it is necessary that some preliminary levelling works are to be done. These include that (i) the axis of rotation of the telescope shall remain vertical, (ii) the axis of the collimator and the telescope shall remain horizontal and (iii) the top of the prism-table shall remain horizontal. Levelling of the apparatus means making these adjustments properly.

Levelling of telescope : A spirit-level is placed on the telescope axis along its length, and the telescope tube is made parallel to the line joining the two base screws (other than that below the collimator). If the bubble of the spirit level is not at the centre of the level, it is brought there halfway towards the centre by turning the two base screws by equal amounts in opposite directions. For the other half, the screws attached below the telescope is rotated in the same direction. The telescope is then turned through 180° so that it goes on the opposite side and remains parallel to the first position. It may be found that the air bubble has moved away from the centre. It is again brought back to the centre by the same way as was done in the first position of the telescope. It may be necessary to repeat the operations several times till the bubble is found to remain at the centre for both positions of the telescope. When done, the axis of rotation of the telescope lies in a vertical plane perpendicular to the line joining the two base screws.

It may now be found that the bubble may not be at the centre for intermediate positions of the telescope. Then the telescope is placed in line of the collimator and the bubble is brought at the centre by turning the third levelling screw lying below the collimator. This makes the axis of the telescope horizontal and the axis of rotation vertical. It is advisable that the first adjustment be checked after making the second adjustment. The work becomes easier if the following clue is remembered.

(i) For parallel positions, adjustment to be done half by the base screws and half by the screws attached to the telescope.

(ii) For the perpendicular position, adjustment to be done by the third screw lying below the collimator.

Levelling of collimator : The spirit level is placed on the collimator along its length and the bubble is brought to the centre by adjusting the screws attached to the collimator body. This makes the collimator horizontal.

Levelling of prism-table : The prism-table has three separate levelling screws attached to its base. The ruled lines are etched on its surface parallel to a line joining two of these screws. The spirit level is placed paralleled to these lines and the air-bubble is brought to the centre by turning the screws lying at the ends of the parallel lines in opposite directions. Then the spirit level is placed perpendicular to the lines again and the bubble is brought to the centre by turning the third screw. This makes the top of the prism-table horizontal.

(b) Alignment of the source and the slit :

To get result with the spectrometer work, proper alignment of the source and the slit is essential. Whatever may be the source (sodium vapour lamp or sodium flame), it must be placed before the slit properly, so that the slit image can be seen very clearly with brightness when viewed through the telescope and the collimator. For this, the brightest portion of the source should be placed in front of the slit in the same line, at a distance of about 8 or 10 inches from the slit. Once the postion of the source is adjusted, it should not be disturbed further. The opening of the slit is then adjusted by means of screws attached to it and the slit must be vertical. Care should be taken that all the portions of the slit get equal illumination from the source. This can be checked by viewing the slit image through the collimator. The slit width can then be adjusted after viewing it. If necessary, the slit image can be porperly focused by the help of the screw of rack and pinion arrangement attached to the collimator tube. This alignment work is very important in the spectrometer work.

(c) Focusing of the cross-wire :

The telescope is turned towards an illuminated source and when viewed through the eye-piece, it will appear that there is a cross-wire and it may be that the cross-wire is blurred. The eyepiece is adjusted till the cross-wire can be seen distinctly without any strain on the eyes. The cross-wires should be kept horizontal-vertical or turned through 45° with respect to the horizontal.

(d) Adjustment of the slit :

The alignment of the slit has to be done first as discussed earlier. Then the telescope is turned and brought in line with the collimator. When viewed through the telescope, generally a blurred image of the slit is observed. The focusing screw of the telescope and of the collimator, if necessary, should be turned until the image of the slit appears in sharp focus. The width of the slit opening should not be very large and the boundary edges of the slit must be sharply defined when viewed through the telescope.

(e) Focusing for parallel rays :

(i) Direct method

In order to focus the telescope and the collimator for parallel rays by direct method, the telescope is turned towards a distant object whose image is focused on the cross-wire without any parallax. Then the collimator and the telescope are brought in a straight line and the image of the slit attached to the collimator is sharply focused with the help of the screw attached to the collimator. There must not be any parallax between the image of the slit and the cross-wire. Thus in the first arrangement, the telescope is focused for parallel rays, while in the second operation, the collimator produces a beam of parallel rays.

(ii) Schuster's method

When a distant object is not available for focusing the telescope and the collimator for parallel rays (as in a dark room), then focusing is done by Schuster's method.

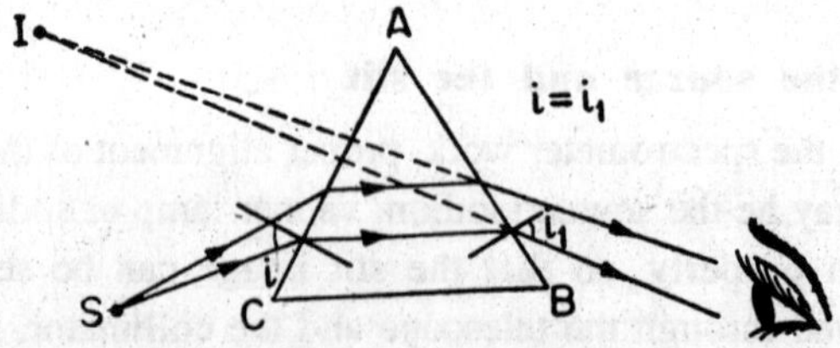

Fig. 5.11 : Position of minimum deviation

The prism is set on the prism-table with its refracting edges vertical and the centre of its base lies at the centre of the table. The refracting face AC is turned towards the collimator and through the face AB, the image of the slit is viewed. With the eyes fixed on the slit image, the table with the prism is slowly rotated in the appropriate direction so that the image (I) is brought as far as possible near the collimator. From this position, in whatever direction the prism is turned, the image (I) will move away from the collimator. When the image is at I (Fig. 5.11), then the prism is at the minimum deviation position and at this position, the angle of incidence ($\angle i$) on the prism is equal to the angle of emergence ($\angle i_1$) of the ray.

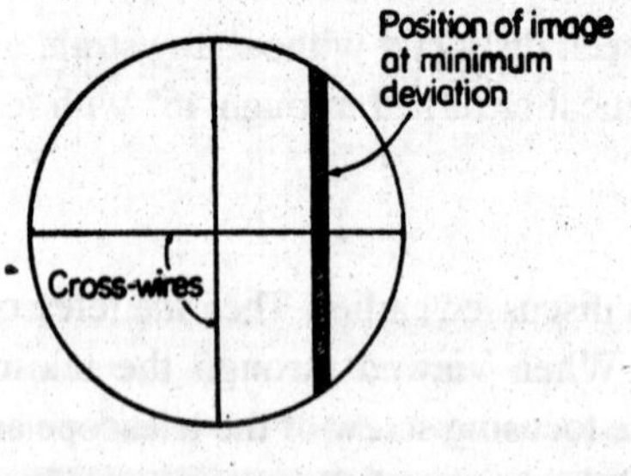

Fig. 5.12 : View of the telescope

Next the prism-table with the prism is rotated towards the prism base (BC) and the slit image is kept at one end of the field of view of the telescope (Fig. 5.12). Keeping telescope fixed, if the prism-table is rotated on either side of this position, the slit-image will be visible twice in the field of view. Once the image will appear broad and once thin. The position of the prism for these two positions are known as *normal position* and *slant position* respectively.

When the prism is brought to the position A_1, B_1 C_1 by rotation, then the angle of incidence ($\angle i$) of the ray on it is greater than the angle of emergence ($\angle i_1$). This position of the prism is known as the *slant position* (Fig. 5.13). In this position, the distance of the image I_1 formed is further away from the position of the image formed at the minimum deviation position. The image I_1 also appears blurred, when it is focused sharply with the aid of the screw attached to the telescope. The image at this position appears to be *thin*.

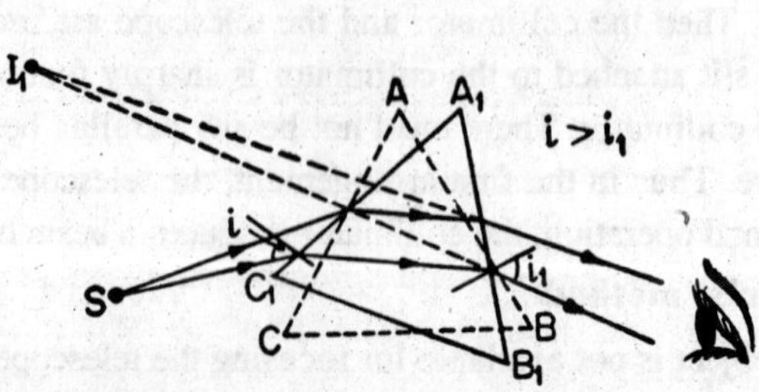

Fig. 5.13 : Slant position of the prism

Next the prism-table is rotated on the other side and the slit image is again viewed through the telescope. This time the image (I_2) appears to be broad (Fig. 5.14). In this position ($A_2B_2C_2$), the angle of incidence ($\angle i$) is less than the angle of emergence ($\angle i_1$) and the position is known as the *normal position*. The distance of the image (I_2) formed is less than that of the image formed at the minimum deviation position. So, again this image also appears blurred, and it is focused with the help of the screw attached to the collimator. The focused image appears to be *broad in size*. If the prism-table is again rotated in the opposite direction to keep the prism in slant position, the image again appears thin and out of focus. As before, the screw attached to the telescope is operated to focus the image. Again, bringing the prism in the normal position, the screw attached to the collimator is operated to focus the image. In general, it will be necessary to repeat these operations several times in succession till the image appears in sharp focus at both positions. Finally,

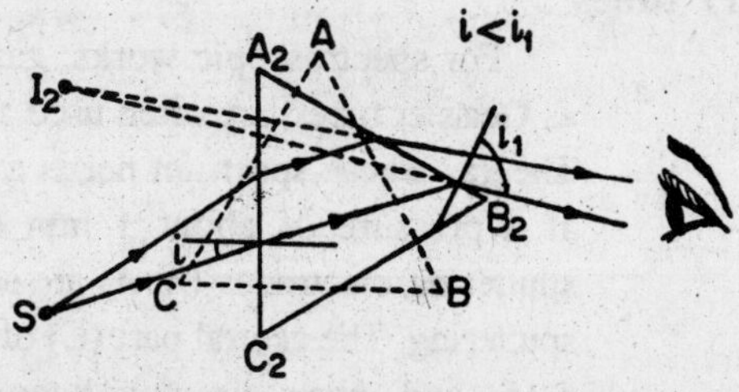

Fig. 5.14 : Normal position of the prism

the image is formed at infinity and the spectrometer is focused for parallel rays. In the final focused position, there must not be any parallax between the image and the cross-wires.

To remember the operations of the focusing screws for Schuster's method, the following *mnemonic* can be remembered.

Slit-image		*Focus by the screw of*
Broad	(B → C)	Collimator
Thin	(T → T)	Telescope

Precautions in handling the prism :

Great care must be taken in handling the prism while working with a spectrometer. Surface of the prism must remain clean. It should never be cleaned by dry cloth or handkerchief, rather a clothpiece soaked in alcohol should be used. While holding a prism, its refracting surfaces should not be touched by hand. It should be held by fingers at the top and the base, and placed on the prism-table. Similar care should be taken while removing it form the table. It should be remembered that if the refracting faces contain any dirts, the slit image does not appear as bright and inconveniences are caused while working. So, all possible cares must be taken to handle the prism during work.

Optical levelling of the prism :

It is essential that the refracting faces of the prism should be vertical during work with it. This can be achieved by the method, already discussed, when the prism-table is levelled. But slight difference of the two refracting faces in being vertical, may exist if the bottom section of the prism is not perfectly perpendicular to the faces. In that case, the difference may be removed by optical levelling. The operation will be as follows :

The prism is placed on the table with its apex at the centre of the prism-table and with one of the refracting surface at right angles to the line joining two of the levelling screws of the prism-table. The telescope is then turned so that its axis makes an angle of about 90° with the axis of the collimator and it is fixed there by the screw. The prism-table is then gradually turned till in the telescope an image of the slit produced by light reflected from the face of the prism kept at right angles to the line joining the two levelling screws, is visible. If the image is not at the centre of the field of view, then it is brought there by turning the above levelling screws in the opposite directions. Next the prism-table is again rotated till the slit image is visible in the fixed telescope when rays of light are reflected from the other face of the prism. If again the image is not at the centre, it is made so by turning the third levelling screw of the prism-table. The first operation makes the first face vertical while the second operation turns the first face in its own plane but does not alter its adjustment, but makes the second face vertical. The two operations together constitute optical levelling.

Discharge (Geissler) tubes :

For spectroscopic works, gas discharge tubes, known as Geissler tubes, are often used as a strong source of light. The gas whose spectrum needs to be studied, fills the tube at a pressure of about 1 mm of mercury. Aluminium sputtering electrodes (A, A) are sealed in the tube for little sputtering. The central part (C) of the tube is capillary (Fig. 5.15) and when direct voltage is applied between the electrodes, light is produced because of gas discharge. The light comes from the positive column of the discharge and it is most intense in the capillary portion where current density is the highest. The tube operates at a high voltage of about 3000 volts. For best results to get, the capillary portion of the tube should be placed in front of the spectrometer slit at an appropriate distance from it. The voltage required for operation, is supplied by an induction coil or a transformer.

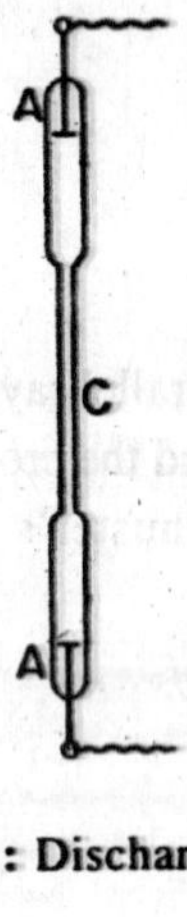

Fig. 5.15 : Discharge tube

5.5 Determination of Refractive Index of the material of a prism by Spectrometer using minimum deviation method

Theory :

If a ray of monochromatic light while passing through a prism, has an angle of minimum deviation δ_m and if A be the angle of the prism, then the refractive index (μ) of the material of the prism is given by

$$\mu = \frac{\sin\frac{A+\delta_m}{2}}{\sin\frac{A}{2}}. \qquad \text{... (5.5a)}$$

Hence, measuring the angle of minimum deviation (δ_m) of the ray passing through the prism and the refracting angle (A) of the prism, the value of μ can be determined.

Apparatus :

Spectrometer, prism, spirit level, source of light.

Method :

(i) At first the levelling, adjustments and focusing of the spectrometer have to be done as discussed in details in Article 5.4.

(ii) Next the angle of the prism (A) has to be determined. To do this, the prism has to be set on the prism-table in such a way that its vertex points towards the collimator and its refracting faces (AB and AC) get almost equally the parallel rays coming out of the collimator (Fig. 5.16). The angle between the rays reflected from these two faces will be $2A$ if the prism angle be A. This is because when the face AC turns and goes to the position of AB, then the angle through which the reflected ray turns, will be equal to $2A$. Hence, if the angle between the rays reflected from these two refracting faces is determined by the spectrometer, then the value of A will be obtained.

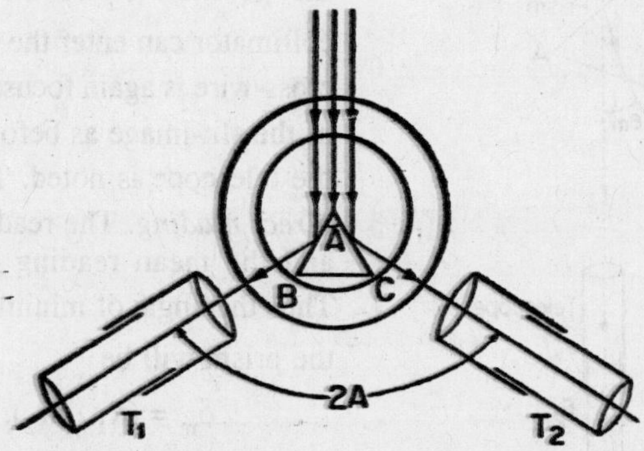

Fig. 5.16 : Determination of the angle of the prism

(iii) The vernier constants of both the circular scales are determined and noted.

(iv) The prism is then properly set on the prism-table (Fig. 5.16) and the slit is properly illuminated by a sodium vapour lamp or a burner.

(v) The telescope is rotated gradually and the image of the slit reflected from the face AB of the prism, is received in it in the position T. The cross-wire is set at a particular end of the slit-image and the reading of the telescope is noted from the circular and the vernier scales. The readings are taken *thrice* and the mean is taken. Let it be M_1.

(vi) The telescope is then rotated to the position T_2 and similarly the slit-image is received in it when light is reflected from the side AC of the prism. Readings of the telescope are taken *thrice* as before and the mean reading is noted. Let it be M_2. Then the angle of the prism (A) will be given by

$$A = \frac{M_1 \sim M_2}{2}. \qquad \text{... (5.5b)}$$

(vii) To determine the angle of minimum deviation (δ_m), the prism is placed on the prism-table in such a way that one of its faces (AC) points towards the collimator (Fig. 5.17). In this position, rays coming out of the collimator will be refracted through the face AC and will finally come out of the face AB. If the prism-table is gradually rotated and the slit-image is viewed through the telescope, it will be found that the image is deviated up to a certain point and then comes back in the same direction, irrespective of the direction of rotation of the prism-table. The point wherefrom the image returns back, is the position of

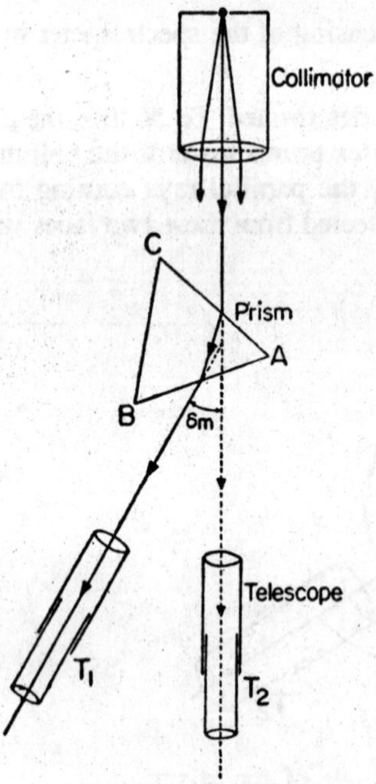

Fig. 5.17 : Determination of the angle of minimum deviation

minimum deviation. Keeping the prism at the position of its minimum deviation, the slit-image there is focused in the telescope in the position T_1 and the telescope reading is taken. This reading is taken at least *thrice* and their mean (R_1) is taken.

(viii) The prism is then withdrawn from the table and the telescope is brought to the position T_2 so that the rays from the collimator can enter the telescope direct. The cross-wire is again focused with the same side of the slit-image as before and the reading of the telescope is noted. This is known as the *direct reading*. The readings are taken thrice and the mean reading (R_2) is determined. Thus the angle of minimum deviation (δ_m) of the prism will be

$$\delta_m = (R_1 \sim R_2). \qquad \text{... (5.5c)}$$

(ix) Hence, determining the values of A and δ_m, the value of the refractive index (μ) of the material of the prism can be determined from Eq. (5.5a).

Results :

(A) Vernier constant determination :

Smallest circular scale division = ...

Vernier constant of the 1st vernier of the prism-table = ... minutes (or seconds)

Vernier constant of the 2nd vernier of the prism-table = ... minutes (or seconds)

Vernier constant of the 1st vernier of the telescope = ... minutes (or seconds)

Vernier constant of the 2nd vernier of the telescope = ... minutes (or seconds)

Table 1 (See next page)

Table 2 (See next page)

Calculation :

From Table 1 and Table 2, the values of A and δ_m are obtained and substituting them in Eq. (5.5a), the value of the refractive index of the material of the prism is obtained as

$$\mu = \frac{\sin \dfrac{A+\delta_m}{2}}{\sin \dfrac{A}{2}} = \text{........}$$

(B) Determination of the angle of the prism (A) :

Table 1

Vernier	Readings for first image at T_1				Readings for second image at T_2				Difference of the mean readings for the two ($M_1 \sim M_2$)	Mean of the differences ($2A$)	Angle of the prism (A)
	Main Scale	Vernier	Total	Mean (M_1)	Main Scale	Vernier	Total	Mean (M_2)			
	...	...	...		...	...	...				
First	...	...	...	...	...	...	...	...	...		
	...	...	...		...	...	...				
	...	...	...		...	...	...			...	...
Second	...	...	...	...	...	...	...	...	...		
	...	...	...		...	...	...				

(C) Determination of the angle of minimum deviation (δ_m) :

Table 2

Vernier	Readings for minimum deviation				Readings for direct rays				Angle of the minimum deviation ($\delta_m = R_1 \sim R_2$)	Mean value of the angle of minimum deviation (δ_m)
	Main Scale	Vernier	Total	Mean (R_1)	Main Scale	Vernier	Total	Mean (R_2)		
	...	...	...		...	...	...			
First	...	...	...	...	...	...	...	...	...	
	...	...	...		...	...	...			...
	...	...	...		...	...	...			
Second	...	...	...	...	...	...	...	...	...	
	...	...	...		...	...	...			

Precautions :

(i) There should not be any parallax between the cross-wires and the slit-image.

(ii) The vertical axis passing through the apex of the prism must pass through the centre of the prism-table, while determining the angle of the prism.

(iii) During rotation of the telescope, if the vernier zero crosses the zero mark of the main circular scale, then the latter should be considered as 360° and calculations should be made accordingly.

(iv) One particular end of the slit-image must be focused at all times during measurements and this end should never be changed during experiment.

Questions and Answers

[Questions on Spectrometers have been discussed at the end of the Experiment 5.6]

5.6 Measurement of the Dispersive Power of the material of a prism

Theory :

The dispersive power (ω) of the material of a glass prism is given by

$$\omega = \frac{\mu_F - \mu_C}{\mu_D - 1}, \quad \text{... (5.6a)}$$

where μ_F and μ_C are the refractive indices for *F*-lines (bluish-green 4861 Å) and *C*-lines (red 6563 Å) of hydrogen spectrum and μ_D is the refractive index for *D*-lines (5893 Å) of sodium respectively.

The refractive indices μ_F, μ_C and μ_D of the material of the prism for the different lines can be obtained from the formula,

$$\mu = \frac{\sin\frac{A+\delta_m}{2}}{\sin A/2} \quad \text{... (5.6b)}$$

where A and δ_m are the angle of the prism and the angle of minimum deviation respectively for the line concerned.

Apparatus :

Spectrometer, prism, spirit level, sodium light source and hydrogen discharge tube.

Method :

(i) The spectrometer is as usual levelled, focused and adjusted for working. This has to be done in the manner discussed in Article 5.4.

(ii) The angle of the prism (A) is determined by the method discussed in the previous experiment (Article 5.5).

(iii) With the sodium light placed in front of the slit, the angle of minimum deviation (δ_m) of the prism for sodium light is obtained by the method discussed in the previous experiment (Article 5.5).

(iv) The sodium light is then replaced by a hydrogen discharge tube which is excited by means of an induction coil or a transformer. The tube is held vertically and the slit is illuminated by it directly. On viewing through the telescope with the prism set properly, well defined lines of hydrogen spectrum are visible. The strong *C*-line in the red is identified and it is focused on the cross-wire at the minimum deviation position of the prism. Telescope readings are noted. Similarly, the strong F-line in the bluish-green is identified and focused at the minimum deviation position of the prism.

(v) Finally, the direct reading of the slit-image is taken from which the values of the angles of minimum deviation corresponding to the red, bluish-green and yellow lines are obtained.

(vi) Substituting the values of δ_m for different lines in Eq. (5.6b), the values of μ_F, μ_C and μ_D are determined, from which the value of the dispersive power (ω) of the material of the prism is obtained using Eq. (5.6a).

Results :

(A) Vernier constants of the spectrometer :

(i) Vernier constant of the prism-table vernier = ...

(ii) Vernier constant of the telescope vernier = ...

(B) Determination of the angle (A) of the prism :

[Make a table similar to Table 1 of Article 5.5]

(C) Determination of the angle of minimum deviation (δ_m) :

(a) For yellow light (Sodium light) :

[Make a table similar to Table 2 of Article 5.5]

(b) For C-lines (red) of Hydrogen spectrum :

[Make a table similar to Table 2 of Article 5.5]

(c) For F-lines (bluish-green) of Hydrogen spectrum :

[Make a table similar to Table 2 of Article 5.5]

Calculations :

Using Eq. (5.6b), the refractive indices μ_D, μ_F and μ_C of sodium D-lines and F- and C-lines of hydrogen are respectively found out from the values of corresponding angles of minimum deviations (δ_m) obtained. Then substituting the values of μ_D, μ_F and μ_C in Eq. (5.6a), the value of the dispersive power (ω) of the material of the prism is determined. Thus,

$$\text{dispersive power of glass, } \omega = \frac{\mu_F - \mu_C}{\mu_D - 1} = \ldots\ldots\ldots$$

Precautions :

(i) In all cases of focusing, the minimum deviation position of the prism corresponding to each line has to be ascertained and the telescope must be set there.

(ii) Several independent settings of the cross-wire are to be made for each reading, avoiding thereby any possible errors in taking readings of the vernier.

(iii) The discharge tube should be held vertically in front of the slit properly, so that the brightest part of the illumination remains in front of the slit.

[For other precautions, Experiment in Article 5.5 may be consulted.]

Error Calculations :

The errors δA and $\delta(\delta_m)$ in A and δ_m are given by half the sum of the errors in the values of M_1, M_2 and R_1, R_2 respectively (See Eqs. 5.6b and 5.6c). Since an error in A alters both numerator and denominator in the expression for μ (Eq. 5.6a) in the same direction though not exactly equally, a reasonable estimate of the error in μ would be about twice the cahange produced in μ substituting $[\delta_m + \delta(\delta_m)]$ for δ_m and re-calculating. Thus finding the error $\delta\mu$ in each value of μ, the maximum error $\delta\omega$ in ω is found out by reworking Eq. (5.6a), having increased μ_F by δ_μ and having decreased μ_C and μ_D by δ_μ.

Hence, maximum percentage error in ω

$$= \frac{\delta\omega}{\omega} \times 100\% = \frac{\delta\mu_F + \delta\mu_C}{\mu_F - \mu_C} + \frac{\delta\mu_D}{\mu_D - 1}$$

$$= \frac{2\delta\mu}{\mu_F - \mu_C} \text{ (approx.)}$$

This is true when the errors in μ_C, μ_D and μ_F are approximately equal and ω is small.

Questions and Answers

[General questions on spectrometer are discussed at the end of the experiment given in Article 5.8.]

Q.1. What do you mean by dispersion of light?

A. : When a ray of white light passes through a prism, it is separated into rays of its constituent colours. The incident ray is said to have been *dispersed* and the phenomenon is known as *dispersion*.

Q.2. Apart from the prism, can any other device disperse the light?

A. : Yes, the diffraction grating can do that.

Q.3. What is the difference between the actions of them in dispersing light?

A. : The dispersion by a prism arises because the refractive index of the material of the prism varies with the wavelength of light. Each colour of light thus undergoes deviation by different amount—the red being the least and the violet the most. But the dispersion by a grating arises because each colour is diffracted by a different amount while passing through it—the violet being the least and the red the most.

Q.4. What is meant by dispersive power of a material?

A. : The dispersive power of a material is its ability to disperse various colours constituting the incident light.

Q.5. On what factors does it depend?

A. : It depends on the material and on the wavelength for which it is determined.

Q.6. What is meant by angular dispersion?

A. : Angular dispersion for any two colours (or wavelengths) is the difference between the deiviation of those wavelengths.

Q.7. Can the dispersive power of the material of the prism for wavelengths λ_1 and λ_2 be taken as $\frac{(\delta_1 - \delta_2)}{\delta}$ where δ is the mean deviation?

A. : The quantity $\frac{(\delta_1 - \delta_2)}{\delta}$ varies with the angle of the prism and hence it is not taken as a measure of the dispersive power.

Q.8. Can any other expression for the dispersive power of a prism is available except that given in Eq. (5.6a)?

A. : Yes, the dispersive power of the material of the prism for wavelengths λ_1 and λ_2 can be represented by $\omega = \frac{\mu_1 - \mu_2}{\mu - 1}$, where μ_1 and μ_2 are the refractive indices for λ_1 and λ_2 and μ is the mean refractive index. But the internationally accepted expression for the dispersive power is that given in Eq. (5.6a).

Q.9. Why are the *F*- and *C*-lines of hydrogen spectrum and *D*-lines of sodium light selected?

A. : The *F*- and *C*-lines lie respectively in the bluish-green and red regions of the hydrogen spectrum while the *D*-line lies in the yellow region. The refractive index for the *D*-line is very nearly equal to the mean of the refractive indices for the *F*- and *C*-lines. These lines are selected because they cover that region of the spectrum to which the eye is most sensitive.

Q.10. What is the best source of light for determining the dispersive power?

A. : Neon lamp is the best source for this purpose because it gives spectral lines in all the regions of the visible spectrum.

5.7 Calibration of a spectrometer and hence to find the wavelength of an unknown line from the (D-λ) curve

Theory :

By setting a prism on a spectrometer table, a series of spectral lines due to one or more sources can be produced and the receiving telesocope will have definite position on the spectrometer scale corresponding to each particular wavelength. Calibration of a spectrometer means to find the values of the wavelengths corresponding to the different deviations of the lines, caused by the prism, as recorded by the telescope. A curve in which the deviations are plotted against the corresponding wavelengths, is called a calibration curve. Such a curve is determined by the material, angle and position (relative to the spectrometer) of the prism. It is customary to keep the prism in its position of minimum deviation for sodium light. Normally a helium or a hydrogen discharged tube is used to draw the calibration curve, using the different spectral lines produced by them. If the deviation produced by the same prism at the same position be found out for some unknown lines, then from the calibration curve, the wavelength of that unknown line can be found out.

Apparatus :

Spectrometer, spirit level, sodium light, discharge tubes, prism (dense flint glass), induction coil, etc.

Method :

(i) The spectrometer is levelled, adjusted and focused for parallel rays by the method described in Article 5.4.

(ii) The vernier constant of the spectrometer is determined and expressed in minutes or seconds.

(iii) After properly adjusting the slit, sodium light is placed in front of the slit and focusing the cross-wire on the slit-image, reading for the direct ray is taken from the scale.

For purposes of calibration, only one vernier of the circular scale may be used but it should be the same vernier all through.

(iv) The prism is set on the prism-table and it is levelled preferably optically (vide Article 5.4). The refracted image of the slit is seen through the prism set at minimum deviation for sodium light and the readings are taken. The prism is re-set and the reading is repeated. The prism must not be moved or displaced from this position for the rest of the experiment. The prism-table is clamped firmly.

(v) The sodium flame is then replaced by a helium discharge tube placed as close as possible to the slit. The tube is excited and the spectral lines will be visible at the focal plane of the telescope already focused. The cross-wires are focused on the successive lines from the red end and ending in violet. For each position of the telescope coinciding with a given line (Fig. 5.18), the readings of the vernier and the scale are noted. The lines are identified and their wavelengths are recorded corresponding to the lines respectively.

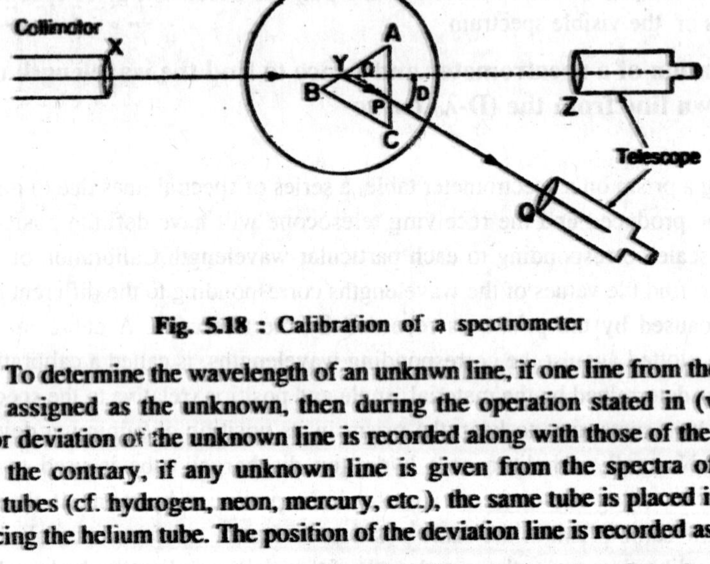

Fig. 5.18 : Calibration of a spectrometer

(vi) To determine the wavelength of an unknwn line, if one line from the spectrum of helium is assigned as the unknown, then during the operation stated in (v) above, the reading for deviation of the unknown line is recorded along with those of the other known lines. On the contrary, if any unknown line is given from the spectra of some other discharge tubes (cf. hydrogen, neon, mercury, etc.), the same tube is placed in front of the slit, replacing the helium tube. The position of the deviation line is recorded as before.

(vii) The wavelengths of the known lines of helium are ascertained from a table, and a graph is drawn by plotting wavelengths (λ) of those lines along *X*-axis and their corresponding deviations (*D*) along Y-axis. The nature of this curve (*D*-λ curve) will be as shown in Fig. 5.19. From this graph, the wavelength (λ) of the unknown line is determined by noting its deviation (at *P*) on the graph.

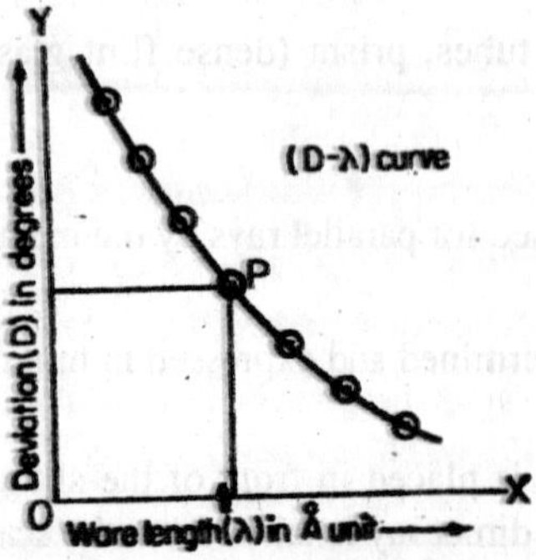

Fig. 5.19 : (*D*-λ) curve→

(viii) The prism is withdrawn and the telescope is rotated to receive the direct light from the collimator whose slit is now illuminated by sodium light. This is to check the reading of the direct beam again for any change of displacement of the step-up.

Results :

(A) Vernier constant of the spectrometer :

[Write in details the method of finding the vernier constant.]

Vernier constant = ... min or sec.

(B) Direct reading of the telescope :

Table 1

	Vernier	Circular scale reading (C)	Vernier reading	Value of vernier reading (V)	Total reading (T) = C + V	Mean reading	Mean reading of
Before experiment	A	...	...	...	...	... (M_1)	Vernier A
		...	...	...	...		= ......
		...	...	...	...		(X)
	B	...	...	...	...	... (M_2)	
		...	...	...	...		
		...	...	...	...		Vernier B
After experiment	A	...	...	...	...	... (M_3)	= ......
		...	...	...	...		(Y)
		...	...	...	...		
	B	...	...	...	...	... (M_4)	
		...	...	...	...		
		...	...	...	...		

Drawing of Graph :

A graph is plotted with the deviations (D) of the spectral line along Y-axis and corresponding wavelengths (λ) along X-axis. This is the calibration curve. The nature of the curve will be as shown in Fig. 5.19. On this graph, the deviation of the unknown line is put at the point P, from which the value of the wavelength (λ) corresponding to the unknown line, is obtained. Another graph $\left(D - \frac{1}{\lambda^2}\right)$ can also be drawn, which would be a straight line.

Hence, wavelength of the unknown line as found from the calibration curve = ... A.U.

Precautions :

(i) The slit should be very narrow and it should be properly illuminated by keeping the brightest part (capillary portion) of the discharge tube in front of it.

(ii) Parallax betweeen the images of the lines and the cross-wires must be avoided.

(iii) During the entire experiment, the prism must not be disturbed once it is set initially at its position of minimum deviation corresponding to the sodium light.

(iv) The known spectral lines of the particular discharge tube must be properly identified.

(C) Deviations and wavelengths of the lines of Helium :

Table 2

Colour of the line	Wave-length in A.U.	Vernier	Telescope readings						Mean Deviation (D)
			Circular scale reading (C)	Vernier reading	Value of Vernier reading (V)	Total reading ($T = C + V$)	Mean reading	Deviation	
		A	...	...	...	...			
			...	...	...	...	...	...	
			...	...	...	...	(X_1)	($X_1 \sim X$)	
Red	6678								....
		B	...	...	...	...			
			...	...	...	...	...	...	(D_1)
			...	...	...	...	(Y_1)	($Y_1 \sim Y$)	
Yellow	5876	A							.....
		B							(D_2)
Green	5016	A							.....
		B							(D_2)
Green-Blue	4922	A							.....
		B							(D_4)
Blue (I)	4713	A							.....
		B							(D_5)
Blue (II)	4471	A							.....
		B							(D_6)
Violet	4026	A							.....
		B							(D_7)
Unknown	λ	A							.....
		B							(D_λ)

Error Calculations :

The only measurement done in this experiment is the deviation of the spectral lines caused by the prism due to dispersion. The maximum error in measuring the deviation (D), by taking the difference of the readings for direct and deviated rays, may be at most 2 vernier divisions, that is 2 × (vernier constant). Since the values of D are quite appreciable, the proportional error in its measurement is significant. However, smaller the value of the vernier constant of the spectrometer, the less will be the proportional and hence, the percentage error in this experiment.

Questions and Answers

[For questions and answers on this experiment, see at the end of the next experiment (Article 5.8), where questions and answers on spectrometer, deviation produced by a prism and allied phenomenon have been discussed.]

5.8 To draw the μ-λ curve for a prism and hence to determine the wavelength of some unknown line.

Theory :

When the spectral lines of a source (say, helium light) are produced by a prism and if the angles of minimum deviations (D_m) of each line of the spectrum and the angle of the prism are determined by a spectrometer, then the refractive index of the material of the prism for light of the given wavelength is obtained from the relation,

$$\mu = \frac{\sin\frac{1}{2}(A+D_m)}{\sin\frac{1}{2}A} \qquad \ldots (5.8a)$$

Using special lines of know wavelengths, the refractive index for each can be determined in the above way. A graph may be drawn with the wavelengths (λ) of the lines as abscissa and the corresponding refractive indices (μ) of the prism as the ordinate. This graph (μ-λ curve) is known as the dispersion curve of the material of the prism. From this curve, the wavelength of some unknown line can be determined from the knowledge of the refractive index of the material of the prism for that line.

The relation between μ and λ is given by Cauchy's relation,

$$\mu = a + \frac{b}{\lambda^2} + \frac{c}{\lambda^4}, \qquad \ldots (5.8b)$$

where a, b and c are constants and characteristics of any particular substance. Retaining the first two terms of the equation, one gets the relation as

$$\mu = a + \frac{b}{\lambda^2}, \qquad \ldots (5.8c)$$

Thus, if μ is plotted against $\frac{1}{\lambda^2}$, the curve should be a straight line, which is more convenient to find the wavelength of the unknown line. The intercept on the μ-axis will give a and the slope of the straight line will give b. These are known as Cauchy's constants.

Apparatus :

Spectrometer, spirit level, sodium light, discharge tube, prism, induction coil, etc.

Method :

(i) The spectrometer is levelled, adjusted and focused for parallel rays by the method described in **Article 5.4**.

(ii) The procedure for work is the same as that described in the previous experiment (**Article 5.7**). The operations (ii) to (vi) and (viii) of the previous experiment have to be performed in this experiment also, only keeping in mind that the prism should be set at the position of minimum deviation for each of the lines.

(iii) From the knowledge of the values of D_m obtained for each line, Eq. (5.8a) is used to find the values of μ of the prism for each line, knowing the value of the angle of the prism.

(iv) Finally, a consolidated table should be prepared to find μ for different values of λ, from which graphs of (μ-l) and (μ-1/λ^2) are drawn. From the latter graph, the values of Cauchy's constants are determined.

Results :

(A) Vernier constant of the spectrometer :

[Write in details the method of finding the vernier constant.]

Vernier constant = ... min or sec.

(B) Direct reading of the telescope :

Table 1

[Make a table just like Table 1 of Expt. No. 5.7]

(C) Deviations (minimum) and wavelengths of the lines of helium :

Table 2

[Make a table here just like Table 2 of Expt. No. 5.7.]

(D) Determination of μ for different values of λ :

Table 3

Source	Wavelength in A.U. (λ)	Reciprocal of the square of wavelength ($1/\lambda^2$) cm^{-2}	Minimum Deviation in degree (D_m) from Table 2	Angle of the prism in degree (A)	Refractive index μ $= \dfrac{\sin\frac{1}{2}(A+D_m)}{\sin\frac{1}{2}A}$
	......	....	....		.....
	......	....	....		.....
Helium	....	....	....	....	.....
	....	....	....		.....
	....	....	....		.....

Calculations :

From Eq. (5.8a) :

$$\mu = \frac{\sin\frac{1}{2}(A+D_m)}{\sin\frac{1}{2}A}$$

Substituting the values of the angles of minimum deviations (D_m) obtained from Table 2 for different values of λ and also the angle of the prism (A), the values of the refractive indices (μ) for different values of λ are obtained.

Drawing of Graphs :

Two graphs are plotted. The first one is a (μ-λ) curve where the values of λ are plotted along the X-axis and the corresponding values of μ along the Y-axis (Fig. 5.20). In this graph, the refractive index corresponding to the unknown wavelength is noted (point Q) wherefrom the value of the unknown wavelength is given by the abscissa of the point Q.

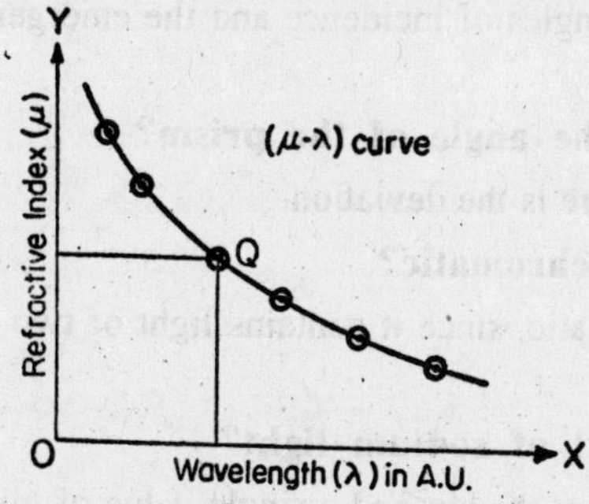

Fig. 5.20 : (μ-λ) curve

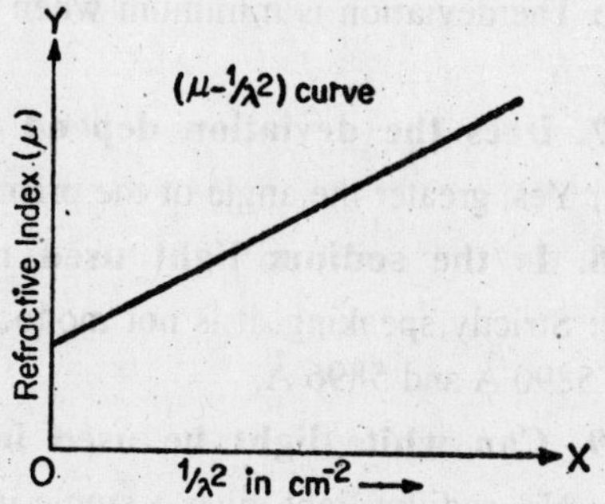

Fig. 5.21 : (μ - $1/\lambda^2$) curve

Another graph (Fig. 5.21) can be plotted with μ along Y-axis and ($1/\lambda^2$) along X-axis. This graph will be a stright line, wherein the refractive index corresponding to the unknown wavelength is noted (point P), wherefrom the value of $1/\lambda^2$ is obtained. From the slope of the curve and its intercept on the Y-axis, the values of b and a are respectively obtained.

Precautions :

(i) While taking readings, the prism should be set at its minimum deviation position for each line. For this, prism-table should be made free from the scale, so that the rotation of the prism-table may not change the direct reading.

(ii) The vertical line of cross-wire must be made coincident with the same edge of each line while taking the readings.

[Other precautions mentioned under Expt. 5.7 are to be adopted as usual.]

Error Calculation :

For calculation of errors, the same discussions made at the end of Expt. 5.7 are to be followed.

Questions and Answers

[General questions on Spectrometer and on Expts. 5.7 and 5.8.]

Q.1. What do you mean by principal section, refracting edge, base, face and angle of the prism?

A. : See any textbook on 'Light'.

Q.2. What are the functions of the collimator and the telescope?

A. : See Article 5.4.

Q.3. How does the ray pass through a prism at minimum deviation?

A. : See any textbook on 'Light'.

Q.4. How does the deviation of a ray vary with its angle of incidence?

A. : The deviation is minimum at a particular angle of incidence but the former increases both for greater and smaller angle of incidence than that particular value.

Q.5. Does the deviation depend on the colour of light?

A. : Yes, the deviation is less for red than for violet light.

Q.6. What is the condition for obtaining minimum deviation?

A. : The deviation is minimum when the angles of incidence and the emergence are equal.

Q.7. Does the deviation depend on the angle of the prism?

A. : Yes, greater the angle of the prism, more is the deviation.

Q.8. Is the sodium light used monochromatic?

A. : Strictly speaking, it is not monochromatic, since it contains light of two wavelengths 5890 Å and 5896 Å.

Q.9. Can white light be used instead of sodium light?

A. : No, sodium light gives a single image of the slit and a single value of minimum deviation is obtained while white light gives a band spectrum.

Q.10. Why are there two verniers provided with the telescope?

A. : To avoid the error which may arise out of the non-coincidence of the axis of rotation of the telescope with the centre of the circular scale.

Q.11. Why is the centre of the cross-wire always focused on one and the same edge of the slit?

A. : Since the slit and the cross-wires have definite widths, the cross-wire centre is always focused on the same edge of the slit to avoid any possible error.

Q.12. What kind of image is produced by the telescope?

A. : It is a virtual image formed at infinity.

Q.13. What are the differences between normal dispersion, anomalous dispersion and irrational dispersion?

A. : See any textbook on 'Light'.

Q.14. What is the difference between emission spectra and absorption spectra?

A. : Light emitted from the excitation of a substance when passed through a dispersive medium, gives rise to emission spectra, while when light emitted from a source is made to pass through a cold substance, the substance absorbs selectively some wavelengths from the incident light and the rest is allowed to pass through. When this transmitted light is dispersed, a spectrum is obtained known as absorption spectrum.

Q.15. What is the difference between a particular gas (cf. helium) spectrum and solar spectrum?

A. : The spectrum arising out of the discharge of a gas is a *line* spectrum while that of the sun is a *band* spectrum.

Q.16. How does a discharge tube emit spectral lines?

A. : [See the action of a discharge tube from any textbook.]

Q.17. Is there any limit for the value of the angle of the prism?

A. : Yes, if the refracting angle of the prism exceeds twice the critical angle for the material of the prism and air, no ray will emerge out of it.

Q.18. Why does the deviation differ with the colour of light?

A. : The deviation depends on the refractive index of light which in turn depends on the colour of light. Hence, the deviation is different for different colours of light.

Q.19. Do the light rays of different colours travel with the same velocity while passing through the prism?

A. : No, since different colours have different values of refractive index in regard to the same prism, they travel with different velocities through the prism. Violet rays travel with less velocity than that of red.

Q.20. Does the deviation depend on the length of the base of the prism?

A. : No, it is independent of the length of the base.

Q.21. Then, why some prisms are made of larger base?

A. : Only to increase their resolving power.

Q.22. On what factors does the value of μ depend?

A. : It depends on the nature of the pair of media and also on the colour (wavelength) of light used.

Q.23. Will the value of μ change if the experiment of (μ-λ) is performed with the prism being placed in water?

A. : Yes, the value will change from that obtained when the prism is placed in air.

Q.24. When a beam of light falls on one face of the prism, is the whole of it refracted?

A. : No, although the major portion of the incident light is refracted, some fractions are reflected and some are absorbed at the face.

Q.25. What important informations are obtained by studying the spectrum of a substance?

A. : By studying a spectrum, an idea about the constitution of the substance and the temperature at which it is emitting lght, can be obtained

Q.26. Why should there be no parallax between the slit-image and the cross-wires?

A. : If there be any parallax, then with the movement of the eye, coincidence of the cross-wire image and the slit-image will change continuously, resulting the error in the recording of the readings.

Q.27. How does the (μ-λ) curve change with prisms of different materials like crown glass, flint glass, etc.?

A. : At any given wavelength, the (μ-λ) curve will be more and more steeper, the larger the refractive index of the dispersive substance.

Q.28. How does the dispersion depend on wavelength?

A. : The dispersion is given by $\frac{d\mu}{d\lambda}$, where μ is the refractive index. Dispersion varies, to a close approximaton, inversely as the cube of the wavelength.

Q.29. What are the limitations of Cauchy's equation?

A. : Cauchy's equation fits the curve of normal dispersion of many transparent substances in the visible region. If refractive index measurements are extended into infrared region, marked deviation of results from Cauchy's equation are observed as one approaches the spectral region of the absorption lines of the dispersive medium.

[Apart from these questions, general descriptions of the parts of a spectrometer and their functions should be known.]

5.8(A) To find the relation and to plot a graph betweeen 'dispersion' $\left(\frac{d\mu}{d\lambda}\right)$ and 'wavelength' (λ)

Theory :

The variation of refractive index μ of a prism with the wavelength (λ) is known as the *dispersion* and it is denoted by $\frac{d\mu}{d\lambda}$. Normal dispersion is that when the refractive index (μ) increases with the decrease of wavelength (λ). But the dispersion $\left(\frac{d\mu}{d\lambda}\right)$ of the material of the prism becomes greater at shorter wavelength. Normal dispersions are obtained for those wavelengths of incident light which are far away from the absorption bands of the material of the prism.

From the μ-λ curve (Fig. 5.20 of Expt. No. 5.8) we get the variation of refractive index (μ) with the wavelength (λ) of the incident light. In this graph, if a tangent is drawn at any point of the graph, then the slope of the curve at that point will be obtained. The value of this slope will give the 'dispersion' $\left(\frac{d\mu}{d\lambda}\right)$ at that particular value of wavelength (λ), where the tangent is drawn.

Now, if successive tangents are drawn at different points of the graph (for different values of the wavelength λ), the respective values of $\frac{d\mu}{d\lambda}$ corresponding to respective wavelengths will be obtained. In this way, several values of $\frac{d\mu}{d\lambda}$ are obtained for different values of λ (say, five points, i.e., five particular values of λ). From these values of $\frac{d\mu}{d\lambda}$, another graph may be drawn with $\frac{d\mu}{d\lambda}$ along Y-axis and corresponding values of λ against X-axis. This graph $\left(\frac{d\mu}{d\lambda} \text{ vs. } \lambda\right)$ will give the dependence of $\frac{d\mu}{d\lambda}$ on λ.

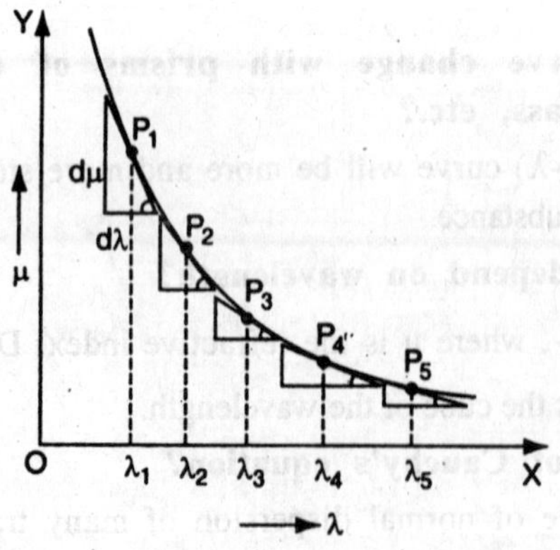

Fig. 5.22(a) : [Finding ($d\mu/d\lambda$)]

Apparatus :

[Same as in Expt. No. 5.8]

Method :

[Same as in Expt. No. 5.8]

Results :

The same data as in Expt. no. 5.8 are to be taken and shown in tabular forms as (A), (B), (C) and (D) of Expt. No. 5.8. The values of μ of the prism for different values of λ are to be obtained and tabulated as in Table 3 (Expt. No. 5.8)

Drawing of Graphs :

As in Expt. No. 5.8, a curve showing variation of μ against λ (Fig. 5.20) is drawn. From this graph (μ-λ), tangents are drawn at different values of recorded wavelength (λ) and in each case, the slope of the curve at that particular point, is calculated [Fig. 5.22(a)]. This slope gives the vlaue of $\left(\frac{d\mu}{d\lambda}\right)$ at that particular value of λ. In this way, at least five values of λ are chosen and values of $\left(\frac{d\mu}{d\lambda}\right)$ at those five points are obtained. Thus we get the variation of $\left(\frac{d\mu}{d\lambda}\right)$ with λ.

Next another graph is drawn with λ along X-axis and absolute value (mod. value) of $\left(\frac{d\mu}{d\lambda}\right)$ along Y-axis.

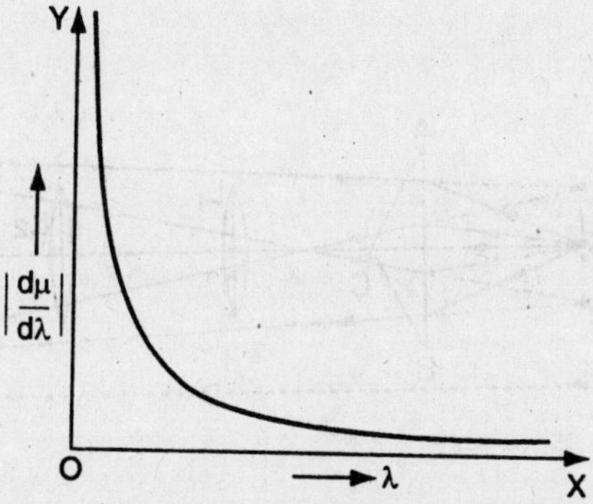

Fig. 5.22(b) : (|dμ/dλ| vs. λ graph)

From Eq. 5.8 (c),

$$\mu = a + \frac{b}{\lambda^2}$$

$$\therefore \quad \left(\frac{d\mu}{d\lambda}\right) = -\frac{2b}{\lambda^3}.$$

Since, λ is always +ve, hence the values of $\left(\frac{d\mu}{d\lambda}\right)$ for all values of λ, will always be negative. So, while plotting the graph of $\left(\frac{d\mu}{d\lambda}\right)$ vs. λ, the absolute value or the mod. of $\left(\frac{d\mu}{d\lambda}\right)$ is taken. The nature of this graph will be as shown in [Fig. 5.22(b) (asymptotic)].

Precautions and Discussions :

[Same precautions as mentioned in Expt. No. 5.8, are to be adopted].

In addition to the graph of $\left|\left(\frac{d\mu}{d\lambda}\right)\right|$ vs. λ, another graph showing $\left|\left(\frac{d\mu}{d\lambda}\right)\right|$ against $\left(\frac{1}{\lambda^3}\right)$ may be drawn. This graph will be a straight line with a negative slope.

[For Questions and Answers, see Expt. No. 5.8]

5.9 Determination of the wavelength of a monochromatic light using Fresnel's biprism

Theory :

If a beam of monochromatic light of wavelength, λ, after passing through a narrow, slit, is made to be incident on a symmetrically placed biprism, with its refracting edge paralleled to the slit, then the beam of light after transmission through the two halves of the biprism, overlap and interfere under suitable conditions. If a screen *Sc* is placed (Fig. 5.23) with its plane perpendicular to the plane containing the slit and the common base of the biprism, then on the screen, interference fringes (alternately dark and bright) will appear in the region where the rays from the two virtual sources overlap. If d is the distance between the two virtual sources of light (S_1 and S_2), D being the actual distance of the screen from the source and λ being the wavelength of the light used, then the distance (x) between two consecutive dark or bright fringes is given by

$$x = \frac{D\lambda}{d}. \qquad \text{... (5.9a)}$$

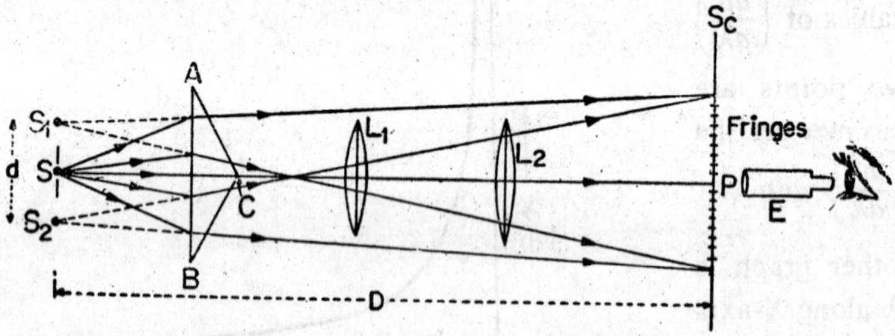

Fig. 5.23 : Biprism arrangement

To measure d, the distance between the two coherent virtual sources S_1 and S_2, a convex lens of focal length f (where $D > 4f$) is interposed between the biprism and the screen with its optic axis coinciding with the line SCP. For two positions of the lens along SCP, real images of S_1 and S_2 may be focused on the screen provided the distance SP is greater than $4f$. If d_1 and d_2 be the distances between the images of S_1 and S_2 for two positions L_1 and L_2 of the lens, then the value of d is obtained from the relation,

$$d = \sqrt{d_1 d_2}\,. \qquad \text{... (5.9b)}$$

Thus, measuring x, d and D, the value of λ, the wavelength of the light used, may be obtained from Eq. (5.9a).

Apparatus :

Optical bench, biprism, sodium light, an electric bulb (white light), index-rod.

The biprism consists of two acute-angled prisms joined base to base. The plane face of the prism is directed towards the slit. It is used on a long optical bench having suitable stands for holding the slit, the biprism, a lens and a micrometer eyepiece which serves the purpose of the screen. All stands have usually vernier attachments. The biprism can be rotated about a vertical axis. The slit is mounted near one end of the bench. Its height and width are adjustable. A tangent screw attached to the slit, can turn the slit in its own plane to make it vertical. The biprism is placed next to the slit being held in an upright. Its height is also adjustable and it also can be turned in its own plane like the slit.

The third upright carries a micrometer eyepiece, usually of Ramsden type. The height of the eyepiece is also adjustable and it can also turn about a vertical axis. The eyepiece can be moved perpendicular to the bench by a micrometer screw having a linear and a circular scale. Cross-wire in the eyepiece are mounted in a drawn-tube beyond the field lens. Before using the eyepiece, the cross-wires are brought to a focus. With the help of the micrometer screw, the shift of the eyepiece perpendicular to the bench, can be determined accurately.

A fourth stand can hold a lens and can be set on the bench when necessary. It also has adjustable height and arrangement for rotation about a vertical axis. When not needed, this upright is removed from the bench.

Method :

The experimental procedure involves some preliminary adjustments and the entire method can be divided into the following stages :

(a) Obtainings fringes
(b) Ensuring to obtain d_1 and d_2
(c) Improvement of the fringes obtained
(d) Alignment of the fringes
(e) Measurements of d, fringe width and the index correction.

(a) Obtaining fringes :

The slit and the biprism are set on the optical bench on their respective uprights and their centres are placed at the same height. The slit must be made vertical and set near one end of the bench. The slit is then illuminated with sodium light, and the brightest part of the source is placed in front of the slit. The plane face of the biprism must be at right angles to the optical bench. By adjusting the tangent screw of the biprism stand, the slit and the base of the biprism are rendered parallel. The slit-width is reduced to a very low value.

The eyepiece is then mounted on the bench and its height adjusted so that the axis passes through the centre of the biprism. The cross-wire in the eyepiece is focused and it is moved near the biprism. On viewing through the eyepiece, indistinct fringes will be visible. If no fringe system or a patch of light is visible, then the transverse screw of the biprism is turned until they are found. By adjusting the tangent screw of the biprism further, the fringes are made as distinct as possible. Formation of well defined fringes of moderate width must be made. The biprism should be set, relative to the slit, in such a way that the lens can be brought close enough to the biprism so that magnified images of the virtual sources can be focused subsequently.

(b) Ensuring to obtain d_1 and d_2 :

After obtaining well-defined fringes, a rough value of the convex lens' focal length is obtained, which is to be used for the determination of d, the distance between the two virtual sources. The eyepiece is then slowly moved away from the biprism and the tangent screw of the biprism is continuously adjusted to keep the fringes in the field of view of the eyepiece. The eyepiece is placed at a distance of about $5f$ from the slit, where f is the focal length of the convex lens supplied.

Next the lens is mounted on the optical bench and its height is adjusted to the same as that of the eyepiece. If is made coaxial with eyepiece lens. The lens is shifted along the bench, so that for two positions of the lens, focused images of the virtual source are obtained in the field of view of the eyepiece. If the two positions of the lens for the images of the virtual source are not obtainable for any reason, then the distance between the slit and the eyepiece is altered suitably by changing the position of the eyepiece, so as to ensure getting the images at the two positions.

(c) Improvement of the fringes obtained :

To improve the visibility of the fringes, first the lens is removed and the eyepiece is slowly moved away from the biprism. The transverse screw of the biprism is continuously adjusted so that the fringes do not go out of field of view in the eyepiece during its motion. When the eyepiece is near the other end of the bench, the slit-width and the source are adjusted to get well-defined fringes. If the fringes are too wide, the biprism is moved away from the slit to narrow it down, but the fringes should never be made too close. Care should also be taken to see that the fringes never go out of the field of view.

(d) Alignment of the fringes :

To make the fringes properly aligned and the line joining the slit and the biprism parallel to the bench, an electric lamp is placed behind the slit removing the yellow source. Coloured fringes, of which the central one is almost white, are visible. The central fringe is brought on the cross-wires. To align the fringes,

(i) the eyepiece is moved towards the biprism turning the micrometer screw of the eyepiece continuously so as to keep the central fringe on the cross-wire. This is done till the eyepiece is brought fairly close to the biprism.

(ii) the eyepiece is then moved away from the biprism. The central fringe is now kept on the cross-wires by turning the transverse screw of the biprism.

The operations (i) and (ii) are repeated several times till it is found that for all positions of the eyepiece, the central fringe remains on the cross-wires. It is then ensured that the line joining the slit and the centre of the biprism is parallel to the optical bench.

This adjustment is very important.

(e) Measurement of d :

After removing the white light, sodium light (flame or vapour lamp) is placed in proper position. The eyepiece is set at a distance of about $5f$ from the slit. The lens is placed on the upright and set in between the biprism and the eyepiece. On moving the upright along with the lens along the bench, focused images of the virtual sources S_1 and S_2 are visible for two positions of the lens. The lens should be correctly adjusted to get the images of both S_1 and S_2 focused simultaneously at each of the two positions of the lens. After aligning the biprism and the slit, the micrometer screw should not be disturbed.

Next the vernier constant of the bench and the least count of the micrometer screw are determined. Keeping the eyepiece fixed in its position, the distances d_1 and d_2 between the two images of S_1 and S_2 for the two positions of the lens by setting the cross-wires on the same edge (right or left) of the images are determined. Care should be taken to avoid parallax between the images and the cross-wires. Backlash error of the micrometer screw should also be avoided. To do this, it is advisable not to take the linear scale reading attached to the micrometer but to take the initial and final readings of the micrometer screw and the number of its complete rotations. For the same position of the eyepiece, three sets of readings of the micrometer screw and the number of its complete rotations. For the same positin of the eyepiece, three sets of readings are taken for three independent settings of the lens. If possible, the observations are repeated for two other positions of the eyepiece.

Measurement of fringe width :

The lens is removed from the bench and the eyepiece is shifted to one end of the bench. The positions of the slit, biprism and the eyepiece are kept unaltered and their positions are recorded. The crosswires of the eyepiece are focused on one of the end fringes (either left or right end of the fringe). The micrometer screw is then gradually rotated in the opposite direction and the cross-wires focused on every alternate 3rd and 4th fringes and the corresponding readings of the micrometer screw are noted. The number of complete rotation of the micrometer screw during two successive readings are to be computed very carefully. The process is continued till the cross-wires reach the other end of the fringe pattern. While taking these readings, care must be taken to avoid the backlash error, and the screw must be turned always in the same direction. Thus, the width of 3 or 4 fringes are obtained and the screw is then turned back again from the extreme end reaching the initial end, taking as before the readings for every 3 or 4 fringes. In this way, widths of 3 or 4 fringes are obtained from forward and backward motion of the screw, from which the mean width of a single fringe is obtained.

Index correction :

To find the index correction, the biprism and the tube carrying the cross-wires of the eyepiece are removed from the socket. An index rod is taken and held parallel to the axis of the bench. Its one end is touched with the slit. The position of the eyepiece on the bench is adjusted until the other end of the rod is clearly visible through the eyepiece. The readings of the slit and the eyepiece are noted from the scale of the bench. The length of the index rod is measured directly by a metre scale. If l be the actual length of the index rod, M_1 and M_2 be the readings of positions of the slit and the eyepiece as obtained from the bench-scale, then the index error will be given by

$$p = l - (M_1 \sim M_2) \qquad \text{... (5.9c)}$$

Therefore, measuring D, x and d, the value of the wavelength (λ) of sodium light can be obtained from Eq. (5.9a)

Results :

(A) Determination of vernier constants of the bench stands :

[Usual records are to be made here.]

Vernier constant = ... cm

(B) Least count of the micrometer screw of the eyepiece :

[Usual records, as in case of spherometer, are to be made here.]

Least count = ... cm

(C) Measurement of distance between the virtual sources (d) :

Table 1 (See next page)

(D) Measurement of fringe-width (x) :

Table 2 (See next page)

(C) Measurement of distance (d) between the virtual sources :

Table 1

Position of eyepiece	No. of Obs.	Position of lens (cm)	C.S.R. for left image (r_1)	C.S.R. for right image (r_2)	No. of complete rotations (N)	Total [$N \times$ pitch + $(r_1 \sim r_2) \times l.c.$] = d_1 (cm)	Mean (cm)	Value of d from	Mean value of $d\left(=\sqrt{d_1 d_2}\right)$
	Readings for d_1								
I		......							
	(a)	($l \to r$)	......	......	......	......	......	I = ... cm	
	(b)	($r \to l$)	......	......	......	......	(d_1)		
II		....							
	(a)	($l \to r$)	......	......	......	......	......	II =... cm	
	(b)	($r \to l$)	......	......	......	......	(d_1)		
III		......							
	(a)	($l \to r$)	......	......	......	......	......		...... cm
	(b)	($r \to l$)	......	......	......	......	(d_1)......		
	Readings for d_2								
I		......							
	(a)	($l \to r$)	......	......	......	......	......	III = ... cm	
	(b)	($r \to l$)	......	......	......	......	(d_2)		
II		......							
	(a)	($l \to r$)	......	......	......	......	......		
	(b)	($r \to l$)	......	......	......	......	(d_2)		
III		......	......	......	......	......	......		
	(a)	($l \to r$)	......	......	......	......	......		
	(b)	($r \to l$)	......	......	......	......	(d_2)		

(D) Measurement of fringe-width (x) :

Table 2

No of obs.	Stand of position			App. value of D (cm)	Fringes serially from left	Initial C. Scale reading (r_1)	Final C. Scale reading (r_2)	No. of complete rotations (N)	Total value of width of 3 fringes $[N \times \text{pitch} + (r_1 \sim r_2) \times l.c] = R$ (cm)	Mean value of R (cm)	Fringe width $x = R/3$ (cm)
	Slit (cm)	Biprism (cm)	Eyepiece (cm)								
I.	......	......	......	...... (D_1)	1—4	...	...	...	......	......	$x_1 = \ldots$ cm
					5—8	...	...	...	......		
					9—12	...	...	...	......		
					13—16	...	...	...	......		
					...						
					...						
					17—14	...	...	...	......		
					13—10	...	...	...	......		
					9—6	...	...	...	......		
					5—2	...	...	...	......		
II.	......	......	......	...... (D_2)	1—4	...	...	...	......	......	$x_2 = \ldots$ cm
					5—8	...	...	...	......		
					etc.						
					...						
					...						
					9—6	...	...	...	......		
					5—2	...	...	...	......		
III.	......	......	......	...... (D_3)	1—4	...	...	...	......		$x_3 = \ldots$ cm
					5—8	...	...	...	......		
					etc.						
					...						
					...						
					9—6	...	...	...	......		
					5—2	...	...	...	......		

[I, II and III correspond to three different distances (D) between the slit and the eyepiece. For this only the position of the eyepiece is changed, keeping the slit and the biprism undisturbed. This is very important.]

(E) Index correction (*p*) between the slit and eyepiece :

Table 3

Length of the index rod (l) cm	Readings on the bench for		Index error $p = l - (M_1 \sim M_2)$ cm
	Slit position in cm (M_1)	Eyepiece position in cm (M_2)	

(F) Determination of wavelength (λ) of sodium light

Table 4

No. of Obs.	Apparant distance between slit and eyepiece (D cm)	Index error (p cm)	Corrected value of D in cm (= App. value + p)	Value of d in cm	Friange width (x) in cm	Wavelength (λ) in cm $= \frac{d}{D}x.$	Mean value of λ in cm
	...... (D_1)	......	......	......	...... (x_1)	......	......
	...... (D_2)	......	......	......	...... (x_2)	......	......
	...... (D_3)	......	......	......	...... (x_3)	......	......

Calculation :

From Eq. (5.9a),

$$x = \frac{D\lambda}{d}$$

or, $$\lambda = \frac{d}{D}.x.$$

Substituting the values of d, D and x, the value of the wavelength of the sodium line (λ) is determined.

Precautions :

(i) The fringes should be made neither too narrow nor too wide. By adjusting the distance of slit-biprism and slit-eyepiece, the proper width of the fringes should be made.

(ii) Focal length of the lens used should be of focal length greater than four times the distance between the slit and the eyepiece.

(iii) Proper alignment of the instrument is a must before taking readings.

(iv) For different sets of readings, the value of D should not be changed by a very large amount.

(v) Source should be placed properly in front of the slit to get the maximum illumination.

(vi) Backlash error of the movement of the micrometer screw should always be avoided.

Error Calculation :

From Eq. (5.9a),

$$\lambda = \frac{d}{D} \cdot x = \frac{\sqrt{d_1 d_2}}{D} x.$$

If R is the distance between n bands ($n = 3$ in the experiment), the

$$\lambda = \frac{\sqrt{d_1 d_2}}{D} \times \frac{R}{n}$$

The proportional error in λ is given by

$$\frac{\delta\lambda}{\lambda} = \frac{1}{2}\frac{\delta d_1}{d_1} + \frac{1}{2}\frac{\delta d_2}{d_2} + \frac{\delta D}{D} + \frac{\delta R}{R}.$$

The values of d_1, d_2 and R are measured by a micrometer screw and thus the error in their measurements is only the least count of the micrometer screw—or, twice the value, since their values are obtained from the difference of two readings. The maximum error in the value of D will be 2 × vernier constant of the bench scale.

$\therefore$ percentage error $= \frac{\delta\lambda}{\lambda} \times 100 =$ .... %.

Since x is small, the maximum contribution of error in the experiment is in the measurement of R. Hence, to find R, readings for fringes in two directions are taken and also the values of n should be as large as possible. The micrometer having a small value of least count also reduces the error.

Questions and Answers

Q.1. What is meant by 'interference of light' and what are the conditions of interference? How are the conditions fulfilled in the case of biprism?

A. : For first part of the question, see any textbook on Light. In biprism experiment, the virtual sources S_1 and S_2 are produced from the same source S by refraction through the two halves of the biprism. Hence there are two sources from which waves interference are coherent.

Q.2. Are there two prisms in a biprism?

A. : No, actually it is a single prism one of whose angles is about 179° and the other two of about 30′ each. It is regarded as a combination of two prisms placed base to base.

Q.3. What type of fringes are produced here?

A. : The fringes are non-localised and formed in space as hyperboloids of revolution with the sources S_1 and S_2 as foci. As the screen (eyepiece) is held parallel to the line joining the two sources, we get straight fringes on the screen by division of wavefronts.

Q.4. Why should the slit be narrow?

A. : The slit is made narrow to obtain a good and distinct fringe pattern with contrasts. With wide or extended slit, blurred fringes are obtained with no contrast.

Q.5. Why is the distance between the slit and the biprism made small while that between the slit and the eyepiece large?

A. : This is done to obtain broad fringes and thus to reduce the error in the measurement of fringe-width.

Q.6. How does the distance between the slit and the biprism affect the fringe-width?

A. : The distance d between the two virtual sources is related to the distance a between the slit and the biprism by the relation, $d = 2a(\mu - 1)\alpha$, where μ and α are the refractive index and angle of the biprism respectively. Thus on increasing the value of a, the value of d increases making the fringes narrower.

Q.7. What happens when the slit is illuminated by a white light?

A. : All the fringes will appear as coloured except the central fringe which is white.

Q.8. What happens when the angle of the biprism is increased?

A. : As the angle of the prism (α) increases, d increases and the fringes become narrower.

Q.9. Will the fringe-width remain the same for all colours of light?

A. : No, the fringe width is proportional to the wavelength (λ) of the light used.

Q.10. What will happen if both the slit and the edge of the prism are made horizontal?

A. : When both of them are horizontal, fringes will appear but they will extend along a vertical line and their width cannot be measured by the eyepiece moving perpendicular to the bench.

Q.11. Are the fringes equispaced?

A. : Yes, they are all equispaced.

Q.12. Is it necessary to set the cross-wires on the bright fringes?

A. : No, they can be fixed on dark fringes also but for practical conveniences, the cross-wires are fixed on the bright fringes.

Q.13. Can a convex lens of any focal length be used?

A. : No, it should be of short focal length because the slit-eyepiece distance must be greater than four times the focal length of the lens.

Q.14. What is index-correction and why is it done? Can it be eliminated?

A. : See 'Index Correction'. It can be eliminated by measuring the fringe-width at two different positions of the eyepiece upon the bench. Then subtraction of one fringe-width from the other eliminates the index error.

Q.15. Can the central fringe be located?

A. : For white light, the central fringe is white but for sodium light, all the bright fringes are of the same colour, thus making it difficult to locate the central one. However, in comparison with the white light fringes, the central fringe can be detected.

Q.16. Why only a limited number of fringes are obtained?

A. : The inhomogeneity of light sets a limit to the number of fringes. The more homogeneous a light is the greater number of fringes are obtained and thus it extends across the whole field of interference.

Q.17. Can you apply this method to measure a very small thickness, say of a mica film?

A. : Yes, if a thin mica film is introduced in the path of only one of the interfering beam using white light, then the thickness t of the mica sheet is given by $t = \frac{x.d}{D(\mu - 1)}$, where μ is the refractive index of mica and x is the displacement of the pattern on introducing the mica sheet; D and d have their usual significances.

Q.18. How do the biprism fringes differ from Newton's rings?

A. : [Discussed at the end of Newton's ring Experiment.]

[Apart from these, questions on adjustments, alignments and focusing of the biprism are to be prepared.]

5.10 Determination of the wavelength of light by Lloyd's mirror

Theory :

If a beam of monochromatic light of wavelength λ after passing through a narrow slit (S_1) is made to be incident at a grazing angle on the surface of an optically plane, front-silvered mirror M_1M_2, while the rest reaches the screen directly, then the reflected light appears to diverge from S_2—the virtual image of the source slit S_1—formed by reflection at the mirror. (Fig. 5.24). The virtual source thus acts as a coherent source and with the original source, is able to produce interference fringes in the region where the rays from the two sources overlap (shaded portions). If d be the distance between the source and its virtual image (S_1 and S_2), D being the actual distance of the screen from the source and λ be the wavelength of light used, then the fringe width (separation between two consecutive fringes) x, as shown by Lloyd, will be given by

$$x = \frac{D\lambda}{d} \text{ or, } l = \frac{d}{D}.x. \qquad \text{... (5.10a)}$$

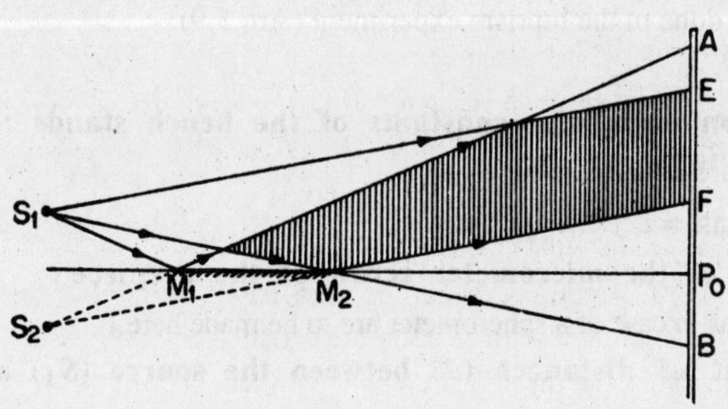

Fig. 5.24 : Lloyd's mirror

The values of d, D and x being measured by an optical bench, the wavelength of the source of light can be determined from Eq. (5.10a).

Apparatus :

Optical bench, a hightly polished and optically plane front-silvered mirror, sodium light, index rod, etc.

Lloyd's mirror is a highly polished front-silvered mirror mounted vertically on an upright set in an optical bench similar to one used in biprism experiment. A vertical slit, illuminated with the light whose wavelength is to be determined, is mounted on another upright and placed in front of the slit as in biprism experiment. The slit is adjusted close to the line of the mirror so that the light may be reflected as far as possible at grazing incidence. The fringes produced are observed in the focal plane of the focused micrometer eyepiece.

Method :

(i) Since the line joining the object and its image formed due to reflection in a plane mirror is always perpendicular to the reflecting surface, hence the reflecting mirror here must be adjusted eactly perpendicular to the plane of cross-wires of the eyepiece. The mirror must also be adjusted exactly parallel to the length of the vertical source slit.

(ii) To make the mirror (M_1M_2) normal to the plane of the cross-wires of the eyepiece, a convex lens of suitable focal length is introduced between the mirror and the eyepiece and adjusted so that its centre lies opposite to the edge M_2 of the mirror, and on a level with the centre of the slit S_1. Finally the lens is fixed in a position so as to form the real images of S_1 and S_2 in the plane of cross-wires. Both images will be in focus in the plane of the cross-wires for one setting of the lens, provided S_1S_2 is parallel to plane of the cross-wires. This is achieved by rotating M_1M_2 slightly about the vertical axis of rotation until the two images are exactly in focus simultaneously.

(iii) Then on removing the lens, fringes are observed through the eyepiece. The mirror M_1M_2 is rotated about its horizontal axis to get the position when the fringes become well defined and brightest. The adjustment of M_1M_2 normal to the plane of the cross-wires may then again be tested by the absence of the lateral shift in the fringes as the eyepiece is moved backward or forward along the optical bench.

(iv) With the preliminary setting of the mirror being thus completed, arrangements for measuring the distance (d) between the source and its virtual image, the slit-eyepiece distance (D) and the fringe-width (x) are being done.

(v) To measure the above three quantities, methods are carried out exactly in the same manner as are done in the biprism experiment (Art. 5.9).

Results :

(A) Determination of vernier constants of the bench stands :

[Usual records are to be taken here.]

∴ vernier constant = ... cm

(B) Least count of the micrometer screw of the eyepiece :

[Usual records as in case of a spherometer are to be made here.]

(C) Measurement of distances (d) between the source (S_1) and its virtual image (S_2) :

Table 1

[Same table as Table 1 of Expt. No. 5.9 (C) is to be made here.]

(D) Measurement of fringe-width (x) :

Table 2

[Same table as Table 2 of Expt. No. 5.9 (D) is to be made here.]

(E) Index correction (p) between the slit and the eyepiece :

Table 3

[Same table as Table 3 of Expt. No. 5.9(E) is to be made here.]

(F) Determination of wavelength (λ) of sodium light :

Table 4

[Same table as Table 4 of Expt. No. 5.9 (F) is to be made here.]

Calculation :

From Eq. : (5.10a), $\lambda = \frac{d}{D}.x$

Substituting the value of d, D and x, the value of the wavelength of sodium light (λ) is determined.

Precautions :

(i) The most important precaution in this experiment is the proper setting of the mirror to satisfy the working formula. These are as discussed in (i), (ii) and (iii) under 'Method' and those operations must be very carefully done.

(ii) The mirror taken must be one having silvering on its front surface to avoid multiple internal reflections.

(iii) When the eyepiece is kept at a distance from the mirror, the point P_0 of the screen, equidistant from S_1 and S_2 receives only direct light from S_1 and not the reflected light. Thus the central fringe of zero geometrical path difference is not visible at that position. Even less than half the interference pattern is visible with the arrangement as shown in Fig. 5.13. Of course, half the central fringe can be brought into view by displacing the screen towards the mirror and just touching its end (M_2). Then the edge M_2, which is equidistant from S_1 and S_2, passes through the centre of the fringe of zero path difference. The fringe appears to be dark instead of being a bright one. This is only because the reflected beam suffers a sudden phase change of π on reflection. Before taking readings, appearance of a central dark fringe on touching the eyepiece with the edge M_2 of the mirror, must be ascertained.

[Other general precautions discussed at the end of the biprism experiment (Art. 5.9) are to be adopted also.]

Error Calculation :

The same calculations as are done at the end of the biprism experiment (Art. 5.9) are to be done here and the percentage error is calculated.

Questions and Answers

[Basic questions on interference are to be discussed apart from the particular questions on this experiment given below.]

Q.1. How fringes are produced in Lloyd's mirror?

A. : See 'Theory' of the experiment.

Q.2. Do the two interfering beams have equal amplitudes?

A. : Interference takes place between the direct and the reflected beams. These beams have almost nearly equal amplitudes, because at grazing incidence almost the entire incident light is reflected.

Q.3. What steps are taken to have a good contrast of the fringes?

A. : Since the amplitudes are nearly equal, automatically good contrast is obtained. However, keeping the slit narrow, more contrast is obtained.

Q.4. What steps are taken to obtain a large fringe-width?

A. : The slit is kept close to the mirror and the eyepiece at a large distance from the slit.

Q.5. Why the cmoplete band is not obtained?

A. : See 'Precautions' (iii).

Q.6. Is it possible to get the complete system?

A. : Yes, by introducing a thin lamina of glass or mica in the path of direct beam, the centre of the pattern is shifted upwards and the complete system becomes visible.

Q.7. Is the central fringe observable?

A. : No. See 'Precaution' (iii).

Q.8. Why the central fringe is dark when it is made visible by introducing a thin sheet of glass or mica in the path of the direct beam?

A. : See 'Precaution' (iii).

Q.9. What happens with a white light instead of sodium light?

A. : With white light, the central fringe remains dark as before. As one goes outwards, a few coloured fringes are obtained and then a general illumination.

Q.10. What are the differences between biprism fringes and these fringes?

A. : In biprism, complete pattern is obtained, the central fringe is bright and less sharp, the distances between the different pairs of coherent sources are the same; while in this experiment, only a few fringes on one side of the central fringe is visible, the central fringe is dark and sharp, and the distances between the different pairs of coherent sources being different, the fringe-width is different for different pairs of sources unlike biprism fringes.

Q.11. What is the difference between the coherent sources of biprism and Lloyd's mirror experiments?

A. : In biprism the coherent sources are the virtual images of an actual source but in this experiment, the actual source and its virtual image constitute the coherent sources.

Q.12. Is this property of fringes used anywhere?

A. : Yes, in the production of achromatic fringes.

Q.13. What important conclusion can be drawn from this experiment?

A. : The central fringe in this experiment being drak instead of being bright, it proves conclusively that a phase change of π is created by reflection from the surface of a denser medium like the metallic surface of the mirror.

5.11 Determination of the Radius of Curvature of the convex surface of a plano-convex lens by Newton's Ring method

Theory :

When a parallel beam of monochromatic light of wavelength λ coming from an extended source is incident on a wedge-shaped film of air enclosed between a glass plate G and a plano-convex lens L, then a part of the beam is reflected from the surface of the lens along ABC (Fig. 5.25) and the other part after refraction through the film is reflected back from a point (D) on the plate surface. These two parts of the beam are in a condition to interfere and give rise to a system of dark and bright rings alternately. Whether the point A will appear bright or dark depends on the actual phase difference between the beams. The point of contact (O) between the lens and the plate will be the centre of the rings formed. With O as the centre and OA as the radius, if a circle is drawn then all the points lying on the circumference of this circle will appear either dark or bright. Thus a bright or a dark ring will be formed with OA as its radius. These rings are known as 'Newton's rings'. These are formed alternately as dark and bright rings.

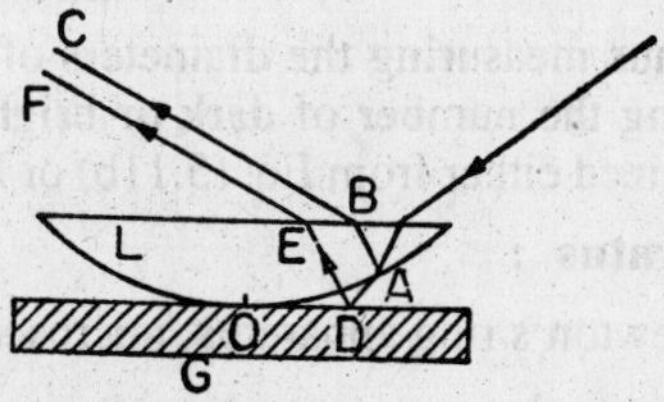

Fig. 5.25 : Formation of Newton's rings

If θ be the angle of incidence of the rays falling on the plate, μ be the refractive index of the film between the lens and the plate and D_n be the diameter of the n-th bright or dark ring, then

$$\frac{D_n^2}{4} = \frac{n\lambda R}{\mu \cos\theta} \qquad \text{...... (for dark rings)}$$

and $$\frac{D_n^2}{4} = \frac{(2n+1)}{2} \times \frac{n\lambda R}{\mu \cos\theta}, \qquad \text{...... (for bright rings)}$$

where R is the radius of curvature of the lower surface of the lens. For normal incidence, $\theta = 0$ and with $\mu = 1$, (for air) in between the lens and the plate, we have

$$\left.\begin{aligned} D_n^2 &= 4n\lambda r & \text{......(for dark rings)} \\ \text{and} \quad D_n^2 &= 2(2n+1)\,\lambda R. & \text{......(for bright rings)} \end{aligned}\right\} \quad \text{... (5.11a)}$$

Thus knowing the value of μ and measuring the diameter of the n-th dark or bright ring, the value of R, the radius of curvature of the convex surface of the lens in contact with the plate, may be detemined.

Now, since the rings very close to the central dark point are not perfectly circular, Eq. (5.11a) is slightly modified and used in practice.

Let D_{m+n} and D_n be the diameters of the $(m + n)$-th and n-th rings (dark or bright) respectively.

Thus from Eq. (5.11a),

$$\left.\begin{aligned} D^2_{m+n} &= 4(m+n)\lambda R \\ \text{and} \quad D^2_{m+n} &= 2[2(m+n)+1]\lambda R \end{aligned}\right\}$$

$$\left.\begin{aligned} \text{Again,} \quad D_n^2 &= 4n\lambda R \\ \text{and} \quad D_n^2 &= 2(2n+1)\lambda R. \end{aligned}\right\}$$

Hence, for dark rings,

$$D_{m+n}^2 - D_n^2 = 4(m+n)\lambda R - 4n\lambda r$$
$$= 4m\lambda R \qquad \dots (5.11b)$$

and for bright rings.

$$D_{m+n}^2 - D_n^2 = 2[2(m+n)1]\lambda R - 2(2n+1)\lambda R$$
$$= 4m\lambda R. \qquad \dots (5.11c)$$

Thus measuring the diameters of the $(m + n)$-th and the n-th bright or dark rings and counting the number of dark or bright rings lying between them, the value of R may be determined either from Eq. (5.11b) or Eq. (5.11c) provided the value of λ is known.

Apparatus :

Newton's ring apparatus, microscope, source of light.

Newton's ring apparatus consists of a plano-convex lens (L) placed on an optically plane glass plate (G) with its curved surface being in contact with the plate. The radius of curvature of this curve surface has to be determined, which is normally very large (Fig. 5.26). A bunsen flame illuminated with sodium salt or a soidum vapour lamp (S) is placed at the focal plane of another lens L' thereby making the beam parallel. Another clean glass plate (P) is placed in the path of a beam, inclined to the vertical at an angle of 45°. The rays after reflection from this plate, fall on the combination of the lens L and the glass plate G normally and after reflection, again pass through P and are viewed through a travelling microscope M. The microscope magnifies the system of rings which forms in the space in between the lens L and the plate G.

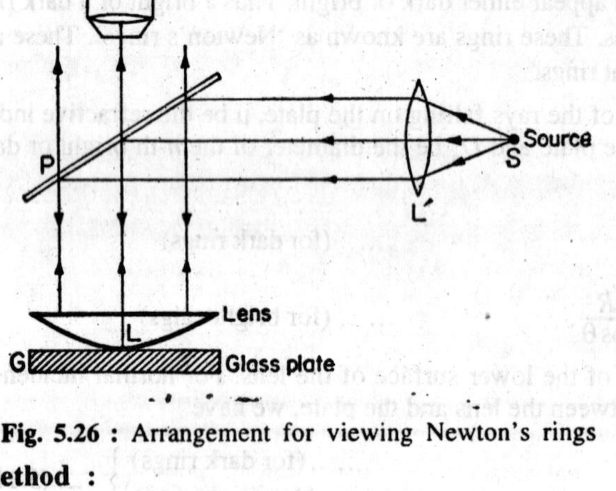

Fig. 5.26 : Arrangement for viewing Newton's rings

Method :

(i) The travelling microscope is levelled and its axis is made vertical. It is focused in the usual manner and the vernier constant of the vernier scale attached to it is determined.

(ii) Both the lens (L) and the glass plate (G) are cleaned and they are set in the manner, already stated, in such a way that clear and distinct Newton's rings are visible. The rings are focused within the field of view of the microscope.

(iii) The rings are made distinct and equally bright by adjusting the source (S), the lens (L') and the glass plate P.

(iv) The cross-wires of the microscope are at first set with the central dark spot of the ring system and they are set in such a manner that the vertical cross-wire remains as a tangent to the rings formed.

(v) Next the microscope is moved on one side so that the vertical cross-wire touches tangentially a bright ring far away from the centre. The number of this bright ring, counted from the centre, is also ascertained. At the focused condition, readings of the microscope (main scale and the vernier scale) are noted.

(vi) The microscope is now gradually moved towards the centre of the ring system, and readings for three or four consecutive rings are taken. Next, a number of intermediate rings are passed off and again readings for three or four consecutive rings are taken. In this way, readings are taken gradually sliding the microscope towards the centre of the ring but readings for a few rings around the centre should not be taken.

(vii) The movement of the microscope is continued till it passes over to the other side of the central spot. Avoiding three or four rings near the centre, again readings are taken for rings as before as the microscope gradually moves away from the centre. While taking all these readings, the number of the rings whose readings are taken, must be noted down.

(viii) The whole process is repeated bringing back the microscope in the reverse direction and taking readings for the same rings as before.

(ix) While taking these readings, special care has to be taken to avoid backlash error of the microscope screws.

(x) Determining thus the diameters (D) of the rings and the difference of ring numbers (m), the value of R can be found out employing Eq. (5.11c).

Result :

(A) Determination of the Vernier constant :

One small main scale division	= .... cm
Number of vernier divisions	= .... cm
Vernier constant	= .... cm

(B) Wavelength of the light used (λ) = ... cm

Caltulation :

Wavelength of the light used (λ)	= .... cm
Difference of ring numbers (m)	= ....
Value of $D_{m+n}^2 - D_n^2$	= .... sq. cm
Radius of curvature = (R) = $\dfrac{D_{m+n}^2 - D_n^2}{m}$	= .... cm

(C) Measurement of the diameters of the rings :

Table 1

[Given ring numbers are hypothetical]

Ring Number	Observation	Microscope reading when the cross-wire touches						Diameter of the ring ($D = X_1 \sim X_2$) cm	Mean diameter (D) cm
		Left end of the ring			Right end of the ring				
		Main Scale (cm)	Vernier (cm)	Total (X_1 cm)	Main Scale (cm)	Vernier (cm)	Total (X_2 cm)		
28th	$l \to r$	...	...	...	...	...	...	...... (left)	......
	$r \to l$	...	...	...	...	...	...	...... (right)	
27th	$l \to r$	...	...	...	...	...	...	...... (left)	......
	$r \to l$	...	...	...	...	...	...	...... (right)	
26th	$l \to r$	...	...	...	...	...	...	...... (left)	......
	$r \to l$	...	...	...	...	...	...	...... (right)	
etc.									
...									
...									
8th	$l \to r$	...	...	...	...	...	...	...... (left)	......
	$r \to l$	...	...	...	...	...	...	...... (right)	
7th	$l \to r$	...	...	...	...	...	...	...... (left)	......
	$r \to l$	...	...	...	...	...	...	...... (right)	
6th	$l \to r$	...	...	...	...	...	...	...... (left)	......
	$r \to l$	...	...	...	...	...	...	...... (right)	

(D) Determination of the Radius of Curvature (R) of the lens :

Table 2

Ring Number	Diameter of the ring (D) (cm)	Square of the distance (D^2) (sq. cm)	Difference of the ring numbers (m)	$\left(D_{m+n}^2 - D_n^2\right)$ (sq. cm)	Mean value of $D_{m+n}^2 - D_n^2$	Radius of Curvature $R = \frac{D_{m+n}^2 - D_n^2}{4m\lambda}$ (cm)
28th	...	...	...	...		
27th	...	...	...	...		
26th	...	...	...	...		
etc.		$\left(D_{m+n}^2\right)$	...	...	...	...
...			...	...		
...			...	...		
...			...	...		
8th	...	...	...	...		
7th	...	...	...	...		
6th	...	... $\left(D_n^2\right)$	...	...		

Determination of R from graph :

A curve can be drawn with the square of the diameters of the rings (D^2) as ordinate and the corresponding ring number as abscissa. The graph according to the Eq. (5.11a) will be a stright line as shown in Fig. 5.27. Taking any two points P and Q on the graph as far apart

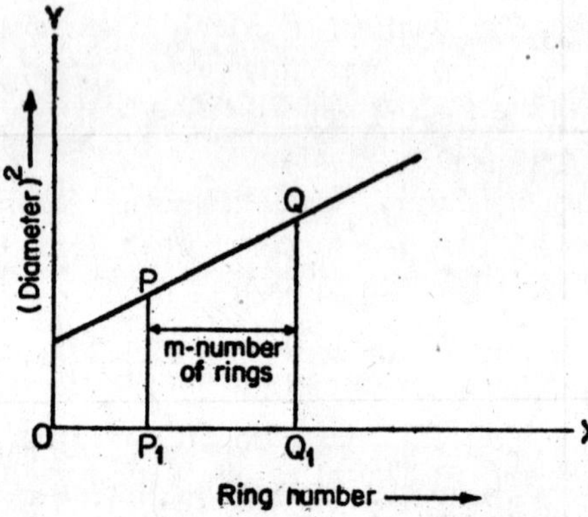

Fig. 5.27 : Ring No.-(Diameter)^2 graph

as possible, the values of D^2_{m+n} and D^2_n will be obtained from their ordinates. If there are m number of rings in between these two points ($m = P_1Q_1$), then the slope (M) of this straight line will be given by

$$M = \frac{D^2_{m+n} - D^2_n}{m}$$

and thus the radius of curvature (R) will be

$$R = \frac{M}{4\lambda}.$$

Finding the value of M from the graph and knowing the value of λ, the value of R can, therefore, be found out.

Precautions :

(i) The set-up must not be disturbed in any way during experimentation.

(ii) The microscope must always be moved in one direction in a particular set while taking the readings, so as to avoid any backlash error.

(iii) A few rings near the centre of the pattern should be avoided while taking readings.

(iv) It is convenient to take the difference or ring numbers (m) as 10.

(v) The lens (L) and the glass plate (G) should be set in such a way that the central point appears as dark.

(vi) The success of this experiment depends on the proper illumination arrangement. Accordingly, the source, the plate P and the lens L' should be appropriately set to have the maximum and proper illumination incident on the lens-plate system.

Error Calculations :

From the relation $R = \dfrac{D^2_{m+n} - D^2_n}{4m\lambda}$, the maximum proportional error in R is given by

$$\frac{\delta R}{R} = \frac{\delta y}{y} + \frac{\delta \lambda}{\lambda}, \text{ where } y = D_{m+n}^2 - D_n^2.$$

Since there is least possibility in having any error in the value of λ of the light used, the only source of error in this experiment is in the measurement of diameters of the ring. In order to reduce that error, the value of y should be large, i.e., the expression should be large. This can be achieved by taking the difference of ring numbers (m) large, and making n as small as practicable.

Thus the percentage error in the measurement of R is

$$\frac{\delta R}{R} \times 100 = \ldots\ldots\ldots\ \%.$$

Questions and Answers

Q.1. What is meant by 'interference of light'?

A. : See any textbook on 'Light'.

Q.2. What are Newton's rings and why are they so called?

A. : The rings formed by the arrangement so described in the experiment are called Newton's rings. Although those were first observed by Hooke in 1665 but it was Newton who studied them in details and measured their diameters. Hence, those are called Newton's rings.

Q.3. Why is the central spot dark?

A. : The central ring with reflected light is dark due to the change of phase of π by reflection from the denser medium like the glass plate.

Q.4. What happens if white light is used in the experiment?

A. : With white light only a few coloured fringes are visible fading into general illumination.

Q.5. Where are the rings formed?

A. : Rings are formed in the air film enclosed between the glass plate and the lens.

Q.6. Why are the fringes circular?

A. : Each fringe is the locus of points at which the thickness of the air film is constant. This thickness increases as we go outwards from the centre, so that points of equal thickness lie on circles with the point of contact as the centre. Thus the fringes are circular.

Q.7. Is it possible to get the central spot bright?

A. : Yes, with transmitted light, the centre of the rings will appear as bright.

Q.8. What type of fringes are obtained here?

A. : These are the fringes of equal thickness.

Q.9. Sometimes the fringes do not appear as perfectly circular but elliptical. Why?

A. : If the glass plate (G) is not perfectly flat or there are dust particles on the lens and plate surfaces, then the fringes may not appear as circular.

Q.10. Why do you hold the glass plate (P) at 45° to the vertical?

A. : So that the rays may be incident normally on the glass plate-lens system.

Q.11. Why should a lens of large radius of curvature be used in this experiment?

A. : There are two reasons. First, greater the radius of curvature of the lens, greater would be the diameters of the rings making less percentage error in the experiment. Secondly, with a large value of R, the angle enclosing the air-film is very small, as assumed in the theory. The lens must also be very thin.

Q.12. Why is an extended source necessary in this experiment?

A. : That enables each point of the film being seen by light from a different point of the sources so that the entire film can be seen for one position of the eye.

Q.13. On what factors do the diameters of the ring depend?

A. : They depend on (i) the wavelength of light used, (ii) refractive index of the film, (iii) the radius of curvature of the lens surface in contract with the plate.

Q.14. Are the rings equispaced?

A. : No, with the increase of the diameters, the rings get closer and closer.

Q.15. What happens if a little liquid is introduced between the lens and the plate?

A. : The rings contract in the ratio of $\frac{1}{\sqrt{\mu}}$, where μ is the refractive index of the liquid.

Q.16. What will happen if the light is incident obliquely and not normally?

A. : The diameters of the rings will increase because of the decrease of path difference at a given point.

Q.17. What will happen if the lens is placed on a plane mirror instead of a glass plate?

A. : The fringes will disappear because of the reflection of the transmitted rays from the lower surface of the mirror and for superposition on the fringes caused by the original condition.

Q.18. If the plane side of the plano-convex lens is placed on the glass plate, what will happen?

A. : Straight bands, instead of rings, will be obtained for the imperfectness of the contact surface.

Q.19. Can the radius of curvature be measured by a spherometer in the usual way?

A. : No, because the radius of curvature being very large in this case, the elevation of the central screw (h) of the spherometer will be very small, causing an appreciable error in the measurement.

Q.20. Can the refractive index of a liquid be measured by this experiment?

A. : Yes, by introducing the liquid in the space between the lens and glass plate (a few drops), diameters of the rings are measured once with air as medium and once with the liquid being interposed. The value of μ of the liquid will be given by

$$\mu = \frac{\left(D_{m+n}^2 - D_n^2\right) \text{air}}{\left(D_{m+n}^2 - D_n^2\right) \text{liquid}}.$$

Q.21. What are the differences between Newton's rings and biprism fringes?

A. : Biprism fringes are formed by division of wavefront, while Newton's rings are formed by division of amplitude.

5.12 Measurement of the width of a narrow slit by the diffraction bands produced by it

Theory :

When a beam of monochromatic light of wavelength λ, is incident normally on the surface of a narrow slit (SS) of width a, then diffraction of the rays takes place by the slit.

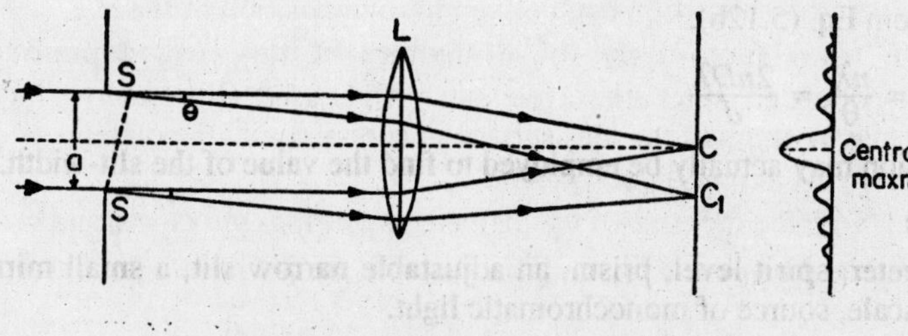

Fig. 5.28 : Diffraction through single slit

If the transmitted light is viewed through a telescope (*T*) the diffraction bands will be visible which consist of alternative bright and dark fringes (Fig. 5.28). These bands will be formed on the either side of a central diffraction maximum (*C*). If θ be the angle between the direct beam and the *n*-th dark band on either side of the central maximum, *a* be the width of the slit (*SS*) and λ be the wavelength of light used, then

$$a \sin \theta = n\lambda. \qquad \dots (5.12a)$$

As θ is very small, then θ ≃ sin θ and hence,

$$a\theta = n\lambda$$

$$\text{or,} \quad a = \frac{n\lambda}{\theta}. \qquad \dots (5.12b)$$

Hence, knowing λ and measuring θ for the *n*-th order, the value of the slit-width (*a*) can be obtained.

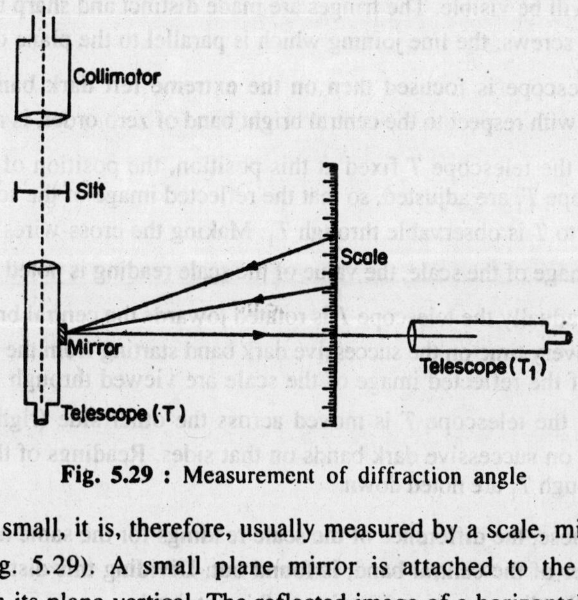

Fig. 5.29 : Measurement of diffraction angle

As θ is very small, it is, therefore, usually measured by a scale, mirror and telescope arrangement (Fig. 5.29). A small plane mirror is attached to the telescope of the spectrometer with its plane vertical. The reflected image of a horizontal scale, placed at a distance *D* from the mirror, is viewed through a second telescope (T_1). If *d* be the linear shift of the scale division coincident with the cross-wires of T_1 when the telescope *T* is turned through an angle θ, then

$$\theta = \frac{d}{2D}. \qquad \text{... (5.12c)}$$

Hence, from Eq. (5.12b),

$$a = \frac{n\lambda}{\theta} = \frac{2nD\lambda}{d}. \qquad \text{... (5.12d)}$$

This relation may actually be employed to find the value of the slit-width.

Apparatus :

Spectrometer, spirit level, prism, an adjustable narrow slit, a small mirror, telescope with a metre scale, source of monochromatic light.

Method :

(i) The spectrometer is levelled, focused and adjusted for experiment in the usual manner. The source is also properly adjusted.

(ii) The slit whose width is to be measured is generally attached to a frame. The frame is set on the prism-table in such a way that the rays coming from the collimator are incident normally on one of its face. The slit and the frame must remain in a vertical plane and the central axis of the slit passes through the centre of the prism-table.

(iii) A small circular plane mirror is attached to the axis of the telescope T by means of some wax in such a way that its reflecting plane remains vertical. A metre scale, a telescope (T_1) and a source of light are placed at a distance from the axis of the telescope T in such a way (Fig. 5.29) that the readings of the illuminated scale, after reflection from the mirror, enter the field of view of the telescope T_1.

(iv) Narrowing down the collimator slit, if viewed through the telescope T, diffraction fringes will be visible. The fringes are made distinct and sharp by turning one of the two prism-table screws, the line joining which is parallel to the plane of the slit.

(v) The telescope is focused then on the extreme left dark band and the order number of the band with respect to the central bright band of zero order, is noted.

(vi) Setting the telescope T fixed at this position, the position of the lamp, metre scale and the telescope T_1 are adjusted, so that the reflected image of the scale reading from the mirror attached to T is observable through T_1. Making the cross-wires of T_1 coincident with the reflected image of the scale, the value of the scale reading is noted.

(vii) Now gradually the telescope T is rotated towards the central bright band and its cross-wires are set every time on the successive dark band starting from the left end and each time the readings of the reflected image of the scale are viewed through T_1 and are noted down. In this way, the telescope T is moved across the other side (right) of the central bright band and set on successive dark bands on that sides. Readings of the scale for each band as viewed through T_1 are noted down.

(viii) From these, the difference of the scale readings for the same number (order) of band (n-th) on either of the central band, is found out. Dividing this distance by twice the order number ($2n$), the distance corresponding to θ is obtained.

(ix) Taking different values of n and for each case, by finding the distance between two consecutive bands (θ), the mean value of the difference of scale readings (d) for any two consecutive dark bands, is obtained.

(x) Finally, the value of D is measured thrice by a scale and its mean value is determined. Knowing the value of λ, thus the value of a can be obtained from Eq. (5.12d).

(xi) The experiment can be repeated by taking another set of readings while the telescope T is initially set at the extreme right end dark band and moving it gradually towards left, crossing the central bright band and reaching the extreme left dark band.

Results :

Wavelength of light used (λ) $= \ldots\ldots\ldots$ cm

Distance of scale from the mirror $= \dfrac{D_1 + D_2 + D_3}{3}$

$= D$ cm

(A) Slit-width measurement from diffraction bandwidths :

Table 1 (See next page)

(B) Slit-width by microscopic measurements :

Table 2 (See next page)

Vernier constant of the travelling microscope = ... cm

Calculation :

From Eq. (5.17d),

$$a = \frac{2nD\lambda}{d}.$$

But since from the experiment, the value of d is obtained for the difference of one band, the vlaue of $n = 1$ in Eq. (5.12d). Thus putting $n = 1$, we get

$$a = \frac{2D\lambda}{d} = \ldots\ldots\ldots \text{ cm}$$

Substituting the values of D, λ and d, the slit-width a is obtained spectroscopically. The value of a so obtained is verified by microscopic measurement as shown in Table 2.

Precautions :

(i) The sources, collimator slit and the given slit must be properly aligned and the source must be sufficiently bright in order to get sharp bands.

(ii) The line joining the scale and the mirror must be perpendicular to both of them.

(iii) The value of D should be fairly large (about 2 metres or more) so as to give an appreciable shift of scale reading.

(iv) Readings must be taken for dark bands, not for the bright bands. Otherwise the theory will not be valid.

(v) It may be noted that spectrometer scale readings need not be taken in this method.

Error Calculation :

From equation, $a = \dfrac{2D\lambda}{d}$ and since $d = \dfrac{y}{6}$ (see Table 1) we have proportional error in a as

$$\frac{\delta a}{a} = \frac{\delta D}{D} + \frac{\delta y}{y}.$$

(A) Slit-width measurement from diffraction bandwidths :

Table 1

Order number of dark bands with central bright band as zero order	Scale readings in telescope T_1 when telescope T is placed on the dark bands and moved			Difference of mean scale readings as seen through T_1 for 6 dark bands (y cm)	Mean value of y (cm)	Width of a single dark band $\left(d = \frac{y}{6}\right)$(cm)	Slit-width $a = \frac{2D\lambda}{d}$
	from left to right (M') (cm)	from right to left (M'') (cm)	Mean $M = \frac{M' + M''}{2}$(cm)				
5	...	...	...... (M_1)	..... $(M_7 - M_1)$			
4	...	...	...... (M_2)	..... $(M_8 - M_2)$			
3	...	...	...... (M_3)	..... $(M_9 - M_3)$			
2	...	...	...... (M_4)	..... $(M_{10} - M_4)$			
1	...	...	...... (M_5)	..... $(M_{11} - M_5)$			
Central bright band					......	......	......
1	...	...	...... (M_7)				
2	...	...	...... (M_8)				
3	...	...	...... (M_9)				
4	...	...	...... (M_{10})				
5	...	...	...... (M_{11})				

(B) Slit-width by microscopic measurements :

Table 2

Vernier constant of the travelling microscope = ... cm

No. of Obs.	Microscope readings for						Width of the slit ($a = A_1 \sim A_2$) cm	Mean width of the slit from	
	Left edge of slit			Right edge of slit				Microscopic measurements (cm)	Spectroscopic measurements (cm)
	Main Scale (cm)	Vernier (cm)	Total (A_1 cm)	Main Scale (cm)	Vernier (cm)	Total (A_2 cm)			
1.	...	...	...	...	...	...	......		
2.	...	...	...	...	...	...	......		
3.	...	...	...	...	...	...	......		
4.	...	...	...	...	...	...	......		

As D is fairly large, hence $\frac{\delta D}{D}$ is very small and its contribution to the error is insignificant. The only source of error lies in the measurement of y, the difference of two readings for a certain number (n) of bands. So the proportional error is minimum in the experiment as y is made larrge.

$\therefore$ percentage error $= \frac{\delta a}{a} \times 100\ \% = \ldots\ldots\ldots\ \%$.

Questions and Answers

Q.1. What is meant by diffraction of light?

A. : See any textbook.

Q.2. What is the difference between interference and diffraction?

A. : See any textbook.

Q.3. What is the difference between Fresnel type and Fraunhoffer type of diffraction?

A. : In Fresnel type, both the source and the screen lie near the diffracting edge, that is at finite distance. But in Fraunhoffer type, both of them lie at infinite distance and as such the rays are incident on the diffracting obstacle parallel. In this case, the diffraction is of Fraunhoffer type.

Q.4. If the slit is of large width, what will happen?

A. : Then the angular separation between the dark bands will decrease and finally diffraction bands will not be visible.

Q.5. If white light is used instead of a monochromatic beam, what will happen?

A. : Then the bands will appear coloured, the central one being white.

Q.6. Why do we measure the separation between the dark bands and not between the bright bands?

A. : The angular separation between the adjacent dark bands remain equal but not between the bright bands. Hence, for width between the bright bands, the working formula used will not remain valid.

Q.7. What happens when the collimator slit-width is increased?

A. : The fringes become indistinct.

Q.8. What is the approximate width of the collimator slit permissible?

A. : If the collimator lens has a focal length f, the slit-width to be measured be a and λ be the wavelength of light used, then the maximum width of the collimator slit for the fringes to remain distinct is given by the expression, $\frac{\lambda f}{4a}$.

5.13 Measurement of the slit-width and the separation of two identical narrow slits by observing the diffraction and interference bands produced by them

Theory :

If a parallel beam of monochromatic light of wavelength λ coming out of the collimator of a spectrometer, falls normally on a plane containing two identical slits (AB and CD) of width a, separated by an opaque space of width b (BC), then the transmitted light when viewed through the telescope focused for infinity, will show a number of diffraction fringes similar to that due to a single slit (Fig. 5.30). On the diffraction bands will be superposed an interference system (fringes) of equal width. If θ be the angular separation between two successive dark bands of diffraction pattern and ϕ be the angular separation between two successive interference, maxima or minima, then

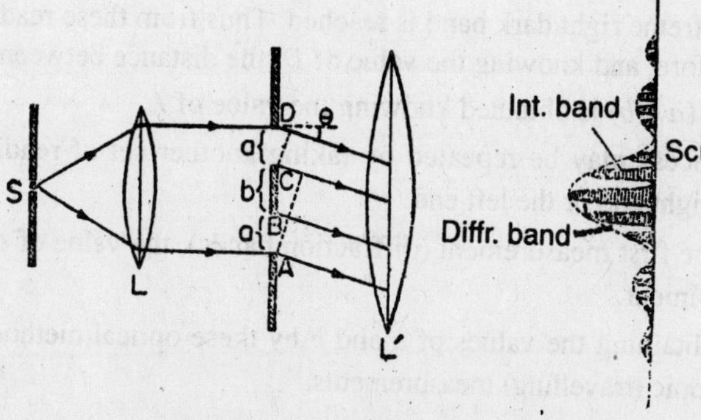

Fig. 5.30 : Double slit fringes

$\lambda = a \sin \theta$ for diffraction fringes

and $\lambda = (a + b) \sin \phi$ for interference fringes.

As both θ and ϕ are very small, $\sin \theta = \theta$ and $\sin \phi = \phi$, and hence,

$$\lambda = a\theta \text{ and } \lambda = (a + b)\,\phi. \quad \text{... (5.13a)}$$

Now, if D be the distance between the scale and the mirror attached with the telescope of the spectrometer (the arrangement as done in the single slit experiment), and if d and d_1 be corresponding displacements of the scale-image when the cross-wires coincide with the successive dark bands of the diffraction and interference patterns respectively, then

$$\theta = \frac{d}{2D} \text{ and } \phi = \frac{d_1}{2D} \quad \text{... (5.13b)}$$

$$\left.\begin{aligned} \text{and thus, } a &= \frac{\lambda}{\theta} = \frac{2D\lambda}{d} \\ \text{and } (a+b) &= \frac{\lambda}{\phi} = \frac{2D\lambda}{d_1}. \end{aligned}\right\} \quad \text{... (5.13c)}$$

Hence, measuring the values of D, d and d_1 and knowing λ, the values of, a, $(a + b)$ and hence, of b can be calculated from Eq. (5.13a).

Apparatus :

Spectrometer, spirit level, prism, double slit, small mirror, telescope with a metre scale, a source of monochromatic light.

Method :

(i) All adjustments to be made are exactly the same as are done in the experiment with single slit (**Article 5.12**). Only here the double slit with the frame has to be placed on the prism-table instead of the single slit.

(ii) As in case of single slit experiment, the angular separation between successive dark bands of the diffraction pattern is measured with the help of the scale-telescope arrangement (Fig. 5.29).

(iii) To measure the value of $(a + b)$, the interference fringes visible in the region occupied by the central diffraction band are used. The telescope of the spectrometer is focused on the extreme left dark band (interference) and the cross-wires are set on it. Taking this as the first order, the telescope is successively focused on the second, third, etc. dark bands and scale readings as seen through the second telescope, are noted. The process is continued till the extreme right dark band is reached. Thus from these readings, the value of d_1 is obtained as before, and knowing the value of D, the distance between the scale and the mirror, the value of $(a + b)$ is obtained knowing the value of λ.

(iv) The process may be repeated by taking another set of readings, rotating the telescope from the right end or the left end.

(v) From the first measurement (diffraction bands), the value of a is obtained as in the single slit experiment.

(vi) After obtaining the values of a and b by these optical methods, the results are verified by microscopic (travelling) measurements.

Results :

Wavelength of the light used (λ) = ... cm

Distance of the scale from the mirror (D) = $\dfrac{D_1 + D_2 + D_3}{3}$ = ... cm

(A) Slit-width (a) measurement from diffraction pattern :

Table 1

[A table similar to Table 1 in item (A) of Expt. 5.12 is to be made here.]

(B) Measurement of $(a + b)$ from interference pattern :

Table 2 (See next page)

(C) Comparison of measurements by vernier microscope :

Table 3 (See next page)

Vernier constant of the microscope = ... cm

Calculations :

Width of the slit from Eq. (5.13c),

$$a = \frac{2D\lambda}{d} = \text{........ cm}$$

and width of $(a + b) = \dfrac{2D\lambda}{d_1}$ = ... cm

Hence, width of the obstacle, $b = (a + b) - (a)$ = ... cm

Precautions :

(i) To get appreciable displacement of scale-image, the distance of the scale from the mirror (D) should be fairly large (about 3 metres or more).

(ii) To get a good number of bands, the source used must be very strong and properly set. Either sodium vapour lamp or good quality coloured glass filters in combination with a high power electric filament lamp may be used.

(iii) The opening of the slits may be of the order of 0.3 mm and the width of the obstacle about 0.8 mm. These values give a good number of bands with appreciable separations.

(B) Measurement of $(a + b)$ from interference pattern :

Table 2

Order number of dark Int. bands from left	Scale readings in telescope T_1 when telescope T is placed on the dark int. fringes and moved			Difference of mean scale readings as seen through T_1 for 5 dark fringes (y) (cm)	Mean value of y (cm)	Width of a single dark fringe $\left(d_1 = \frac{y}{5}\right)$ (cm)	Value of $(a+b)$ $= \frac{2D\lambda}{d_1}$ (cm)
	from left to right (M') (cm)	from right to left (M'') (cm)	Mean $M = \frac{M'+M''}{2}$ (cm)				
1	...	...	...... (M_1)	$(M_6 \sim M_1)$ = ...			
2	...	...	...... (M_2)	$(M_7 \sim M_2)$ = ...			
3	...	...	...... (M_3)	$(M_8 - M_3)$ = ...			
4	...	...	...... (M_4)	etc.	......	......	......
5	...	...	...... (M_5)				
6	...	...	...				
...	...	...	...				
...	...	...	...				

[It is not essential to take 5 bands while taking the difference. This will depend on the number of bands available.]

(C) Comparison of measurements by vernier microscope :

Table 3

Vernier constant of the microscope = ... cm

	No. of Obs.	Microscope reading in cm for the						Width ($= T_2 \sim T_1$) (cm)	Mean width (cm)	Width in cm	
		Left edge			Right edge					From microscopic measurement	From spectroscopic measurement
		Main Scale (M)	Vernier (V)	Total ($T_1 = M + V$)	Main Scale (M)	Vernier (V)	Total ($T_2 = M + V$)				
1st Slit	1.	......	......	......	......	......	......			$\frac{a_1 + a_2}{2}$	$a = ...$
	2.	......	......	......	......	......	......	......	......	= ...... and	(from Table 1)
	3.	......	......	......	......	......	......	......	(a_1)		
Obstacle	1.	......	......	......	......	......	......	......			
	2.	......	......	......	......	......	......	......	......	$b = ...$	$b = ...$
	3.	......	......	......	......	......	......	......	(b)		(from Table 2)
2nd Slit	1.	......	......	......	......	......	......	......			
	2.	......	......	......	......	......	......	......			
	3.	......	......	......	......	......	......	......	(a_2)		

Error Calculation :

The proportional error in the measurement of slit-width is given by

$$\frac{\delta a}{a} = \frac{\delta D}{D} + \frac{\delta y}{y} \text{ (see 'Error Calculation' at the end of Expt. 5.12).}$$

Similarly, for measurement of width of obstacle,

$$\frac{\delta(a+b)}{(a+b)} = \frac{\delta D}{D} + \frac{\delta y}{y}.$$

Calculations are to be made as discussed at the end of the 'Single Slit Experiment' (Art. 5.12)

Questions and Answers

[In addition to the questions discussed at the end of the Single Slit Experiment (Art. 5.12), the following questions are also to be studied.]

Q.1. What happens when the width of the collimator slit is increased?

A. : Normally the collimator slit width must be very very small in comparison with those of the double slits. If not, the interference fringes become indistinct. As the collimator slit-width increases, finally the interference fringes disappear.

Q.2. If the width of the obstacle is increased, what changes are observed in the nature of the fringes?

A. : If the width of the double slits remains unaltered, then the diffraction pattern remains unaltered but the number of the interference fringes increases.

Q.3. How the angular widths of the fringes can be increased?

A. : The widths of the double slit being decreased, the angular width of the diffraction dark bands increases, while the value of $(a + b)$ should be made smaller to increase the angular width of the interference fringes.

Q.4. When a particular order of bright interference fringe may be missing?

A. : A particular order of an interference maxima may be missing when the condition for a maximum of the interference pattern due to double slit is fulfilled by the condition for minimum of the difraction pattern due to a single slit for the same value of θ. Therefore, if $(a + b) = 4a$, the 4th, 8th, 12th, etc. orders will be missing.

Q.5. If the width of the obstacle is very small so that the two slits almost join together, then what will happen?

A. : Then the system will behave like a single slit whose width is double the width of each slit, resulting in the disappearance of the interference pattern and making the diffraction bands closer.

Q.6. Is the principle of double slit utilised for any practical purposes?

A. : The principle is employed to determine the angular width of a source like distant starts, etc.

5.14 Determination of the wavelength of an unknown line with a plane transmission grating

Theory :

If a beam of light of wavelength λ coming parallel out of the collimator lens of a spectrometer, be incident normally on a plane transmission grating placed vertically on the

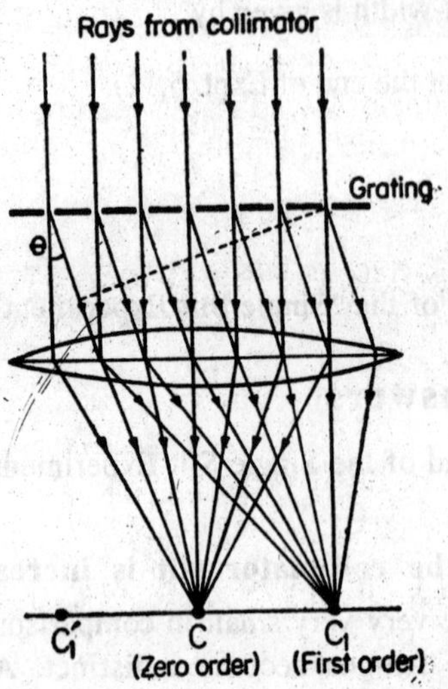

Fig. 5.31 : Diffracted images by a grating

prism-table and having its rulings parallel to the collimator slit, then the transmitted rays after diffraction through the grating, will form at the focal plane of a properly focused telescope, a number of diffracted images of the slit. If viewed through the telescope, a number of primary maxima of different orders will be visible (Fig. 5.31) on both sides of the central maximum (C) of zero order. If θ be the angle of diffraction of the nth order primary maximum, and N be the number of lines per centimetre ruled on the grating, then the wavelength λ of the incident light is given by

$$\lambda = \frac{\sin\theta}{nN}. \qquad \text{... (5.18a)}$$

The value of the number of lines per centimetre (N) of the grating may be determined by employing a light of known wavelength like sodium light and by measuring θ for a particular order n of the spectrum (from Eq. 5.18a). Once the value of N is known, the wavelength of any other radiation can easily be determined by measuring the angle of diffraction θ for a particular order n of the spectrum of the unknown radiation, employing Eq. (5.18a).

If the incident light contains light of several wavelengths (heterogenous), then the diffraction maxima corresponding to different colours will be formed at different angles for the same order of spectra. Thus a series of illuminated lines will be obtained on either side of the central maximum.

Apparatus :

Spectrometer, spirit level, prism, sodium light, light of unknown wavelength (Helium or Hydrogen discharge tube) and a plane transmission grating.

A plane transmission diffraction grating is generally a plane glass plate on which is ruled a large number of parallel, equidistant grooves of the same width. Such a piece being very costly, normally a photographic *replica* of the same is used in the laboratory. Care should be taken neither in trying to clean the surfaces of this replica grating nor in trying to touch its surfaces. It should always be handled by holding it by the edges.

Methods :

Preliminary work for this experiment are namely, (a) Adjustment of the spectrometer and (b) Adjustment of the grating.

(a) Adjustment of the Spectrometer :

The spectrometer is levelled and the prism-table is also levelled by a spirit level. With the help of a sodium flame and a prism, the spectrometer adjustments are done in the usual way as discussed earlier (Art. 5.4).

(b) Adjustment of the Grating :

The proper adjustment of the grating involves several steps and those are successively done in the following manner :

(i) *Making the telescope axis perpendicular to that of the collimator*

The slit is illuminated by sodium light and the direct beam is received by the telescope whose cross-wires are focused on the image. The reading of the telescope vernier is noted and if this value be A, then the telescope is rotated through 90° so that its reading becomes $(90° + A)$. By this the axis of the telescope is made perpendicular to that of the collimator.

(ii) *Mounting of grating and making its surface vertical*

Next the grating is placed on the prism-table so that its vertical surface passes through that diameter of the prism-table which is perpendicular to the line joining the screws X and Z of the prism-table (Fig. 5.32). Then an image of the slit is thrown into the telescope by reflection from one face of the grating and the screws are adjusted to bring the image into the same part of the field as that occupied by the direct image. This makes the faces vertical.

(iii) *Making the grating surface normal to the rays*

In order to set the grating surface normal to the incident rays, the collimator and the telescope are adjusted at right angles to each other and the telescope is fixed there. Next the prism-table with the grating on it, is turned until the slit-image is on the cross-wires after being reflected from one of the grating faces. The position of the table is noted. If its value is B, then the table with the grating is rotated either through 45° or 135° in the appropriate direction, so that the *unruled* surface of the grating faces the collimator. (The ruled and the unruled surfaces of the grating should be identified prior to this, by the usual method). This makes the grating surface normal to the incident rays with the ruled surface directed towards the telescope.

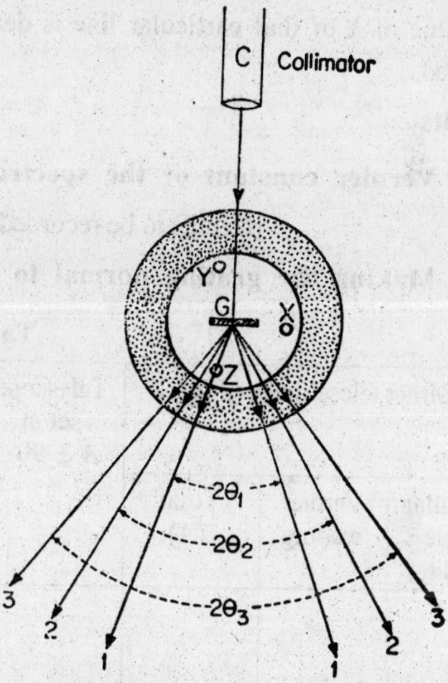

Fig. 5.32 : Grating arrangement

(iv) *Making the grating vertical*

Though the plane of the grating is vertical, its lines may not be. To make sure that lines are vertical, the diffracted orders of the images are viewed through the telescope, making the slit as narrow as is convenient. If the diffracted images on either side of the direct image are not coincident with the centre of the cross-wires, then the screw X is rotated (Fig. 5.32) until the best images are obtained on both the sides. The rotation of the screw X makes the grating vertical and the lines and the slit parallel.

Finally, the collimator slit is properly illuminated by appropriately setting the source in front of it and the slit is made as narrow as possible to get sharp and equally bright fringes on both sides of the central band.

(c) Determining the grating constant :

In order to determine the number of rulings per centimetre of the grating (grating constant), the telescope is set for the highest order of the diffracted image, first on one side and then on the other. The readings of the telescope for the same order on either side of the central band are taken with both the verniers. The difference of the two readings gives twice the angle (2θ) of diffraction for that order of the image (Fig. 5.32). The order number is also noted. Next the angle of diffraction for the lower order is measured similarly and in this way, the diffraction angles are measured for successive orders. Thus from Eq. (5.18a), knowing the order number (n), the diffraction angle (θ) and the wavelength (λ) of sodium light, the value of N (number of rulings per centimetre) can be calculated.

(d) Measuring wavelength of unknown line :

Having known the value of N, the sodium light is replaced by the source whose one of the radiation of unknown wavelength has to be determined. Normally a discharge tube (hydrogen or helium), the unknown wavelength of a line which needs to be determined, is taken. After properly setting the discharge tube very close to the slit, the bright diffraction spectra are visible on either side of the central band. Fixing up a particular line (red, green or violet), the diffraction angles are measured for various orders as before. Using Eq. (5.18a) the value of λ of that particular line is determined, substituting the value of N already obtained.

Results :

(A) Vernier constant of the spectrometer :

[Details to be recorded as in Expt. No. 5.13(A).]

(B) Making the grating normal to the rays :

Table 1

Direct telescope reading			Telescope set at $A \pm 90°$	Prism-table reading when light reflected from grating surface enters the telescope			Prism-table fixed at ($B \pm 45°$ or $135°$)
Circular scale reading	Vernier reading	Total (A)		Circular scale reading	Vernier reading	Total (B)	

(C) Determination of grating constant and wavelength of the unknown line :

Table 2

Wavelength	Order No. (n)	Vernier	Telescope readings for diffracted images: Left — Main Scale reading	Left — Vernier reading	Left — Total	Left — Mean	Right — Main Scale reading	Right — Vernier reading	Right — Total	Right — Mean	Difference of left and right readings (2θ)	Mean value of 2θ	Value of θ	Value of N = sin $\theta/n\lambda$	Mean Value of N
5896 (A.U.)	1	A	...	...	...	...	...	...	...	...	...				
						(M_1)				(M_1')	($M_1 - M_1'$)	...	...	...	...
		B	...	...	...	...	...	...	...	...	...				
			...	...	...	(M_2)	...	...	...	(M_2')	($M_2 - M_2'$)	...			
	2	A	...	...	...	...	...	...	...	...	...				
			...	...	...	(M_1)	...	...	...	(M_1')		...	...	...	
		B	...	...	...	...	...	...	...	...	...				...
			...	...	...	(M_2)	...	...	...	(M_2')					
	3		etc.	etc.	...	...	etc.	etc.	...	...	...				
												...	...	...	
			etc.	etc.	...	...	etc.	etc.	...	...	...				
	etc.														
	etc.														
														Value of λ = sin θ/n.N	Mean Value of λ in (A.U.)
Unknown line	1	A	...	...	...	...	...	...	...	...					
												...	...		
		B	...	...	...	...	...	...	...	...	...				
	2	A	...	...	...	...	...	...	...	...		...	...		
		B	...	...	...	...	...	...	...	...	...				
	etc.	A													
		B													

Calculation :

From Eq. (5.18a),

(i) $N = \dfrac{\sin\theta}{n\lambda}$.

Substituting the value of θ for different order number from Table 2, the values of N are obtained and the mean value of N is thus obtained.

(ii) Putting this mean value of N so obtained in the same equation the value of λ for the unknown line is calculated as

$$\lambda = \frac{\sin\theta}{nN}.$$

Values of θ for different orders are substituted and corresponding values of λ are obtained. From these, the mean value of λ for a given unknown line is thus determined.

Precautions :

(i) The grating should be placed in such a way that incident rays fall on its unruled surface normally.

(ii) Collimator slit should be made as narrow as possible to get bright bands for different orders.

(iii) Surface should be properly set so as to give the same number of visible orders on both sides of the central band having equal brightness.

(iv) The grating should be so placed on the prism-table as to have the maximum area of the ruled surface exposed to the incident light.

(v) Before taking all readings, the telescope should be fixed rigidly by the fixing screw attached to it.

(vi) While rotating the telescope, it should be moved always in the same direction so as to avoid any backlash error.

Error Calculation :

From equation, $\lambda = \dfrac{\sin\theta}{nN}$

$$\frac{\delta\lambda}{\lambda} = \frac{\delta(\sin\theta)}{\sin\theta} + \frac{\delta N}{N} = \cot\theta.\,\delta\theta + \frac{\delta N}{N}.$$

But, $N = \dfrac{\sin\theta'}{n\lambda'}$, where λ' is constant.

$$\therefore \frac{\delta N}{N} = \cot\theta'.\,\delta\theta'.$$

$\therefore$ proportional error $= \cot\theta.\,\delta\theta + \cot\theta'.\delta\theta'$.

Maximum error in measuring $\delta\theta$ or $\delta\theta'$ would be 1 vernier division of spectrometer reading, and the percentage error in the measurement of l will be $\dfrac{\delta\lambda}{\lambda} \times 100 = \ldots\ldots\ \%$.

The percentage error will increase if the order number is decreased when θ becomes smaller. So, in order to get minimum error, higher order spectra are always preferred for computation of the result.

Questions and Answers

Q.1. What is a diffraction grating and how is it constructed?

A. : See any textbook on Light.

Q.2. What is 'grating element'?

A. : If 'a' be the width of the slit and 'b' be the width of an opaque space, then $(a + b)$ is known as the grating element or grating constant and any two points in the consecutive slits separated by a distance of $(a + b)$ is known as 'consecutive points'.

Q.3. What are different types of grating?

A. : Transmission grating, Reflection grating and Concave grating.

Q.4. Which type of grating is better?

A. : Concave grating is better since it produced self-focused spectra and is usable in the U-V region of spectra.

Q.5. What happens when the number of rulings are changed?

A. : For increased number of rulings, a few order numbers with large angle of separation are obtained while for less number of rulings, large number of orders are obtained with small angle of separation.

Q.6. What are Ghost lines?

A. : If the rulings are not precisely parallel and the consecutive points are not equidistant, then additional lines sometimes appear along with the actual spectral lines. These additional lines are known as 'Ghostlines'.

Q.7. Why is it essential to have the unruled surface facing the collimator?

A. : Otherwise the incident rays will be first diffracted at the ruled surface and then refracted by the glass surface (unruled end), making thereby θ not due to diffraction alone. The observed θ will then be due to both diffraction and refraction as well.

Q.8. What happens if white electric light is used?

A. : The central band will be white while the other bands will be coloured with red in the outermost position and violet in the innermost.

Q.9. What is the difference between spectra of grating and that of prism?

A. : Prism gives only one spectrum but grating gives a number of spectra on both sides of the centre. Prism spectrum is a band spectrum but grating spectra are line spectra. The former is much more intense than the latter. The order of the colours are reverse in one another. Prism spectrum shows less details and are slightly curved. Grating spectra are practically straight. There is no chance of 'ghost spectra' in case of prism.

Q.10. If the incident light is not strictly monochromatic, what will happen?

A. : Two lines of wavelength λ and $(\lambda + d\lambda)$ will be found to be more and more resolved as higher orders are approached.

Q.11. What are the requisites of a good grating?

A. : Lines should be exactly similar in form; they should be exactly parallel and should be equispaced over the entire ruled surface.

Q.12. How is grating element related to the number of rulings per centimetre?

A. : Grating element is the reciprocal of the number of rulings per centimetre.

Q.13. Are there any secondary maxima formed?

A. : Yes, but they are not visible with a grating having a large number of rulings.

Q.14. Are the spectra of different orders of the same intensity?

A. : No, they are of different intensities. The intensity decreases with the increase of order number.

Q.15. On what factors does the available number of order depend?

A. : It depends on the grating element $(a + b)$ and the maximum number obtainable is $\frac{(a+b)}{\lambda}$.

Q.16. What type of grating is preferable—with a large number of rulings or a small number?

A. : A grating with a small number of rulings give a large number of orders but then the angle of diffraction becomes small for a given order, making thereby greater percentage error in the measurements. Moreover, with more number of orders, intensity rapidly fall off. Hence, grating with a large number of rulings per centimetre is always preferable at the cost of non-availability of higher order numbers.

Q.17. What happens if the width of the ruled lines on the grating is the same as that of the clear space in between them?

A. : Then the spectra of the even orders (2, 4, 6, etc.) will be found to be missing.

Q.18. What are meant by 'dispersive power' and 'resolving power' of a grating?

A. : The dispersive power is defined as $\frac{d\theta}{d\lambda}$ (rate of change of angle of diffraction with wavelength), while the resolving power is defined as $\frac{\lambda}{d\lambda}$ ($d\lambda$ being the wavelength difference that can just be resolved at wavelength λ.)

Q.19. What are the different adjustments to be made in this experiment?

A. : [See 'Adjustment of the grating'].

5.14(A) Measurement of wavelength difference between D_1-and D_2-lines of sodium using a slit of adjustable width

Theory :

In the previous experiment (Expt. No..5.14), if sodium light is used and the D_1 (5896Å) and D_2 (5890Å) lines are resolved in the order n, the wavelength λ of the D_2-line can be obtained from Eq.5.14(a) [$\lambda = \frac{\sin\theta}{nN}$] from the known value of N and by measuring θ.

If the wavelengths of the D_2- and D_1-lines of sodium be λ and $(\lambda + d\lambda)$ respectively, the resolving power of the grating is

$$\frac{\lambda}{\Delta\lambda} = nm, \qquad \text{... [5.14(A) a]}$$

where n is the order number in which D_1 and D_2 lines are just resolved by adjusting the width of the slit and m is the total number of lines effective in the production of the diffraction pattern. This is nothing but the total nummber of lines exposed to the adjustable slit. If x is the width of the adjustable slit for which the nth order fringe is just resolved, then $m = Nx$. Hence Eq. [5.14(A)a] gives

$$\frac{\lambda}{\Delta\lambda} = nNx$$

or, $\Delta\lambda = \frac{\lambda}{nNx}$. ... [5.14(A) b]

By measuring x, the difference in wavelength $\Delta\lambda$ can be obtained from Eqn. [5.14(A) b].

Apparatus :

Same as in Expt. No. 5.14. In addition, a slit of adjustable width is necessary, microscope to measure x.

Method :

(i) Adjustments of the spectrometer and the grating are to be made as in Expt. No. 5.14

(ii) To obtain N, the procedures mentioned in (C) under Method of Expt. No. 5.14 are to be followed. For accuracy, angles of diffraction are preferably taken for higher orders.

(iii) The collimator is illuminated with sodium light. The telescope is rotated to the left or right to get D_1 and D_2 lines of a given order. If the lines are not resolved, the next higher order is reached. The angle of diffraction for the D_2-line is obtained following step (ii) and its wavelength λ is calculated. The process is repeated for different orders for the D_2-line. The mean wavelength is calculated.

(iv) To measure the wavelength difference between D_1- and D_2-lines of sodium, the slit of adjustable width is placed exactly behind the grating. The plane of the slit is made vertical. The collimator is properly illuminated with sodium light.

(v) The telescope is turned to the left to observe the resolved D_1- and D_2-lines of a given order. The slit is first made wide and narrowed slowly so that D_1- and D_2-lines of the order in which they are resolved, approach each other and just merge together. When the two lines are just resolved at this stage, readings are taken from the microscope.

(vi) The slit is then made narrower and then wider till the D_1- and D_2-lines are just resolved again. Readings in microscope are taken again.

(vii) The process is repeated two or three more times by decreasing and increasing the slit-width alternately. The telescope is turned to the right and the operations are repeated for the lines of the same order.

(viii) The mean of the readings are taken to find x for the given order.

(ix) For higher values of the order n, step (v) above is repeated.

(x) From the measured values of N, λ (for D_2-lines), and x, the wavelength difference $\Delta\lambda$ between D_1- and D_2-lines for different orders from Eq. [5.14(A) b] is calculated. Mean value of $\Delta\lambda$ is then obtained.

Results :

[Tables (A), (B) and (C) of Expt. No. 5.14 are made as usual to obtain vernier constant of the spectrometer, to make grating normal, to get the grating constant, to get angle of diffraction for the bands and the unknown wavelength λ of D_2-lines.]

(D) Determination of the width of the adjustable slit (x) when nth order fringe is just resolved.

Table 4

Order No. (n) just resolved	Reading at one edge of slit when nth order fringe is just resolved				Corresponding readings at the other edge of the slit				Slit-width $x = (x_1 \sim x_2)$ (cm)
	Main scale (cm)	Vernier (cm)	Total ($= x_1$) (cm)	Mean x_1 (cm)	Main scale (cm)	Vernier (cm)	Total ($= x_2$) (cm)	Mean x_2 (cm)	
	……	……	……	……	……	……	……	……	
……									……
	……	……	……	……	……	……	……	……	
	……	……	……	……	……	……	……	……	
……									……
	……	……	……	……	……	……	……	……	
	……	……	……	……	……	……	……	……	
……									……
	……	……	……	……	……	……	……	……	

Table 5

(E) Determination of $\Delta\lambda$ (Difference of D_1- and D_2-lines)

Wavelength λ from Table 2 (Expt. No. 5.14)	Lines per cm (N) of grating [from Table 2]	Order No (n)	Slit-width from Table 4 (cm)	$\Delta\lambda = \dfrac{\lambda}{nNx}$ (Å)	Mean $\Delta\lambda$ (Å)
		……	……	……	
……	……	……	……	……	……
		……	……	……	
		……	……	……	

Calulation :

From Eq. [5.14 (A) b],

$$\Delta\lambda = \frac{\lambda}{nNx}.$$

The values of λ, n, N and x are substituted and value of $\Delta\lambda$ is obtained using different order numbers and slit-widths.

Precautions :

(i) The adjustable slit should be placed symmetrically about the central line of the grating.

(ii) The D_1- and D_2-lines may not be resolved in the lower orders. Then, higher orders are to be looked upon.

(iii) Backlash error in readings of the microscope should be avoided.

Error Calcualtion :

From equation, $\Delta\lambda = \frac{\lambda}{nNx}$, the proportional error in $\Delta\lambda$ is

$$\frac{\delta(\Delta\lambda)}{\Delta\lambda} = \frac{\delta\lambda}{\lambda} + \frac{\delta N}{N} + \frac{\delta x}{x},$$

where δx = twice the vernier constant of the microscope.

The values of $\frac{\delta\lambda}{\lambda}$ and $\frac{\delta N}{N}$ are obtained as shown in Expt. No. 5.14. If λ of D_2-line is supplied, then $\frac{\delta\lambda}{\lambda}$ term is not necessary. To get minimum error, higher order spectra are always preferred for computation of the result.

Questions and Answers

[The same as in Expt. No. 5.14]

5.15 Calibration of a polarimeter and hence to determine the specific rotation and the concentration of a sugar solution

Theory :

When a beam of monochromatic light of wavelength λ is transmitted through a column of solution of length l cm ($\frac{l}{10}$ decimetre) at temperature t°C and if the solution contains m gm of active substance per cubic centimetre, which can produce a rotation of the plane of polarisation, and if ϕ be the amount of rotation of the plane of polarisation of a plane polarised light, then

$$\phi = \frac{Sml}{10}, \qquad \text{... (5.15a)}$$

where S is known as the specific rotation of the substance depending upon the wavelength (λ) of radiation and temperature of the solution. This quantity is defined as the rotation produced by a column of solution of one decimetre in length containing 1 gm of active substance per cc of the solution.

If 100 cc of the solution contains c gm of active substance, then

$$m = \frac{c}{100} \text{ gm/cc} \qquad \text{... (5.15b)}$$

Hence, Eq. (5.15a) becomes

$$\phi = \frac{S.c.l}{1000} \qquad \text{... (5.15c)}$$

Now, with the help of a polarimeter, if the angles of rotation (ϕ) for different values of known concentrations (c) of a solution are determined, and if a graph be plotted with ϕ against c, then the graph will be a straight line. The polarimeter is thus calibrated. Taking then the solution of an unknown concentration, rotation of its plane of polarisation can be measured and from the graph, its unknown concentration can be determined.

Again, using Eq. (5.15c), the value of the specific rotation of the solution can also be determined.

Apparatus :

Polarimeter, monochromatic source of light, sugar, distilled water, balance, measuring cylinder.

The polarimeter employed to find the concentration of an active solution is a half-shade polarimeter (or biquartz type), the principle of the which is shown in Fig. 5.33. In this, an analyser and a polariser are used for polarising the light and to analyse the same after it passes through the active solution. The light from the source after being rendered parallel by

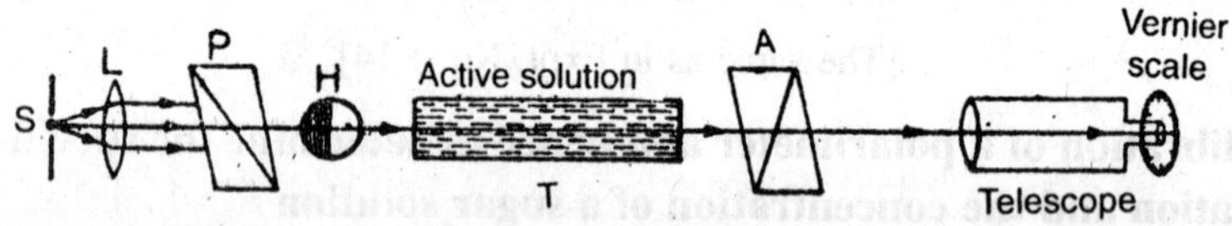

Fig. 5.33 : Polarimeter

the lens L passes through the polarising Nicol P and then passes out through the half-shade plate H (or, biquartz). After this the light passes through the active solution contained in a tube T whose two ends are closed by optically plane glass plates. Light emerging out of the solution is then made to pass through the analysing Nicol A and finally observed through a telescope attached with the vernier scale. As the telescope rotates along with the analyser, the angle of rotation of the system is obtained from the scale. What is required to be done, is to set the analyser so that the two halves of the field appear equally bright, once before the insertion of the active material and again after the insertion, using the position of lower intensity. Normally, the condition of the field of equal brightness of the two halves, can be obtained for two positions of the vernier-readings which are usually 180° apart. The axes of the tube, telescope, source, etc. are all in one horizontal line.

Method :

(i) The smallest reading of the circular scale and the vernier constant of the vernier attached to the analyser tube are detertmined. The length of the glass tube (l) is measured thrice between the end plates, and their mean value is noted.

(ii) The tube is then completely filled with the pure distilled water and placed in its position in such a way that the axes of the analyser, tube and the polariser lie in the same horizontal line.

(iii) The eyepiece of the telescope is focused on the field of view of the analyser and the latter is rotated till minimum light in the field of view is visible, matching the intensities in both parts of the field. The setting of the analyser is noted reading both the verniers, and the analyser is reset twice more and readings are taken for each. This gives the initial reading.

(iv) Some amount of dry sugar is taken and about 15 gm of it is dissolved in 80 cc of water taken in a measuring cylinder. When the sugar completely dissolves

20 cc of water is then added to the solution. At this state, 15 gm of sugar remains dissolved in 100 cc of water making thereby a 15% sugar solution. This would be the solution to start with.

(v) About half of the amount of the above solution is kept away as the stock solution and from the remaining portion, the tube is filled with this solution in place of distilled water, Care should be taken to see that there remains no air bubble inside the tube.

(vi) Keeping the tube in its proper position, the analyser is reset to obtain minimum illumination and matched fields. The new setting is noted from the circular scale and the vernier. From this reading and that obtained during operation (iii), the angle of rotation of the plane of polarisation is determined.

(vii) Next, the concentration of the solution has to be changed. The stock solution is taken and a portion from it is taken in the measuring cylinder. Let x cc of this solution be taken. If the new concentration to be made c_1% and the existing solution be of strength c% (which was actually 15%), then the amount of water (y cc) that has to be added to x cc to convert c% to c_1% concentration strength, will be given by

$$y = \frac{c - c_1}{c_1} \times x \text{ cc} \qquad \text{... (5.15d)}$$

This gives the method of converting a certain amount of the solution taken of known strength (c%) to a new strength (c_1%).

(viii) In this way, the experiment is started with a 15% strength solution and gradually changing the strength to 12.5%, 10%, 7.5% and 5% respectively the observations are made. For each concentrations, the angles of rotation (ϕ) of the plane of polarisation are determined and finally a graph is drawn with the concentration (c%) along the X-axis and the corresponding angle of rotation (ϕ) along the Y-axis. The graph will be a straight line (as shown in Fig. 5.34), passing through the origin.

(ix) Taking a point (P) on the graph well away from the origin, the value of the specific rotation (S) is determined, using the values of θ and c corresponding to that point. The temperature at which the measurements were done and the wavelength of the light used are also noted. From the same graph, noting the angle of rotation for a solution of an unknown strength (c_x%), the value of the strength can also be obtained.

To find the molecular rotation, the specific rotation is multiplied by the molecular weight.

Result :

(A) Determination of the vernier constant :

Value of one smallest scale division = ...

Number of vernier divisions = ...

Value of vernier constant = ...

(B) Determination of the length of the tube :

Length of the tube (l) = $\frac{l_1 + l_2 + l_3}{3}$ = ... cm

(C) Readings for the distilled water :

Table 1

No. of Obs.	Reading in degree for							
	First vernier				Second vernier			
	Circular Scale	Vernier	Total (X_0)	Mean (X_0)	Circular Scale	Vernier	Total (Y_0)	Mean (Y_0)
1. ($l \rightarrow r$)	…							
2. ($r \rightarrow l$)	…	…	…		…	…	…	
				… degree				… degree
3. ($l \rightarrow r$)	…	…			…	…	…	
4 ($r \rightarrow l$)	…	…	…		…	…	…	

(D) Conversion of one strength to another strength :

Table 2

Serial No.	Volume of stock solution taken (x cc)	Strength of the solution taken	Final strength to be made	Quantity of water (y cc) to be added to the stock solution $\left(y = \frac{c - c_1}{c_1} \times x \text{ cc}\right)$	Temperature of the solution in °C
1.	…… (x_1)	c% (15%)	c_1% (12.5%)	$y_1 = \frac{c - c_1}{c_1} x_1$	…
2.	…… (x_2)	c_1% (12.5%)	c_2% (10%)	$y_2 = \frac{c_1 - c_2}{c_2} x_2$	…
3.	…… (x_3)	c_2% (10%)	c_3% (7.5%)	$y_3 = \frac{c_2 - c_3}{c_3} x_3$	…
4.	…… (x_4)	c_3% (7.5%)	c_4% (5%)	$y_4 = \frac{c_3 - c_4}{c_4} x_4$	…
5.	…… (x_5)	c_4% (5%)	c_5% (2.5%)	$y_5 = \frac{c_4 - c_5}{c_5} x_5$	…

(E) Readings for solutions of different strengths :

Table 3

No. of Obs.	Strength of solution (c %)	Vernier		Readings in degrees: Circular Scale	Readings in degrees: Vernier	Readings in degrees: Total (X or Y)	Readings in degrees: Mean (X or Y)	Rotation of the vernier (in degree)	Mean rotation of the vernier (ϕ in degree)
1.	15%	First :	$t \to r$	...	...	...	... (X_1)	$\phi'_1 = (X_1 \sim X_0)$	$\phi_1 = \frac{\phi_1 + \phi_1''}{2}$
			$r \to l$	...	...	...			
		Second :	$l \to r$	...	...	...	... (Y_1)	$\phi_1'' = (Y_1 \sim Y_0)$	
			$r \to l$	...	...	...			
2.	12.5%	First :	$l \to r$	...	...	...	... (X_2)	$\phi'_2 = (X_2 \sim X_0)$	$\phi_2 = \frac{\phi_2' + \phi_2''}{2}$
			$r \to l$	...	...	...			
		Second :	$l \to r$	...	...	...	... (Y_2)	$\phi_2'' = (Y_2 \sim Y_0)$	
			$r \to l$	...	...	...			
etc.	etc.	......		...	...	...	...	...	...
etc.	etc.	......		...	...	...	...	...	...
6.	2.5%	First :	$l \to r$	...	...	...	... (X_6)	$\phi'_6 = (X_6 \sim X_0)$	$\phi_6 = \frac{\phi_6' + \phi_6''}{2}$
			$r \to l$	...	...	...			
		Second :	$l \to r$	...	...	...	... (Y_6)	$\phi_6'' = (Y_6 \sim Y_0)$	
			$r \to l$	...	...	...			
7.	c_x% (unknown)	First :	$l \to r$	...	...	...	... (X_7)	$\phi'_7 = (X_7 \sim X_0)$	$\phi_7 = \frac{\phi_7' + \phi_7''}{2}$
			$r \to l$	...	...	...			
		Second :	$l \to r$	...	...	...	... (Y_7)	$\phi_7'' = (Y_7 \sim Y_0)$	
			$r \to l$	...	...	...			

Drawing of Graphs :

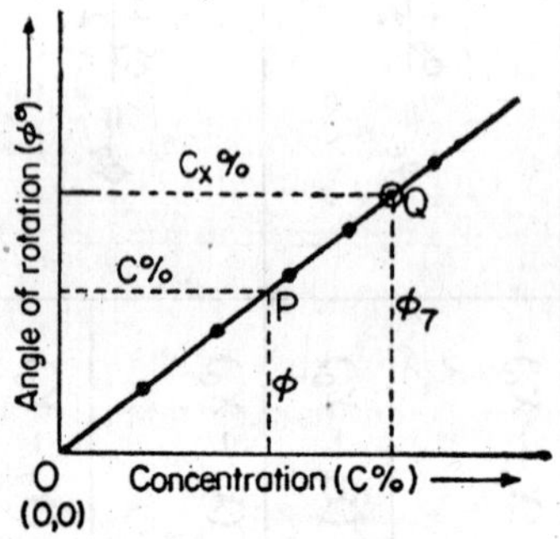

Fig. 5.34 : (c–ϕ) curve

Plotting the strengths (concentrations) of the solutions along the X-axis and the corresponding rotations (ϕ) of the plane of polarisation along the Y-axis of a graph paper, the curve obtained will be a straight line passing through the origin (Fig. 5.34). From this graph, taking any arbitrary point (P), the corresponding values of the concentration (c) and the angle of rotation (ϕ) are noted and these values when substituted in Eq. (5.15c), will give the value of the specific rotation (S) of the solution taken. Again, substituting the value of ϕ_7 on the graph (for the unknown concentration) at the point Q, the corresponding value of abscissae gives the value of the unknown concentration (c_x%) of the unknown solution taken.

Calculation :

From the graph, noting the amount of rotation (ϕ_7) for the solution of unknown strength (c_x%), the value of the latter can be had and again taking any arbitrary point P on the graph, the values of c% and ϕ corresponding to this point are noted and substituting them in Eq. (5.15c), one gets

$$S = \frac{1000\phi}{cl},$$

which gives the value of the specific rotation of the given solution at room temperature.

Hence, specific rotation of the given solution

at temperature .... °C = ...

and strength of the unknown solution = ... %

Precautions :

(i) The vessel in which the solution is prepared, should be thoroughly cleaned with distilled water beforehand.

(ii) The experimental tube should not contain any air bubble.

(iii) Temperature of water before and after the experiment should be noted and their mean taken, because the value of specific rotation depends very much on the temperature of the solution.

(iv) The quantity of brightness of the two halves of the half-shade plate should be carefully judged and made equal as far as practicable.

(v) The rotation depends on wavelength of light employed. Hence it is better to use a known monochromatic light (sodium light) for the experiment.

Error Calculations :

From Eq. (5.15c), $c = \dfrac{1000\phi}{Sl}$

or, $\dfrac{\delta c}{c} = \dfrac{\delta\phi}{\phi} + \dfrac{\delta S}{S} + \dfrac{\delta l}{l}$

Again, to find S, M gm of an active substance is dissolved in V cc of solvent, so that the concentration of the known solution is

$$c' = \frac{100M}{V}\ \%.$$

$$\text{Hence, } S = \frac{1000\phi'}{c'l'} \text{ or, } \frac{\delta S}{S} = \frac{\delta\phi'}{\phi'} + \frac{\delta c'}{c'} + \frac{\delta l'}{l'},$$

$$\text{but, } \frac{\delta c'}{c'} = \frac{\delta M}{M} + \frac{\delta V}{V}$$

$$\text{or, } \frac{\delta S}{S} = \frac{\delta\phi'}{\phi'} + \frac{\delta M}{M} + \frac{\delta V}{V} + \frac{\delta l'}{l'}$$

$$\text{or, } \frac{\delta c}{c} = \frac{\delta\phi}{\phi} + \frac{\delta\phi'}{\phi'} + \frac{\delta M}{M} + \frac{\delta V}{V} + \frac{\delta l'}{l'} + \frac{\delta l}{l}.$$

Out of the quantities in the R.H.S., the contributions to the error is maximum by the measurements of the angle of rotation ϕ. All other terms contribute very little to the error. Thus the measurement of the angle of rotation (ϕ) should be done carefully, and repeated for several times for one position of the vernier taking readings for both the verniers.

Hence, percentage error in measurement of the unknown concentration is

$$\frac{\delta c}{c} \times 100 = \ldots\ldots\ldots\ \%.$$

Questions and Answers

Q.1. Define the terms : Polarisation, Plane of polarisation, Double refraction, Optic axis, Dextro and Laevorotatory substances.

A. : [See any textbook on Optics.]

Q.2. What types of polarisation of light are possible?

A. : Lights may be plane-polarised, circularly polarised and elliptically polarised.

Q.3. What is meant by 'optically active' substance?

A. : A substance which can rotate the plane of polarisation of a polarised light.

Q.4. What conclusion can be drawn from the phenomenon of polarisation of light?

A. : It proves the transverse nature of the light waves.

Q.5. What is the difference between *ordinary* and *extra-ordinary* rays?

A. : While passing through a double-refracting crystal, the ray which obeys the law of refraction is the ordinary ray while the other does not obey.

Q.6. How does the optical activity depend on the wavelength of light employed?

A. : The amount of rotation is approximately inversely proportional to the square of the wavelength.

Q.7. Apart from sugar solutions, does any other substance rotate the plane of polarisation?

A. : Yes, tartaric acid, quartz crystals, etc. can rotate the plane of polarisation.

Q.8. How does the sugar solution behave?

A. : It is of dextro-rotatory type (rotates the plane of polarisation towards right).

Q.9. On what factors does the specific rotation depend?

A. : It depends on temperature of the solution and the wavelength of light employed. For solutions, it also depends to some extent, on the properties and concentration of the solute and the solution.

Q.10. Can this experiment be done with white light instead of sodium light?

A. : In that case, biquartz has to be used in place of half-shade plate when rotation of the yellow light can be measured very precisely.

Q.11. What is meant by *molecular rotation*?

A. : It is the product of the specific rotation and the molecular weight of the substance.

Q.12. How does the rotation change with temperature?

A. : The rotation ϕ changes with temperature t according to Gernez's relation, $\phi = A + Bt + Ct^2$, when B and C are positive for solids like quartz, etc. and negative for liquids like turpentine, tartaric acid, etc.

[Apart from these questions, construction of Nicol prism and functions of analyser-polariser are to be prepared.]

5.16 Determination of the nodal points of a lens system by Nodal Slide

Theory :

The positions of the nodal points of a lens system can be determined from its properly that an incident ray directed towards the first nodal point emerges in a parallel direction appearing to emerge from the second point.

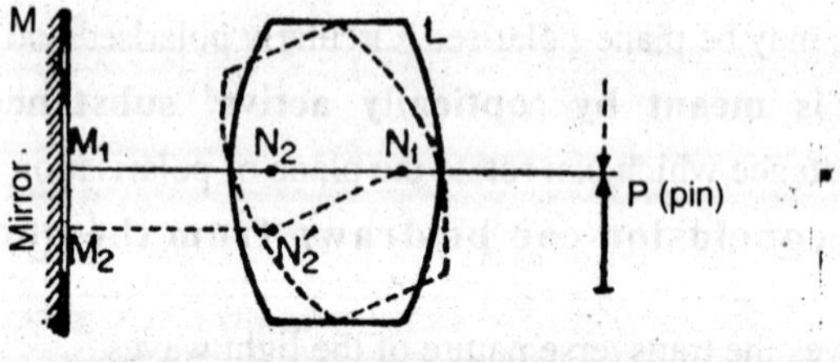

Fig. 5.35 : Rays through nodal slide

In (Fig. 5.35) P is a pin which is placed by trial at the first focal point of the lens system L. The rays emergent from the lens system are parallel. These rays are reflected back normally from a plane mirror M which is set normal to the axis of the lens. The image of the pin, therefore, coincides with the pin itself. If the lens is now turned about an axis perpendicular to its principal axis, the image will be displaced. But, if the axis of this rotation passes through the first nodal point (N_1), then the image will not be displaced. The reason for this is understandable from the following.

Let the second nodal point of the system be at N_2, and let N'_2 be its position when the lens is displaced. A ray from the pin P directed towards N_1 passes along $N_1N_2M_1$ and is reflected back along the same path after reflection from the mirror M_1. When the lens system is slightly rotated about N_1, PN_1 is refracted and emerges from the lens parallel to its original direction, appearing to come from N'_2. The path of this ray is thus along $PN_1N'_2M_2$, which is reflected back by the mirror along $M_2N'_2$ and emerges along.

The images of the pin is thus formed at the same postion. Under these circumstances, the point of intersection of the rotation axis with the principal axis, will be the position of the *first nodal point.*

By rotating the lens system by 180°, and following the same procedure, the position of the *second nodal point* is located. In the first case, the distance between the object pin *P* from the vertical axis of rotation gives the *second focal length* of the lens system, while in the second case, the distance between the pin *P* from the vertical axis of rotation gives the *first focal length* of the lens system.

Apparatus :

Nodal slide, plane mirror, reference pins, lenses, etc.

Nodal slide is the apparatus used for determination of the nodal points of a lens system. It consists of a turntable mounted on a tripod and kept horizontal with the levelling screws (Sc_1, Sc_2, Sc_3). The table can rotate about a vertical axis. It has a scale (S_1S_2) mounted on it, which can be moved parallel to itself (Fig. 5.36). The lens system (L_1, L_2) whose nodal points are to be determined, are mounted on two rails fixed on a frame which can be made to slide on the table along the scale. The optical axis of the lens system should be horizontal and parallel to the scale. A pin can be placed on one side of the lens system and a plane mirror (*M*) can be fixed vertically on the other side. The tip of the pin lies in the horizontal line passing through the centre of the lenses and cutting the mirror normally. The separation between the lenses can also be altered.

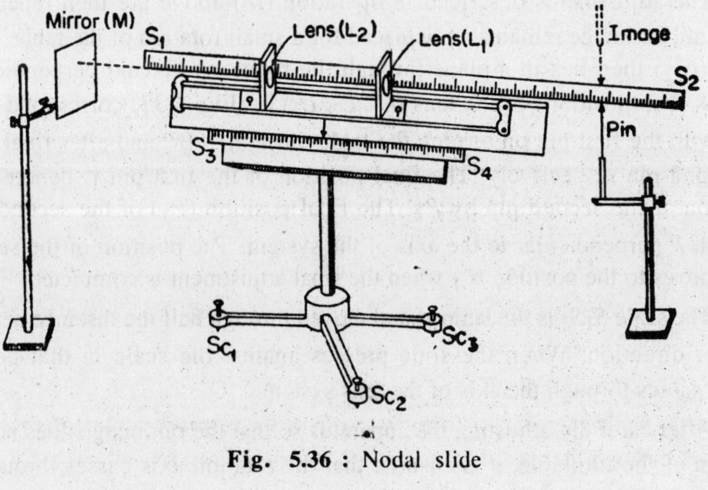

Fig. 5.36 : Nodal slide

Method :

(i) The turntable is properly levelled and the mirror is made vertical. The pin (*P*) is adjusted to lie at the proper height (axis of the lens system).

(ii) The lenses are mounted properly and the position of the pin (P) is adjusted so that its image formed by transmision through the lens system and reflection at the mirror, coincides with itself.

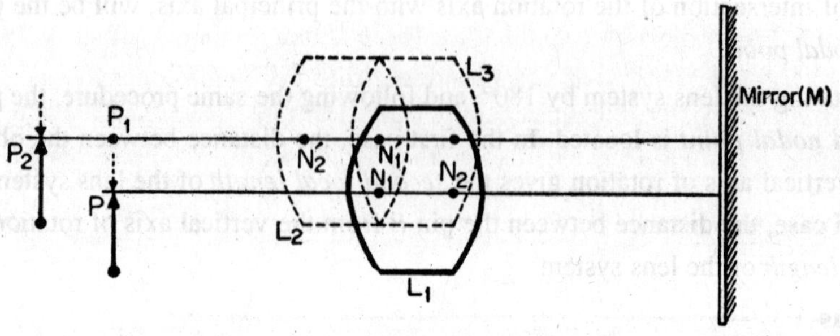

Fig. 5.37 : Determination of nodal points

(iii) Next the position of the first nodal point and the rotation axis of the table is how far from the principal axis of the lens system, have to be determind. To find the former, the table is turned by about 10° and the image is viewed. If it is displaced, the slide is moved until with focusing exactly made for each position of the slide, a small rotation of the table causes no displacement between the pin and its image. When this occurs, the first nodal point (N_1) lies in a plane ($N_1N'_1$) through the rotation axis of the table and perpendicular to the principal axis. This position of the pin (P) must not be disturbed (Fig. 5.37).

(iv) To determine the distance of the rotation axis from the principal axis of the lens system, the table is turned through 180° so that the lens L_1 turns about the axis of rotation and swings over to the position L_2. A second pin P_2 is so set that its image is formed at its own point. The adjustments described in operation (iii) above are then repeated for the second pin until its image remains undisplaced for a small rotation of the table. The second nodal point (N'_2) then lies in a plane through the rotation axis and perpendicular to the principal axis. The positions of the lens (L_1, L_2, L_3) in (Fig. 5.37) correspond to the final adjustment with the first pin on turning the table through 180°, and after final adjustment with the second pin respectively. The final position of the first pin is denoted by P, the initial position of the second pin by P_2. The final position (P_1) of the second pin is in a plane through P perpendicular to the axis of the system. The position of the second nodal point (N'_2) moves to the position N'_1 when the final adjustment is completed.

(v) The scale S_1S_2 is unclamped and moved through half the distance between P_1P_3 in the proper direction. When the slide presses against the scale in that position, the rotation axis passes through the axis of the lens system.

(vi) After carefully adjusting the apparatus so that the pin image does not move for small rotation of the turntable, it is ensured that the rotation axis passes through the first nodal point and the pin lies at the position of the first principal focus. The distance PN_1 is then the *first principal focal length.* To find the actual focal length, a metre scale is placed on the turntable, close to the fixed scale and parallel to it. The reading of this scale corresponding to the pin P is noted. The table is then turned through 180° and the new scale reading is again taken for the same position of the pin. The difference of these two scale readings is twice the focal length (f_1).

(vii) Putting back the table in the original position, the distance of the pin (P) from the pole of the first surface of the lens is measured. This distance subtracted from f_1 gives the distance (r) of the first nodal point from the pole of the first surface of the lens. Some fixed reference mark (X) on the fixed scale is noted at this position.

(viii) The table is then again rotated through 180° so that N_2 (the second nodal point) now lies between P and N_1. The lens system is moved away from P until its image remains undisplaced for small rotation of the table. The second nodal point (N_2) will then be on the axis of rotation. The position of the reference mark (X) of the slide is again noted on the fixed scale. The displacement of this mark (X) gives the distance between the two nodal points or the nodal separation (x).

(ix) As before, the distance of the second nodal point from the pole of the second curved surface is again determined.

(x) The entire observations are repeated by changing the separation between the lens, and graphs may be drawn with (a) focal length (f) against separation (d) between the lenses and also (b) nodal separation (x) against separation (d) between the lenses.

Results :

(A) Determination of positions of focal and nodal points :

Table 1 (See next page)

(B) Determination of focal length and nodal points separation :

Table 2

No of Obs.	Separation between the lenses (d) (cm)	Mean value of		Focal length $f = \frac{f_1+f_2}{2}$ (cm)	Mean value of		Nodal separation $x = \frac{x_1+x_2}{2}$ (cm)	Mean value of	
		f_1 (cm)	f_2 (cm)		x_1 (cm)	x_2 (cm)		r_1 (cm)	r_2 (cm)
1.	......	... ...	... ...	......	... ...	...	....	... ...	... ...
2.	......	... ...	... ...	......	... ...	...	....	... ...	... ...
3.	......	... ...	... ...	......	... ...	...	....	... ...	... ...
etc.	etc.	etc.	etc.	etc.	etc.	etc.	etc.	etc.	etc.

Drawing of graphs :

Two graphs may be plotted. The first graph is plotted with lens-separation (d) along X-axis and the focal length (f) along Y-axis. This curve (f vs. d) would be a rectangular hyperbola. It is of course better to plot d versus $\frac{1}{f}$ when the graph would be a straight line. Another second graph is plotted with lens-separation (d) along X-axis and nodal-separation (x) along Y-axis, The second graph shows the variation of nodal separation with the change of distance between the lenses.

Calculations :

The values of the focal distances, nodal distances, nodal separation, etc., are all calculated from the readings shown in Table 2.

Results :

(A) Determination of positions of focal and nodal points

Table 1

No. of Obs.	Separation between the lenses (d) (cm)	Position of the pin on the scale (R_1) (cm)	Mean value of R_1 (cm)	Position of the pin on the scale on reversing the table R'_1 (cm)	Mean value of (R'_2) (cm)	Focal length of the lens system, $f = \frac{R_1 \sim R'_1}{2}$ (cm)	Distance (r) of the pin from the nearest surface of the lens (cm)	Mean value of (r) (cm)	Position (x) of the reference mark (X) on the fixed scale when rotation axis passes through the nodal point	Mean Value (x)
1.	......	For N_1	...	... ...	...	$f_1 = \dots\dots$	... ...	... (r_1)	....	...... (x_1)
		For N_2	... ...	... ...	...	$f_2 = \dots\dots$	... ...	... (r_2)	....	...... (x_2)
2.	......	For N_1		... ...	...	$f_1 = \dots\dots$	... ...	... (r_1)	....	.... (x_1)
		For N_2	...	... ...	... ...	$f_2 = \dots\dots$	... ...	... (r_2)	....	.... (x_2)
etc.	etc.	etc.	etc.	etc.	etc.	etc.	etc.	etc.	etc.	etc.

Precautions :

(i) The separation between the lenses (*d*) should not be very large. If the value of *d* equals the sum of the focal lengths of the two lenses, then the system will become telescopic and this method of finding the focal points and nodal points will fail as all of them will be located at infinity then.

(ii) While setting, parallax must be avoided for the images.

Error Calculations :

In this experiment, measurements are only of those of distance and separation measurements. These measurements are done from the scale attached to the instrument. Sufficient errors are, therefore, not obtained from these measurements. Rather, errors may creep in while setting the pins and rotating the table, etc., which should be done very carefully.

Questions and Answers

Q.1. What are the nodal points of a lens?

A. : The nodal points are a pair of conjugate points on the principal axis of the system, such that a ray of light directed towards one passes after refraction through the other in a direction parallel to itself. These are points of unit positive angular magnification.

Q.2. What are the cardinal points of a lens system?

A. : There are six cardinal points of an optical system, viz., (i) two focal points, (ii) two principal points and (iii) two nodal points.

Q.3. In this experiment, are the principal points and the nodal points the same?

A. : Yes, since the lenses are surrounded by the same medium (air) on all sides.

Q.4. Are the two focal lengths of the system equal?

A. : Yes, since they are surrounded by the same medium on two sides.

Q.5. What is the importance of the knowledge of the cardinal points of an optical system?

A. : If the cardinal points can be located, then the size and the position of the image formed by a coaxial system can be located without considering refraction through each component of the system individually.

Q.6. If the lens system produces an image of a distant object, will the image be displaced if the system is rotated a little about its second nodal point?

A. : No, it will not be displaced.

Q.7. With the increase of value of *d* (lens-separation), whether the focal length and nodal separation increase or decrease?

A. : Both of them increase with the increase of the value of *d*.

Q.8. What is the focal length of the mirror used and what part does it play here?

A. : Since it is a plane mirror, its focal length is infinity and it forms image of the object pin coincident with the pin itself.

Q.9. Sometimes one gets more than one image, why?

A. : It is the image of the pin formed by the reflection at the surface of the lens besides the desired image formed by the reflection at the plane mirror.

Q.10. How can the desired image be ensured?

A. : The desired image moves when viewed, on rotating the plane mirror, while the undesired one remains fixed.

Q.11. Why should the rotation of the lens systems about the vertical axis be small?

A. : A large rotation results in the blurring of the images.

Q.12. Why observations are made by rotating the nodal slide through 180°?

A. : This gives the other focal length of the system which of course, is the same as the first one, since the lenses are surrounded by the same medium (air) on both sides.

Q.13. What are equivalent focal length and equivalent lens?

A. : A lens is said to be equivalent to a number of lenses when it produces an image at the same plane and of the same size as that formed by the combination of lenses. The focal length of such an equivalent lens is called equivalent focal length.

Q.14. What is a telescopic system?

A. : It is a system whose cardinal points are at infinity. For such a system, the distance of separation of the two lenses will be numerically equal to the sum of their focal lengths.

5.17 Verification of Fresnel's equation of reflection and refraction of e.m. waves with the help of spectrometer, prism, two polaroids and sodium light

Theory :

From Brewster's relation, the polarising angle

$$\phi = \tan^{-1} \mu, \quad \text{... (5.17a)}$$

where μ is the refarctive index of the prism material for a monochromatic light (say, sodium light of $\lambda = 5893$ Å). The angle of refraction r is obtained from the relation

$$\sin r = \frac{\sin i}{\mu}, \quad \text{... (5.17b)}$$

Let A and B be the amplitudes of the electric vector of an e.m. wave in the plane and perpendicular to the plane of incidence respectively. If A be made equal to B, then the ratio of the reflected components A_r and B_r is given by

$$\frac{B_r}{A_r} = \frac{\cos(i+r)}{\cos(i-r)}, \quad \text{... (5.17c)}$$

where i and r are the angles of incidence and refraction respectively. The condition that vector $A = B$, may be achieved by polarising a light at an angle of 45° to the plane (horizontal) of incidence. From electomagnetic theory of light, the ratios of vectors A and A_r in the incident plane and B and B_r in the plane perpendicular to the plane of incidence, are given as

$$\frac{A_r}{A} = \frac{\tan(i-r)}{\tan(i+r)} \quad \text{(First Law)} \qquad \text{...(5.17d)}$$

The condition for complete transmission i.e. $A_r = 0$ is that $\tan(i + r) = \infty$, or $(i + r) = \frac{\pi}{2}$. In that case, no light will be reflected.

$$\text{Also, } \frac{B_r}{B} = \frac{\sin(i+r)}{\sin(i+r)} \quad \text{(Second Law)} \qquad \text{...(5.17e)}$$

This equation shows that B_r can never be zero. If the condition $(i + r) = \frac{\pi}{2}$ is set for light, the part in the plane of incidence will be completely transmitted but the reflected part will always be present because of the vectors B and B_r in a plane perpendicular to the plane of incidence. It shows that the reflected light in such a case is always plane-polarised in a plane perpendicular to the plane of incidence. The expressions are the mathemtical forms of Fresnel's laws. However, if $A = B$, we obtain from Eqs. (5.17d) and (5.17e),

$$\frac{B_r}{A_r} = \frac{\cos(i+r)}{\cos(i-r)}. \qquad \text{... (5.17f)}$$

Relation given in Eq. (5.17f) is experimentally verified.

The angles of incidence, reflection and refraction are shown in (Fig. 5.38). It is evident from the figure that $\sin r = \cos i$. Hence, $\mu = \frac{\sin i}{\sin r} = \tan i$. At $i = \phi$ (polarising angle), the reflected light becomes plane-polarised. Hence ϕ and i are found using Brewster's law given in Eq. (5.17a).

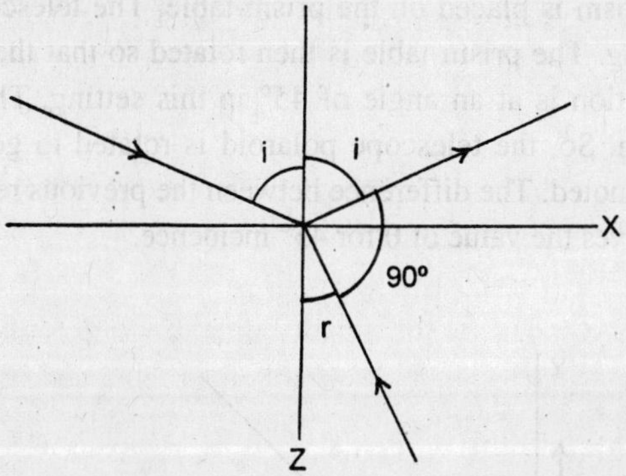

Fig. 5.38 : Angle of incidence, reflection and refraction

The angle r is obtained by using Snell's law which gives $r = \frac{\sin i}{\mu}$. Knowing i and r, the value of $\frac{B_r}{A_r} = \frac{\cos(i+r)}{\cos(i-r)}$ is obtained. The ratio $\frac{B_r}{A_r} = \tan\theta$ is also found experimentally and a graph (Fig. 5.39) is plotted with $\tan\theta$ and $\frac{\cos(i+r)}{\cos(i-r)}$, which is a straight line passing through the origin and inclined at an angle of 45° to either axes.

Apparatus :

Spectrometer, two polaroids, p: ism, sodium vapour lamp.

Method :

(i) The spectrometer is set up in the usual manner and the refractive index of the material of the prism for monochromatic light (sodium) is determined as in Expt. No. 5.5.

(ii) Using Brewster's law ($\mu = \tan \theta$), the Brewster's angle (ϕ) is thus obtained from the value of μ. It usually would come out nearly 58°. The light incident at this angle is reflected as plane-polarised light from the face of the prism.

(iii) The direct slit is then focused and the vernier reading is noted. The prism is placed on the table and the prism-table is rotated so that the slit image falls on the telescope cross-wires. The light is reflected from the prism face at angle 45° in this setting. In order to make the reflection at Brewster's angle (58°), the prism-table is clamped and the telescope rotated by an angle equal to double the difference of ϕ and 45° (58° – 45°) = 13° × 2 = 26°. The telescope is clamped and the prism-table rotated so that the slit is again focused on cross-wires. The reflection is at Brewster's angle in this setting and the reflected light is plane- polarised.

(iv) Next a polaroid is fitted on the telescope objective. The polaroid is rotated so that the intensity of the slit appears minimum. This is known as tint of passage. The polaroid pointer reading is noted.

(v) The prism is then removed. The direct slit is observed. Another polaroid is fitted on the collimator lens and it is rotated to get the minimum intensity of the slit. The collimator polaroid is rotated at an angle of 45° from this setting. This setting makes the vibration of the incident light at an angle of 45° to the plane of incidence. This condition is necessary to make $A = B$. This position of the collimator polaroid must not be disturbed.

(vi) Next the prism is placed on the prism-table. The telescope is rotated through 90° from the first setting. The prism-table is then rotated so that the slit is focused on the cross-wires. The refraction is at an angle of 45° in this setting. The intensity of the slit would not be minimum. So, the telescope polaroid is rotated to get minimum intensity. The polaroid reading is noted. The difference between the previous reading and this reading of telescope polaroid, gives the value of θ for 45° incidence.

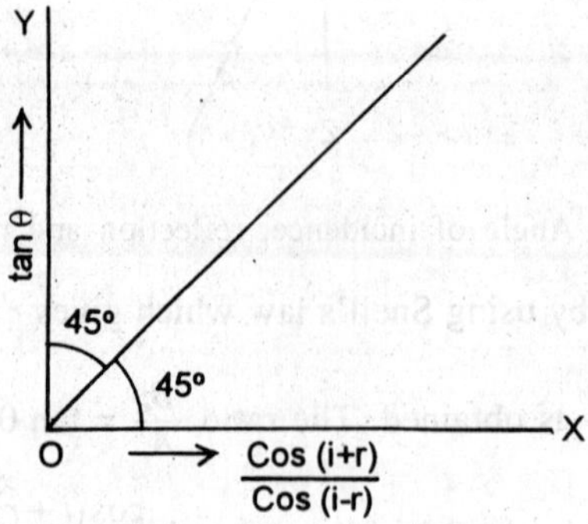

Fig. 5.39 : tan θ - $\left(\frac{B_r}{A_r}\right)$curve

(vii) The telescope is rotated by 10°. The prism-table is then rotated to focus the slit or the cross-wires. The telescope polaroid is then rotated to get the minimum intensity. The difference between this and the first reading of telescope polaroid, gives the value of θ for 50° incidence. This procedure is repeated for other angles of incidence.

(viii) The value of r (the angle of refraction) for each value of i (45°, 50°, ...) is calculated from Snell's law.

(ix) A graph is plotted with $\frac{\cos(i+r)}{\cos(i-r)}$ against tan θ, which is a straight line (Fig. 5.39) passing through the origin and inclined at 45° to either axes. It establishes and verifies Fresnel's law.

Results :

[Tables (A), (B) and (C) of Expt. No. 5.5 are recorded as usual to determine the refractive index of the material of the prism.]

(D) Determination of the polarising angle :

Standard wavelength used (λ)	Prism angle (A)	Angle of minimum deviation for wavelength λ (δ_m)	Ref. Index of prism $\mu = \frac{\sin\frac{A+\delta_m}{2}}{\sin\frac{A}{2}}$		Polarising angle $\phi = \tan^{-1}\mu$
...... cm	...... degree	....	....	Mean	......degree

(E) Polaroid readings for polarising incident light at 45° to the plane of incidence :

Telescope polaroid for minimum intensity on receiving reflected light at Brewster's angle (a)	Collimator polaroid for minimum intensity in direct observation with telescope polaroid set to position (a). (b)	Collimator polaroid after rotating it through 45° from its setting (b) (c) → fixed	Remarks
......degree	......degree	......degree	......degree

(F) Verification of Fresnel's laws :

Serial No.	Angle of incidence	Telescope Polaroid				Angle of Refraction		
		Initial	Final	Difference θ	tan θ	sin i	$\sin r = \frac{\sin i}{\mu}$	r
1.	45°	...	...	...	...	...	...	...
2.	50°	...	...	...	...	...	...	...
3.	55°	...	...	...	...	...	...	...
4.	...	...	...	...	...	...	...	...
.....	...	...	...	...	...	...	...	...
.....	...	...	...	...	...	...	...	...
.....	...	...	...	...	...	...	...	...

(G) Readings for Graph-plotting :

Telescope polaroid		Angle of incidence	Angle of refraction	$(i+r)$	$(i-r)$	$\cos(i+r)$ $= B_r$	$\cos(i-r)$ $= A_r$	$\frac{B_r}{A_r}$
θ	tan θ	(*i*)	(*r*)					
...	...	...	...	...	...	...	...	...
...	...	...	...	...	...	...	...	...
...	...	...	...	...	...	...	...	...
...	...	...	...	...	...	...	...	...
...	...	...	...	...	...	...	...	...
...	...	...	...	...	...	...	...	...

Calculations :

$$\mu = \frac{\sin\frac{A+\delta_m}{2}}{\sin\frac{A}{2}}$$

Polarsing angle, $\phi = \tan^{-1}\mu$,

Angle of refraction, $\sin r = \frac{\sin i}{\mu}$; $r =$ ....degree.

$B_r = \cos(i+r) =$ ....; $A_r = \cos(i-r) =$ ....

$$\frac{B_r}{A_r} = \frac{\cos(i+r)}{\cos(i-r)} = \text{.........}$$

Drawing of Graphs :

As stated earlier, a graph is plotted with tan θ against $\frac{B_r}{A_r}$. The graph will be a straight line passing through the origin and inclined at 45° with either axes. This graph obviously, if a straight line passing through origin and inclined at 45° with the axes, verifies Fresnel's laws (Fig. 5.39).

Precautions :

(i) Parallax error must be avoided.

(ii) One particular end of the slit-image should always be viewed.

(iii) During rotation of the prism-table, telescope and collimator with polaroids, care must be taken to avoid any slipping.

(iv) Other general precautions for spectrometric work should be taken as indicated in experiments involving spectrometers.

5.18 To study the diffraction pattern of a crossed grating with the help of a LASER source

Theory :

A combination of two plane transmission gratings having rulings of one perpendicular to those of the other, is known as a crossed grating. When a laser beam of wavelength λ is incident normally on such a crossed grating, a pattern of two dimensional diffraction spots is formed on the screen placed parallel to the plane of the grating at a distance D from it (Fig. 5.40). If θ_n be the angle of diffraction of the nth order, the grating equation gives

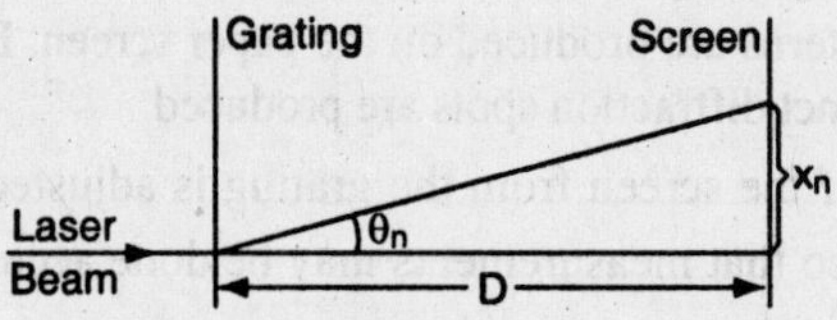

Fig. 5.40

$$d \sin \theta_n = n\lambda, \qquad \dots (5.18\ a)$$

where d is the grating constnát.

If the nth order maximum occurs at a distance x_n from the central zero maximum, then

$$\sin \theta_n = \frac{x_n}{\sqrt{D^2 + x_n{}^2}} = \frac{x_n}{D}, \qquad \dots (5.18\ b)$$

when $x_n << D$.

Putting the value of $\sin \theta_n$ from Eq. (5.18a), we get

$$d.\frac{x_n}{D} = n\lambda, \text{ or, } x_n = \frac{n\lambda D}{d}. \qquad \dots (5.18\ c)$$

Hence, if a curve is plotted with x_n versus n, then a straight line is obtained, the slope of which $m = \frac{\lambda D}{d}$. From this slope, knowing the values of D and λ, the grating constant can be obtained.

Hence, grating constant, $d = \frac{\lambda D}{\text{slope}}$. ... (5.18 d)

Apparatus :

A crossed grating, laser source, screen with a graph paper attached to it or a photo-detector, mounting accessories.

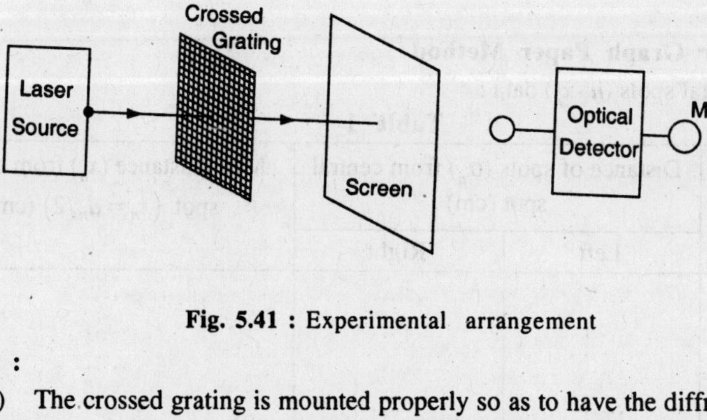

Fig. 5.41 : Experimental arrangement

Method :

(i) The crossed grating is mounted properly so as to have the diffraction elements in horizontal and vertical planes. At about 3 metres away, a white paper screen is placed properly. The laser source is placed in front of the grating.

(ii) The laser source is switched on with density fitters, height from the laser must be incident normally on the grating.

(iii) Diffraction patterns are produced on the paper screen. By properly adjusting the grating and the screen, distinct diffraction spots are produced.

(iv) The distance of the screen from the grating is adjusted to make the separation between the spots distinct so that measurements may be done accurately. The experimental set-up is shown in (Fig. 5.41).

(v) The distance of the screen from the grating (D) is measured accurately and carefu!ly.

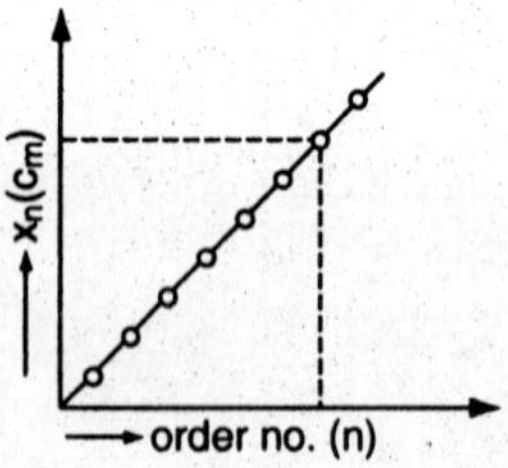

Fig. 5.42 : $(x_n - n)$ graph

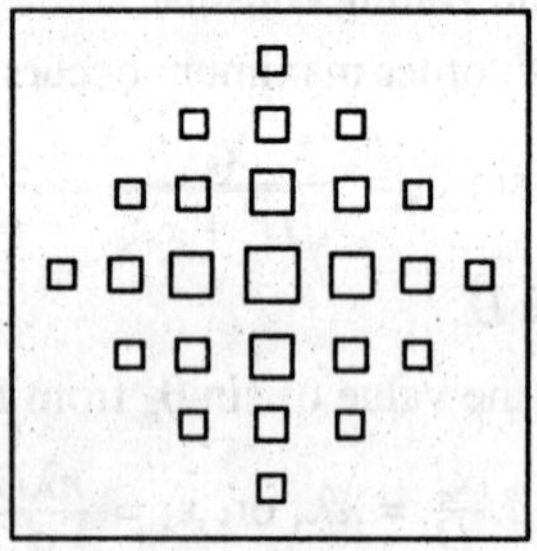

Fig. 5.43 : Difraction spots

(vi) The distances of the bright diffraction spots of different orders on the screen may be measured either by using an 'optical detector' or by using a mm graph paper.

(vii) A graph is drawn finally with order number (n) along X-axis and x_n along Y-axis for horizontal spots. The graph is a straight line passing through the origin (Fig. 5.42). The slope of the graph is measured and from it, the grating constant (d) can be had using Eq. (5.18d).

The pattern of the diffraction spots on the screen is shown in (Fig. 5.43).

(viii) Step (vii) above has to be repeated for the vertical spots also.

Results :

(A) Millimeter Graph Paper Method

For horizontal spots (n -x_n) data :

Table 1

Order number of the spot (n)	Distance of spots (α_n) from central spot (cm)		Mean distance (x_n) from central spot $(x_n = d_n/2)$ (cm)
	Left	Right	
1.	......	......	......
2.	......	......	......
3.	......	......	......
......	......	......	......
......			
......			

Table 2

For measurement with vertical spots :

[Same Table as of Table 1 is to be made]

Determination of slope (m) and Grating constant (d) :

Table 3

Calculations for	Value of slope (m) from the graph	Value of D (cm)	Wavelength used (λ) (cm)	Grating Constant (d)
Horizontal spot	......	......	......	......
Vertical spot	......	......	......	......

(B) Optical Detector Method

Readings for horizontal spots :

Table 4

Order No. of spots (n)	Detector readings for movement from				Mean Positon $r_n = \frac{a_1 + a_2}{2}$ (cm)	Mean Current/voltage $\left(\frac{\alpha_1 + \alpha_2}{2}\right)$ (mA/mv)
	Right to Left		Left to Right			
	Position (α_1) cm	Maxm. Current/voltage (mA/mV) (α_1)	Position (α_2) cm	Maxm. Current/Voltage (mA/mV) (α_2)		
8	......	......	......	......	... (r_8)	......
7	......	......	......	......	(r_7)	......
6	......	......	......	......		......
...						
...						
0	......	......	......	......	... (r_0)	......
1	......	......	......	......	... (r'_1)	......
2						
...						
...						
10	......				(r'_8)	......

Table 5

[For horizontal spots (n -x_n) data]

Order No. (n)	From Table 4 positions		Difference ($r_n \sim r'_n$) (cm)	Distance x_n of nth order from the central spot $\left[x_n = \frac{r_n - r'_n}{2}\right]$ cm
	One side (r_n) (cm)	Other side (r'_n) (cm)		
1.	......	......	......	......
2.	......	......	......	......
3.	......	......	......	......
...	......	......	......	......
...	......	......	......	......
...	......	......	......	......
8.	......	......	......	......

For measurement of Vertical Spots

Table 6

[Make a Table similar to Table 4]

(n-x_n) data for Vertical Spots

Table 7

[Make a Table similar to Table 5]

Determination of Slope and Grating constant (d)

Table 8

[Make a Table similar to Table 3]

Error Calculation :

[See 'Error Calculation' of Expt. No. 5.14]

Precautions :

(i) Alignments of the whole set-up are to be done properly.

(ii) Laser should be operated at low power level.

(iii) Grating should be uniformly illuminated.

(iv) The value of D should remain fixed and large throughout.

(v) Laser beam should not be viewed directly as it may damage the eyes.

(vi) If D is not large, then instead of x_n, one should plot $\sin \theta_n$ against n.

Questions and Answer

Q.1. What is LASER?

A. : It is an abbreviated form of Light Amplification by Stimulated Emission of Radiation. It is a device that generates an intense narrow beam of coherent monochromatic light by stimulating the emission of photons from exicted atoms or molecules.

Q.2. What are its characteristic features?

A. : It has intensity, directionality, coherence and monochromaticity.

Q.3. What are different types of lasers?

A. : Ruby laser, Semiconductor laser, He-Ne laser, quantum well lasers, etc.

Q.4. What are laser's range of wavelength and power covered?

A. : It has range from ultraviolet to radio frequency. Output power varies from some milliwatts to megawatts.

Q.5. What are the wavelengths emitted by He-Ne lasers?

A. : It has three wavelengths. Red light (λ = 6328 Å) and two in the infrared region. Of these, λ = 6328 Å is widely used.

Q.6. What do you mean by 'crossed grating'?

A. : See Theory.

Q.7. How does it differ from ordinary plane grating?

A. : A plane grating consists of a series of parallel equidistant lines for diffraction. The pattern obtained is bright and dark lines. The crossed grating consists of a large number of rectangular meshes distributed two-dimensionally. The pattern here is made of a number of bright spots on a two-dimensional screen.

Q.8. Can this experiment be done with a Na-vapour lamp instead of a Laser?

A. : In that case, beam will not be so intense and not so monochromatic.

Q.9. What type of diffraction is obtained here?

A. : It is Fraunhoffer type of diffraction.

Q.10. Where is the possible maximum error in this experiment?

A. : It is in the measurement of x_n.

[Also see the Questions/Answers under Expt. No. 5.14]

(A) Study of Divergence of Laser Beam

(B) Study of Laser as a Monochromatic Beam.

To study Laser Beams as regards to its divergence property and monochromatic nature, separate and independent experiments are beyond the scope of this book as those experiments are non-conventional and non-classical. However, that laser beams are very much non-diverged and unidirectional, can be easily realised from the experiment stated above (Expt. No. 5.18) and that it is very much monochromatic in nature, is also evident from the above experiment which gives diffraction pattern as very bright and intense.

Chapter 6

Magnetism

6.1 Determination of the horizontal component of the earth's magnetic field by reflection and vibration magnetometers

Theory :

If a bar magnet (*NS*) is placed near a magnetic needle (*ns*) in the end-on position and the axis of the bar magnet is at right angles to the earth's magnetic field, then the magnetic needle will be deflected by an angle θ, and in equilibrium, it will be under the action of two equal and opposite couples (Fig. 6.1). One couple will be due to the earth's horizontal field H acting along the meridian and the other couple will be due to the field F of the bar magnet *NS* whose moment is *M* and magnetic length 2*l*. If *d* be the distance between the centres of the magnet and the needle, then

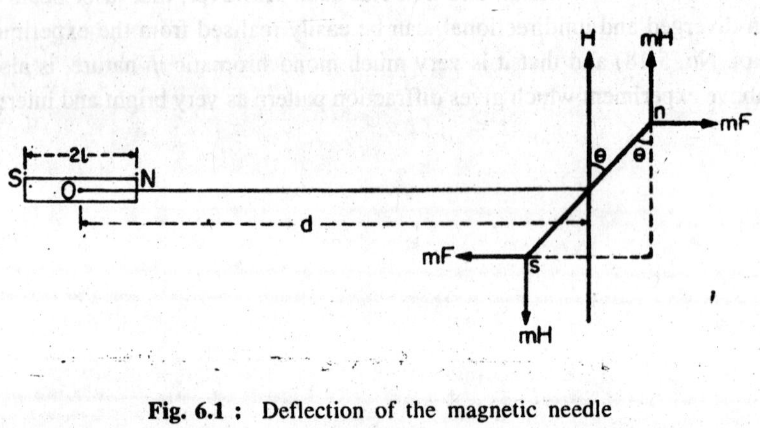

Fig. 6.1 : Deflection of the magnetic needle

$$\frac{M}{H} = \frac{\left(d^2 - l^2\right)^2}{2d} \tan\theta, \qquad \text{... (6.1a)}$$

where $2l$ = magnetic length of the magnet (*NS*)
= actual length of the magnet × 0.85.

If the same bar magnet be allowed to oscillate at the same place in a vibration magnetometer in the earth's magnetic field with its axis horizontal, then neglecting the moment of inertia of the stirrup and assuming the suspension fibre to be torsionless, the time period (*T*) of oscillation of the bar magnet will be given by

$$T = 2\pi\sqrt{\frac{I}{MH}}, \qquad \text{... (6.1b)}$$

where I is the moment of inertia of the bar magnet about the axis of suspension. If the magnet be of rectangular shape and the axis of suspension passes through the centre of the magnet and is normal to the broad face of the magnet, then

$$I = m\,\frac{l'^2 + b'^2}{12}, \qquad \text{... (6.1c)}$$

where m is the mass of the magnet and l', b' are its geometrical length and breadth respectively.

From Eqs. (6.1a) and (6.1b) we have,

$$H = \sqrt{\frac{MH}{M/H}} = \sqrt{\frac{4\pi^2 I}{T^2} \times \frac{2d}{\left(d^2 - l^2\right)^2 \tan\theta}} \qquad \text{... (6.1d)}$$

and $$M = \sqrt{MH \times \frac{M}{H}} = \sqrt{\frac{4\pi^2 I}{T^2} \times \frac{\left(d^2 - l^2\right)\tan\theta}{2d}}. \qquad \text{... (6.1e)}$$

Since, all the quantities on the right-hand side of these equations can be measured, the value of H or M can be computed.

Apparatus :

Reflection and vibration magnetometers, bar magnet, slide calliper, magnetic needle, stopwatch, lamp and scale arrangement, balance and weight box, spirit level, metre scale.

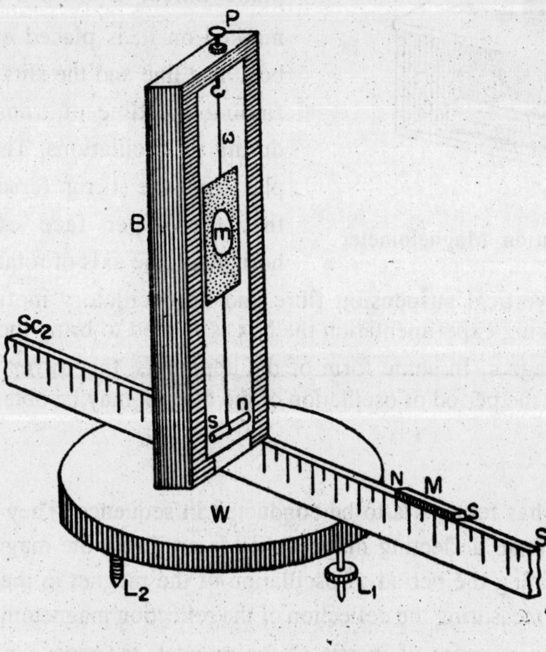

Fig. 6.2 : Reflection magnetometer

Reflection magnetometer :

The essential components of a reflection magnetometer is a rectangular box (B) mounted on a wooden base W provided with leveling screws (L_1, L_2). There is a provision for suspending a small magnetic needle (ns) by a torsionless fibre (w) from a torsion head

(P). This is enclosed in a non-magnetic case with a glass front. The wooden base is provided with two long wooden arms in a straight line and at right angles to the length of the freely suspended magnetic needle. These arms contain a wooden scale extending on both sides of the case (Fig. 6.2). The zero mark of the scale lies on the prolongation of the axis of the suspension of the magnet system. The scale is placed magnetic east-west direction and the deflecting magnet (NS) is placed on the scale (Sc_1 and Sc_2) in the end-on position. The deflection of the needle is recorded usually by a telescope applying lamp and scale arrangements.

Vibration magnetometer :

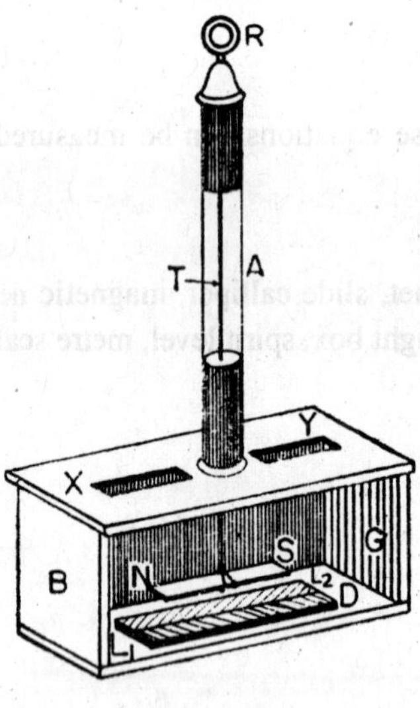

Fig. 6.3 : Vibration Magnetometer

It consists of a hollow wooden rectangular box (B) whose front surface is made of glass G. The top surface has two openings (X and Y) which are fitted with glass. A cylindrical glass tube A having a torsion head R at the top, is fixed to the middle of the top surface of the box (Fig. 6.3). A magnet NS can be placed on a light non-magnetic stirrup suspended by a torsionless fibre (T) from the torsion head. A plane mirror D with a central line L_1L_2 marked on it, is placed on the floor of the box. The line and the slits at the top, help in finding the time of transit of the magnet during its oscillations. The magnet is to be placed on the stirrup (cradle) in such a way that the upper face of the magnet is horizontal; the axis of rotation of the system coincides with the vertical suspension fibre and no oscillatory motion occurs for the suspension fibre. During experimentation the box is rotated to bring the line L_1L_2 parallel to the axis of the magnet. In some form of the apparatus, the stirrup may have a small mirror fixed to it and the period of oscillation of the magnet may be obtained by a lamp and scale arrangement.

Method :

The experiment has four parts to be conducted in sequence. They are (i) finding the moment of inertia of the deflecting magnet; (ii) determining the magnetic length of the magnet; (iii) determining the period of oscillation of the magnet in the earth's horizontal field and finally, (iv) measuring the deflection of the reflection magnetometer.

(i) To find the moment of inertia of the magnet, the mass (m) of the magnet is determined by a balance while its length (l') and breadth (b') are determined by a slide calliper. The moment of inertia (I) is then calculated from Eq. (6.1c).

(ii) The deflecting magnet (NS) is generally short and strong. Its magnetic length will be 0.85 times its geometrical length. The latter of the magnet, being already measured (l') in operation (i), when multiplied by 0.85, gives the magnetic length ($2l$) of the magnet.

If the magnet is small (of the order of 10 cm or less) and the value of d is large (of the order of 25 cm or more), then the contribution towards the error in the measurement, if geometrical length is employed instead of magnetic length, is very small compared to other sources of error. However, it is adivisable to use the magnetic length for all practical purposes.

(iii) In determining the period of oscillation of the magnet, all magnets and magnetic materials are removed from the working table. The vibration magnetometer (Fig. 6.3) is placed on the table in such a way that the line L_1L_2 in it lies along the magnetic meridian previously determined by a magnetic needle. If the suspension fibre of the vibration magnetometer exerts negligible torsional couple, then a brass bar should first be placed on the stirrup and the system is brought to rest, till the suspension fibre is made torsion-free. The torsion head (R) is rotated till in the rest position, the brass bar is parallel to L_1L_2. The bar is then replaced by the deflecting magnet, its north pole pointing north, and the axis of suspension passing through the centre of gravity. The magnet is brought to rest and then it is slightly deflected with the help of another magnet from a distance. Oscillations are set up and time for 30 oscillations are noted. Three such readings are taken from which the mean time period T is obtained. If the torsional couple of the suspension fibre and moment of inertia of the stirrup are not negligible, then MH has to be calculated from the knowledge of T_C (for cradle only), T_B (with the bar) and T_M (with magnet).

(iv) To measure the deflection of the magnetometer, after removing all magnets and magnetic substances from the working table, the magnetometer is set on the table with its two arms perpendicular to the magnetic meridian, i.e., perpendicular to the magnetic needle. This is further tested by observing equal deflections of the needle when the S-pole and the N-pole of the deflecting magnet alternately points towards the needle. The instrument is then levelled by the screws at the base. A scale-telescope arrangement is set to measure the deflection of the magnetometer mirror (m). The scale should be perpendicular to the mirror and the central reading of the scale must coincide with the vertical line of the cross-wire of the telescope. A scale with zero mark at the centre is preferred here.

(v) The deflecting magnet is then placed on the magnetometer scale in the end-on position. It is ensured that its axis is at the same height as that of the magnetometer needle. If the deflections are not the same on two sides of the central zero mark of the scale, when the deflecting magnet is turned end for end in its own position, necessary rectifications of error have to be done.

(vi) The deflecting magnet position on the scale is recorded and readings are taken for both the ends of the magnet. The mean of these two readings marks the position of the centre of the magnet, which gives the value of d in Eq. (6.1a). The deflection in the telescope is noted (x). The magnet is then turned upside down and the value of x is again noted. Then the magnet is turned end for end, so that its poles interchange their places and again upside down, taking readings (x) for all the cases.

(vii) The deflecting magnet is then placed on the other side of the scale at the same distance as before and observations made in operation (vi) above are all repeated.

(viii) The mean of the observations (deflections) made by the telescope in all these cases, is used in computing the value of θ for the particular value of d taken. The value of x

is related to θ in Eq. (6.1a) by the expression $\theta = \frac{x}{2D}$, where D is the perpendicular distance between the scale and the mirror (m).

(ix) Two other values of d are taken, each about 5 cm greater than the previous value and the observations are all repeated for both.

(x) The distance (D) is measured from the scale to the mirror (m) with the help of a scale or a thread.

(xi) Computing thus the values of I, T, d, l and, θ, the values of M and H can be calculated employing the Eqs. (6.1d) and (6.1e).

Results :

(A) Determination of length, breadth and the moment of inertia of the bar magnet :

Vernier constant of the slide calliper = ... cm.

Table 1

Mass of the bar magnet (m) gm	Length of the magnet (l') cm	Mean length (l') cm	Breadth of the magnet (b') cm	Mean breadth (b') cm	Moment of Inertia $I = m \times \frac{l'^2 + b'^2}{12}$ (gm cm²)	Magnetic length of the magnet $2l = l' \times 0.85$ (cm)
... (gm)	(i) ...... (ii) ...... (iii)	...... (cm)	(i) ...... (ii) ...... (iii)	...... (cm)	I = ... (gm. cm²)	...... (cm)

(B) Determination of the period of oscillation of the magnet in the earth's field and to find MH :

Table 2

Body oscillating	No. of obs.	Time for 30 oscillations	Mean time for 30 oscillations	Time period (T) in secs.	$MH = \frac{4\pi^2 I}{T_M^2}$
* Stirrup only	1. 2. 3.	... min ... sec ... min ... sec ... min ... sec	...... sec	...... sec (T_C)	
* Stirrup + Brass bar	1. 2. 3.	... min ... sec ... min ... sec ... min ... sec	...... sec	...... sec (T_B)	.... (dyne-cm)
Stirrup + Magnet	1. 2. 3.	... min ... sec ... min ... sec ... min ... sec	...... sec	...... sec (T_M)	

[* These two readings need be taken only, if necessary, i.e., if the stirrup has an appreciable moment of inertia and the suspension fibre is not torsionless.]

(C) Measurement of the deflection of the magnetometer and to find $\frac{M}{H}$:

Deflections :

Table 3

No. of Obs.	Position of the bar magnet			Position of the magnetic needle					Mean deflection (x) cm	Distance of the scale from the mirror (D) cm	$\theta = \frac{x}{2D}$	Value of $\frac{M}{H}$
	End I	End II	Mean		East		West					
	(d_1)	(d_2)	$d = \frac{d_1 + d_2}{2}$		Pole facing needle		Pole facing needle					
					North	South	North	South				
I	... cm	... cm	...... cm	Face up	...... cm	...... cm	...... cm	...... cm	...... (cm)	...... (cm)	......	......
				Face down	...... cm	...... cm	...... cm	...... cm	...... (cm)	...... (cm)		
II	... cm	... cm	...... cm	Face up	...... cm	...... cm	...... cm	...... cm	...... (cm)	...... (cm)	......	......
				Face down	...... cm	...... cm	...... cm	...... cm				
III	... cm	... cm	...... cm	Face up	...... cm	...... cm	...... cm	...... cm	...... (cm)	...... (cm)	......	......
				Face down	...... cm	...... cm	...... cm	...... cm				

∴ mean value of $\frac{M}{H}$ = ... dyne-cm/oersted^2.

Calculations :

From Tables 2 and 3, the values of MH and $\frac{M}{H}$ are respectively obtained. Again, from Eq. (6.1d) and (6.1e),

$$H = \sqrt{\frac{MH}{M/H}} = \sqrt{\frac{4\pi^2 I}{T^2} \times \frac{2d}{\left(d^2 - l^2\right)^2 \tan\theta}}$$

$$= \frac{2}{T\left(d^2 - l^2\right)} \sqrt{\frac{2Id}{\tan\theta}} = \ldots\ldots \text{ oersted}$$

$$\text{and } M = \sqrt{MH \times \frac{M}{H}} = \sqrt{\frac{4\pi^2 I}{T^2} \times \frac{\left(d^2 - l^2\right)^2 \tan\theta}{2d}}$$

$$= \frac{2\pi\left(d^2 - l^2\right)}{T} \sqrt{\frac{I \tan\theta}{2d}} = \ldots\ldots\ldots \text{ dyne-cm/oersted.}$$

Precautions :

(i) All magnets and magnetic materials should be removed from the vicinity of the magnetometers; otherwise, the earth's field may be distorted there due to their presence.

(ii) The suspension fibre should be almost torsionless.

(iii) The axes of the deflecting magnet and of the needle must be at the same horizontal plane.

(iv) During oscillation of the vibration magnetometer, the amplitude should be small (less than 8°).

(v) The magnetic needle should be made free by adjusting the levelling screws at the base of the deflection magnetometer.

(vi) The bar magnet should be properly set on the stirrup, so that the axis of the suspension passes through its centre of gravity and the upper surface of the magnet should remain horizontal.

(vii) If the moment of inertia of the cradle is comparable to that of the magnet and if the suspension fibre is fully torsionless, then necessary corrections are to be made accordingly while determining the moment of inertia of the bar magnet. (See the foot-note of Table 2.)

Error Calculation :

From Eqs. (6.1d) and (6.1e),

$$\frac{M}{H} = \frac{\left(d^2 - l^2\right)^2}{2d} \tan\theta = \frac{\left(d^2 - l^2\right)^2}{2d} \theta = \frac{\left(d^2 - l^2\right)^2}{2d} \times \frac{x}{2D} \text{ (when } \theta \text{ is small)}$$

$$\text{and } MH = \frac{4\pi^2 I}{T^2}.$$

$$\therefore \quad H^2 = MH \div \frac{M}{H} = \frac{16\pi^2 IdD}{T^2\left(d^2 - l^2\right)^2 x} = \frac{16x^2 IdD}{T^2 x . y^2} \text{ [where } y = (d^2 - l^2)]$$

Hence the proportional error in the measurement of H is

$$\frac{2\delta H}{H} = \frac{\delta I}{I} + \frac{\delta D}{D} + \frac{\delta d}{d} + \frac{2\delta T}{T} + \frac{\delta x}{x} + \frac{2\delta y}{y}.$$

Since, $I = \dfrac{m(l'^2 + b'^2)}{12}$, $\dfrac{\delta I}{I} = \dfrac{\delta m}{m} + \dfrac{2\delta l' + 2\delta b'}{l'^2 + b'^2}$.

Again, the time period T is obtained by noting time (t) for 30 oscillations. Hence, $T = \dfrac{t}{30}$.

or, $\dfrac{\delta T}{T} = \dfrac{\delta t}{t}$.

Thus, $$\frac{2\delta H}{H} = \frac{\delta m}{m} + \frac{2(\delta l' + \delta b')}{(l'^2 + b'^2)} + \frac{\delta D}{D} + \frac{\delta d}{d} + \frac{2\delta t}{t} + \frac{\delta x}{x} + \frac{2\delta y}{y}.$$

From individual calculations of contributions of errors for each term, it comes out that the value of H obtained is reliable up to two places of decimals. The above calculation of proportional errors indicate that the percentage error $\left(\dfrac{\delta H}{H} \times 100\right)$ in H can be minimised by taking care in measuring t and x which contribute greatest to the errors. Hence, t (i.e., T) and the displacement x of the scale image should be measured very carefully and quite a number of times to reduce the errors contributed by them.

Questions and Answers

Q.1. Define the terms : Horizontal component of earth's intensity, Magnetic moment, Magnetic intensity and Pole strength.

A. : [See any textbook.]

Q.2. Are the values of MH and $\dfrac{M}{H}$ same everywhere on the earth's surface?

A. : No, for the same magnet, their values are different at different places because of the variations of the value of H at different places.

Q.3. Does this experiment give the value of the pole strength?

A. : Yes, since the magnetic moment = pole strength × length of the magnet. The value of magnetic moment (M) being measured and the length of the magnet being known, the value of the pole strength can be known.

Q.4. What are the magnetic elements of the earth? Are they all constants at all places?

A. : Declination, dip and the horizontal intensity are the magnetic elements which completely specify the earth's magnetic field at a place. Their values are different at different places of the earth.

Q.5. What happens if any magnet or magnetic substance lies near the magnetometers during the experiment?

A. : In that case, the magnetic needle will not move freely and the bar magnet and the needle will both be affected by their presence.

Q.6. How does the earth's magnetic field run?

A. : Earth's magnetic field runs from south to north along a vertical plane known as the magnetic meridian, inclined to the geographical meridian by a certain angle called **declination**. The total field runs along the magnetic meridian by making a certain angle with the horizontal, which is known as the **dip** at the place.

Q.7. Can the experiment be performed by placing the axis of the deflecting magnet along the magnetic meridian?

A. : No; in that case, the field due to the magnet and the earth's field become parallel to each other and there will be no deflecting couple acting on the magnetic needle.

Q.8. Why is the deflection of the needle kept around 45°?

A. : To minimise the error in the measurement of $\frac{M}{H}$.

Q.9. Can this experiment be performed in the broad-side-on position and of the two, which position is preferred and why?

A. : It can be done in the broad-side-on position but the end-on position is preferred because for the same distance of the magnet from the needle, the field due to the magnet is greater in the end-on position than in the broad-side-on position increasing thereby the accuracy of the measurements.

6.2 To study the nature of dependence of dipolar field of a short bar magnet on distance with the help of a deflection magnetometer and to determine the horizontal component of the earth's magnetic field

Theory :

According to the theory of the previous experiment [Expt. No. 6.1], the value of $\frac{M}{H}$ as per Eq. (6.1a) is

$$\frac{M}{H} = \frac{(d^2 - l^2)}{2d} \tan\theta.$$

For a very short magnet (approximately a dipole), when $l << d$ this equation reduces to,

$$\frac{M}{H} = \frac{d^3}{2} \tan\theta \qquad \text{... (6.2a)}$$

or, $$\tan\theta = 2\left(\frac{M}{H}\right) \times \frac{1}{d^3} \qquad \text{... (6.2b)}$$

So, a plot of $\tan\theta$ against $\frac{1}{d^3}$ will be a straight line passing through the origin. The slope of this straight line graph will be $\frac{2M}{H}$ from which the value of $\frac{M}{H}$ can be obtained,

i.e., slope $(m) = \frac{2M}{H}$. ... (6.2c)

If M is given or obtained, then the value of H can be had from the graph. Alternatively, the value of $\left(\frac{M}{H}\right)$ and MH being obtained as in the previous experiment, the value of H may be obtained following Eq. (6.1d) of the previous experiment.

Apparatus :

[Same as in Expt. No. 6.1]

Method :

[Same as in Expt. No. 6.1]

Drawing of Graph :

For small values of d (when the bar magnet is close to the deflecting magnet), the graph will not be a straight line. But when d is large enough, the graph of $\left(\tan\theta - \frac{1}{d^3}\right)$

will be a straight line. From this straight portion of the graph, the slope of the graph is obtained (m) from Eq. (6.2c) and hence the value of H is obtained knowing M (knowing the mass of the magnet and hence its moment $M = 2ml$).

Results :

[Prepare same Tables 1, 2, 3 as in Expt. No. 6.1]

Table 4

Determination of $\frac{M}{H}$ using the graph :

Distance (d) of the dipole magnet (cm) [From Table 3]	Mean Deflection of the magnetic needle (θ)	Value of $\tan\theta$	Value of $\left(\frac{1}{d^3}\right)$ cm^{-3}
......	......	......	......
......	......	......	......
......	......	......	......

Table 5

Data for Graph Drawing and Slope determination :

Value of $\left(\frac{1}{d^3}\right)$ from graph (cm^{-3})	$\tan\theta$ from graph	Slope of the line, $m = \frac{\tan\theta}{(1/d^3)}$	Value of $\frac{M}{H}$ from graph ($m = 2M/H$) (dyne-cm/oersted²)
......	......	......	......
......	......	......	......

Calculations :

[Same as in Expt. No. 6.1]

Only calculations for obtaining the slope of the curve of $\left(\tan\theta - \frac{1}{d^3}\right)$ is to be done as shown.

Precautions :

[Same as in Expt. No. 6.1]

Error calculation :

[Same as in Expt. No. 6.1]

Questions and Answers

[In additions to the Questions-Answers of Expt. No. 6.1]

Q.1. When the bar magnet is considered as a dipole?

A. : When the distance between the two poles of the magnet is very short compared to other distances measured in the particular problem.

Q.2. Can this experiment be performed with the broad-side-on position (Tangent-B position)?

A. : Yes, but tan-A position is performed over tan-B position because the field due to the bar magnet is greater in tan-A configuration.

Chapter 7

Electrical Accessories and Instruments

7.1 Resistances and their different types

In an electrical circuit, it is often required to vary the current strength and for this, resistances are inserted in the circuit. With the increase of the circuit resistances, the current decreases and *vice versa*. Normally, the resistances inserted in the circuit are variable in nature and sometimes fixed resistances are also used. The different types of resistors used in electrical circuits are discussed below briefly.

(A) Fixed resistance coils

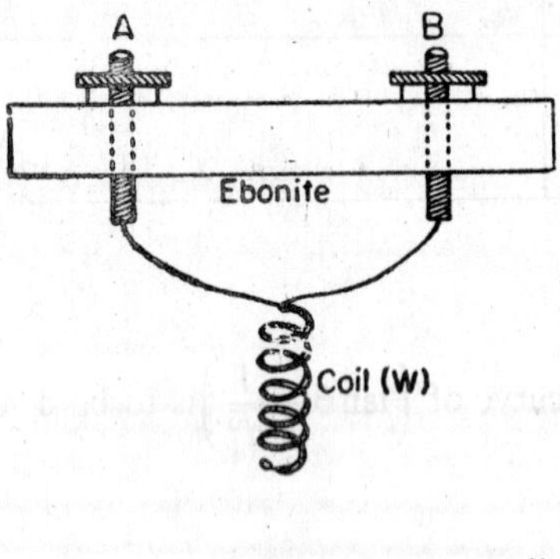

Fig. 7.1 : Fixed resistance coil

Often in an electrical circuit, coils of wire having fixed known resistances are used. It consists of two brass screws (A, B) fixed on an ebonite plate (Fig. 7.1). The two ends of a piece of enamelled wire, are soldered to the screws. The wire (W) has a fixed resistance and it is wound on a wooden bobbin after being doubled. This type of winding reduces self-induction to a minimum. The wire is generally thick and of short length for low value resistances, while for larger values, it is of long length and thin. The wire is generally made of manganin (manganese—12%, copper—84% and nickel—4%). Its specific resistance is very high and temperature coefficient almost nil. The completed coil is furnished with a protective non-conducting coating of varnish or wax.

(B) Resistance box

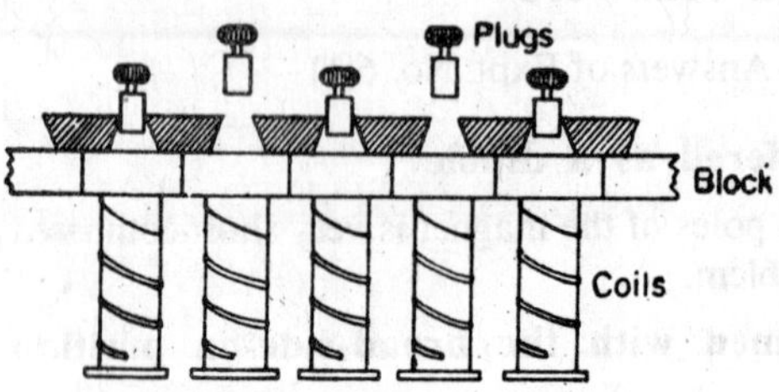

Fig. 7.2 : Resistance box

It consists of a wooden box in which a number of fixed resistance coils are so mounted that any number of them can be used at a time in series. The coils are constructed in the same manner as described above. The ends of the coil are soldered to two pins connected to two brass blocks mounted on a sheet of ebonite. Each block is separated from each other by a small air gap and a conical hole inside which a conical brass plug with an ebonite top can be fitted in (Fig. 7.2). When the plug is put inside the hole, then the current passes through the plug instead of through the coil, as if the coil is being shortcircuited. The coil thus becomes inoperative then. When the plug is taken out of the hole, the current passes from one block to the other through the resistance coil, which means the value of the resistance assigned to the particular coil is then introduced in the circuit. In this way, if more than one plug are taken off simultaneously, then more than one resistance are introduced in the circuit. Since the coils are in series the total effective resistance put in the circuit will be the sum of the resistances whose plugs are taken off (Fig. 7.3). When all the plugs are kept in positions, the effective resistance applied is zero. The points A and B are the terminals of the resistance box. The value of the particular resistance is written down near the corresponding hole on the box.

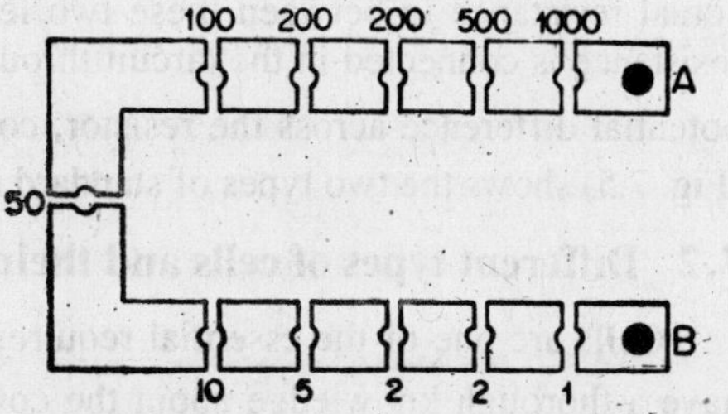

Fig. 7.3 : Arrangement in a resistance box

Care should be taken to see that *while using a resistance box, all plugs other than those are opened for use, must be kept failrly tight*; otherwise, extra resistances would be introduced in the circuit which are not taken into account.

(C). Rheostat

This is a form of a variable resistance, which when connected in a circuit, enables to change the current *continuously* in the circuit as necessary. It consists of a coil of bare manganin wire would in a thin, narrow groove cut uniformly on a porcelain cylinder. The ends of the wire (A and B) are fixed to two screws (Fig. 7.4). An insulated rod PQ is fixed above the cylinder. On this lies a sliding contact (R), which always touches the base wire. There is also another screw (S) fixed at one end of this rod. If the rheostat is connected in the circuit by the terminals, A and B, the whole of the resistance of the coil is included in the circuit and no variation of resistance is obtained by sliding R. But if one end of the circuit is connected to the screw S and the other end either to A or B, then current flows through A to R and then R to S (or, B to R and then R to S) and only a portion of the coil becomes effective as the resistance put in the circuit. The actual value of the resistance included in the circuit, depends on the position of R set on the coil.

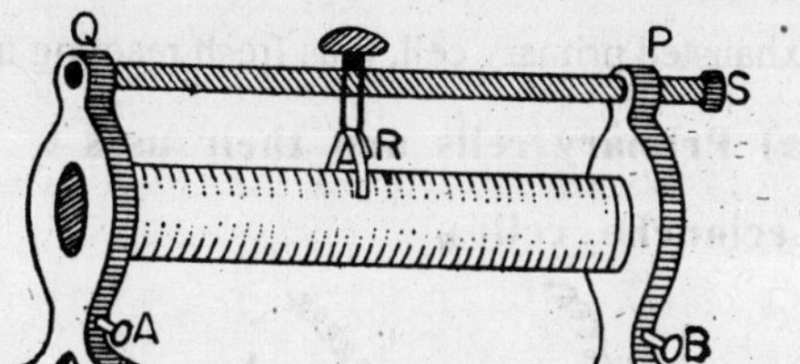

Fig. 7.4 : Rheostat

(D) Standard resistance

It is generally a resistance of fixed value of 1 ohm or less. It consists of an ebonite plate

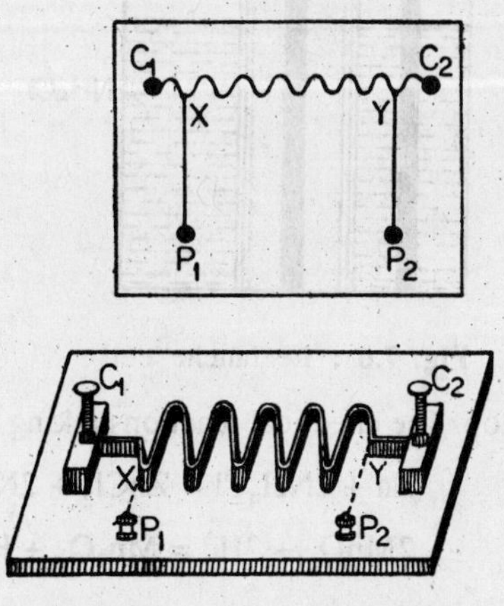

Fig. 7.5 : Standard resistance

on which a metal strip or a coil is fixed by two binding screws (C_1 and C_2). Through one of them current enters and leaves through the other (Fig. 7.5). These terminals are, therefore, called the *current leads*. The value of the resistance assigned to this standard resistor is slightly less than the actual resistance in between these two screws. Just beside the above terminals on the inner side, there are two more screws (P_1 and P_2) attached to two points (X, Y) of the copper strip. These two screws are known as the *potential leads*. The actual resistance in between these two leads is the value specified on the resistor. The resistance is connected in the circuit through the current leads (C_1, C_2) and to measure the potential difference across the resistor, connections are to be made to the potential leads. (Fig. 7.5) shows the two types of standard resistances.

7.2 Different types of cells and their uses

Cells are one of the essential requirements in electrical experiments. Thus, one must have a thorough knowledge about the construction, action and uses of different types of cells, before employing them in electrical circuits.

Electrical cells can be classified under two types : (a) *Primary cells* and (b) *Secondary cells*. A primary cell is a device from which electrical energy is available by chemical reactions. The common examples of this type of cells are the Daniell cell, the Leclanche cell, the Standard cell etc. The amount of electrical energy available from these types of cells depends upon the total amount of the reacting material contained inside the cell. The cell thus ceases to function when the reacting materials are exhausted. To restore back the exhausted primary cell, thus fresh reacting materials are to be added.

(a) Primary cells and their uses :

Leclanche cell

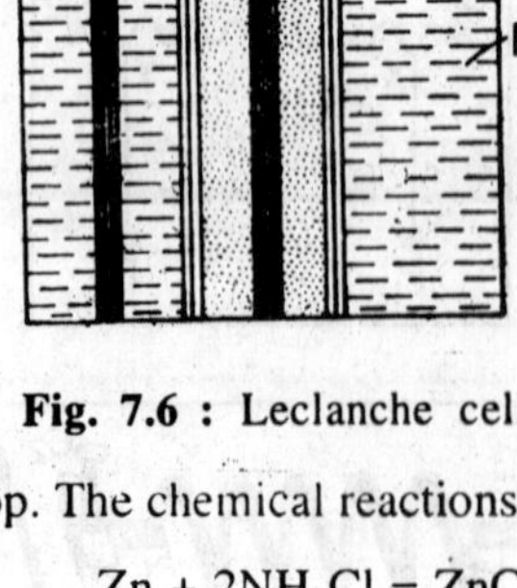

Fig. 7.6 : Leclanche cell

In this type of cell, a porous chinaclay pot (P) is placed inside a glass jar which contains saturated ammonium chloride solution. This solution is the electrolyte. An amalgamated zinc rod (Z) is immersed in the electrolyte (Fig. 7.6) and this rod behaves as the negative pole. It has a screw fixed at its top. Within the chinaclay pot, remains a carbon rod (C) immersed in it and a mixture of powdered charcoal and MnO_2 are placed around the rod. The carbon rod is the positive pole. The mouth of the pot is closed with pitch. The carbon rod also has a screw attached to its top. The chemical reactions taking place are as follows :

$$Zn + 2NH_4Cl = ZnCl_2 + 2NH_3 + 2H^+$$

$$2MnO_2 + 2H^+ = Mn_2O_3 + H_2O.$$

The e.m.f. of this cell is about 1.45 volt.

Daniell Cell

This cell consists of a cylindrical copper vessel which serves as the positive pole. It contains a saturated solution of $CuSO_4$ which also acts as depolariser. This solution is kept saturated by allowing it to come in contact with copper sulphate crystals placed on a perforated shelf *S*. The negative plate is a amalgamated zinc rod (*Z*) which is kept immersed (Fig. 7.7) partly in dilute sulphuric acid contained in the porous pot *P* placed inside the copper vessel. The acid is the exciting liquid. The chemical reaction taking place inside the cell is as follows.:

$$Zn + H_2SO_4 = ZnSO_4 + 2H^{++}$$

$$2H^{++} + CuSO_4 = H_2SO_4 + Cu^{++}$$

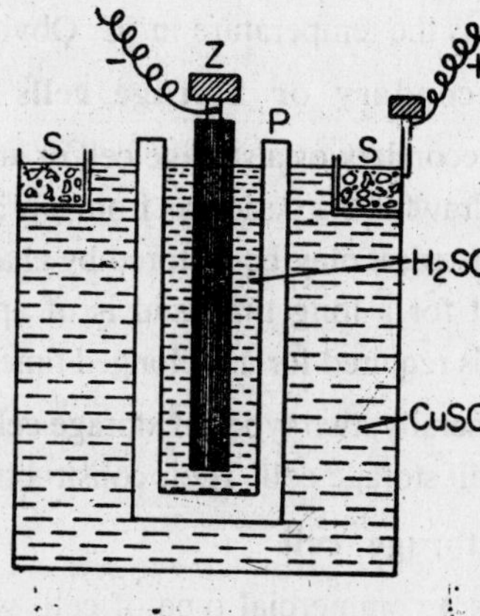

Fig. 7.7 : Daniell cell

The e.m.f. of this cell is 1·08 volt and it remains steady even when the current is drawn from it. It cannot be employed in those cases where large current is necessary, since this cell has a high internal resistance. It is most preferably used where small but constant current is required.

Standard cell

It is a basically a primary cell and is used for providing a standard for measuring a potential difference. It should never be used as a source of current and should not deliver a current greater than 10^{-4} ampere. Two types of standard cells are in use, which are (i) Clark cell and (ii) Weston Cadmium cell. The latter is commonly used, in which the electrodes are placed in the opposite sides of an H-shaped glass vessel (Fig. 7.8) which contains the electrolyte ($CdSO_4$ solution). The positive and the negative poles are connected to the base of the two arms (A_1 and A_2) by means of platinum wires. The positive-end arm (A_1) contains mercury at its base upon which lies a mixture of Hg_2SO_4 and $CdSO_4$. Above it lies the electrolyte ($CdSO_4$ soln.). The other arm (A_2) contains a mixture of Hg and Cd at its base upon which lies the $CdSO_4$ solution. The two arms are connected by the arm *R*. The $CdSO_4$ solution has to be kept saturated. The platinum wire from the two arms are connected with two very high resistances (R_1, R_2) in series. The internal resistance of this cell is very high and so current can never be drawn from this cell, rather it is used as a standard, when making comparison of e.m.f. of other cells or sources. It should never be handled roughly. The container of the

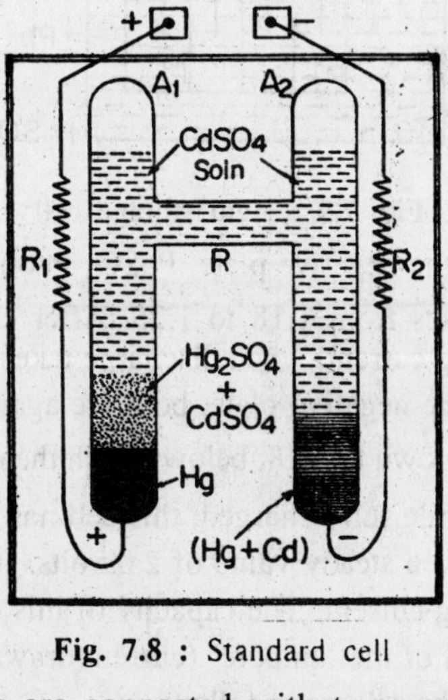

Fig. 7.8 : Standard cell

cell is hermetically sealed and the cell is covered by a lightproof cover since it is sensitive to light.

The e.m.f. of this cell is

$$E_t = 1.0183 - 0.0000406\ (t - 20)\ \text{volt},$$

where t is the temperature in °C. Obviously, the e.m.f. of the cell at 20 °C is 1.0183 volt.

(b) Secondary or Storage cells :

A secondary or a storage cell or an accumulator is a device from which electric current can be drawn for a period as from a primary cell but unlike the latter, when the accumulator is exhausted, it may be restored by charging it again. The e.m.f. of this type of cell remains constant for a long time and is of appreciable high value. For this reasons, when high current is required for a prolonged time, then these cells are only used.

Generally, two types of storage cells are used in the laboratory, (i) Acid storage cell and (ii) Alkali storage cell. Their constructions and actions are as follows :

Acid storage cell

It is a commercial type of cell, which when charged is ready to deliver a current of moderate strength for a prolonged time. In this cell, both the positive and the negative plates are made of lead having the structure of a grid or network. The plates are immersed in sulphuric acid contained in a polythene or glass vessel. The Pb plates react with H_2SO_4 and produce $PbSO_4$. The capacity of the cell is increased in increasing the working surface of the plates. Thus the capacity and the current output will be greater, if the number of plates and their surface areas become larger. The cell is charged from a D.C. source properly before use. During charging, in the positive plate, $PbSO_4$ is converted into PbO_2 and in the negative plate, $PbSO_4$ into metallic Pb (Fig. 7.9). During this, the sp. gravity of acid rises from 1.18 to 1.25. After charging, the cell can be employed in a circuit for drawing current, when discharge takes place. During discharge, PbO_2 in positive plate and Pb in the negative plate, both are again converted into $PbSO_4$. The sp. gravity of the acid comes down to 1.18, below which the cell should not be used.

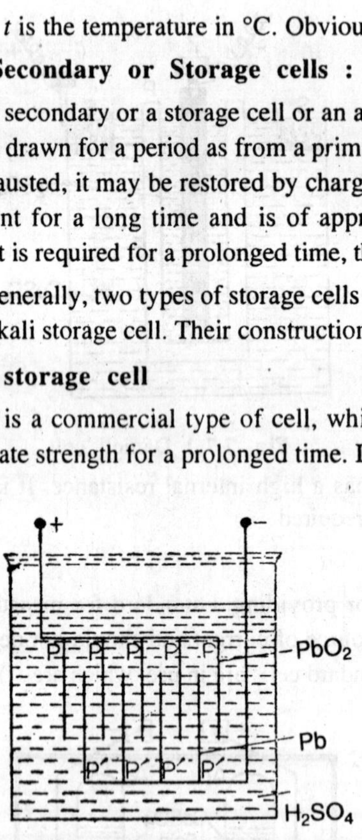

Fig. 7.9 : Acid storage cell

While fully charged, this cell has the e.m.f. of 2.20 volts but after use for some time, it shows a steady value of 2.0 volts. If it comes down below 2.0 volts, it is better to stop drawing current. The capacity of this cell is expressed in 'ampere-hour', which gives the product of the 'ampere' (current drawn) and the 'hour' of use. The reactions taking place inside the cell are as follows :

During charging,

at positive plate : $PbSO_4 + H_2O + O = PbO_2 + H_2SO_4$

at negative plate : $PbSO_4 + 2H = Pb + H_2SO_4$.

During discharge,

at positive plate : $PbO_2 + 2H + H_2SO_4 = PbSO_4 + 2H_2O$

at negative plate : $Pb + O + H_2SO_4 = PbSO_4 + H_2O$.

Alkali cell

In this cell, invented by Edison, the electrolyte is a 21% solution of KOH in distilled water and a small amount of Li(OH) of about 1%. Thus, it is known as 'alkali cell' or 'Edison cell' or 'Ni-Fe cell' as the negative plate consists of finely divided mixture of Fe and FeO contained in a perforated nickel-steel frame and the positive plate is of some higher oxides and hydroxides of Ni contained in a similar frame. During charging, $Ni(OH)_3$ is formed at the positive plate and in the negative plate, the mixture of Fe and FeO remains. During discharge, $Ni(OH)_3$ is converted into $Ni(OH)_2$ and Fe is oxidised to FeO. During charge or discharge, the electrolyte does not undergo any change (Fig. 7.10). During discharge, the e.m.f. of the cell remains steady at 1.20 volt and when this value goes down to 1.10, it is better then not to use the cell. The strength of the electrolyte remains steady during discharge and its sp. gravity remains at 1.19. The reactions taking place are as follows :

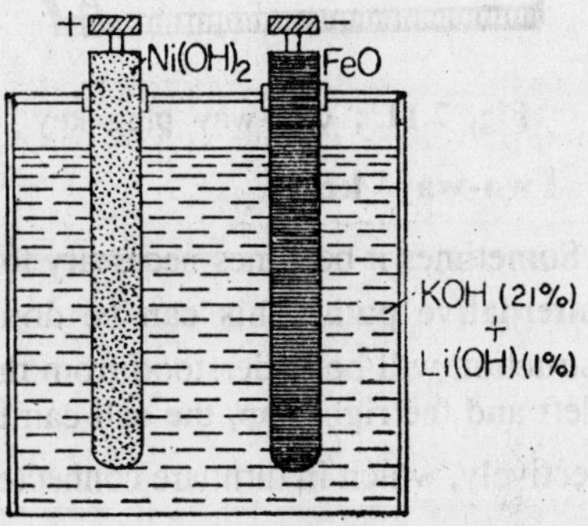

Fig. 7.10 : Alkali cell

$$Fe + 2Ni(OH)_3 \underset{\text{charge}}{\overset{\text{discharge}}{\rightleftharpoons}} Fe(OH)_2 + 2Ni(OH)_2$$

In all respects, this type of cell is advantageous and preferable for uses as compared to an acid storage cell, as this cell is not spoilt by accidental reverse charging and does not become damaged after prolonged use. It can, in fact, withstand violent mechanical treatment and handling, overcharging and discharging above the normal rate. It can also be left unused for a long time after being charged.

7.3 Keys and their uses

Keys are the simple devices for making or breaking a circuit. There may be different types of keys used for different purposes. They are briefly discussed below :

(a) One-way plug key

With the help of this key, circuits can be made close or open and it acts just like a switch. It has an ebonite base (*A*) on which are mounted two brass blocks (*M* and *N*) separated by a small air gap (Fig. 7.11). A brass plug (*P*) with an ebonite top can fit into this hole. When the plug (*P*) is inserted in this hole metallic connection is established between the blocks *M* and *N* which are connected to the circuit. Thus, on inserting the plug, the circuit is closed and on taking out the plug, the circuit is broken. This key, however, should never be connected in a circuit powered by supply mains, because the air gap being small, an arc across the gap may start on breaking the circuit.

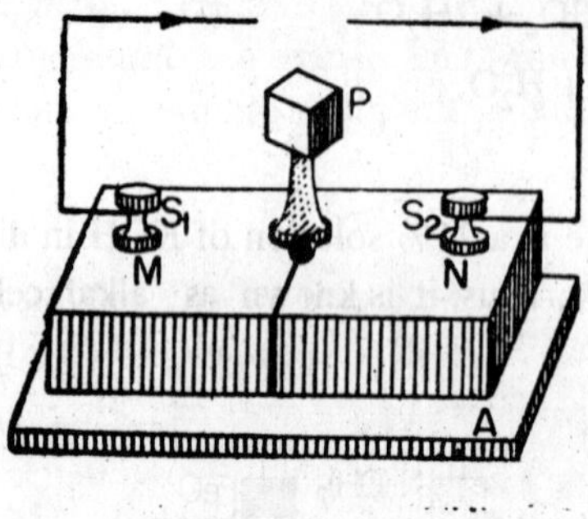

Fig. 7.11 : One-way plug key

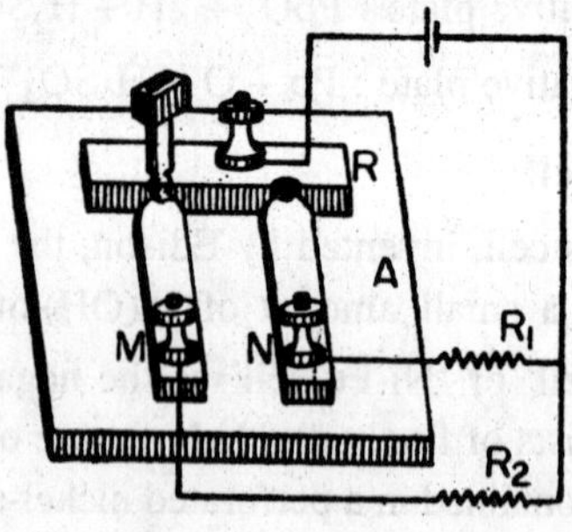

Fig. 7.12 : Two-way key

(b) Two-way key

Sometimes it becomes necessary to send a current when desired through a particular or an alternative path. This can be done by employing a two-way key (Fig. 7.12). Its construction will be understood from the figure and here by placing the plug alternately in the left and the right gap, the cell can be connected either through the resistance R_2 or R_1 respectively, which in turn, are connected to the binding screws *M* and *N*.

(c) Four-way key

This is also a key like a two-way type and its principle of construction is the same. But in this key, arrangement is there to pass current through any of the four alternative paths. The principle of action of the key will be evident from the (Fig. 7.13). The main circuit can be connected to either of the path 1, 2, 3, 4 by properly placing the key.

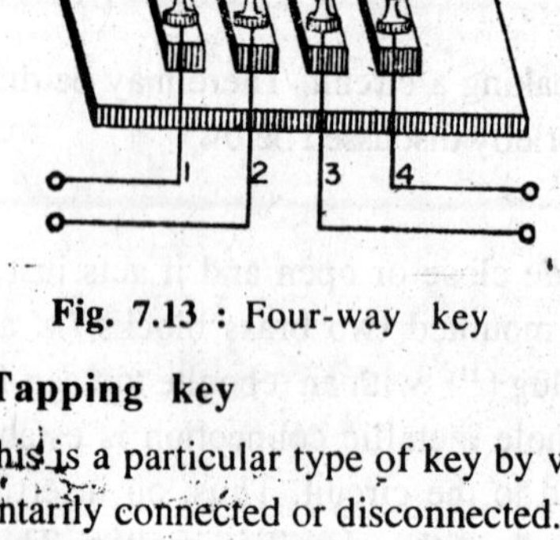

Fig. 7.13 : Four-way key

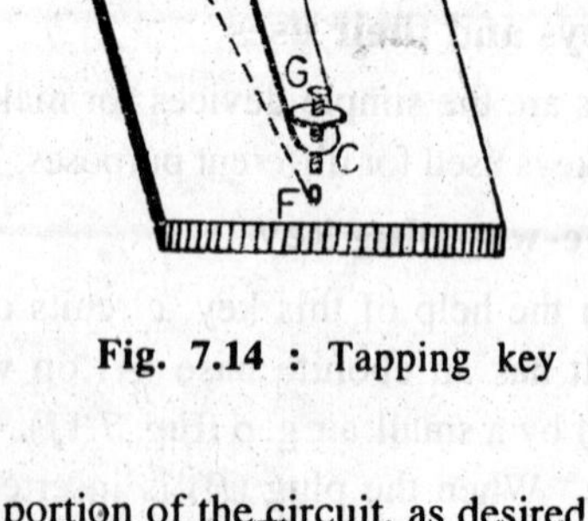

Fig. 7.14 : Tapping key

(d) Tapping key

This is a particular type of key by which any portion of the circuit, as desired, may be momentarily connected or disconnected. The principle of its action will be understood from the (Fig. 7.14). When the top portion (*G*) of the end *C* is pressed, the point *F* is connected to *C* and also to *A*, making the circuit closed through *A* and *B*. On releasing the pressure on *G*, the strip *M* rises up and the circuit is disconnected at *C*.

Commutators :

A commutator is a device by which the direction of a current through a circuit or a part thereof, can be reversed without changing the connections. There are different types of commutators of which the most commonly used are (i) the plug commutator and (ii) the Pohl's commutator.

(a) Plug commutator

It consists of four blocks of brass, making up a circle or a square, mounted on an ebonite base (*E*) and separated from each other by a small air gap (Fig. 7.15). Binding screws are attached to each block. Adjacent blocks may be put into metallic contact by inserting a brass plug (*P*) into the hole. The portion of the circuit in which the direction of current has to be reversed, is connected to two diagonally opposite blocks (like 1, 3, or 2, 4). The other two blocks are connected to the cell or battery. When the keys are placed face to face, the current flows in one direction while when the keys are placed on the other two holes (face to face), the direction of the current is reversed.

Care should be taken that while reversing the direction of the current, *both the plugs should be removed first* and then inserted on the other holes, otherwise there will be a risk of the battery being short-circuited.

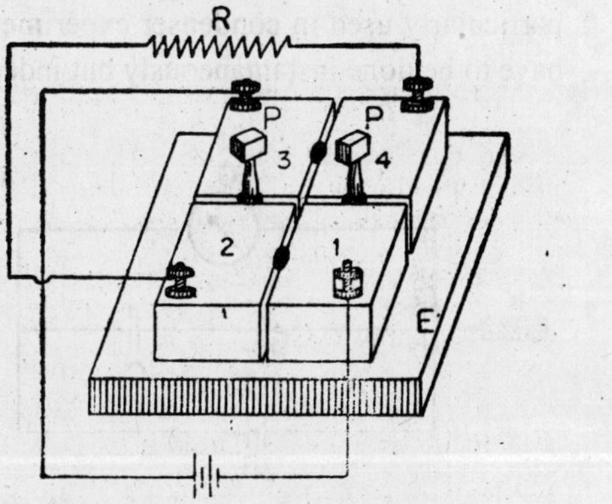

Fig. 7.15 : Plug type commutator

(b) Pohl's commutator

In this type of commutator, on an ebonite base, there are six small holes or cups containing mercury. Each cup is metalically connected to a binding screw (Fig. 7.16). The

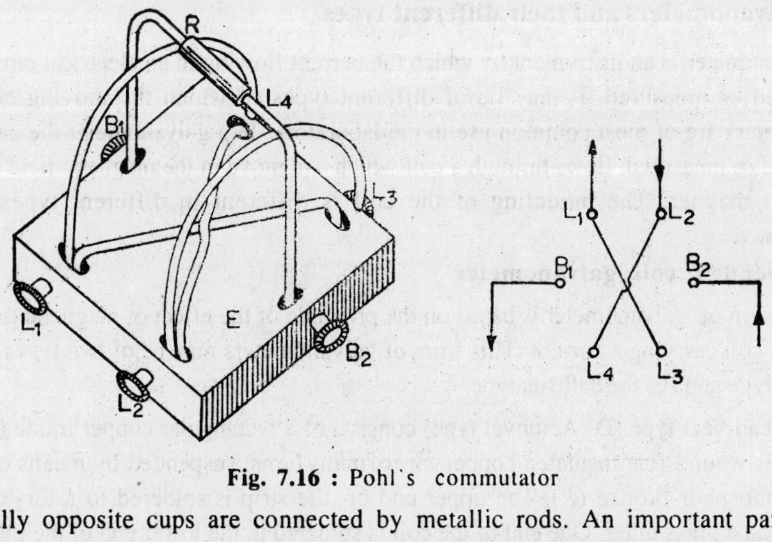

Fig. 7.16 : Pohl's commutator

diagonally opposite cups are connected by metallic rods. An important part of this commutator is the 'rocker' (*R*) which has a horizontal handle connecting two sets of legs, three on each side. The outer legs are a little shorter than the middle legs. So, when two legs are on one side remain in contact with mercury, other two legs on the other side are above the mercury surface. The screws B_1 and B_2 are connected to the battery circuit and the portion of the circuit in which the direction of the current is to be reversed, is connected to the pair of the screws (either L_1, L_2 or L_3, L_4). The method of connecting the circuit will be clear from Fig. 7.16. When the rocker is moved from one side to the other, the direction of the current is automatically reversed in the required portion of the circuit.

Charging and Discharging key

Sometimes it becomes necessary to close a particular portion or section of a circuit and to switch over to another portion of the circuit, disconnecting the former one. To achieve this, a special type of key is used which is known as 'charging-discharging key'. This is particularly used in condenser experiment, where charging and discharging of a condenser have to be done instantaneously but independently. Fig. 7.17 shows the schematic diagram of such a key. By pressing the arm *K* down, the point (*b*) is in contact with *P* containing the capacitor *C* and it is charged. When the key *K* springs back in the upward direction, automatically the point *b* is cut off, and contact is established with the point (*a*) making the galvanometer circuit operative, and discharging of the condenser takes place. Arrangement is also made in this type of key, to keep the handle in the neutral position, connecting neither the point *a* nor the point *b*.

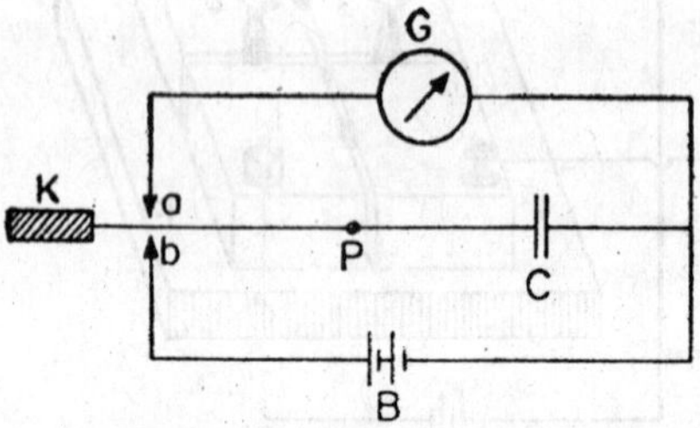

Fig. 7.17 : Charging-discharging key connection

The use of this key will be mainly in experiments involving condensers, particularly in determining the value of the capacitance of a condenser or in the measurement of a high resistance by the method of leakage, etc.

7.4 Galvanometers and their different types

Galvanometer is an instrument by which the current flowing in an electrical circuit may be detected or measured. It may be of different types of which the moving coil type galvanometers are of most common use in the laboratory. In a galvanometer the current to be detected or measured, flows through a coil which can move in the magnetic field due to a permanent magnet. The mounting of the coil is different in different types of the galvanometer.

(a) Suspended coil galvanometer

This form of galvanometer is based on the principle of the effect of magnetic field on a suspended coil carrying a current. This from of galvanometers may be of two types : (i) the dead-beat type and (ii) the ballistic type.

The dead-beat type (D' Arsonval type) consists of a rectangular copper frame (*ABCD*) on which is wound fine insulated copper wire of many turns, suspended by means of a fine piece of phosphor-bronze (*P*). The upper end of the strip is soldered to a torsion head supported on a brass pillar. One end of the coil is soldered to the lower end of the phosphor-bronze, while its other end is attached to a fine spiral spring of phosphor-bronze (*Q*) and connected to a screw (Fig. 7.18). The current thus passes through the suspension fibre, the coil and the spiral spring. Normally, the coil with the frame hangs freely in a cylindrical cavity between the concave pole-pieces (*N* and *S*) of a permanent horse-shoe magnet. Within the pole pieces, a soft iron cylinder (*I*) is placed coaxially, so that the field produced is radial. The coil lies outside the cylinder and can rotate freely in the annular gap between it and the pole faces.

A small miror is attached to the suspension wire from which light spot is reflected and studied by a lamp and scale arrangement.

If i amount (e.m.u.) of current flows through the coil having n number of turns and cross-sectional area A, and if this causes the rotation of the coil by an amount θ radian, then

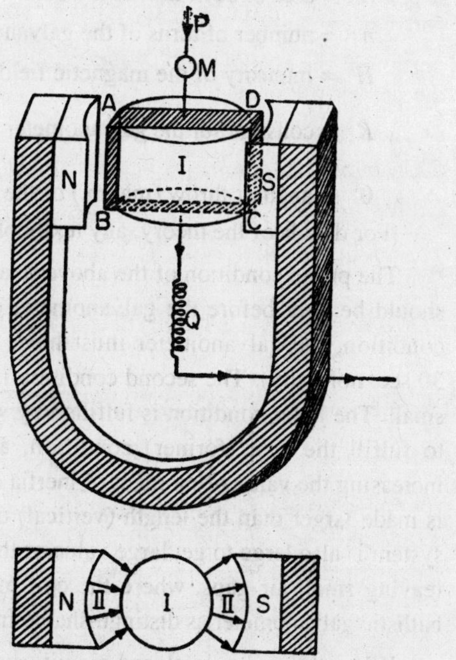

Fig. 7.18 : Moving-coil galvanometer

$$i = \frac{c\theta}{nAH} = K\theta,$$

where H = strength of the magnetic field

c = torque per unit twist of the suspension

K = Galvanometer constant = $\dfrac{c}{nAH}$.

The whole instrument is caged in a brass cover with a glass window and it has three levelling screws, by adjusting which the coil can be made to rotate freely in the air gap between the magnet and the softiron piece without touching any one of them.

(b) Ballistic galvanometer

This form of the galvanometer is used for measuring the total charge that moves through a conductor in a circuit when a current passes for a very short duration. Normally, a moving coil type of this galvanometer is used in the laboratory. Its construction is similar to that of a current measuring D' Arsonval type of galvanometer but with some structural differences giving it a large periodic time of oscillation and a small dumping.

When an instantaneous current passes through a ballistic galvanometer, an angular impulse exerted on the moving system gives the system an angular momentum. The coil is deflected and the maximum angle of throw is proportional to the quantity of charge passing through the coil.

If q is the charge which passes through the galvanometer for a transient current, then it is given by

$$q = \frac{cT}{2n\pi AH}\theta = K\theta,$$

where T = periodic time of oscillation of the coil in the open circuit.

c = torque per unit twist of the suspension.

A = area of each turn of the coil.

n = number of turns of the galvanometer coil.

H = intensity of the magnetic field in which the coil moves.

K = constant for the galvanometer $\left(= \dfrac{cT}{2\pi nAH}\right)$

θ = angular throw (maxm.) of the coil.

[For details of the theory, any textbook on Electricity may be cunsulted.]

The prime condition of the above relation is that the impulse arising out of the current should be over before the galvanometer coil moves appreciably from rest. To fulfil this condition, the galvanometer must have a long time period of oscillation (about 20 to 30 sec. normally). The second condition is that the electromagnetic damping must be very small. The latter condition is fulfilled by winding the coil in a non-conducting frame while to fulfill the first (former) condition, a delicate suspension of small value of c and increasing the value of moment of inertia of the coil, are employed. The breadth of the coil is made larger than the length (vertical) of the coil. The vertical length of the suspension system is also large to get large value of the period and the soft iron pole pieces are broader leaving small air gaps, where the coil moves. These are the characteristic features of a ballistic galvanometer as distinguishable from a current measuring galvanometer.

When the coil is in closed circuit, some electromagnetic damping as well as damping due to air resistance and viscosity of the suspension fibre are present. As such, the coil deflection does not attain its full value (θ) as required by the formula. The observed maximum deflection θ' is related to the ideal value θ by the relation.

$$\theta = \theta'\left(1 + \frac{\lambda}{2}\right),$$

where λ is called the *logarithmic decrement*. It can be calculated by observing successive maximum deflections on the two sides as the coil oscillates. If $\theta_1, \theta_2, \theta_3, \ldots, \theta_n$, etc. are successive amplitudes of the oscillations alternately to the right and left, then logarithmic decrement,

$$\lambda = \frac{1}{n+1}\log_e \frac{\theta_1}{\theta_n} = \frac{2.303}{n-1}(\log_{10}\theta_1 - \log_{10}\theta_n).$$

The value of λ depends on the total resistance of the galvanometer circuit and its value should be determined under the conditions in which the throw occured.

An oscillatory system subject to damping can be made dead-beat by suitably increasing the damping. The point at which the transition from oscillatory to dead-beat character is obtained, is known as *critical damping*. This is achieved by inserting an *additional resistance* in series with the coil externally. For a particular value of this resistance R_c, the critically damped condition is attained. This value (R_c) of the resistance is known as the *external critical damping resistance* (*XCDR*). This value is normally supplied by the manufacturer. For values higher than R_c, the coil motion is oscillatory, while for values smaller than R_c, the motion is *dead-beat*.

Sensitivity of a Galvanometer

If a weak current flowing through a galvanometer, produces a large deflection in it, then the galvanometer is said to be *current sensitive*. While if a small voltage applied across its terminals, produces a large deflection, it is said to be *volt-sensitive*.

The *current sensitivity* or the *figure of merit* of a galvanometer is defined to be the current in ampere required to produce a deflection of 1 millimetre on a scale placed at a distance of 1 metre from the galvanometer mirror. Lower this value, the greater is the sensitivity of the galvanometer. It increases by increasing the number of turns of the coil and by making the suspension system delicate. Normally, the galvanometers used in the laboratory, have a figure of merit of the order of 10^{-8} ampere per division.

The volt-sensitivity of a galvanometer is defined as the potential difference in volt to be applied across the galvanometer terminals to produce a deflection of 1 mm on a scale placed at a distance of 1 metre from the mirror. Thus volt-sensitivity = figure of merit × galvanometer resistance.

A volt-sensitive galvanometer should have a *low resistance* while a current-sensitive galvanometer should have a *high resistance*.

7.5 Shunts and their uses

Shunt is a very low resistance connected across a galvanometer in *parallel* so as to protect the galvanometer from being damaged due to passage of a heavy current through it. It is normally a resistance of value less than 1 ohm which means a fractional resistance box connected in parallel with the galvanometer. Sometimes a standard low resistance also may be used as a shunt but by using a box, the value of the shunt can be altered at will and the strength of the current desired to be passed through the galvanometer may be adjusted. The arrangement of the shunt connection is shown in Fig. 7.19. If the value of the shunt resistance is S ohms and the resistance of the galvanometer is G ohms, then the ratio of the currents flowing through the galvanometer (C_g) and through the shunt (C_s) will be given by

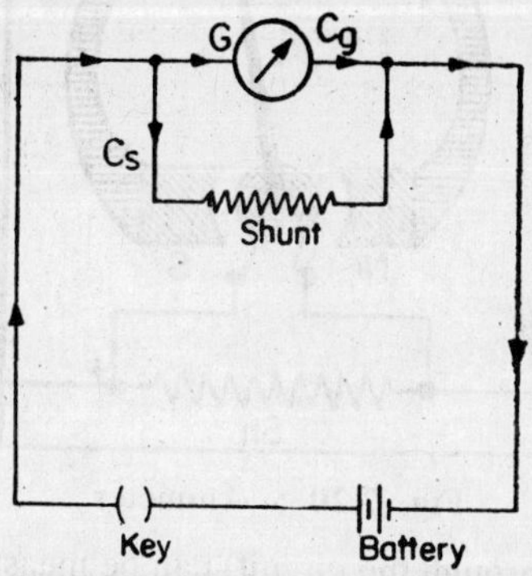

Fig. 7.19 : Shunt connection

$$\frac{C_g}{C_s} = \frac{S}{G}.$$

If the main and the total current in the circuit be C, then $C = C_g + C_s$. Hence, current through the shunt will be

$$C_s = \frac{G}{S+G}.C$$

and current through the galvanometer will be

$$C_g = \frac{S}{S+G}.C.$$

Naturally, since the value of S is much smaller than G, a small fraction of the total current flows through the galvanometer protecting it thereby from being damaged.

One should be careful to understand that high values of resistance should never be inserted in a shunt resistance box. In that case, there is every possibility of the galvanometer being damaged.

7.6 Ammeters and Voltmeters

Ammeter is used to measure directly the current flowing through a circuit, while *Voltmeter* is used to measure the potential difference between any two points in a circuit. In fact, both the ammeter and the voltmeter are the modified forms of a moving coil galvanometer. These are generally light and portable and in both the cases, the deflection of the coil is measured by a pointer moving across a scale. These scales are graduated in *ampere* (in case of an ammeter) and in *volts* (in case of a voltmeter).

Ammeter

It is a short form of a suspended coil type galvanometer in which a low-resisance shunt is connected (*Sh*) in parallel to the coil (Fig. 7.20). As such, the equivalent resistance across the two main connecting screws is very small. The scale is graduated either in *amperes* or in *milliamperes* and calibrated previously. From the position of the pointer on this scale, the magnitude or the strength of the current passing through the instrument (and

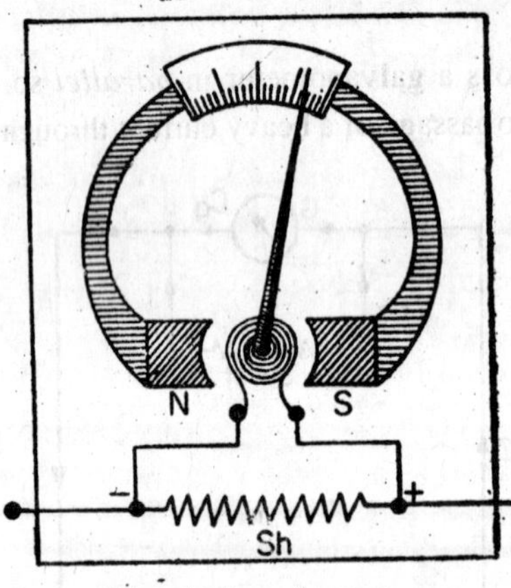

Fig. 7.20 : Ammeter

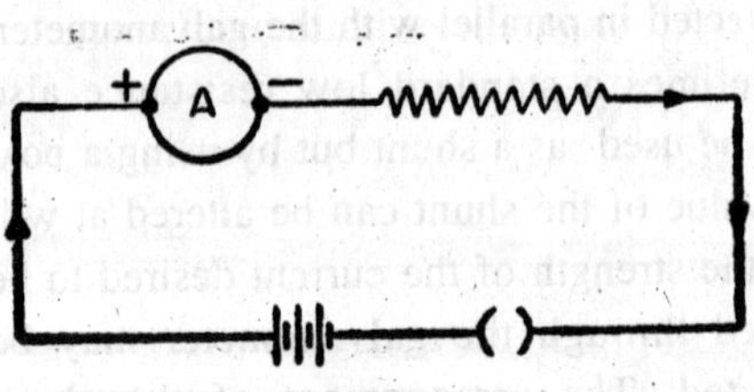

Fig. 7.21 : Ammeter connection

hence through the circuit) can be measured directly. The shunt carries the major part of the current and reduces the effective resistance of the instrument. That the resistance of the instrument should be low, is a necessity, otherwise its insertion in the circuit would alter the value of the current which needs to be measured. To measure a current in a circuit by an ammeter, the latter must be connected in *series* in the circuit (Fig. 7.21).

Voltmeter

It is also a compact portable model of a moving coil galvanometer in which a very *high* resistance is connected in *series* with the coil (Fig. 7.22). Its construction is exactly similar to that of an ammeter, the only difference being that while the ammeter has a low resistance in parallel with the coil, the voltmeter has a high resistance in series with the coil incorporated within the instrument. The points between which the potential difference has to be measured, the voltmeter has to be connected in *parallel* across those two points. Of these two points, the higher potential point has to be connected to that screw of the voltameter marked (+). Care should be taken in this regard. Because of a high resistance in-

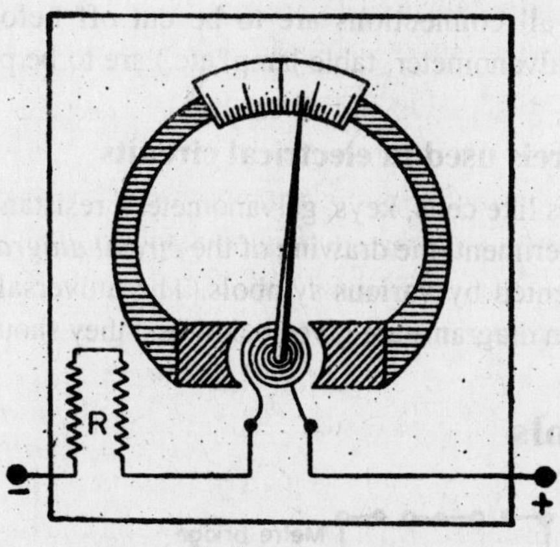

Fig. 7.22 : Voltmeter

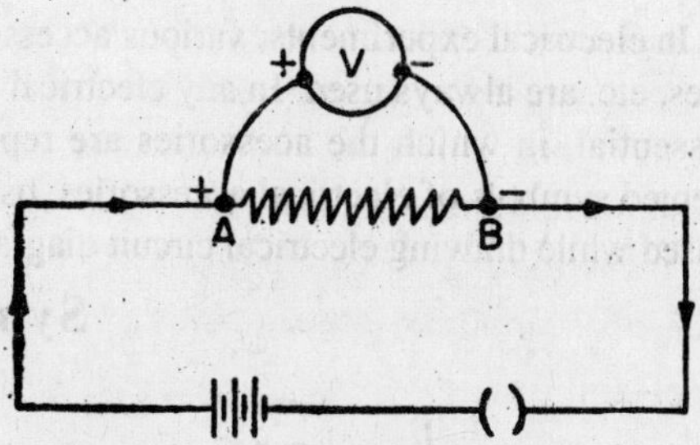

Fig. 7.23 : Voltmeter connection

inserted in the instrument in series, the effective resistance of the instrument is very high and for this, when the instrument is connected in parallel across a certain portion of the circuit, the resistance of the circuit does not change practically, and the value of the potential difference across those two points does not effectively change because of the insertion of the instrument. This scale is graduated either in volts or in millivolts.

The way a voltmeter is connected in a circuit is shown in Fig. 7.23

By altering the values of the shunt or the series resistances in an ammeter and a voltmeter respectively, the range of the particular instrument may be extended. Care must be taken in connecting these instruments properly in the circuit.

7.7 General instructions for electrical experiments

In performing electrical experiments, some general instructions are to be followed and necessary precautions are to be adopted. Those are discussed briefly below :

(i) To ensure good electrical contacts, the ends of the connecting wires must be made clean by sand papers and junction screws must be made tight properly.

(ii) In resistance boxes or P.O. Boxes, the plugs must be inserted tightly by properly rotating and pushing them after insertions.

(iii) A tapping or a pressing key must be used in parallel to a sensitive galvanometer, so that the oscillation of the galvanometer coil can be stopped quickly by pressing the key at the right moment.

(iv) For a suspended coil galvanometer introduced in the circuit, either a low resistance shunt or a higher resistance in series, should be used to protect the galvanometer.

(v) For a steady current in the circuit, a storage cell (preferably an alkali cell) should be used but in Wheatstone bridge experiments Leclanche's cell may be employed.

(vi) Before inserting plugs in the keys, the connections of the circuit must be checked to ensure whether proper connections are being done. Current in the circuit should be allowed to flow as long as it is required. Continuous flow of current in the circuit should never be allowed, unless it is required in the experimental procedure. Continuous flow of current will heat the circuit.

(vii) After completing the experiment, all connections are to be cut off before leaving the working table. Switches of lights (galvanometer, table lamp, etc.) are to be put off also.

7.8 Conventional symbols of the accessoreis used in electrical circuits

In electrical experiments, various accessories like cells, keys, galvanometers, resistance boxes, etc. are always used. In any electrical experiment, the drawing of the *circuit diagram* is essential, in which the acessories are represented by various symbols. The universally accepted symbols of electrical accessories, used in diagrams are given below and they should be used while drawing electrical circuit diagrams.

Symbols

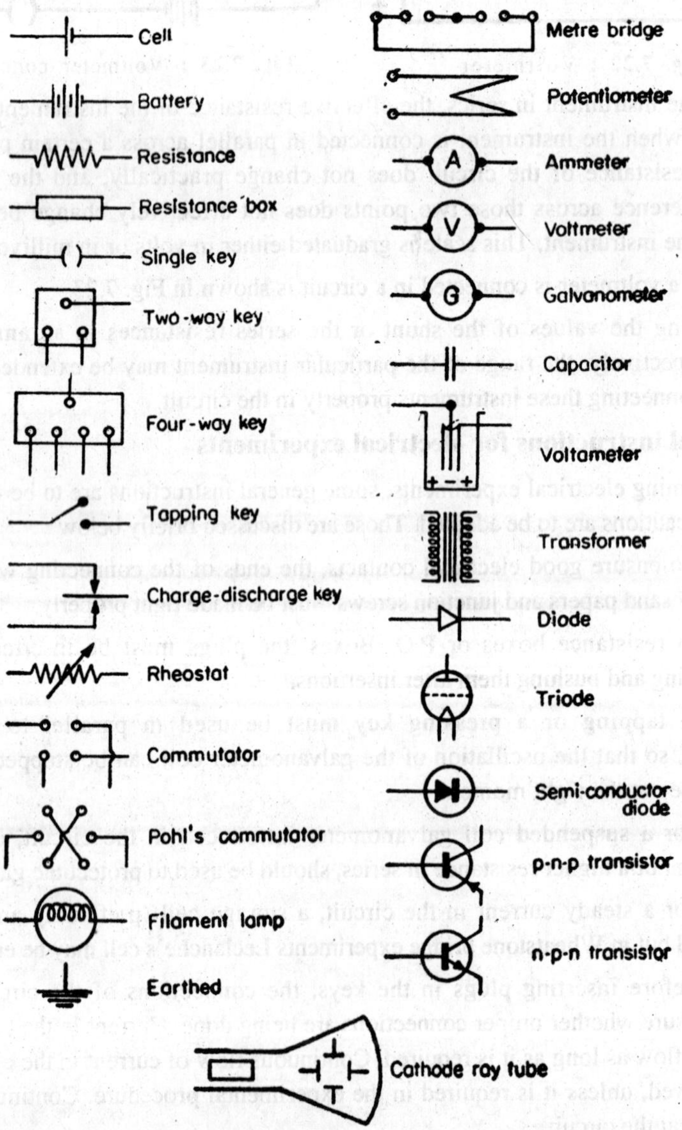

Fig. 7.24

Chapter 8

Electrical Experiments

(Based on Wheatstone's principle)

8.1 Principle of Wheatstone Bridge

In order to determine the resistance of a conductor in an electrical experiment in the laboratory, the method usually employed is based on a very important principle known as *Wheatstone Bridge principle*. It is a special type of circuit arrangement in the form of a closed network (Fig. 8.1). It consists of four resistances P, Q, R and S arranged as shown. Of these P and Q as well as R and S are in series. The PQ arrangement remains in parallel to the RS arrangement. The points A and C are connected to the terminals of a battery (E) while the other two points (B and D) of the bridge are connected to a galvanometer. The values of the four resistances P, Q, R and S are so arranged that no current flows through the galvanometer, i.e., the potentials of B and D are made the same. Under this condition the directions of the current flowing through other portions of the circuit are shown in the diagram. If V_A, V_B, V_C, V_D are the potentials of the points A, B, C, D respectively, then

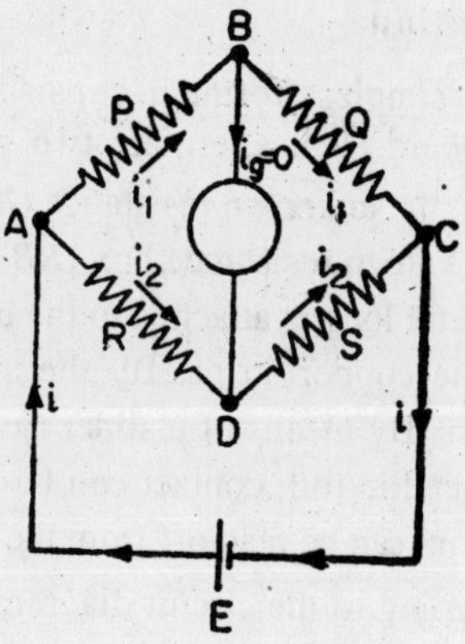

Fig. 8.1 : Wheatstone Bridge Circuit

$$V_A - V_B = i_1 P; \qquad V_A - V_D = i_2 R$$

$$V_B - V_C = i_1 Q; \qquad V_D - V_C = i_2 S.$$

But, $\quad V_A - V_B = V_A - V_D$ (since $V_B = V_D$)

and $\quad V_B - V_C = V_D - V_C$

$$\therefore \quad \frac{V_A - V_B}{V_B - V_C} = \frac{V_A - V_D}{V_D - V_C}$$

$$\text{or,} \quad \frac{i_1 P}{i_1 Q} = \frac{i_2 R}{i_2 S}; \text{ or, } \frac{P}{Q} = \frac{R}{S}.$$

This relation is known as *Wheatstone's balanced condition*. From this relation, if the values of the three resistances (P, Q, R) are known, then the value of the fourth (S) resistance can be determined.

This principle is known as Wheatstone's principle and this is employed in the instruments like P.O. Box, Metre bridge, etc.

It is to be noted that if $\frac{P}{Q}$ is made equal to $\frac{R}{S}$, there will be no current through the galvanometer whether it is connected across B and D and the battery across A and C or vice versa. Hence, the points B, D and A, C are said to be conjugate. The condition for balance, however, does not depend in any way on the resistances in series with the battery and the galvanometer. Any alternation of these resistances simply affects the sensitivity of the arrangement.

8.2 Determination of the end-corrections of a Metre Bridge

A metre bridge, constructed and based on Wheatstone's principle, is usually employed in the laboratory to measure the resistance of a conductor.

Construction

In its simplest form, it consists of three thick copper strips (A, B, C) fixed on a wooden board (W). There are two gaps (G_1 and G_2) in between the strips (Fig. 8.2). In these gaps, the unknown resistance (X) whose value is to be determined is connected in one gap and a known resistance box (RB) in the other. A uniform length of manganese wire of about 1 metre long is attached to the board in a stretched condition, and is fixed between the ends of the copper strips. By the side of the wire is fixed a metre scale graduated in millimeters. By means of a slider (jockey J) moving through a groove parallel to the wire or on a metallic rod, contact can be made at any point of the wire and the reading of the contact point can be read off from the scale.

According to the circuit diagram shown in (Fig. 8.2), a cell E is connected through a key (K) and a rheostat (Rh) to the corner points of the strips A and C. A galvanometer (G) is

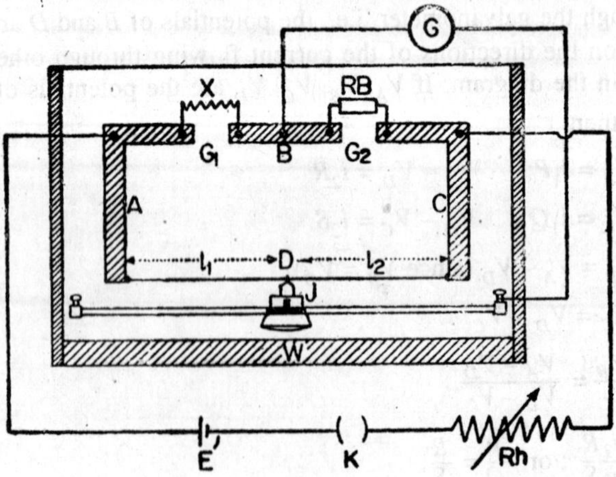

Fig. 8.2 : Metre bridge

connected between the strip B and the jockey (J). When the unknown resistance X is placed in its proper position, if a certain known resistance R is inserted from the box RB, it will be found that by sliding the jockey on the wire, a point (D) will be obtained when there will be no deflection of the galvanometer. At this stage, the two segments of the wire divided by the jockey and the two resistances (X and R), act as the four arms of the Wheatstone bridge. According to the Wheatstone bridge principle, under the condition,

$$\frac{\text{Resistance between } A \text{ and } B}{\text{Resistance between } B \text{ and } C} = \frac{\text{Resistance between } A \text{ and } D}{\text{Resistance between } D \text{ and } C}.$$

Now, the resistance between A and B is X and that between B and C is R (known). Since the wire is uniform, its resistance (ρ) per unit length is constant. If the distance of the point D is l_1 from the left end, then the resistance of the left segment (AD) of the wire is $l_1\rho$ and that of the right segment (DC) is $l_2\rho$.

Hence, $$\frac{X}{R} = \frac{l_1\rho}{l_2\rho} = \frac{l_1}{l_2} = \frac{l_1}{(100 - l_1)} \text{ (since, } l_2 = 100 - l_1)$$

or, $$X = R.\frac{l_1}{100 - l_1}. \qquad \text{... (8.2a)}$$

Since the value of R is known and l_1 is known from the scale, the value of the unknown resistance (X) can be determined.

End-correction of the metre bridge

In deducting the working principle (Eq. 8.2a) of a metre bridge, it has been assumed that the copper strips and the junctions have no resistances. Further it is assumed that the end points of the copper wire coincide with the zero and 10 marks of the scale. But in reality, these are not the cases, for which the end points of the wire possess some resistances whose values are equal to λ_1 and λ_2 cm of wire-lengths. These resistances are known as the 'end-corrections' of the bridge.

Theory for determination of end-corrections

If P and Q are the two known resistances inserted in the two gaps of the metre bridge and if null point is obtained when the jockey is placed at D_1 (Fig. 8.3), then

$$\frac{P}{Q} = \frac{l_1 + \lambda_1}{(100 - l_1) + \lambda_2}, \qquad \text{... (8.2b)}$$

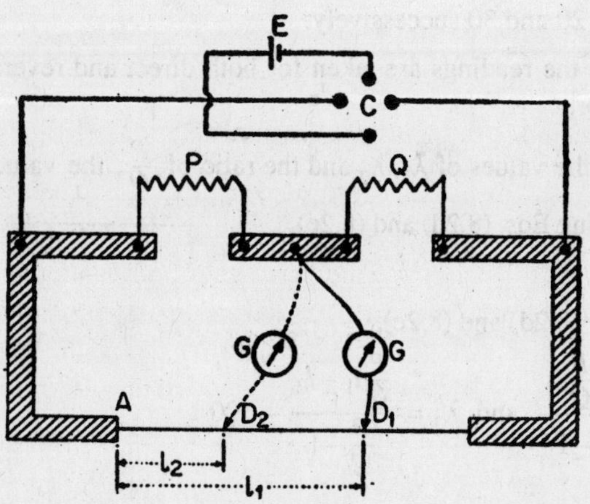

Fig. 8.3 : Determination of the end-correction of a metre bridge

where λ_1 and λ_2 are the end-corrections of the bridge. Again, after changing the positions of P and Q, if the null point is obtained when the jockey is placed at D_2, then

$$\frac{Q}{P} = \frac{l_1 + \lambda_1}{(100 - l_1) + \lambda_2}. \quad \text{... (8.2c)}$$

From Eqs. (8.2b) and (8.2c), $\lambda_1 = \dfrac{l_1 - \frac{P}{Q} l_2}{\frac{P}{Q} - 1}$ ... (8.2d)

and $$\lambda_2 = \frac{\frac{P}{Q} l_1 - l_2}{\frac{P}{Q} - 1} - 100. \quad \text{... (8.2e)}$$

From these equations, the values of the end-corrections can be determined.

The circuit connection is made as shown in (Fig. 8.3). The values of P and Q are so chosen that their ratio $\left(\frac{P}{Q}\right)$ is 10 or greater. The cell (E) is connected through a commutator (C) as shown.

Method :

(i) The jockey is at first brought towards the end point at the left and then at the right. If the galvanometer shows an opposite deflection in these two cases, then the circuit connection appears to be alright.

(ii) The jockey is then slided along the wire from the left side and the null point is determined (say, at D_1). The distance of this point (D_1) from the left end is measured from the scale. Let it be l_1 cm ($AD_1 = l_1$).

(iii) Interchanging the positions of P and Q, again the jockey is slided over the wire and the second null point at D_2 is obtained similarly. Let the distance AD_2 be equal to l_2 cm.

(iv) The experiment is repeated by changing the value of the ratio of $\left(\frac{P}{Q}\right)$ and making it equal to 10, 20 and 30 successively.

(v) Each time the readings are taken for both direct and reverse currents with the help of the commutator.

(vi) Knowing the values of λ_1, λ_2 and the ratio of $\frac{P}{Q}$, the values of λ_1 and λ_2 can be determined employing Eqs. (8.2d) and (8.2e).

Calculation :

From Eqs. (8.2d) and (8.2e),

$$\lambda_1 = \frac{l_1 - \frac{P}{Q} l_2}{\frac{P}{Q} - 1} \quad \text{and} \quad \lambda_2 = \frac{\frac{P}{Q} l_1 - l_2}{\frac{P}{Q} - 1} - 100$$

From the above table, taking the values of $\frac{P}{Q}$ and of l_1 and l_2, the values of λ_1 and λ_2 can be determined.

Results :

Table 1

No. of Obs.	Resistance in		Ratio	Distance of null point from the point A			Value of end correction (left) λ_1 cm	Value of end correction (right) λ_2 cm	Mean value of λ_1 (cm)	Mean value or λ_2 (cm)
	Left gap P (or Q) ohm	Right gap Q (or P) ohm	$\frac{P}{Q}\left(\text{or } \frac{Q}{P}\right)$	Direct current (cm)	Reverse current (cm)	Mean l_1 (or l_2) (cm)				
1.	10	1	10	...	...	...	...	...		
2.	20	1	20	...	...	...	...	...		
3.	30	1	30	...	...	...	...	...		
4.	40	1	40	...	...	...	...	...	...	...
5.	1	10	10	...	...	...	...	...		
6.	1	20	20	...	...	...	...	...		
7.	1	30	30	...	...	...	...	...		
8.	1	40	40	...	...	...	...	...		

Precautions :

(i) The value of the ratio of $\frac{P}{Q}$ should not be less than 10. In that case, the difference of the values of l_1 and l_2 should be small.

(ii) Short thick wires are to be used to connect P and Q.

[Questions on Metre Bridge are discussed at the end of Expt. No. 8.3.]

8.3 Determination of the specific resistance of the material of a wire

Theory :

Specific resistance of a substance is defined as the resistance of the material having unit length and unit cross-section. Thus, if a conducting wire has a length of L cm and diameter d cm, and if the resistance of that wire be R ohm, then the specific resistance (ρ) of the material of the wire will be given by

$$R = \rho \,.\, \frac{L}{\frac{\pi d^2}{4}} \qquad \text{... (8.3a)}$$

or, $$\rho = \frac{\pi d^2 R}{4L} \text{ ohm- cm} \qquad \text{... (8.3b)}$$

If the specimen sample wire is connected across the right gap of a metre bridge (Fig. 8.4) and if a known resistance Q is inserted from a resistance box connected across the left gap of the bridge and if the null point is obtained at a distance of l cm on the wire from the left end, then from Wheatstone's principle,

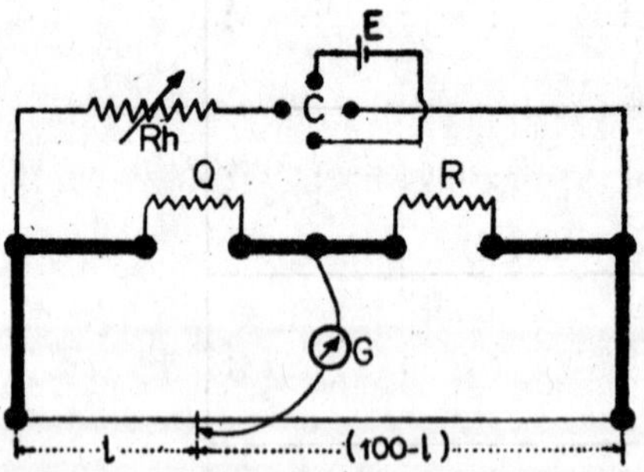

Fig. 8.4 : Specific resistance of a wire

$$\frac{Q}{R} = \frac{l}{100-l}$$

or, $$R = Q.\frac{(100-l)}{l}. \qquad \text{... (8.3c)}$$

Thus, measuring R, d and L, the value of ρ can be obtained.

Apparatus :

Metre bridge, cell, resistance box, metre scale, screw gauge, rheostat, commutator, galvanometer and the experimental wire (at least of 80 cm length).

Method :

(i) The connection of the apparatus is done as shown in (Fig. 8.4). The experimental wire is taken and it is folded 1 cm at each end at right angles to the rest of the wire. These folded portions are inserted within the binding screws at the gap.

(ii) This experimental wire (R) is connected on one gap of the bridge and a resistance box is connected in the other gap, from which a suitable resistance (Q) is employed as shown in (Fig. 8.4). Under this condition, the null point is obtained on the brdige wire. Let the distance of the null point be l_1 cm.

(iii) The direction of the current is reversed with the aid of the commutator and readings are taken. The mean value of the null point readings is noted.

(iv) In this way, three different values of the resistances are employed from the resistance box and for each case, the null points are detected on the bridge wire. Values of these resistances are to be so chosen that the positions of the null points lie within 40 cm and 60 cm of the bridge wire.

(v) Next the positions of the sample wire and the resistance box are interchanged in their gaps, and for the same values of Q from the box, the operations (ii), (iii) and (iv) are repeated. Let the distance of the null point in the second case be l_2 cm.

(vi) From Eq. (8.3c), the value of the resistance of the wire is determined for each case and the mean value of its resistance is hence calculated.

(vii) The length (L) of the experimental wire, excluding the bent portions at the both ends, is determined by a metre scale and its diameter (d) is determined by a screw gauge, taking readings at least at 5 different places and at perpendicular directions at each place.

(viii) Knowing thus resistance (R), length (L) and the diameter (d) of the wire, its specific resistance (ρ) can be calculated employing Eq. (8.3b).

Results :

(A) Determination of length and diameter of the wire :

Table 1

No. of Observation	Length of the wire (L) in cm	Mean length of the wire (L) in cm
1.	……	
2.	……	……
3.	……	
4.	……	

Table 2

Pitch of the screw gauge = ……cm

No. of circular scale divn. = ……

Least count of the screw gauge = …… cm

No. of Obs.	Main scale reading (cm)	Vernier reading (cm)	Total (diameter) in cm	Mean diameter (d in cm)
1. (a)	…	…	…	
(b)	…	…	…	
2. (a)	…	…	…	
(b)	…	…	…	……
3. (a)	…	…	…	
(b)	…	…	…	
4. (a)	…	…	…	
(b)	…	…	…	

[(a) and (b) correspond to readings at prependicular directions at the same place.]

(B) Determination of resistance of the wire :

Table 3

No. of Obs.	Resistance		Position of the null point for current			Value of the unknown resistance in ohm	Mean value of the unknown registance (R) in ohm
	Left gap (ohm)	Right gap (ohm)	Direct (cm)	Reverse (cm)	Mean (cm)		
1.	Sample wire (R)	Q	...	...	...	...	
2.	"	Other value of Q	...	...	...	...	
3.	"	"	...	...	... (l_1)	...	
4.	Q	Sample wire (R)	...	...	...	...	...
5.	Other value of Q	"	...	...	...	...	
6.	"	"	...	...	... (l_2)	...	

Calculations :

When the sample wire is in the right gap,

$$\frac{Q}{R} = \frac{l_1}{100 - l_1} \quad \text{or,} \quad R = Q \times \frac{(100 - l_1)}{l_1} \text{ ohms}$$

and when the sample wire is in the left gap,

$$\frac{R}{Q} = \frac{l_2}{100 - l_2} \quad \text{or,} \quad R = Q \times \frac{l_2}{(100 - l_2)} \text{ ohms}$$

∴ mean value of R = ... ohms

Again, from Eq. (8.3b),

$$\text{specific resistance } \rho = \frac{\pi d^2 R}{4L} = \text{,..... ohm-cm}$$

Precautions :

(i) The null point should not be far from the middle; it should preferably be within 40 cm and 60 cm of the wire.

(ii) None of the plugs in the resistance box Q should be kept loose.

(iii) Current should not be allowed to flow while not taking readings.

Questions and Answers

Q.1. Why is the instrument called a 'metre bridge'?

A. : Since, in this arrangement of Wheatstone bridge, a one-metre long wire is used in the apparatus.

Q.2. Should the wire be of uniform cross-section?

A. : Yes, otherwise the value of ρ will be different at different places of the wire.

Q.3. If the positions of the cell and the galvanometer be interchanged, will the null point position be affected?

A. : No, in that case the working principle would not be affected.

Q.4. When is this arrangement most sensitive?

A. : When the four resistances are of the same order.

Q.5. Can the metre bridge method be employed to determine the values of very high or very low resistances?

A. : No, in that case, the resistances in the different arms are not of the same order and hence the sensitivity of the bridge is lost. This method is suitable only for measurement of resistances of moderate values.

Q.6. What is meant by 'specific resistance'?

A. : See Theory of Expt. No. 8.3.

Q.7. What are meant by 'conductance' and 'conductivity'?

A. : The reciprocal of resistance $\left(\frac{1}{R}\right)$ is known as 'conductance' while the reciprocal of specific resistance $\left(\frac{1}{\rho}\right)$ is known as the 'electrical conductivity' of a material.

Q.8. On what factors does the specific resistance depend?

A. : It depends upon the temperature and material of the substance.

Q.9. Why is the null point taken near the middle of the wire length?

A. : In order to minimise the proportional error in the measurement of length of the wire.

Q.10. Why are the resistances interchanged in the gaps?

A. : In order to eliminate the error due to the non-coincidence of the mark on the jockey with the actual point of contact of the wire.

Q.11. Do the resistance of the cell or the galvanometer affect the position of the null point?

A. : No, the null point position is independent of them.

8.4 Measurement of the average resistance per unit length of a metre bridge wire by Carey Foster's method and hence to determine the value of an unknown resistance

Theory :

If in a Carey Foster's bridge, two almost equal resistances P and Q are inserted in the two inner gaps (1 and 2) and in one of the outer gap (3), a resistance Y and in the other gap (4), another resistance X are inserted and if the first null point is obtained at a distance l_1 cm (Fig. 8.5), then from Wheatstone's principle,

$$\frac{P}{Q} = \frac{Y + l_1\rho + \lambda_1}{X + (100 - l_1)\rho + \lambda_2}, \quad \text{... (8.4a)}$$

where λ_1 and λ_2 are the end-corrections, and ρ is the average resistance per unit length of the bridge wire.

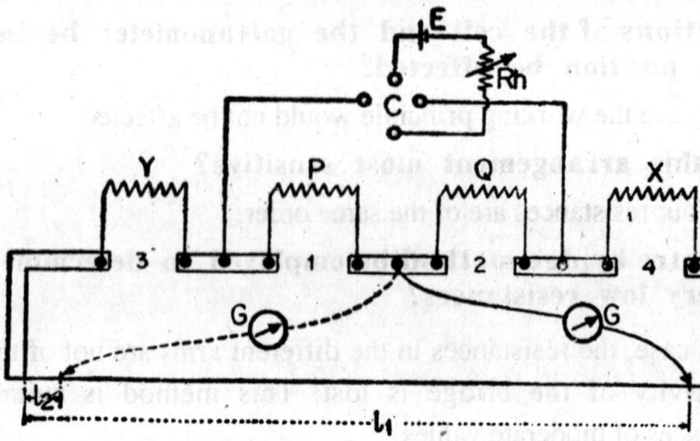

Fig. 8.5 : Carey Foster's bridge

Again, if the positions of X and Y are interchanged without changing P and Q and if the null point is then obtained at a distance l_2 cm on the bridge wire, then

$$\frac{P}{Q} = \frac{X + l_2\rho + \lambda_1}{Y + (100 - l_2)\rho + \lambda_2}. \qquad \text{... (8.4b)}$$

Hence, from Eq. (8.4a), $\frac{P}{P+Q} = \frac{Y + l_1\rho + \lambda_1}{X + Y + 100\rho + \lambda_1 + \lambda_2}$ ... (8.4c)

and from Eq. (8.4b), $\frac{P}{P+Q} = \frac{X + l_2\rho + \lambda_1}{Y + X + 100\rho + \lambda_1 + \lambda_2}$ ... (8.4d)

Equating the two (8.4c) and (8.4d),

$$Y + l_1\rho + \lambda_1 = X + l_2\rho + \lambda_2$$

or, $\rho = \frac{X - Y}{l_1 - l_2}$. ... (8.4e)

If one of the outer gap is fitted with a thick copper strip, then $Y = 0$ and Eq. (8.4e) reduces to

$$\rho = \frac{X}{l_1 - l_2} \qquad \text{... (8.4f)}$$

Hence, knowing the value of X and finding the positions of two null points (l_1 and l_2) on the bridge wire, the value of ρ can be calculated.

Apparatus :

Carey Foster's bridge, two nearly equal resistances (about 1 ohm each), resistance box with fractional resistances, rheostat, galvanometer, cell, commutator and connecting wires, unknown resistance.

Procedure :

(i) The circuit connections are made as shown in the diagram (Fig. 8.5).

(ii) A thick copper strip is inserted in gap number 3 of the bridge (Y) and the resistance box is inserted in the gap 4 of the bridge. On keeping all the plugs of the resistance box tight, if the null point is obtained almost at the mid-point of the wire, with two one-ohm resistances inserted in each of the gaps 1 and 2, then the circuit connections appear to be perfectly alright.

(iii) Suitable resistances are then to be inserted from the resistance box (X) so that the null point is obtained towards one end of the bridge wire. Readings are taken both for direct and reverse currents under this circumstance.

(iv) In this way, at least five sets of readings are to be taken for five different values of the resistances inserted from the box. But care should be taken to see that the null points are always either above 80 cm or below 20 cm length of the bridge wire.

(v) Next the positions of the copper strip and the resistance box are interchanged and readings are again taken for the same values of the resistances inserted in the first set of readings. Both the direct and the reverse currents readings are taken.

(vi) Finally, substituting the values of X in Eq. (8.4f), the values of ρ are obtained for each case and their mean value is thus calculated.

Results :

Determination of resistance per unit length :

Table 1

Value of P and Q : $P = Q = 1$ ohm

No. of Obs.	Resistance in the Left gap (ohm)	Resistance in the Right gap (ohm)	Positions of null point in cm for Direct current	Positions of null point in cm for Reverse current	Positions of null point in cm for Mean (l_1 or l_2)	Value of ($l_1 \sim l_2$) (cm)	Resistance per unit length (ρ) of the wire (ohm/cm)	Average value of ρ (ohm/cm)
1.	O	X_1	...	...	...	...	...	
	X_1	O	...	...	...			
2.	O	X_2	...	...	...	...		
	X_2	O	...	...	...			
3.	O	X_3				...	...	...
	X_3	O	...					
4.	O	X_4	...	...	...	...		
	X_4	O	...	...	...			
5.	O	X_5	...	...	...			
	X_5	O	...	...	...			

Calculations :

From Table 1 the values of ($l_1 \sim l_2$) corresponding to the individual values of X are taken and from Eq. (8.4f), the value of ρ is calculated as

$$\rho = \frac{X}{l_1 \sim l_2} = \text{......} \text{ ohm/cm}$$

Precautions :

(i) Whether the values of P and Q are almost the same should be tested first. If they are, the null point corresponding to (ii) of the Procedure, should be at about 50 cm position of the bridge wire.

(ii) The values of X should be so chosen that the null points must be at positions less than 20 cm or greater than 80 cm of the bridge wire.

(iii) While taking readings, the plugs not in use in the resistance box, should be kept tight.

Questions and Answers

Q.1. Why are the null points taken near the ends of the wire?

A. : Because the value of $(l_1 \sim l_2)$ then becomes large and more accurate value of ρ will be obtained.

Q.2. If the values of P and Q are not the same or nearly equal, what will happen?

A. : In that case, the experiment cannot be performed. It is essential that the ratio of P and Q should be 1 or very nearly equal to 1, having each individual resistance equal (nearly) to that of 50 cm length of the bridge wire. Then only the bridge remains sensitive.

Q.3. Why the value of ρ obtained is called the average value?

A. : Since the value of ρ is calculated for a length of $(l_1 \sim l_2)$ cm of the bridge wire.

Q.4. On what factors, does the value of ρ depend?

A. : It depends upon the material and diameter of the wire.

Q.5. Can the value of specific resistance be obtained from the knowledge of the value of ρ?

A. : Yes, if the diamter (d) of the wire is known, then $\rho = \dfrac{4S}{\pi d^2}$, where S is the specific resistance.

Q.6. Could the value of ρ be obtained by determining the total resistance of the wire and then dividing the same by the length of the wire?

A. : No, that would have been a rough method and this experiment measures the values of ρ at different positions which on averaging, gives a more reliable value.

Q.7. Does the presence of any end-error affect the result?

A. : No, the method is independent of end-errors as is evident from the theory.

Q.8. Can two nearly equal resistances be compared by this method?

A. : Yes, from the theory $(X - Y) = \rho\,(l_1 - l_2)$ and hence knowing ρ and determining l_1 and l_2, the small difference between two resistances X and Y can be obtained. Hence this method can standardise a given resistance. Of course, the difference between the two resistances must be less than the total resistance of the bridge wire.

8.5 Determination of the value of an unknown resistance by a Post Office Box

Theory :

In a Post Office Box, which is constructed on the basis of Wheatstone's principle, if the ratio arms contain resistances P and Q respectively and if R is the resistance inserted in its third arm, then the value of the unknown resistance S connected in its fourth arm (Fig. 8.6) will be given by

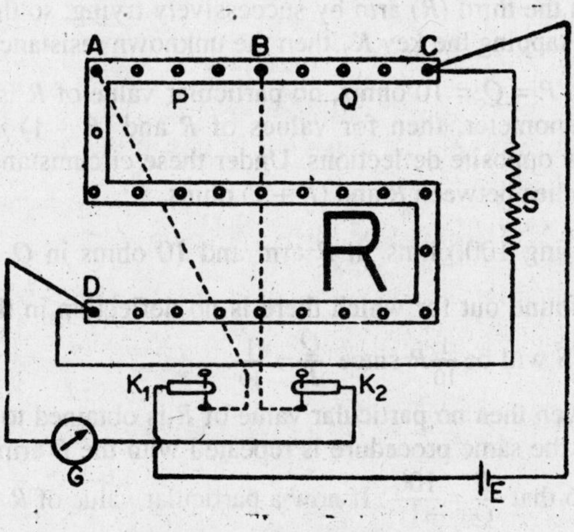

Fig. 8.6 : Post Office Box

$$\frac{P}{Q}=\frac{R}{S}, \text{ or, } S = R.\frac{Q}{P}. \qquad \text{... (8.5a)}$$

Knowing the value of R and that of the ratio $\frac{Q}{P}$, the value of the unknown resistance S can be determined.

Apparatus :

Post Office Box, cell, galvanometer, unknown resistance and connecting wires.

The Post Office Box (P.O. Box) is a form of Wheatstone bridge and was originally designed for measuring the resistance of telegraph wires in Post Offices and hence the name. It consists of a wooden box in which a number of coils of known resistances are arranged so as to form the three arms of a Wheatstone bridge. The coils and their manipulations are just like in a resistance box.

There are three sets of resistances in the box (Fig. 8.6) of which in the first set, the arms AB and BC contain 10 ohms, 100 ohms and 1000 ohms each. These arms are known as the *ratio arms*. The third arm AD contains resistances from 1 ohm to 5000 ohms including an infinite resistance. There are four screws at A, B, C and D. The unknown resistance to be measured is connected between the points C and D. The box also incorporates two keys K_1 and K_2, one for the battery circuit and the other for the galvanometer circuit. The key K_1 is placed between the cell and the point A, while K_2 is placed between the galvanometer and the point B. The arms AB, BC and AD serve as the P, Q and R arms of the Wheatstone bridge. A cell is connected between the points A and C through the key K_1 and a galvanometer is connected between the points B and D through the key K_2.

Method :

(i) Connections of the circuit are made as shown in the diagram (Fig. 8.6). To check the correctness of the connection, resistances of 10 ohms are inserted in each of the P and Q arms and once putting zero resistance and once infinite resistance in the third (R) arm, the deflections of the galvanometer are noted by tapping the respective key. If the galvanometer shows opposite deflections, then the circuit connection is alright.

(ii) With 10 ohms resistance in each of the arms P and Q, suitable value of resistance is inserted in the third (R) arm by successively trying, so that the galvanometer shows no deflection on tapping the key K_2, then the unknown resistance $S = R$ since $P = Q$.

(iii) But, if with $P = Q = 10$ ohms, no particular value of R is obtained to get no-deflection in the galvanometer, then for values of R and $(R - 1)$ in the third arm, the galvanometer will show opposite deflections. Under these circumstances, the value of the unknown resistance (S) lies between R and $(R + 1)$ ohms.

(iv) Then inserting 100 ohms in P arm and 10 ohms in Q arm (i.e., $\frac{P}{Q} = \frac{10}{1}$), suitable value of R is found out for which there is no deflection in the galvanometer. At this stage, the value of S will be $\frac{1}{10}R$ since $\frac{Q}{P} = \frac{1}{10}$.

(v) Again, if even then no particular value of R is obtained to get no-deflection in the galvanometer, then the same procedure is repeated with the P arm having 1000 ohms, and Q arm 10 ohms, so that $\frac{P}{Q} = \frac{100}{1}$. If now a particular value of R shows no-deflection in the galvanometer, then the value of S will be $\frac{1}{100}R$. It may be pointed out that the ratio of 100 : 1 is better not to take, since in that case, the sensitivity of the bridge becomes very poor.

[The values of equivalent resistance for two or more resistances connected in *series* or *parallel* may be found out by the same method after connecting them as shown in Fig. 8.7.]

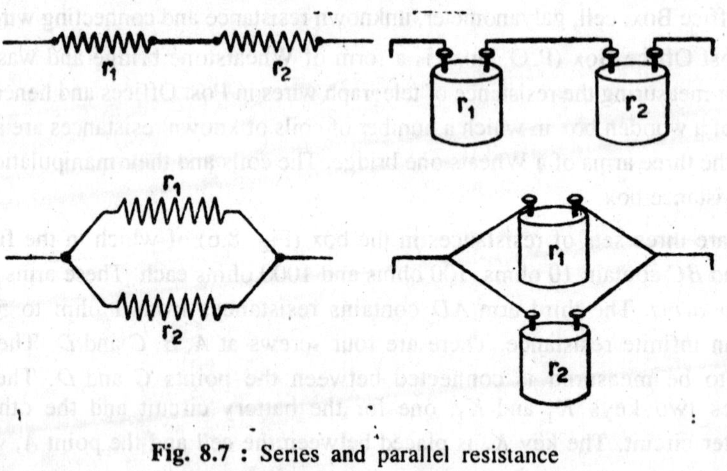

Fig. 8.7 : Series and parallel resistance

Precautions :

(i) The battery key (K_1) should always be pressed first and then the galvanometer key (K_2) to avoid the effect of self-inductance.

(ii) Resistance plugs in the box, not in use, should be kept tight always.

(iii) When for a particular value of R, the null point is obtained in the galvanometer, then the value of R is to be changed to $(R + 1)$ and $(R - 1)$ and it is to be seen whether opposite deflections are obtained.

(iv) It is advised to use a Leclanche's cell in this experiment, otherwise if a storage cell is used, then a rheostat is to be used in series with the cell.

(v) If the value of the resistance determined with a higher ratio does not tally with that determined with a lower ratio, then either the plugs may be loose or the actual resistance of a coil is not exactly the same as written against it. After tightening the plugs, if the discrepancy still exists, then it is better to depend on the higher ratio for the value of the resistance to be determined.

Results :

(A) Determination of the value of the unknown resistance :

Table 1

Resistance in the P arm (ohms)	Resistance in the Q arm (ohms)	Ratio of $\frac{P}{Q}$	Resistance in the third arm (R) (ohms)	Direction of galvanometer deflections	Comment about R in comparison to the unknown resistance (S)	Inference about the unknown resistance S
10	10	1	0	left	too low	
			∞	right	too high	
			100	right	high	
			50	"	"	
			30	"	"	
			10	left	low	
			25	right	high	
			15	left	low	
			20	left	low	S lies
			22	slightly left	slightly high	between 22
			23	slightly right	slightly high	& 23 ohms
100	10	10 : 1	230	right	high	
			220	left	low	
			225	left	slightly low	S lies bet-
			228	slightly right	slightly low	ween 22.7
			227	slightly left	slightly low	and 22.8 ohm
1000	10	100 : 1	2270	left	low	
			2280	right	high	
			2275	slightly right	slightly high	
			2274	slightly right	slightly high	
			2273	No deflection		S = 22.73
			2272	slightly left	slightly low	ohms

[The numerical figures given in the table are hypothetical and are furnished just for an example only.]

Questions and Answers

Q.1. What is a P.O. Box and why is it so called?

A. : See the description of the apparatus.

Q.2. What will be the effect if the positions of the cell and the galvanometer are interchanged?

A. : Since, $\frac{P}{Q} = \frac{R}{X}$, hence there will be no effect if the positions of the cell and the galvanometer be interchanged.

Q.3. Can a very high or a very low resistance be measured by a P.O. Box?

A. : No; when the resistances of P, Q, R and S are of the same order then, the bridge is most sensitive.

Q.4. If the e.m.f. of the cell employed be very high, what will happen?

A. : In that case, the resistances in the box will be heated for high value of current flowing and their values will be thus changed from the specified values.

Q.5. For resistance measurement, whether the P.O. Box or the metre bridge method is preferable?

A. : Since in a P. O. Box, the value of *S* can be determined up to two decimal places and since the metre bridge wire may not be uniform althrough, hence P.O. Box method is more preferable.

8.6 Construction of a one-ohm coil using Carey Foster's bridge

Theory :

A certain length of the specimen wire from which a one-ohm coil is to be constructed, is taken and its two ends are soldered to two thick copper strips. The specimen so prepared is then compared with a standard ohm using a Carey Foster's bridge. The circuit is made as shown in (Fig. 8.8) where P and Q are two equal resistances, each of one ohm, connected in gaps 1 and 2; the specimen prepared (*X*) is placed in gap 3 and a standard one-ohm coil (*Y*) is placed in gap 4 of the bridge.

According to the principle of the Carey Foster's bridge (Art. 8.4), if the null point (J_1) is at a distance of l_1 cm from the left end when the specimen wire is in the left gap, and the null point (J_2) is at a distance l_2 cm (J_2) from the same end when the specimen wire is in the right gap, then

$$X - 1 = \rho\,(l_2 - l_1) \qquad \text{... (8.6a)}$$

where ρ is the average resistance per unit length (cm) of the bridge wire.

If the specimen wire (X) is replaced by a thick copper strip (zero resistance) and by interchanging this strip and the standard one-ohm coil (Y) in the gaps 3 and 4, if the null point shifts by L cm on the wire, then from Eq. (8.6a),

$$\rho L = 1 \text{ ohm or, } \rho = \frac{1}{L} \text{ ohm/cm} \qquad \text{... (8.6b)}$$

Thus ρ can be determined.

Hence, Eq. (8.6a) can be written as

$$X - 1 = \frac{(l_2 - l_1)}{L}$$

$$\text{or,} \quad X = 1 + \frac{(l_2 - l_1)}{L} \text{ ohms} \qquad \text{... (8.6c)}$$

Apparatus :

Carey Foster's bridge, standard one-ohm resistance, two resistances of about one ohm each, cell, commutator, galvanometer, rheostat, copper strip and sufficient length of the specimen wire.

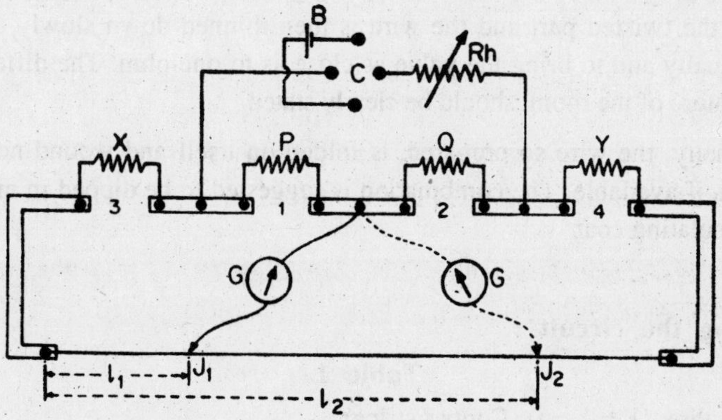

Fig. 8.8 : Construction of one-ohm coil

Method :

(i) The circuit connections are made as shown in (Fig. 8.8). First by closing the gaps 3 and 4 by two thick copper strips and inserting two almost equal (one ohm each) resistances at gaps 1 and 2, the null point is found out. It should be very close to the cenntre of the bridge wire. If not, P and Q are checked for loose contact or inequality.

(ii) The bridge must be made sensitive by adjusting the rheostat (Rh) so that the displacement of the jockey by 1 cm from the true null point, must give an appreciable deflection in the galvanometer.

(iii) Next, to determine the value of ρ, the method used in Carey Foster's bridge (Art. 8.4) is employed. Either the gap 3 or 4 is fitted with a copper strip and the other gap by a standard one-ohm. Null point is obtained. The strip and the one-ohm coil are interchanged and again the null point is obtained. If the shift of null point is by L cm, then ρ is obtained from the expression given in Eq. (8.6b). Three such sets of readings are taken from which the mean value of ρ is determined.

(iv) Next, the actual length of specimen wire required for construction has to be determined. This can be easily done by substitution method, in which the standard ohm (*Y*) is replaced by the specimen wire and its length adjusted until the null point is again restored to the last observed value.

(v) After doing this, the wire is bent at the binding screws so as to mark its effective length. A piece, 2 cm shorter than the length between the bent positions, is cut off and the ends are soldered to two copper strips, including lengths slightly over 1 cm of the wire at each end in the solder.

(vi) This length of the specimen so prepared should have a resistance very close to one ohm. The sample is compared with the standard one-ohm, placing them on two extreme end gaps of the bridge, and observing the null point, interchanging them and noting the null point again.

(vii) It is tested whether the specimen so constructed, has a value lower or higher than one ohm. If lower, it may be rubbed with a sand paper to thin down the radius and thus increase the resistance till it comes to the required value.

(viii) If in any case the value is higher than one ohm, it may be folded on itself and a very small portion at the middle may be twisted by a pair of pliers to reduce the resistance. This has to be continued till the resistance falls slightly below one ohm. A drop of solder is then put on the twisted part and the wire is then thinned down slowly to increase its resistance gradually and to bring the value as close as to one-ohm. The difference, if any, and the temperature of the room, should be clearly stated.

(ix) Finally, the wire so prepared, is folded on itself and wound non-inductively round a bobbin, if available. The combination is suggested to be dipped in molten paraffin to give it an insulating coat.

Results :

(A) Checking the circuit :

Table 1

$P = Q = 1$ ohm; $X = Y = 0$ (Copper strips)

Resistance in the extreme gaps		Null points when current is			Mean value of the mean values obtained (cm)
Right gap (Y) ohms	Left gap (X) ohms	Direct (cm)	Reverse (cm)	Mean (cm)	
0 (Copper strip)	0 (Copper strip)	......	......	.....	
		......			....
		.....	.....	.....	
		.....	.....	.....	

(B) Determination of the value of r of the bridge wire :

Table 2

P = Q = 1 ohm

[Make a similar table as that of Table 1 of Experiment No. 8.4]

(C) Determination of the resistance of the specimen wire prepared :

Table 3

P = Q = 1 ohm

[Make a similar table as that of **Table 1** of Experiment No. 8.4 with the specimen wire and a standard one-ohm being placed at two extreme end gaps. From this, find the value of the resistance of the specimen wire employing Eq. (8.6a) given in the theory.]

N.B. Operations (vii) and (viii) given under Method, may be shown in tabular forms under suitable headings, if possible.

Precautions :

(i) As the specimen wire may not be uniform, it is better to take the difference of $(l_2 - l_1)$ as large as possible to get the average value of ρ more precisely.

(ii) Care should be taken against loose contact and thermal e.m.f. generated out of rubbing the specimen wire.

(iii) After each rubbing of the wire, some time must be allowed for dissipation of the heat generated due to rubbing.

(iv) Rubbing should be done by a very small amount each time so that the change in the value of the resistance takes place very slowly.

(v) All junctions must be kept clean during the experiment.

Questions and Answers

Q.1. How 'one-ohm' is defined?

A. : According to Ohm's law, it is the ratio of one volt to one ampere.

Q.2. What is an International ohm?

A. : It is defined as the resistance offered by a column of mercury of length 106.3 cm and of uniform cross-section, having a mass of 14.4521 gm kept at 0 °C.

Q.3. How is it related to the British Association (B.A.) ohms?

A. : A committee of British Association determined the value of ohm known as B.A. ohm whose value is 0.9866 times the value of the International ohm.

Q.4. What type of material is preferable for construction of standard resistances?

A. : A material having high specific resistance and low temperature coefficient will be preferred.

Q.5. What are the specimens having these properties?

A. : Manganin and constantan have these properties but nichrome is also sometimes used. Though the latter has high specific resistance and hence shorter length is required for construction of a one-ohm coil, but its temperature coefficient is not very low like the other former specimens.

Q.6. What is the order of the resistance of the bridge wire preferred in this experiment?

A. : It should be nearly 2 ohms.

Q.7. Can copper wire be used as the specimen?

A. : No, since copper has low specific resistance and high temperature coefficient.

Q.8. How can the sensitivity in this experiment be increased?

A. : The value of the rheostat resistance should be decreased to get the desired sensitivity.

[Apart from these questions, constructions of resistance boxes, standard resistance and rheostats should be known.]

8.7 Determination of the boiling point of a liquid by a Platinum Resistance thermometer

Theory :

(i) *Temperature measurement :*

Resistance of a pure sample of platinum wire varies with temperature according to the relation,

$$R_t = R_0\,(1 + \alpha t) \qquad \text{... (8.7a)}$$

where R_t = Resistance at t °C, R_0 = Resistance at 0°C

t = temperature

and α = temperature coefficient of resistance.

If R_0 and R_{100} be the resistances of the platinum wire at 0 °C and 100 °C (freezing and boiling points of water under atmospheric pressure), then from Eq. (8.7a),

$$R_{100} = R_0\,(1 + \alpha.100)$$

or,
$$\alpha = \frac{R_{100} - R_0}{R_0 \times 100}. \qquad \text{... (8.7b)}$$

Now, if R_t be the resistance of the platinum wire at an unknown temperature t°C (in the gas scale), then the temperature t_p obtained on the platinum scale, from Eq. (8.7a) is

$$t_p = \frac{R_t - R_0}{R_0 \alpha} = \frac{R_t - R_0}{R_{100} - R_0} \times 100 \qquad \text{... (8.7c)}$$

This temperature t_p obtained on the platinum scale does not agree with the correct temperature t in the gas scale. The relation between t and t_p is given by Callendar's formula,

$$t - t_p = \delta \left\{ \left(\frac{t}{100} \right)^2 - \frac{t}{100} \right\}, \qquad \text{... (8.7d)}$$

where δ is a constant and may be taken as 1.5 for pure platinum. It is evident from Eq. (8.7d) that t and t_p agree only at two temperatures, namely at 0°C and 100°C. Knowing the value of t_p from Eq. (8.7c), the correct temperature t can be obtained from Eq. (8.7d), either by the method of successive approximation or from a curve obtained by plotting correction $(t - t_p)$ along Y-axis and t_p along X-axis.

(ii) *Measurement of resistance :*

Taking a Post Office Box of which the third arm is extended by the inclusion of a uniform wire of resistance greater than half that of the smallest value in the third arm, a null point is first obtained at the room temperature by short circuiting the P, P and the C, C pairs of terminals with the help of two short thick wires. Inserting equal resistances (10 ohms) in the ratio arms and plugging in all the resistances in the third arm, the null point is obtained on the extended wire. This point is called the *electric mid-point*. Let the distance of this point from the end of the third arm be l_0, and let ρ be the average resistance per unit length of this wire.

Next inserting the platinum thermometer and the compensating leads within the circuit as shown in (Fig. 8.9), let a null point be obtained at a distance l cm from the end M, when a resistance R is inserted in the third arm. Under this condition, the resistance of the thermometer R_t is given by

$$R_t = R + 2\,(l - l_0)\rho \qquad \text{... (8.7e)}$$

If the value of R is altered by 1 ohm and then if the null point shifts from l_1 to l_2 cm on the wire, then the value of ρ will be given by

$$\rho = \frac{1}{2(l_1 \sim l_2)}. \qquad \text{... (8.7f)}$$

By using Eqs. (8.7e) and (8.7f), the value of R_t is obtained at any temperature t°C. Knowing R_t, the value of t_p is obtained from Eq. (8.7c), since R_{100} and R_0 are known. Thus t can be calculated from t_p employing Eq. (8.7d).

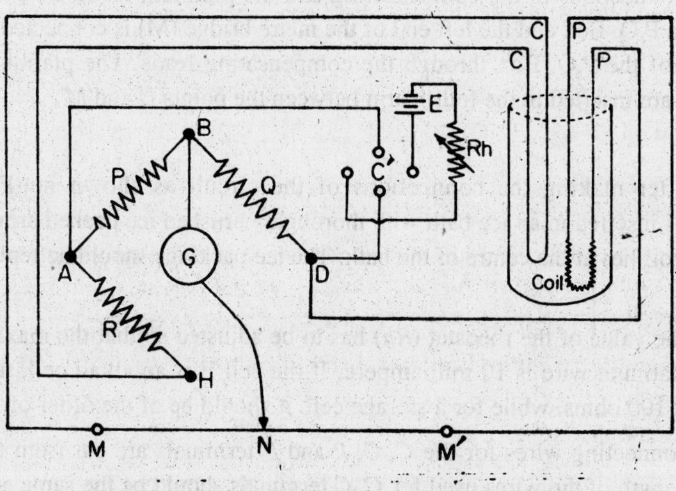

Fig. 8.9 : Connection of resistance thermometer

Apparatus :

Platinum resistance thermometer, P.O. Box, metre bridge, galvanometer, cell, commutator, rheostat, heating device with hypsometer.

A platinum resistance thermometer consists of a platinum coil (Fig. 8.10) wound non-inductively on a mica frame (*m*). The ends of the coil are connected to two screws marked *P*, *P* by means of two thick copper leads (platinum leads). Two other identical leads of the same length are set parallel to the platinum leads and are joined together. These leads are known as the compensating leads and are connected to two other binding screws *C*, *C*. The coil along with the leads are enclosed in a porcelain or hard glass tube, being separated from one another by mica separators (*M*). The resistance of the total length of the platinum leads is made exactly equal to that of the compensating leads. When the glass tube containing the resistance coil is partially introduced into a hot bath, the platinum coil is heated and its resistance increases.

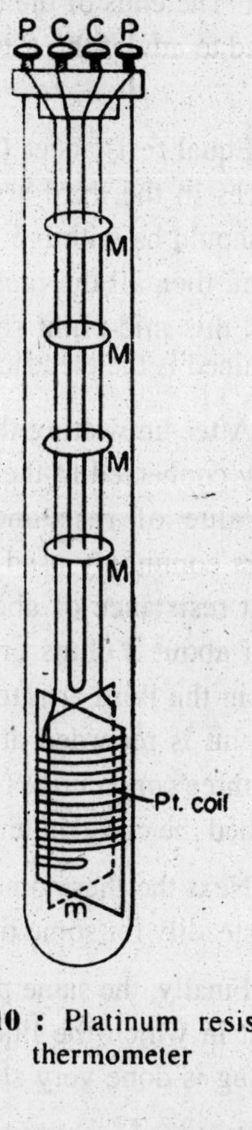

Fig. 8.10 : Platinum resistance thermometer

Connection of the circuit :

Connection of the circuit is done as shown in (Fig. 8.9). One end of the galvanometer is connected to the metre bridge jocky *N*. Four equal and sufficiently long insulated wires are used for connections to the compensating and the platinum leads. *P*, *Q* and *R* are the three arms of a P.O. Box and the left end of the metre bridge (M) is connected to the end of the third arm of the P.O. Box through the compensating leads. The platinum coil along with the leads are inserted in the fourth arm between the points *D* and *M′*.

Method :

(i) After making the connections of the circuit as shown in (Fig. 8.9), the thermometer is inserted in an ice bath with thoroughly crushed ice packed around it, so that the platinum coil lies at the centre of the bath. The ice packings should be replenished from time to time.

(ii) The value of the rheostat (*Rh*) has to be adjusted so that the maximum current through the platinum wire is 10 milliampere. If the cell E is an alkali cell, then the value of *Rh* is about 100 ohms, while for a storage cell, it should be of the order of 150 ohms.

(iii) Connecting wires for the *C*, *C*, *P* and *P* terminals are cut from the same coil and the total length of the wires used for *C*, *C* terminals should be the same as for the *P*, *P* terminals. One end of each of these four wires are connected to the points *M*, *H*, *D* and *M* of the circuit. The ends of the two wires meant for connection to *P*, *P* are twisted together and are bound to any of the *P* terminals. The same is being done also for the pair meant for *C*, *C*.

(iv) Equal resistances (10 ohms each) are put in the ratio arms and plugging in all the resistances in the third arm, the null point is obtained both for direct and reversed currents. It should be within 5 mm of the centre of the wire. But, if it is farther away from the mid-point, then all the plugs in the P.O. Box should be checked against looseness. If properly set, this mid-point should be very close to the middle of the wire and the null point so obtained is the electric mid-point (l_0).

(v) After untwisting the ends of the connecting wires, the four ends are properly and correctly connected to the leads *P*, *P* and *C*, *C*. The null point is again obtained for a particular value of resistance inserted in the third arm. Normally, the platinum thermometers commonly used have a fundamental interval of about 1 ohm, which makes the ice-point resistance of about 2.5 to 2.6 ohms. Thus, if the metre bridge wire has a resistance of about 2 ohms or more, it becomes possible to have three or four different values of *R* in the third arm to get different null points. While observing the null points, each null point is recorded at intervals of three minutes with the current unchanged in direction. If three consecutive readings remain the same, then the thermometer may be taken to have reached the constant temperature of the bath in which it is placed.

(vi) Next the thermometer is inserted in a hypsometer. After steam is found to be coming out steadily for some time, readings for the null points are again taken as before.

(vii) Finally, the same procedure is repeated with the thermometer being introduced in the vessel in which the liquid boils. A condenser arrangement is used to recover the liquid. Boiling is done very slowly keeping the size of the flame low. Sets of null points are taken at the boiling point of the liquid both for direct and reversed currents. The barometric pressure is also recorded.

Result :

(A) Determination of Electric mid-point :

Table 1

P (1st arm) = Q (2nd arm) = 10 ohms

Resistance in 3rd arm = 0 (zero)

Resistance in 4th arm = 0 (zero)

Room temperature = ... °C

No. of Obs. (cm)	Null points for current			Grand mean (l_0 cm)
	Direct (cm)	Reverse (cm)	Mean (cm)	
1.	......	......	......	
2.	......	......	......	......
3.	......	......	......	
4.	......	......	......	

(B) Measurement of resistances at different temperatures and calculation of the value of ρ :

Table 2 (See next page)

P (1st arm) = Q (2nd arm) = 10 ohms

Value of l_0 from Table 1 = ... cm

Barometric pressure (P) = ... cm of mercury.

Boiling point of water at P cm of mercury (T) = ... °C

Thermometer kept in	No. of Obs.	Time in min	Resistance in the 3rd arm (ohms)	Null points when steady temperature is attained for currents			Grand mean (l cm)	Value of ρ (ohms/cm) $= \frac{1}{2(l_1 - l_2)}$	Mean value of ρ (ohms/cm)	Value of $(l - l_0)$ cm	Resistance of thermometer $R_t = R_0 + 2(l - l_0)\,\rho$ (ohms)	Mean value of R_t (ohm)
				Direct (cm)	Reverse (cm)	Mean (cm)						
Ice	1.	0	2	...	...	...	... (l_1)	...	...	...	...	... (R_0)
		3		...	...	...						
		6		...	...	...						
		9		...	...	...						
	2.	0	3	...	...	...	... (l_2)	...		...	...	
		3		...	...	...						
		6		...	...	...						
		9		...	...	...						
	3.	0	4	...	...	...	... (l_3)	...		...	...	
		3		...	...	...						
		6		...	...	...						
		9		...	...	...						

(contd.)

Steam	1.	0	…	…	…	…	…	…	…	…	…	…
		3		…	…	…	…					
		6		…	…	…						
		9		…	…	…						
	2.	0	…	…	…	…	…	…		…	…	… (R_{100})
		3		…	…	…						
		6		…	…	…						
		9		…	…	…						
	3.	0	…	…	…	…	…	…		…	…	
		3		…	…	…						
		6		…	…	…						
		9		…	…	…						
Liquid bath	1.	0	…	…	…	…	…	…	…	…	…	… (R_t)
		3		…	…	…						
		6		…	…	…						
		9		…	…	…						
	2.	9	…	…	…	…	…	…		…	…	
		3		…	…	…						
		6		…	…	…						
		9		…	…	…						
	3.	0	…	…	…	…	…	…		…	…	
		3		…	…	…						
		6		…	…	…						
		9		…	…	…						

(C) Calculation of t_p :

Table 3

Value of R_0 in ohms	Value of R_{100} in ohms	Value of R_t in ohms	$(R_{100} - R_0)$ in ohms	$(R_t - R_0)$ in ohms	Boiling point of water ($T°$ C)	Value of $t_p = \frac{R_{100} - R_0}{R_t - R_0} \times T$ in degrees
...	...	...	...	...	...	......

(D) Calculation of t from t_p :

(a) *By successive approximation :*

To understand this process of calculation, let us suppose that from measurement of R_0, R_{100} and R_t, the value of t_p is found as 81°C. To find the corresponding centigrade temperature t corresponding to this value, the value of t_p obtained is substituted for t in the R.H.S. of Eq. (8.7d). The value of t so obtained is again substituted in the R.H.S. for t, which is known as 2nd approximation. Thus,

1st approximation :

$$t - 81 = 1.5\ \{(0.81)^2 - 0.81\} = -\,0.23085$$

$$\therefore \quad t = 81 - 0.23085 = 80.77$$

2nd. approximation :

$$t - 81 = 1.5\ \{(0.8077)^2 - 0.8077\} = -\,0.233$$

$$\therefore \quad t = 81 - 0{\cdot}233 = 80.767\ °\text{C}.$$

No further approximation is necessary as the correction introduced becomes smaller than the experimental error.

(b) *By drawing correction* $(t - t_p) - t_p$ *graph* :

From Eq. (8.7d), giving arbitrary values of t from 0°C to 110°C by steps of 10°C, the values of $(t - t_p)$— the correction terms, are calculated, assuming the value of δ as 1.5. These values are shown in Table 4. From the knowledge of the values of t and $(t - t_p)$, the corresponding values of t_p are obtained.

Table 4

Values of t	0	10	20	30	40	50	60	70	80	90	100	110
Corrections $(t - t_p)$	0	...	...	...	...	...	...	...	...	...	0	...
Values of t_p	0	...	...	...	...	...	...	...	...	...	100	...

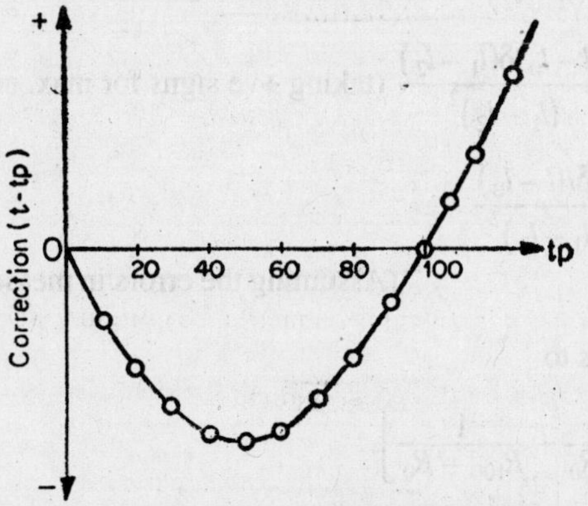

Fig. 8.11 : Correction curve

A graph is then drawn by plotting the correction $(t - t_p)$ along Y-axis and t_p along X-axis. The nature of the graph is shown in Fig. 8.11. Now finding the value of t_p from Table 3, the corresponding correction is obtained from the graph. The correct temperature (t) in the gas scale is obtained thus by algebraically adding the correction to t_p.

Precautions :

(i) The thermometer tube should not touch the water or the liquid surface during their boilings. It should be kept above their surface so that it may be surrounded by their vapours.

(ii) The actual boiling point of water should be calculated from the barometric readings, during the experiment.

(iii) Ice packings should be properly maintained in the flask so that the presence of air in between the ice pieces should be avoided.

(iv) Null points should be observed at intervals of 3 minutes and it should be kept constant at least for three successive intervals.

(v) The lengths of the connecting wires for the platinum leads and the compensating leads must be the same and same current should be passed in the circuit at all temperatures of the thermometer so that the heating effect may remain the same.

Error Calculation :

From Eq. (8.7c), by logarithmic differentiation,

$$\frac{\delta t_p}{t_p} = \frac{\delta(R_t - R_0)}{(R_t - R_0)} + \frac{\delta(R_{100} - R_0)}{(R_{100} - R_0)} = \frac{2\delta R}{R_t - R_0} + \frac{2\delta R}{R_{100} - R_0} \quad \dots (8.7g)$$

[Assuming the errors in measuring the difference of two resistances as $2\delta R$].

Again, from Eq. (8.7e) and (8.7f)

$$R_t = R + 2\,(l - l_0)\,\rho = R + \frac{2(l - l_0)}{2(l_1 - l_2)}$$

$$\therefore \delta R = \frac{(l_1 - l_2)\delta(l - l_0) - (l - l_0)\delta(l_1 - l_2)}{(l_1 - l_2)^2}$$

$$= \frac{\delta(l - l_0)}{(l_1 - l_2)} + \frac{(l - l_0)\delta(l_1 - l_2)}{(l_1 - l_2)^2} \text{ (taking +ve signs for max. error)}$$

$$= \frac{2\delta l}{(l_1 - l_2)} + \frac{2\delta l(l - l_0)}{(l_1 - l_2)^2} \quad \ldots (8.7h)$$

(Assuming the errors in measuring the difference of two lengths as $2\delta l$)

Eq. (8.7g) thus reduces to

$$\frac{\delta t_p}{t_p} = 2\delta R \left[\frac{1}{R_t - R_0} + \frac{1}{R_{100} - R_0}\right].$$

Substituting the value of R_t, R_{100} and R_0 from the observed data and calculating the value of δR from Eq. (8.7h), the proportional error in the measurement of t_p is calculated and hence the percentage error in t_p comes out as $\frac{\delta t_p}{t_p} \times 100\%$. This error normally is not more than 5%, which may further be decreased by taking mean of several, observations.

Questions and Answers

Q.1. Why is this apparatus called a thermometer?

A. : Since studying the variations of resistances of platinum, the temperature of an unknown substance can be determined.

Q.2. Why platinum is chosen instead of any other metal?

A. : Since platinum (pure) has a fixed resistance at a fixed temperature and it has a large and uniform variation of resistance with temperature.

Q.3. What is the range of temperature that can be measured by this type of thermometer?

A. : The range of platinum thermomenter is between –200 °C to 1200 °C and between 0 °C and 700 °C, the readings are accurate within one-hundredth of a degree centrigrade.

Q.4. What is the order of resistance of the platinum wire?

A. : It is about one ohm at the ice-point.

Q.5. What bridge is being used here?

A. : It is known as a Callendar and Griffith's bridge.

Q.6. What is 'platinum scale of temperature'?

A. : The temperature obtained from the relation $t_p = \frac{R_t - R_0}{R_{100} - R_t} \times 100$ is known as platinum scale of temperature. This scale agrees with the gas scale only at 0 °C and 100 °C and at no other temperatures.

Q.7. Why two equal resistances are applied in the ratio arms?

A. : Since the resistances of the platinum leads and the compensating leads are made equal, they cancel out if and only if the resistances in the 1st and the 2nd arms of the P.O. Box are made equal.

Q.8. What are the functions of the compensating leads?

A. : It eliminates the resistance of the platinum leads as they are connected in opposite arms, leaving thereby only the resistance of the platinum wire alone.

Q.9. What is the fundamental interval of this thermometer?

A. : It is the difference of resistance of the platinum wire at steam-point and ice-point. Its value normally is of the order of 1 ohm.

Q.10. What are the advantages and disadvantages of a platinum resistance thermometer?

A. : Advantages : (i) wide range, (ii) small temperature change of the order of $\left(\frac{1}{100}\right)$°C can be measured, (iii) free from charge of zero and (iv) dependable standard.

Disadvantages : (i) takes appreciable time to attain the desired temperature, (ii) large thermal capacity, (iii) instantaneous temperature at a point cannot be measured.

Q.11. Platinum has a positive temperature coefficient. Is there any substance with negative temperature coefficient?

A. : Carbon and electrolytes have negative temperature coefficients.

Q.12. The value of δ has been assumed as 1.5. Can its value be determined?

A. : Yes, by finding t_p at the boilding point of sulphur (444.6°C), the value of δ can be found out in the Callendar's formula.

Q.13. What is the order of degree of purity of platinum used for construction of the thermometer?

A. : The degree of purity would be such that the value of the ratio of R_{100} to R_0 should not be less than 1.40.

Q.14. Can a metre bridge having any value of resistance be used in this experiment?

A. : No; for limitations of the value of the resistance of the metre bridge wire, see (v) under 'Methodd'.

Q.15. Can an alloy be used in place of platinum?

A. : No; since the varition of resistance with temperature in case of alloys is very small.

Q.16. The melting point of platinum is about 1750 °C, then can it not be used up to 1700 °C at least?

A. : No; since at about 1200 °C, the variation of resistance of platinum does not take place linearly and hence the working formula needs to be modified accordingly.

Q.17. On what factors does the accuracy of the experiment depend?

A. : It depends on (i) the uniformity of the bridge wire, (ii) the accuracy of the ratio $\frac{P}{Q} = 1$ and (iii) how far the end-corrections of the bridge wire tend to zero.

Q.18. Why is the metre bridge wire used?

A. : It serves as the continuation of the 3rd arm of the P.O. Box whence a continuous variation of resistance in the 3rd arm can be achieved.

Q.19. What is the significance of determining the electric mid-point?

A. : This checks the equality of the resistances of the compensating and the platinum leads when equal resistances are employed in the ratio arms.

Q.20. Why is not the thermometer immersed directly in hot boiling water or the boiling liquid?

A. : In that case, the platinum wire does not accurately attain the temperature of the bath and hence it must be surrounded by the water/liquid vapour for attaining constancy.

8.8 Determination of equivalent conductivities of a substance in solution at different concentrations by Kohlrausch's method

Theory :

The conductivity of a substance (σ) is defined as the reciprocal of its specific resistance (ρ) as and the equivalent conductivity (λ) of a solution is defined as the ratio of its conductivity to the concentration of the solution expressed in gm-equivalents per cubic centimetre.

Hence, conductivity (σ) $= \frac{1}{\rho}$ ... (8.8a)

and equivalent conductivity (λ)

$$= \frac{\sigma}{\text{conc. of gm - equivalent per cc}} \quad \text{... (8.8b)}$$

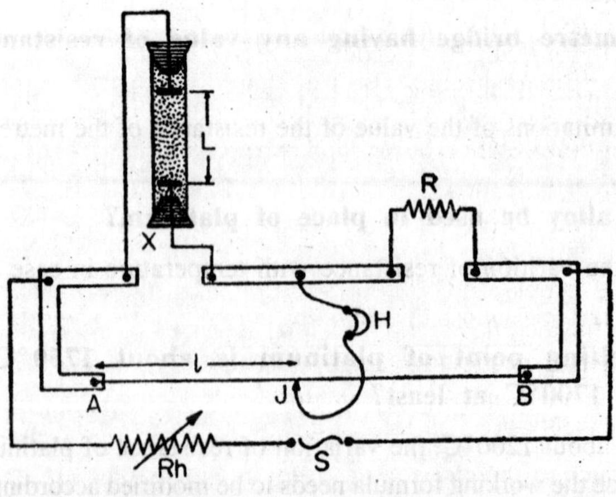

Fig. 8.12 : Kohlrausch's method

To determine conductivity of an electrolyte, at first its specific resistance (ρ) is determined by the usual metre bridge method and knowing (ρ), the values of σ and λ are obtained from Eqs. (8.8a) and (8.8b). In the arrangement shown in (Fig. 8.12) X is the electrolytic cell, R is a resistance of known value, AB is the metre bridge, H is an earphone and S is an A.C. source. If R is the value of the resistance used in the box for obtaining balance (minimum sound in the earphone) on the metre bridge at a distance l on the scale, and if X is the resistance of the electrolytic solution between the electrodes at distance L apart, then

$$X = \frac{l}{100 - l} \cdot R. \qquad \text{... (8.8c)}$$

If X_1 and X_2 are the resistances of lengths L_1 and L_2 of the solution, then the specific resistance (ρ) is given by

$$X_2 - X_1 = \rho \, \frac{L_2 - L_1}{A}, \qquad \text{... (8.8d)}$$

where A is the area of cross-section of the tube between L_1 and L_2 and the conductivity (σ) is given by

$$\sigma = \frac{1}{\rho} = \frac{L_2 - L_1}{A(X_2 - X_1)}. \qquad \text{... (8.8e)}$$

If c be the concentration of the solution in gm equivalent per cc, then equivalent conductivity (λ) is given by

$$\lambda = \frac{\sigma}{c}. \qquad \text{... (8.8f)}$$

A graph may be plotted (Fig. 8.13) with dilution of the solution in cc per gm along X-axis and corresponding equivalent conductivity along Y-axis.

Apparatus :

Electrolytic cell with electrodes, metre bridge, rheostat, A.C. source (buzzer or valve oscillator), earphones for null detection, resistance box, experimental electrolyte.

The electrolytic cell normally used, consists of either a simple glass tube of uniform cross-section or a U-tube having uniform cross-section. Electrodes made of copper or silver or platinum are used and they have diameters slightly less than that of the tube. Their positions can be altered inside the tube. A small induction coil (buzzer) or a low frequency. (1000 Hz) valve oscillator is used as the source of e.m.f. in the birdge. Instead of the galvanometer, a low resistance headphone (H) is connected to the jockey and this headphone serves as the null detector.

Method :

(i) At first the preparation of the solution has to be made with the highest concentration needed for the experiment. For this, solutions of different concentrations by weight have to be prepared. A solution of concentration c_1 by weight means that 100 gm of the solution should contain c_1 gm of the anhydrous salt. Hence, if c_1 gm of the anhydrous salt is dissolved in $(100 - c_1)$ gm of distilled water, then the desired strength of the solution is obtained. If the salt contains water of crystallisation, then from its molecular formula, the amount w gm of the crystalline salt which contains c_1 gm of the anhydrous salt, is calculated. This amount of crystalline salt (w gm) is powdered and dissolved in (100 – w) gm of distilled water. Assuming the density of water as unity, (100 – w) cc of distilled water is taken in a measuring cylinder and w gm of powdered salt is added to it. After completely dissolving the salt by stirring, the volume of the solution is noted. If V is the volume of the solution, then its density (d_1) will be $\frac{100}{V}$.

(ii) The electrolytic cell and the electrodes are cleaned first with distilled water and then with a little of the prepared solution. The cell is then filled with the given solution (strength c_1%) and its temperature is noted. The electrodes are fixed at two ends of the tube and their positions are marked. The cell is held vertically and there should not be any air bubble inside the solution.

(iii) Connections are made as shown in (Fig. 8.12) using straight and short connecting wires. By adjusting the rheostat Rh, the value of the current is controlled so that the bridge may not be heated. By adjusting resistances from the box R, a balance (minimum sound in the earphone) is obtained with the jockey near the middle of the bridge wire. The sound in the earphone may not be totally zero or minimum at a sharp point for which the exact null point may not be precisely sharp.

(iv) For two other values of R, the observations are repeated and for each case, the value of the electrolyte resistance (X) is calculated.

(v) The box (R) and the electrolyte cell are interchanged and the observations are repeated for the same three values of R. From these, the mean value of X is determined (X_1).

(vi) The upper electrode is now shifted slightly so that the effective length of the solution in between the electrodes is changed. The observations are repeated in the same way as before and the resistance (X_2) of the electrolyte of a different length (L_2) is thus determined. The temperature of the solution is again noted.

(vii) The concentration of the solution is to be changed then. To do this, a volume v_1 cc of the stock solution of concentration c_1% is taken. The volume (v_2) of distilled water which should be added to v_1 cc to make a new concentration c_2% from c_1% is given by $\frac{c_1 - c_2}{c_2} \times v_1 d_1$ cc. The density of this new concentration solution is $d_2 = \frac{v_2 + v_1 d_1}{v_2 + v_1}$ gm/cc.

(viii) With this new working concentrations, the experiment is again repeated in the same manner as was done with the first concentration. The entire experiment is thus performed with such five different concentrations.

(ix) If the cell is a unifrom cylindrical tube, its inner diameter is measuremed by using an inner calliper. But, if it is a U-tube, then the inner cross-section is measured by filling it with water with the electrodes set at two fixed positions in the same level and weighing the water (W_1). By removing the electrodes and setting them at a different position in the tube, water is again filled with up to the electrodes and the weight of this water is noted (W_2). Taking density of water as unity, the average cross-section of the tube will be given by $\frac{W_1 - W_2}{l}$, where l is the sum of the lengths between the two sets of position of the electrodes. The lengths between the two sets of positions of the electrodes are measured very carefully.

(x) The values of ρ, σ and λ of each solution at the temperature observed, are calculated and a graph is plotted (Fig. 8.13) with equivalent conductivity (λ) against dilution.

Results :

(A) To prepare stock solution of known concentration (say, 15%) of copper sulphate :

Molecular formula of crystalline copper sulphate is $CuSO_4, 5H_2O$. Thus {(63.5 + 32 + 4 × 16) + 5(2 + 16)} or, 249.5 gm of crystalline salt contains only (63.5 + 32 + 4 × 16) or, 159.5 gm of anhydrous salt.

Therefore,

159.5 gm of anyhydrous salt is present in 249.5 gm of crystalline salt

or, 15 gm of anyhydrous salt is present in 23.5 gm of crystalline salt

So, to prepare a 15% solution by weight, 23.5 gm of crystalline salt is to be dissolved in (100 – 23.5) or, 76.5 cc of distilled water.

(B) Conversion of solution from one strength (c_1%) to another strength (c_2%) :

Table 1 (See Page 288)

(C) Measurement of the resistance of the electrolyte :

Table 2 (See Page 289)

(D) Determination of cross-section of the tube :

Table 3

No. of Obs.	Mass of water with electrodes in the		Mass of water occupying the lengths (l) between the two positions $(W_1 - W_2)$ gm	Cross-section of the tube = $\frac{W_1 - W_2}{l}$ (sq. cm)
	First position (W_1 gm)	Second position (W_2 gm)		
1.	......	......	......	......
2.	......	......	......	......
3.	......	......	......	......

∴ mean cross-section = ... sq. cm

(E) Determination of Equivalent conductivity :

Table 4

Conc. by weight (c%)	Density of soln. (*d* gm/cc)	Value of c.d.	Conc. in gm-equiv. per cc $\left(\frac{c.d}{100 \times 79.75}\right)$	Dilution (100/cd) (cc/gm)	Conductivity $\sigma = \frac{1}{\rho}$ (ohm^{-1}cm^{-1})	Eq. conductivity $\lambda = \frac{\sigma}{\text{conc.in gm-equiv.}}$
15 %	...	...	...	...	...	...
12.5%	...	...	...	...	...	...
10 %	...	...	...	...	...	...
7.5%	...	...	...	...	...	...
5.0%	...	...	...	...	...	...

(F) Drawing of graphs :

Finally a graph is plotted with equivalent conductivity (λ) along *Y*-axis and corresponding dilution along *X*-axis. The nature of the graph will be as shown in (Fig. 8.13). It is evident from the graph that the equivalent conductivity practically remains constant at higher dilutions. Another graph may be plotted with concentration (c%) by weight along *X*-axis and the corresponding conductivity (σ) along *Y*-axis. From these graphs, the conductiviy or the equivalent conductivity of the electrolyte at any concentration or dilution can be determined.

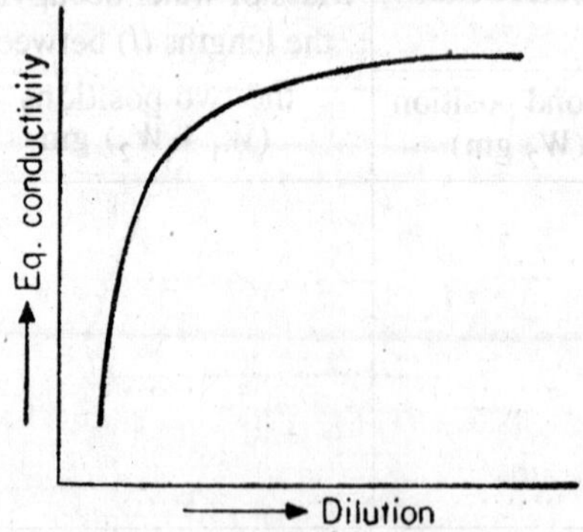

Fig. 8.13 : Equivalent Conductivity-Dilution Curve

(B) Conversion of solution from one strength (c_1%) to another strength (c_2%) :

Table 1

No. of Obs.	Strength of original solution taken (c%)	Volume of original solution taken (v_1 cc)	Density of the original solution (d_1 gm/cc)	Volume of water (v_2) to be added to v_1 to convert to new concentration (cc)	Temp. of the solution in °C	Density of the final solution (d_2 gm/cc)	Strength of final solution by weight	Dilution of final solution $\eta = \frac{100}{cd}$ (cc/gm)
1.	15.0%	...	...	...	...	...	12.5%	......
2.	12.5%	...	...	...	...	...	10.0%	......
3.	10.0%	...	...	...	...	...	7.5%	......
4.	7.5%	...	...	...	...	...	5.0%	......
5.	5.0%	...	...	...	...	...	2.5%	......

(C) Measuremet of the resistance of the electrolyte :

Table 2

No. of Obs.	Conc. of soln. by wt.	Distance between the electrodes (L cm)	Resistance in		Null point at		Resistance of the electrolyte (X ohm)	Mean value of X (ohm)	Value of	
			Left gap (ohm)	Right gap (ohm)	(cm)	Mean (cm)			$(X_2 \sim X_1)$ ohm	$(L_2 \sim L_1)$ cm
			X_1	R_1	... ...	...	...... (X_1)			
			X_1	R_1'	... ...	...	...... (X_1)			
1.	15 %	L_1	X_1	R_1''	... ...	...	...... (X_1)	...... (X_1)		
			R_1	X_1	... ...	...	...... (X_1)			
			R_1'	X_1	... ...	...	...... (X_1)			
			R_1''	X_1	... ...	...	...... (X_1)			
			X_2	R_2	... ...	...	...... (X_2)			
			X_2	R_2'	... ...	...	...... (X_2)			

(contd.)

1.	15 %	L_2	X_2	R_2''	…	…	…… (X_2)			
			R_2	X_2	…	…	…… (X_2)	…… (X_2)		
			R_2'	X_2	…	…	…… (X_2)			
			R_2''	X_2	…	…	…… (X_2)			
2.	12.5 %									
3.	10 %									
4.	etc.		etc.				etc.			

Precautions :

(i) The solution should be freshly prepared and at each time its temperature must be noted, since the conductivity of an electrolyte increases with increase of temperature.

(ii) Since the bridge is not perfectly non-inductive and non-capacitive, perfect silence is never obtained in the earphone and hence, the positions of minimum sound in it should be detected very carefully by taking several readings for a particular observation.

(iii) A valve oscillator as a source of A.C. is always preferable to a buzzer since the former can give a single frequency.

(iv) The electrodes do not fit perfectly in the tube and hence the lines of current flow are not parallel to the tube axis. This end effect does not give precise value of the conductivity by direct determination of the electrolytic resistance. The effect may be eliminated by determining the resistances of two lengths of the same solution provided the lengths are both greater than 1.5 times the diameter of the tube.

(v) A low resistance rheostat should be used to control the current flow and care should be taken to prevent heating of the bridge wire.

(vi) Always a least number of resistance coils should be used from the box R to make up a desired valued of resistance.

Error Calculation :

The proportional error in the measurement of the conductivity (σ) is given by

$$\frac{\delta\sigma}{\sigma} = \frac{\delta M}{M} + \frac{2\delta L}{L} + \frac{\delta l}{l} + \frac{\delta l}{(100 - l)},$$

where $M = (W_1 - W_2)$ ...... (See Table 3)

$L = (L_2 - L_1)$ ...... (See Table 2)

l = distance of null point on the wire.

The maximum contribution to the error is due to measurement of L which thus should be made as large as possible. To minimise error due to M, it is better to find the weights of water directly by weighing and not by measuring its volume by measuring cylinder and compting therefrom weight, taking density of water. The error due to measurement of l is fairly small compared to other errors.

Questions and Answers

Q.1. Define, specific resistance, conductivity, dilution and equivalent conductivity.

A. : See any textbook.

Q.2. Is the equivalent conductivity of an electrolyte constant?

A. : It increases with the increase of dilution and remains fairly constant at higher dilution. If the degree of dissociation remains constant, the equivalent conductivity also remains constant.

Q.3. How does electrolytic conduction differ from metallic conduction?

A. : Matallic conduction is due to the movement of free electrons in the conductor due to imposition of a potential difference, while electrolytic conduction is due to the movement of ions in the electrolyte due to an electrical field produced in between the electrodes.

Q.4. Do the positive and negative ions travel with the same velocity in the field?

A. : No, the ions move with a limiting uniform velocity depending on the charge of the ion, nature of the solvent, intensity of the applied electrical field and on the size of the ions, which are different for different kinds of ions.

Q.5. Why is not a cell used in the metre bridge circuit?

A. : The direct current will produce a change in concentration of solute near the electrode surfaces, causing thereby a change in the resistance of the electrolyte. Hence, an A.C. source is used.

Q.6. As a source of e.m.f., which is preferable—a buzzer or a valve oscillator?

A. : A valve oscillator having a single frequency is preferred since a buzzer produces a current which is not sinusoidal, though periodic. This periodic current can be analysed by Fourier's theorem into a number of sinusoidal current superimposed on a direct current. This d.c. produces polarisation near the electrodes.

Q.7. Why not a galvanometer is used as the null detector?

A. : The deflection of a galvanometer is proportional linearly with the current. Hence with an A.C., it will not show any deflection. Thus either a low-resistance earphone or a thermo-galvanometer has to be employed.

Q.8. Why is the sound in the telephone is minimum and not zero at the null point?

A. : Due to finite inductance and the capacitance of the resistance coils, perfect balance (zero sound) cannot be attained when proportionalities in the four arms of the Wheatstone bridge is maintained.

Q.9. Why are the minimum number of resistance coils used?

A. : To avoid the existence of finite capacitances and inductances of coils, minimum number of plugs are opened to make a certain desired value of resistance.

Q.10. Why should the frequency of the A.C. supply be low?

A. : For low frequency A.C. supply, the effects of the existent inductances and capacitances would be inappreciable.

8.9 Verification of Stefan's Law by electrical method

[Instead of the conventional thermal method of verification of this law, a much more easy and convenient electrical method with accessories easily available, is discussed and given in this experiment.]

Theory :

According to Stefan's law, the total amount of energy radiated by a perfectly black body per second per unit area, is directly proportional to the fourth power of the absolute temperature of the body. That means if E is the energy radiated by a black body per second per unit area maintained at a temperature $T°A$ and if the surrounding temperature be $T_0°A$, then

$$E = \sigma\left(T^4 - T_0^4\right). \qquad \text{... (8.9a)}$$

Even if the body is not black, then the total energy emitted, $E = C\left(T^n - T_0^n\right)$, where C is some constant depending upon the material and area of the body and n is a numerical constant. If $T \geq T_0$ (T_0 being the room temperature), then the second term within the bracket may be neglected and the equation becomes

$$E = CT^n. \qquad \text{... (8.9b)}$$

If the radiating body be an electrical filament lamp which having a potential difference of V volts across its ends, consumes a current of I ampere, then at the steady state, it dissipates a power P given by

$$P = V.I. \qquad \text{... (8.9c)}$$

Let us take, according to Stefan's law, an arbitrary relation in this case,

$P \propto T^n$, where T is in Kelvin scale.

or, $$P = CT^n, \qquad \text{... (8.9d)}$$

where C is a constant. If this constant value comes out to be very close to 4, then Stefan's law would be verified.

Taking logarithm of Eq. (8.9d),

$$\log P = \log C + n.\log T. \qquad \text{... (8.9e)}$$

Thus, if a graph be plotted with $\log P$ against $\log T$, it would be a straight line graph and from the slope of the line, the value of n can be found out, whence Stefan's law would be verified. According to the law, the value of n would be nearly about 4.

To measure the value of P, Eq. (8.9c) is employed while to measure the temperature (T) of the filament, the relation

$$R_t = R_0\,(1 + \alpha t + \beta t^2) \qquad \text{... (8.9f)}$$

is employed where R_0 and R_t are the resistances of the filament at 0°C and t°C respectively and α and β are constant for tungsten filament ($\alpha = 5.24 \times 10^{-3}$ per °C and $\beta = 7 \times 10^{-7}$ per °C^2). From the knowledge of R_t and R_0, the value of t in °C can be obtained and hence the corresponding value in degree Kelvin (T) is ascertained.

To find the value of T, instead of finding the value of R_0, the following method may be adopted.

It is known that a tungsten filament bulb just starts glowing at 800Å (= 527°C) approximately. If R_g be the resistance of the filament at the stage when it starts glowing, then by applying several increasing and decreasing values of current (i) passing through the filament and measuring corresponding voltages (v) across its terminals, the value of R_g may be found out which is the resistance of the filament just at the time of glowing ($t = 527°C$).

Hence, $R_t = R_0\,(1 + \alpha t + \beta t^2)$ at any temperature

and $R_g = R_0(1 + \alpha.527 + \beta.527^2)$.

$$\therefore \quad \frac{R_t}{R_g} = \frac{1+\alpha t+\beta t^2}{1+527\alpha+(527)^2\beta}.$$

Substituting the values of α and β for tungsten,

$$\frac{R_t}{R_g} = \frac{1}{3.95}\left(1+\alpha t+\beta t^2\right). \qquad \text{.... (8.9g)}$$

Since, the value of R_g is already determined, hence, measuring R_t at temperature t°C, the value of t°C (= T°A) is obtained.

Knowing the value of P from Eq. (8.9c) and determining the value of corresponding T, finally a graph is plotted with log P along Y-axis and log T along X-axis, from which the slope gives the value of n, which comes out very close to 4, verifying, therefore, Stefan's law.

Apparatus :

One 6V battery, one rheostat (1 amp, 100 ohm), one voltmeter (10 V range), one ammeter (1 ampere range) and one tungsten filament bulb (6V/6 watts like an auto lamp).

Method :

(i) The connections are made as shown in (Fig. 8.14). It is checked at first that if varying the rheostat permits to go from below glow stage upto a stage when the filament becomes dazzling white.

(ii) Two wires are soldered right to the base points to the bulb. These wires are connected directly to the voltmeter. Extra socket devices are avoided.

(iii) By the help of the rheostat, the value of current is adjusted such that the filament just glows. Several trials with increasing and decreasing currents (i) are made. The corresponding values of the voltages (v) across the bulb are recorded from the voltmeter. From these values, the values of R_g (= $\frac{v}{i}$) and then mean value is obtained.

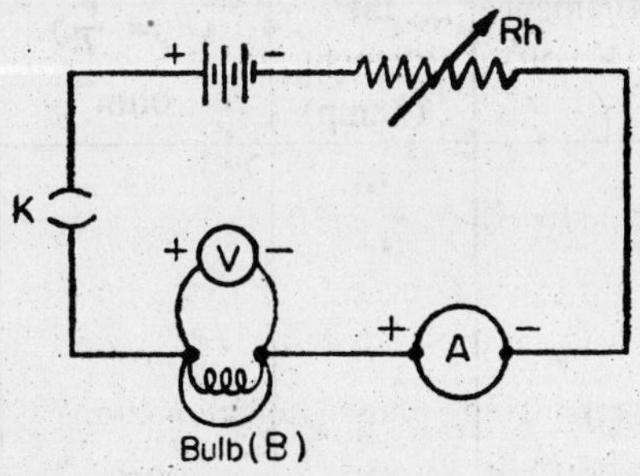

Fig. 8.14 : Stefan's law verification

(iv) Then gradually the value of the current is increased in step when the glow becomes brighter. At each step, from the values of the current and the corresponding voltages, the value of R_i is obtained. Applying Eq. (8.9g), the value of the coresponding temperature (t°C) and hence T°A is found out. (See 'Discussion' also.)

(v) From the above values of currents and voltages employing Eq. (8.9c), each time the power (P) dissipated is also calculated.

(vi) Finally, a graph is plotted (Fig. 8.15) with log P against log T from which the slope of the curve gives the value of n.

Results :

(A) Determination of the value of R_g :

e.m.f. of the battery used = ... volt

Room temperature = ... °C

Glowing temperature of tungsten = 527 °C = 800°K (approx.)

Table 1

No. of Obs.	Current strength (i amp) when the bulb just glows	Voltage across the filament (v volt)	Resistance (R_g) of the filament when the bulb just glows (ohm)	Mean value of R_g (ohm) at T = 800 °K (approx.)
1.	.....	.....	.....	
2.	.....	.....	.....	
3.	.....	.....	.....	.....
4.	.....	.....	.....	
5.	.....	.....	.....	

(B) Evaluation of Resistances (R_t) and Power (P) dissipated at different temperatures (T °K)

Table 2

For tungsten, value of $\alpha = 5.24 \times 10^{-8}$ per °C

value of $\beta = 7.0 \times 10^{-7}$ per °C^2.

No. of Obs.	Voltage across the filament (V volt)	Current through the filament (I amp)	Resistance of the filament $(R_t = \frac{V}{I})$ ohm	Power dissipated in Watts ($P = V \times I$)	Temperature of the filament from Eq. (8.9g) t°C	Corresponding value of the temperature in absolute scale (T°K)
1.	...	...	...	...	...	...
2.	...	...	...	...	...	...
3.	...	...	...	...	...	...
4.	...	...	...	...	...	...
5.	...	...	...	...	...	...
6.	...	...	...	...	...	...
7.	...	...	...	...	...	...
8.	...	...	...	...	...	...
9.	...	...	...	...	...	...
10.	...	...	...	...	...	...

(C) Data for drawing the graph :

Table 3

No. of Obs.	Value of P (watts)	Value of T (°K)	Value of $\log P$	Value of $\log T$	Slope of the curve (= QR/RS)	Value of n
1.	...					
2.	...	...	...	...		
3.	...	...	...	...		
4.	...	...	...	...		......
5.	...	...	...	...	...	

From the results, the value of n would come out close to 4 and hence Stefan's law is verified.

(D) Drawing of graph :

From Table 3, a graph is plotted with $\log P$ along Y-axis and $\log T$ along X-axis. The nature of the curve will be a straight line (Fig. 8.15). From this graph, any arbitrary points Q and S are taken and from the ordinate (QR) and absciassa (RS) values, the slope of the curve is determined. This is equal to n.

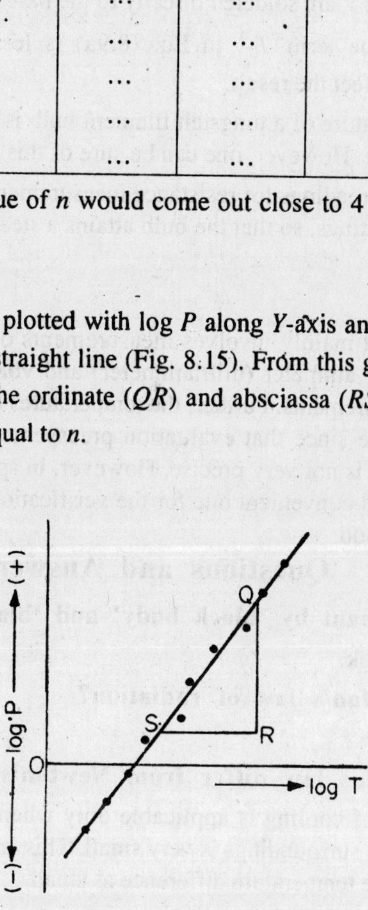

Fig. 8.15 : (log P-log T) graph

Discussions :

(i) Since during the experiment, the room temperature is not far away from 0°C, the value of the filament resistance at room temperature can be approximately assumed to be very close to R_0. S_0, if the resistance of the filament at the room temperature is determined with a very small value of current passing through the filament, this value may be assumed to be equal to R_0. Then applying the eqn. $R_t = R_0(1 + \alpha t + \beta t^2)$, the value of t can be obtained by solving the quadratic equation from the knowledge of R_t, α and β.

This is an alternative way of finding the temperature (t°C) of the filament.

(ii) It is assumed that in the steady state, the electrical power $V.I.$ supplied to the filament should be equal to the radiated power, the loss of energy by conduction along the leads and through the gas in the bulb being negligible compared to the radiation loss. This assumption, however, is not perfectly ture.

(iii) The value, of the resistance (R_0) of the bulb determined at low temperature, is the sum of the actual resistance of the bulb plus the resistance of the leads of the bulb, the latter being possible to be a substantial fractions of the total value, even when the connections for measuring V are soldered directly to the base of the bulb.

(iv) Neglecting the term T_0^4 in Eq. (8.9a) is less justified whose significant presence, if any, would affect the result.

(v) Glow temperature of a tungsten filament bulb is assumed to be equal to 800°K which is not very accurate. However, one can be sure of this value within ± 50°C

(vi) While taking readings for resistance measurements, some time must be allowed each time after all the settings, so that the bulb attains a steady temperature, and the steady state is reached.

Error Calculation :

Since the experiment mainly involves measurements of currents and voltages, hence use of precision types of ammeter (milliammeter) and voltmeter (millivoltmeter) would reduce errors in the measurements. Further, the temperatures calculated from formula are not very precise and accurate since that evaluation preassumes the glow temperature of the filament as 800°K which is not very precise. However, in spite of all these limitations, this method is a very easy and convenient one for the verification of Stefan's law, compared to the rigorous thermal method.

Questions and Answers

Q.1. What are meant by 'black body' and 'black body radiation'?

A. : See any textbook.

Q.2. What is Stefan's law of radiation?

A. : See 'Theory'.

Q.3. How does this law differ from Newton's law of cooling?

A. : Newton's law of cooling is applicable only when the difference of temperature between the body and the surroundings is very small. This law, in fact, can be deduced from Stefan's law assuming the temperature difference as small.

Q.4. Can the value of Stefan's constant be determined from this method?

A. : Yes, taking the value of n as 4, the value of C can be obtained from Eq. (8.9a) or from the value of the intercept of the graph (Fig. 8.15) also, the value C can be obtained (Eq. 8.9e), if the radiating body is not assumed as a black body. Assuming this to be a black body, this value of C so obtained will correspond to σ, the Stefan's constant (Eq. 8.9a).

Q.5. Is this method superior to the conventional thermal method?

A. : This method is though not very precise and accurate, however, it has some points of advantages. The bulb is never truly a black body and at a steady state, the power radiated is never equal to V.I exactly. The working theory in this method is to some extent approximate, nevertheless, the method is very simple and the accessories are easy to procure. It gives an approximate idea about Stefan's law, Stefan's constant and the verification of the law.

Chapter 9
Electrical Experiments

(Potentiometer and Galvanometer Experiments)

9.1 Potentiometer and its working principle

A device by which the potential difference between or potential drop across any two points can be measured, is normally known as a *potentiometer*. The commonly used potentiometer is shown in (Fig. 9.1). It consists of a wooden board (*W*) having an insulted glass plate mounted on it. It generally consists of an even number of wires (usually 10) of identical and uniform nature and made of same material, each one metre long and joined in series by thick copper strips at the ends. They are set on the glass plate parallel to each other. The two ends of the series system (*A* and *B*) are fixed to two screws which serve as the terminals. A jockey (*J*) slides along a metre scale and its two legs can pass through a groove while the third leg rests on a copper strip (*CD*). The jockey can be electrically connected by wires at either of the points *C* or *D*. Over the jockey frame, there is a knife-edge contact arrangement (*T*) fitted with a spring. When *T* is pressed down, contact between the wire and the jockey is established.

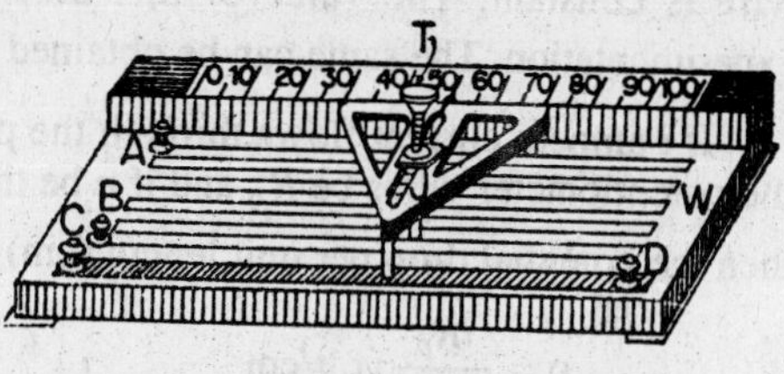

Fig. 9.1 : Potentiometer

The jockey can slide over the wires, and from the wire number and the scale, the actual length of the contact point can be ascertained.

Working Principle :

The working principle of the potentiometer can be understood from (Fig. 9.2). If two independent potential differences V_1 and V_2 (in opposition) are connected in series through a galvanometer so as to form a closed circuit and if further there is no current through the galvanometer, then from Kirchhoff's law, we have $V_1 = V_2$. When the current flows through the potentiometer, then a potential drop is maintained between the ends of the wire *A* and *B*. If the jockey is moved across the wires, a variable potential drop is obtained. In order to measure the value of an unknown drop of potential *V*, the terminal *A* and the jockey are connected across *V* in opposition through the galvanometer *G*. The jockey is slided over the wires and a point is obtained when there is no current through the galvanometer. This point is called the *null point*. Then the drop of potential between *A* and the null point equals *V*. In order to protect the galvanometer in such a circuit from being damaged due to passage of a heavy current, always a high resistance R is connected in series with the galvanometer. This resistance, however, does not affect the value of the potential drop between A and the null point, when the latter is obtained. In order to increase the sensitivity of the circuit, the value of this high protective resistance may be decreased, once the null point is obtained.

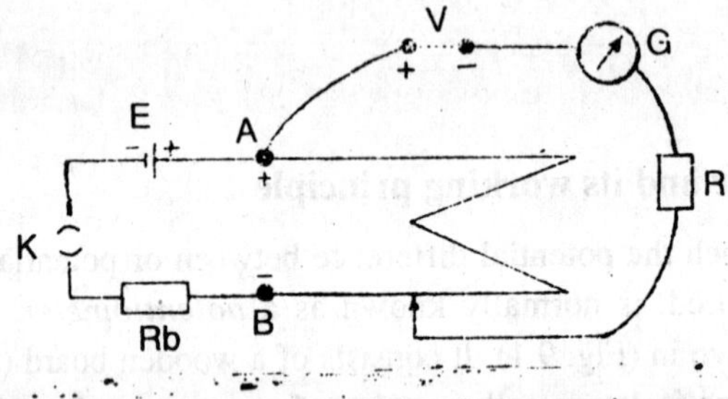

Fig. 9.2 : Working principle of a potentiometer

Care should always be taken to see that the similar terminals (positive or negative) of the driving cell (E) and of the unknown potential difference to be measured, must be connected to the point A of the wire.

Since the potentiometer wires are all uniform, the potential drop per unit length of the wire is constant. The value of this drop per unit length (ρ) has to be known during experimentation. The same can be obtained by the following way.

If i ampere current flows through the potentiometer wires and if the total resistance of the potentiometer wires be R_p and if n be the total number of wires, each of length 1 metre, then the potential drop per unit length (cm) of the wire,

$$\rho = \frac{iR_p}{100n} \text{ volt/cm} \quad \text{... (9.1a)}$$

Again, if the value of the current (i) is not known but the e.m.f. of the cell (E) is known, then

$$i = \frac{E}{R_p + R_b} \text{ and } \rho = \frac{E.R_p}{100n(R_p + R_b)} \text{ volt/cm} \quad \text{... (9.1b)}$$

Hence, if a total length of l cm balances the unknown potential (V) to be measured, then

$$V = \rho . l \quad \text{... (9.1c)}$$

It is to be noted that the potential drop to be measured must always be less than the total potential drop between the terminal ends A and B, otherwise the method will not be valid.

9.2 Measurement of the e.m.f. of a cell by potentiometer, using a milliammeter

Theory :

If a current of i milliampere passes through a potentiometer of resistance R_p, and if it has a total length of wire as L cm, then the drop of potential per centimetre of the wire is

$$\rho = \frac{iR_p}{10^3 \times L}. \quad \text{... (9.2a)}$$

If a cell of e.m.f. E' be balanced on a length l of the potentiometer wire, (Fig. 9.3) then E' is given by

$$E' = \rho.l = \frac{iR_p l}{10^3 L} \text{ volt} \quad \text{... (9.2b)}$$

Apparatus :

Potentiometer of known resistance, battery, milliammeter, resistance box; galvanometer, key, high resistance box, cell whose e.m.f. has to be determined.

Method :

(i) Connections of the circuit are made as shown in (Fig. 9.3). Care must be taken to see that similar polarities of the cells E and E' are connected to the point A.

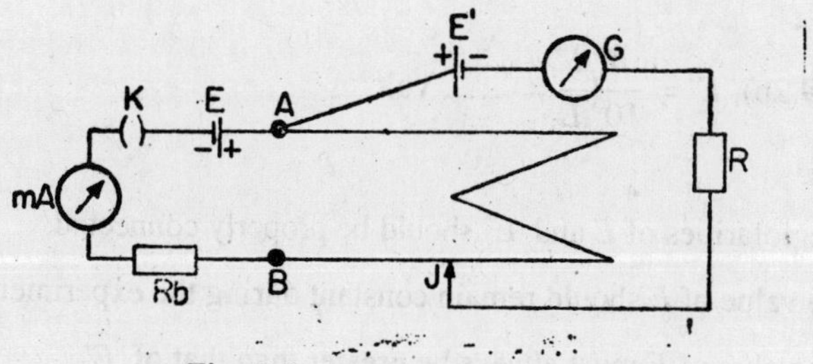

Fig. 9.3 : e.m.f. of a cell

(ii) Making the value of R_b zero and inserting sufficient resistance from R, the jockey (J) is once touched on the first wire and then on the last wire. The galvanometer deflections are observed. If the value of E is greater than E', then the galvanometer deflections in these two cases must be opposite. If not so, then there must have been some fault in the connections.

(iii) When opposite deflections are obtained on the galvanometer, then the value of R_b has to be so adjusted that the null point is obtained somewhere on the tenth wire. Making R zero, accurate position of this null point has to be obtained. The milliammeter reading is noted and the length of the null point is ascertained from the wire number and the jockey position on the scale.

(iv) In this way, the values of R_b are set to other different values and for other values of the current (i) through the milliammeter, null points are obtained at least for two other values on the ninth and the eighth wire. Each time the position of the null point is ascertained from three scale readings.

Results :

Total resistance of the potentiometer wires (R_p) = ... ohm

Total length of the potentiometer wires (L) = ... cm

e.m.f. of the cell (E) = ... volt (before expt.)

e.m.f. of the cell (E) = ... volt (after expt.)

Table 1

No. of Obs.	Milli-ammeter reading (i)	Null point			Total length of wire (l) (cm)	e.m.f. of the cell (E' volt)	Mean value of the e.m.f. (E' volt)
		On wire number	At scale position (cm)	Mean scale reading (cm)			
1.	...	...	... ...	...	...	...	
2.	...	...	... ...	...	...	...	...
3.	...	...	... ...	...	...	...	

Calculations :

From Eq. (9.2b), $E' = \dfrac{iR_p l}{10^3.L} = \ldots\ldots$ volt

Precautions :

(i) The polarities of E and E' should be properly connected.

(ii) The value of E should remain constant during the experiment.

(iii) The value of E must always be greater than that of E'.

(iv) The sensitivity of the circuit has to be maintained so that the galvanometer gives deflections for 1 cm shift of the jockey from the null point either way.

(v) The galvanometer used in this experiment must be of low resistance.

Questions and Answers

Q.1. What is the difference between e.m.f. and potential difference?

A. : They are the same when the cell does not send any current but in a closed circuit, e.m.f. = P.D. in the external circuit + P.D. in the internal circuit, i.e., $E = I.R. + I.r.$

Q.2. Can the e.m.f. of the cell be measured if no milliammeter is provided?

A. : Yes, then by employing Eq. (9.1b), the e.m.f. can be measured.

Q.3. The e.m.f. could also be measured by a voltmeter. Then why is this method adopted?

A. : Voltmeter records only the potential difference in an external circuit. That does not measure e.m.f. of a cell if the internal resistance be very high.

Q.4. What are the conditions of success of this experiment?

A. : The value of E must be greater than E' and the positive polarities of both E and E' must be connected to the same point.

Q.5. What is the significance of introducing the resistance *R* in the circuit?

A. : It protects the galvanometer from being damaged.

Q.6. Why are the null points taken on the 9th or 10th wires?

A. : If the number of wires be higher, the total length of the balance point becomes higher and hence percentage error in the measurement of the balancing length becomes much smaller.

9.3 Measurement of current flowing in a circuit by measuring the potential drop across a resistance using a potentiometer

Theory :

If a current of *i* ampere flows through a standard resistance *r* shown in the circuit (Fig. 9.4), then the potential difference across the resistance will be ir ($= e$) and if this p.d. is balanced by a length *l* cm of potentiometer wire having a potential drop of ρ per cm, then

$$e = i.r. = l\rho. \qquad \ldots (9.3a)$$

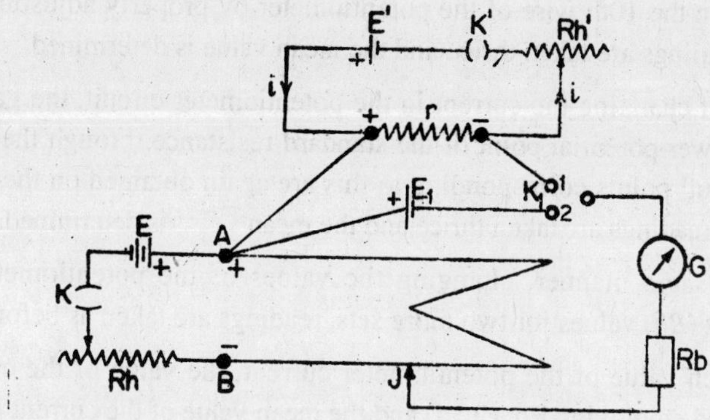

Fig. 9.4 : Current by a potentiometer

The value of *r* can be obtained with the help of a standard or a given cell of known e.m.f. E_1. If this e.m.f. is balanced on a length l' cm of the potentiometer wire under the same arrangement, then

$$E_1 = \rho.l' \qquad \ldots (9.3b)$$

or, $$\rho = \frac{E_1}{l'}$$

From Eq. (9.3a),

$$e = i.r = l.\rho = l.\frac{E_1}{l'} \qquad \ldots (9.3c)$$

$$\therefore \quad i = \frac{E_1}{r}.\frac{l}{l'} \qquad \ldots (9.3d)$$

Thus, knowing the value of the standard low resistance (*r*) and the e.m.f. of the cell (E_1), the value of *i* can be calculated from Eq. (9.3d).

[If the standard/known cell is not given in this experiment, then the potentiometer resistance (R_p) has to be determined by a P. O. Box, the e.m.f. (E) of the driver cell has to be known and instead of the rheostat (Rh), a resistance box (R_b) has to be introduced. From these the value of ρ will be $\rho = \frac{E}{R_b + R} \cdot \frac{R_p}{1000}$ and from Eq. 9.3a, $i = \rho . \frac{l}{r}$.]

Apparatus :

Potentiometer, rheostats, cells, keys, standard resistance, standard cell, two-way key, galvanometer, high resistance box.

Method :

(i) Circuit connections are made as shown in (Fig. 9.4). Care should be taken to see that the positive poles of E, E_1 and E' are all connected to the same point A. The other end of the standard resistance and the negative of E_1 are connected to the points 1 and 2 of the two-way key.

(ii) The value of e.m.f. of E must be greater than that of E_1 and p.d. across the standard resistance (r), otherwise, null points will not be available.

(iii) Connecting the galvanometer with the point 2 of the two-way key, null point for E' is obtained on the 10th wire of the potentiometer by properly adjusting the value of the rheostat Rh. Readings are taken thrice and the mean value is determined.

(iv) Without changing the current in the potentiometer circuit, the galvanometer is connected to the lower potential point of the standard resistance through the point 1 of the two-way key and null points corresponding to this are again obtained on the potentiometer wire. As before the readings are taken thrice and the mean v·· is determined.

(v) In the same manner, changing the values of the potentiometer current by altering the rheostat (Rh) values for two more sets, readings are taken as before.

(vi) For each value of the potentiometer current, the value of the current flowing through r is obtained employing Eq. (9.3d) and the mean value of the current is determined.

Results :

e.m.f. of the standard cell (E_1) = ... volt

Value of the standard resistance (r) = ... ohm

Calculations :

From Eq. (9.3c), $e = E_1 . \frac{l}{l'}$ = ... volt

and from Eq. (9.3d), the value of the current flowing through the resistance (r) is

$$i = \frac{e}{r} = \frac{E_1}{r} . \frac{l}{l'} = \text{........ ampere}$$

Table 1

No. of Obs.	For P. D. across	Null points			Total balancing length (cm)	$e = \left(E.\frac{l}{l'}\right)$(volt)	Mean value of e in volt	Current $i = \frac{e}{r}$ (amp)
		On wire number	At scale position (cm)	Mean scale readings (cm)				
1.	Cell (E_1)	...	... ...	...	...... (l')	...	...	...
	Standard resistance (r)	...	... ...	...	...... (l)			
2.	Cell (E_1)	...	... ...	...	...... (l')	...		
	Standard resistance (r)	...	... ...	...	...... (l)			
3.	......	...	...	...	...	...		
	......	...	...	...	...			

Precautions :

(i) Whether E_1 is greater or $i.r$ is greater, it is not known at first. Hence the value of Rh is so adjusted that null points are obtained for both of them on the potentiometer under the same set-up. Then by adjusting the value of Rh, the null point corresponding to the higher value (E_1 or $i.r$) has to be brought on the 10th wire. Null point corresponding to the other will then automatically be obtained on some lower number wire.

(ii) While readings are taken for E_1, it is advised to keep open the key K', so that current should not then unnecessarily flow through r.

(iii) Current should not be passed through the potentiometer circuit unnecessarily.

9.4 Comparison of two low resistances by a potentiometer

Theory

If i_1 and i_2 be the currents in the potentiometer and the low resistance (auxiliary) circuits respectively and a_1, b_1, c_1 and d_1 be the null points corresponding to the potential terminals P_1, P_2, P_3 and P_4 respectively in (Fig. 9.5), then

$$\left.\begin{aligned} i_2.X &= i_1(b_1 - a_1)\rho = i_1 l_1 \rho \\ i_2.R &= i_1(d_1 - c_1)\rho = i_1 l_2 \rho \end{aligned}\right\} \quad \ldots (9.4a)$$

and

where $l_2 = (b_1 - a_1)$

$l_1 = (d_1 - c_1)$

and p = resistance per unit length of the potentiometer wire.

$$\text{Hence, } \frac{X}{R} = \frac{i_1 l_1 \rho}{i_1 l_2 \rho} = \frac{l_1}{l_2} \quad \ldots (9.4b)$$

$$\text{or, } X = R.\frac{l_1}{l_2} \quad \ldots (9.4c)$$

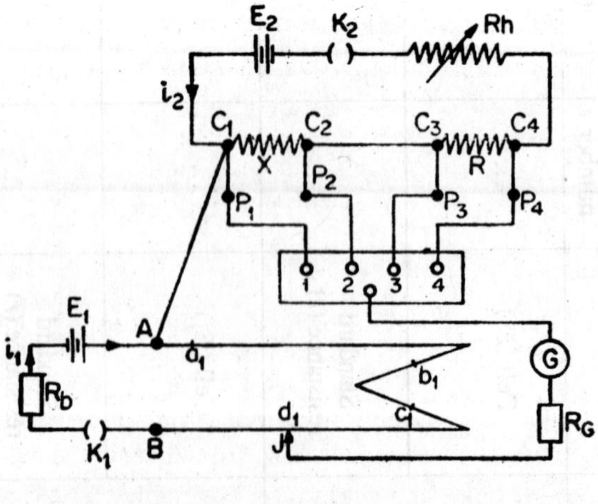

Fig. 9.5 : Comparison of two low resistances

Thus, for comparison of two low resistances, Eq. (9.4b) is employed while if the value of one resistance (R) is known, then the unknown value of the low resistance (X) can be obtained from Eq. (9.4c) if l_1 and l_2 are measured.

Apparatus :

Two low resistances, potentiometer, rheostats, keys, one four-way key, galvanometer, cells, resistance boxes.

Connections of Apparatus :

The arrangement of the apparatus is shown in (Fig. 9.5). The auxiliary circuit consists of two low resistances X and R connected in series along with the rheostat (Rh), key (K_2) and the cells E_2. The low resistances are connected across their current leads (C_1, C_2, C_3, C_4). The potentiometer circuit is the potentiometer in series with the key (K_1), resistance box (R_b) and the cells E_1. The two circuits are joined by a wire connecting the point A (the end of the potentiometer) and the current lead C_1. The *potential leads* P_1, P_2, P_3 and P_4 of the two low resistances are connected to the points 1, 2, 3 and 4 respectively of the four-way key, whose other end is connected to the galvanometer (G). The end of the galvanometer is connected to the jockey (J) through a protective resistance box R_G. Care should be taken to see that the point A of the potentiometer is connected to the positive terminals of both E_1 and E_2.

Method :

(i) The e.m.f.'s of the cells E_1 and E_2 are measured at first before starting the experiment and again they are measured at the end of the experiment to see whether they remain steady during the experiment.

(ii) The auxiliary circuit is closed and by adjusting the rheostat Rh, an appreciable amount of potential drop is created against the low resistances. Closing the potentiometer circuit, the value of R_b is made zero and the null point corresponding to the last potential lead (P_4) is obtained on the bridge wire. Then gradually changing the value of resistance in the box R_b to some suitable one, this null point (corresponding to point 4) is brought in the middle of the last wire (10th) of the potentiometer. The null point is noted thrice and the mean value is determined. Then in the same set-up, without altering anything, null points corresponding to other three points (P_1, P_2, P_3) are obtained successively and in each case, mean of three independent null points are noted. The null point (a_1) corresponding to P_1 should be close to the point A.

(iii) From the positions of the null points (a_1, b_1, c_1, d_1), the values of l_1 ($= b_1 - a_1$) and l_2 ($= d_1 - c_1$) are obtained from which, the value of $\frac{X}{R}$ is obtained employing Eq. (9.4b).

(iv) Keeping the current in the low resistance circuit constant, the potentiometer current is varied to two other values by changing R_b and again the null points for the four different potential leads are obtained in each case. Preferably the values of R_b should be so adjusted that the null points (d_1) are obtained in the 9th and 8th wires respectively. From each set, the value of $\frac{X}{R}$ is obtained for a given value of current in the low resistance circuit.

(v) By changing the current in the low resistance circuit to another value, operations (ii) to (iv) are repeated and another set of $\frac{X}{R}$ is obtained.

(vi) For all the above sets, the resistance X and R are interchanged in their gaps and the experiments are repeated. Thus, for three different values of potentiometer currents and for two different values of current in the low resistance circuit, 12 values of $\frac{X}{R}$ in all are obtained for resistances being interchanged. The mean value of $\frac{X}{R}$ is obtained from these twelve values.

Results :

(A) E.M.F.'s of the batteries :

Table 1

Battery	E.M.F. in volts			Inference
	Before experiment	After experiment	Mean	
E_1	...	...	...	
E_2	...	...	...	

(B) Determination of null points :

Table 2

Current in low resistance	Resistance in		No. of Obs.	Null points corresponding to											
	Right Gap (ohm)	Left Gap (ohm)		d_1			c_1			b_1			a_1		
				Wire No.	Scale position (cm)	Mean (cm)	Wire No.	Scale position (cm)	Mean (cm)	Wire No.	Scale position (cm)	Mean (cm)	Wire No.	Scale position (cm)	Mean (cm)
High	R	X	1	10	... ...	...	...	... ...	...	...	... ...	...	...	... ...	...
			2	9	... ...	...	...	... ...	...	...	... ...	...	...	... ...	...
			3	8	... ...	...	...	... ...	...	...	... ...	...	...	... ...	...
	X	R	1												
			2												
			3												
Moderate	R	X	1												
			2												
			3												
	X	R	1												
			2												
			3												
Low	R	X	1												
			2												
	X	R	1												
			2												

(C) Calculation of the ratio of X and R :

Table 3

Current in low resistance	Resistance in Right Gap (ohm)	Resistance in Left Gap (ohm)	No. of Obs.	Total balancing lengths corresponding to d_1 (cm)	c_1 (cm)	b_1 (cm)	a_1 (cm)	$l_1 = (b_1 - a_1)$ (cm)	$l_2 = (d_1 - c_1)$ (cm)	Ratio $\left(\frac{X}{R} = \frac{l_1}{l_2} \text{ or, } \frac{l_2}{l_1}\right)$	Mean value of $\frac{X}{R}$
High	R	X	1.	...	...	...	...	...	...	...	
			2.	...	...	...	...	...	...	...	
			3.	...	...	...	...	...	...	...	
	X	R	1.	...	...	...	...	...	...	...	
			2.	...	...	...	...	...	...	...	
			3.	...	...	...	...	...	...	...	
Moderate	R	X	1.	...	...	...	...	...	...	...	
			2.	...	...	...	...	...	...	...	
			3.	...	...	...	...	...	...	...	
	X	R	1.	...	...	...	...	...	...	...	
			2.	...	...	...	...	...	...	...	
			3.	...	...	...	...	...	...	...	

Calculations :

The ratio of the two low resistances X and R from Eq. (9.4b) is $\frac{X}{R} = \frac{l_1}{l_2}$ or, $= \frac{l_2}{l_1}$ depending upon the positions of X and R in the left or right gaps. If the ratio has to be determined, then from Eq. (9.4b),

$$\frac{X}{R} = \frac{l_1}{l_2} \left(\text{or} \frac{l_2}{l_1}\right) = \ldots\ldots$$

If one of the resistances (R) is known, then the value of the unknown resistance (X) is obtained from Eq. (9.4c) as

$$X = R.\frac{l_1}{l_2} = \ldots\ldots \text{ ohm}$$

Precautions :

(i) To get an appreciable potential drop across the low resistances, the current in the auxiliary circuit should be sufficient.

(ii) In order to reduce the *dead length* b_1c_1 corresponding to the potential difference between P_2 and P_3, the terminals C_2 and C_3 should be connected by a *short thick* copper wire or by a small brass rod.

(iii) The batteries used should be of large capacity and fully charged.

(iv) The current in the low resistance circuit should not be very high, otherwise heating will be produced at the low resistances, changing there by their values.

(v) Before taking readings, the connections of the circuit is tested by observing opposite deflections in the galvanometer, while the jockey is contacted with the first and last wire, corresponding to the potential leads P_2 or P_3.

(vi) The higher potential leads of E_1 and E_2 must be connected properly at the same point (A) of the potentiometer.

Error Calculations :

From Eq. (9.4b), $\frac{X}{R} = \frac{l_1}{l_2} = x$ (say).

$\therefore$ proportional error in measurement of x is $\frac{\delta x}{x} = \frac{\delta l_1}{l_1} + \frac{\delta l_2}{l_2}$.

The maximum error in the measurement of l_1 or l_2 is 2 times the smallest scale divisions. Since the values of l_1 and l_2 are fairly large, hence the proportional error in the measurement of x is small. This is why the null points are taken on balancing points as large as possible.

$\therefore$ percentage error $= \frac{\delta x}{x} \times 100 = \ldots\ldots\ \%$.

Questions and Answers

Q.1. Why the low resistance value is not determined by the Wheatstone bridge method?

A. : The drop of potential method is a more accurate method since in Wheatstone bridge principle the bridge remains sensitive only when the resistance in the four arms are of the same order. Hence, low resistances cannot be measured accurately by that method.

Q.2. What are low resistances?

A. : Resistances having values less than one ohm are generally called low resistances.

Q.3. What is actually compared here?

A. : Actually the potential drops across the low resistances are compared here.

Q.4. Is there any other method of measuring low resistances?

A. : Kelvin's double bridge method may be employed to measure low resistances.

Q.5. What should be done to ensure maximum accuracy in this experiment?

A. : It is better to connect a rheostat in the potentiometer circuit in place of resistance box and the former varied until the balance point is obtained on the maximum number of wire corresponding to the potential lead P_4.

Q.6. What are the low resistances provided with four binding screws?

A. : The thick pair of the screws are the current leads through which current enters and leaves. Resistance between these two leads is normally slightly greater than that specified on the top. The specified value is actually the resistance between the two inner (thin) leads which are known as potential leads.

Q.7. If the potnetial lead P_1 is directly joined to the point A, what will happen?

A.: There may be some irregularities and resistances at the end A of the potentiometer. Since A is connected to C_1, the null point (first) is brought a little inside the potentiometer (away from A) by connecting A to C and P_1 to the first point of the four-way key.

Q.8. Why the null point corresponding to P_4 is brought towards the last wire?

A. : In that case, the length becomes large which thereby reduces the proportional errors.

Q.9. What happens if the current in the potentiometer circuit only is reversed?

A. : Then the point A will be connected with different polarities of E_1 and E_2 and null points would never be obtained according to the principle of the potentiometer.

Q.10. Can two very high resistances be compared by this method?

A. : No, in that case the potentiometer readings become very much insensitive.

Q.11. What is to be done if the null point is not obtained on any of the wires?

A. : Then, either the resistance of the main circuit has to be altered, or the current in the low resistance circuit has to be changed by changing the number of cells in E_2 or E_1 to a different value.

9.5 Calibration of an ammeter by potential drop method

Theory :

In the circuit shown in (Fig. 9.6), if the ammeter (A) shows a current i' ampere flowing through a standard resistance r, then the potential drop across it is $i'r$. If this drop is balanced by a length l cm of a potentiometer wire in which the rate of fall of potential is ρ volt per cm, then

$$ir = l\rho \qquad \dots (9.5a)$$

or, $i = \frac{l}{r}\rho$, where i is the actual current flowing through r, while i' is the apparent value of the current as observed in the ammeter. The correction $(i - i')$ has to be added algebraically to the observed value i', to get the correct value. Two graphs may be drawn; one, the observed ammeter reading against the calculated current and the other, the observed ammeter reading against the correction. These curves are known as the calibration curves.

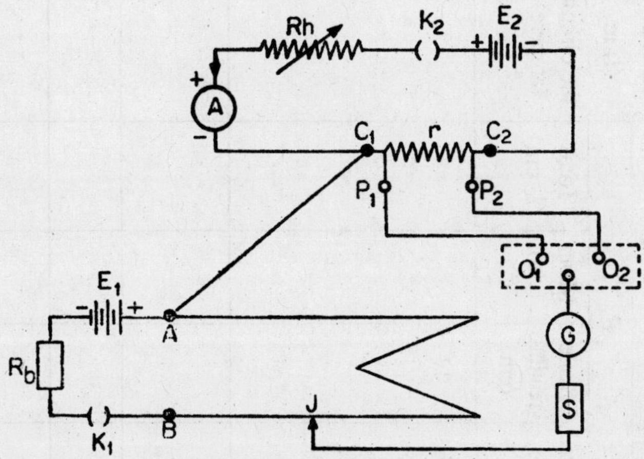

Fig. 9.6 : Calibration of an ammeter by potential drop method

Apparatus :

Potentiometer, ammeter, keys, rheostat, standard resistance, two-way key, galvanometer, resistance boxes, cells, P.O. Box.

The connection of the circuit is done as shown in (Fig. 9.6). The potential leads of the standard resistance are P_1 and P_2, while C_1 and C_2 are the current leads. S is a protective resistance connected in series with the galvanometer. Polarities of the cells (E_2 and E_1) are to be properly connected so that the points A and C_1 correspond to the similar polarities (+ve).

Method :

(i) The resistance R_p of the potentiometer wire is first determined by a P.O. Box. The e.m.f. of the battery E_1, E_2 are measured before and after the experiment and their mean values determined.

(ii) After making the connections properly, the key K_2 is closed and the rheostat Rh is so adjusted as to send the highest current (i') through the ammeter for which correction is to be made. Then current is passed in the potentiometer circuit by closing the key K_1 and connecting the point O_2 of the two-way key to the galvanometer, the value of the resistance in the box R_b is increased gradually from a low value, so that the null point corresponding to the potential lead P_2 of the standard resistnce (r) is obtained somewhere on the 10th wire. This is noted thrice and the mean value (l_2) is noted.

(C) Calculations of currents and corrections :

Table 3

No. of Obs.	Ammeter readings (i' ampere)	Resistance in the potentio-meter circuit (R_b ohm)	Null points corresponding to								Current (i) in the circuit (ampere)	Mean current (i) (ampere)	Corrections $(i - i')$ (ampere)
			P_2				P_1						
			Wire No.	Scale reading (cm)	Mean (cm)	Total (l_1 cm)	Wire No.	Scale reading (cm)	Mean (cm)	Total (l_2 cm)			
1.	1.0	... ...	... ...	... ...	... ...	... ...	... ...	... ...	... ...	... ...	... ...	...	...
2.	0.8	... ...				... ...				... ...	... ...	...	...
etc.	etc.	etc.											
4.	4												
5.	0.2												

(iii) Next the point O_1 of the two-way key is connected to the galvanometer, which corresponds to the potential lead P_1 of the standard reistance. This is done without altering anything in the two circuits. The null point will be obtained somewhere at the beginning of the 1st wire. This is also noted thrice and the mean value (l_1) is determined. Thus, the balancing length (l) corresponding to the potential drop across the standard resistance is $(l_2 - l_1)$.

(iv) The observation is repeated for a different value of the potentiometer current which can be had by changing the value of R_b in such a way that the null point corresponding to P_2 now comes on the 9th wire. From these two sets, the value of the actual current (i) passing through the ammeter is calculated employing the theory. The value of ρ is calculated in the normal way (see Calculations).

(v) Next the ammeter current is changed to some desired lower value for which correction needs to be done and the operations (ii) to (iv) are repeated again.

(vi) By changing the rheostat Rh successively, the observations are made for 5 different values of current passing through the ammeter (cf. 1.0, 0.8, 0.6, 0.4 and 0.2 ampere) and for each value, the corresponding corrections $(i - i')$ are calculated.

(vii) Finally, the calibration curves, as discussed in the Theory, are drawn for the ammeter.

Results :

(A) Determination of the potentiometer resistance (R_p) by a P.O. Box :

Table 1

[Make a table like Table 1 of Experiment No. 8.5]

(B) E.M.F.'s of the cells :

Table 2

Cells	e.m.f.s in volts		Mean e.m.f. (Volt)	Conclusion
	Before Expt.	After Expt.		
E_1	...	...	...	...
E_2	...	...	...	...

(C) Calculations of currents and correstions :

Table 3 (See previous page)

(D) Data for drawing calibration curves :

Table 4

No. of Obs.	1	2	3	4	5
Observed current (i') ampere	...	...	...	...	...
Calculated current (i) ampere	...	...	...	...	...
Corrections $(i - i')$ ampere	...	...	...	...	...

Calculations :

If E_1 is the e.m.f. of the driving cell in the potentiometer circuit, R_p is the potentiometer resistance, R_b is the resistance in the box in series with the potentiometer, then the potential drop across the ends of the potentiometer (*AB*) will be $\frac{E_1 R_p}{R_p + R_b}$ volts. If *L* is the total length of the potentiometer wires, then the potential drop per cm of the potentiometer wire is

$$\rho = \frac{E_1 R_p}{L(R_p + R_b)} \text{ volt/cm} = \ldots\ldots\ldots \text{ volt/cm}$$

Hence, from Eq. (9.5), then the actual current (*i*) passing through the standard resistance is

$$i = \frac{l}{r}\rho \text{ ampere} = \ldots\ldots\ldots \text{ ampere,}$$

where $l = (l_2 - l_1)$ cm

Drawing of Graphs :

From Table 4, two calibration curves are drawn. One with observed ammeter reading (i') along *X*-axis and the calculated current (i) along *Y*-axis. The graph will be as shown in (Fig. 9.7). Another graph is plotted with the observed current (i') along *X*-axis and the correction ($i - i'$) along *Y*-axis. This graph will be as shown in (Fig. 9.8). These curves are known as calibration curves. From these curves, correction for ammeter at any reading may be obtained.

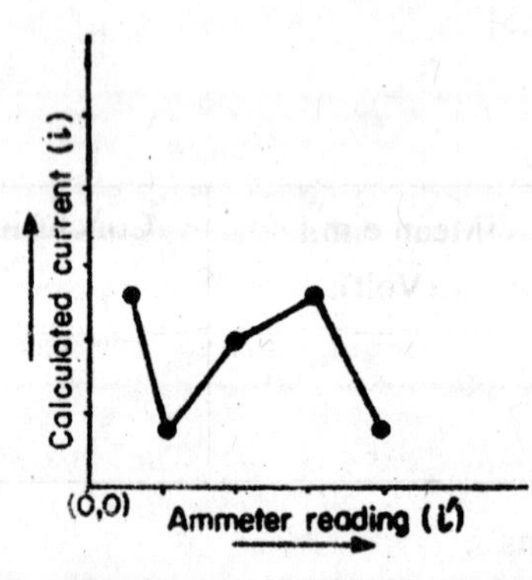

Fig. 9.7 : (i vs. i') curve

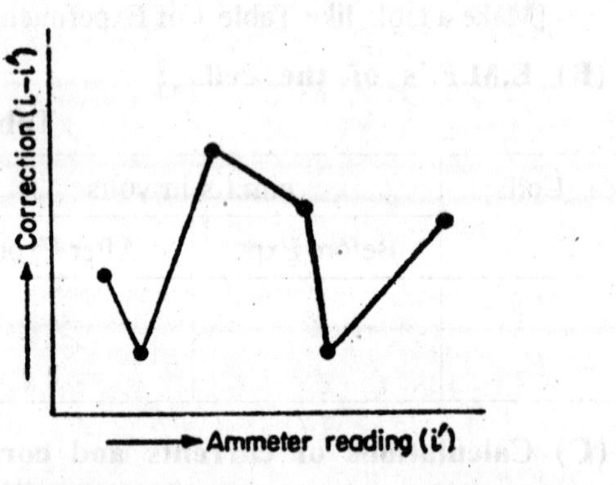

Fig. 9.8 : i' vs. $(i - i')$ curve

Precautions :

(i) The general precautions for potentiometer experiment have to be adopted.

(ii) The ammeter reading should be set at the desired value as precisely as possible and if necessary, in addition to the rheostat *Rh*, a carbon rheostat may be used in series for the purpose.

(iii) For a particular setting of the ammeter reading, the value of R_b should never be disturbed while taking readings for the leads P_1 and P_2.

(iv) While drawing graphs, the consecutive points must be joined by straight lines which indicate the corrections for the ammeter readings between those two points, assuming a linear variation of correction with the change of current between those two points.

Error Calculation :

From the expression if $i = \frac{E_1 R_p l}{L(R_p + R_b).r} = \frac{E_1 R_p (l_2 - l_1)}{L.r(R_p + R_b)}$, it is evident that except $(l_2 - l_1)$ all other quantities in the experiment remain almost constant for a particular set of current value. The only measurable quantity is $(l_2 - l_1)$, which may give rise to some errors. But since the value of $l\ (= l_2 - l_1)$ is made very large (of the order of 900 cm or more), it contributes very little to the error in the experiment. Thus, if the null point corresponding to the leads P_1 and P_2 are taken at a long distance apart, then the proportional error in this experiment will be very small. Hence, great accuracy may be obtained in the calibration of an ammeter.

[Questions and Answers on 'Calibration of Ammeter' are given at the end of the next (9.6) experiment.]

9.6 Calibration of an ammeter by the copper deposition method

Theory :

If a current of i ampere flows through a circuit containing an ammeter (to be calibrated) and a copper voltameter in series and if this current produces a deposition of W gm of copper in a time t seconds on the cathode of the voltameter, then the true value of the current (i) is given by

$$i = \frac{W}{Z \times t}, \quad \dots (9.6a)$$

where Z is known as the electro-chemical equivalent of copper. If the ammeter reads a current i', in the circuit, then the correction for the reading of the ammeter is $(i - i')$, which has to be added algebraically to i' to obtain the correct value of the current corresponding to the observed value of the ammeter. By finding the corrections for different values of the ammeter readings, graphs may be plotted (i) with observed values against the true values and (ii) with the observed values against the corresponding corrections. These graphs are known as the calibration curves.

Apparatus :

Potentiometer, storage cells, plug keys, copper voltameter, ammeter, rheostat, carbon, rheostat, standard low resistance, resistance box, galvanometer, protective resistance and two-way key.

Connection of the circuit is done as shown in (Fig. 9.9). The points A and C must be connected to the similar polaritis of the cells. Care should be taken to connect the cathode of the voltameter carefully to the negative terminals of E_2. The potential leads (P_1 and P_2) of the standard resistance are connected to O_1 and O_2 of the two-way key, while the current leads C_1 and C_2 are connected to the original circuit.

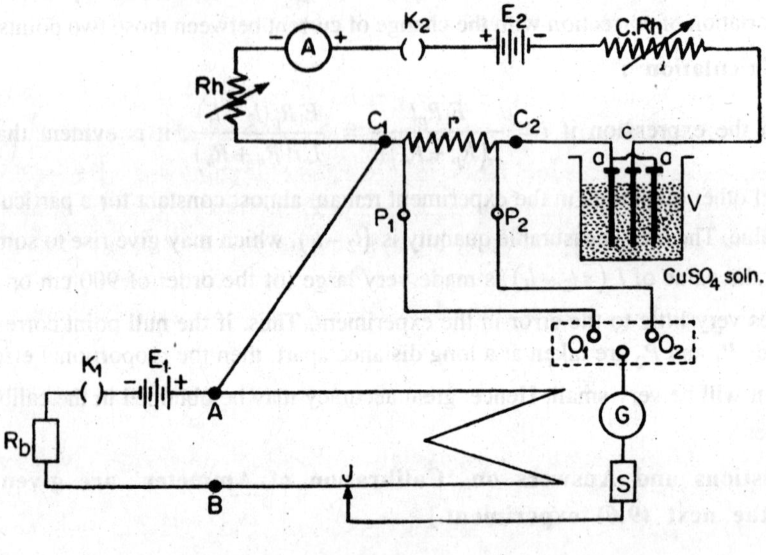

Fig. 9.9 : Calibration by copper deposition method

Method :

(i) The electrolyte (copper sulphate) is first prepared. The prepared solution may be about of 25% to 30% strength. To the solution are added 1cc of strong sulphuric acid and 1cc of ethyl alcohol for every 100cc of the solution.

(ii) Next the total area of the cathode plate (considering both the faces) is measured, that will remain submerged in the solution during electrolysis. The maximum current that may be sent through the voltameter for a good deposit, is one ampere for every 50 sq. cm are of the cathode plate. From this the maximum current that may be used, can be calculated.

(iii) The cathode on which final deposition will be made, is cleaned by rubbing with a fine-grained sand-paper on both the faces. The surface must be clean, free from any grease, on which deposit will be uniform. The cleaned surfaces should never be touched with fingers. The loose granules may be removed by striking the plate on the table without bringing the surfaces in contact with the table. The plate then may be washed in a stream of distilled water and finally dried under a fan. This plate is then weighed carefully by a sensitive balance using a rider. Let the mass be m_1 gm.

(iv) By introducing a dummy cathode in the voltameter (V) the connection of the voltameter circuit is completed by closing the key K_1. By adjusting the rheostat (Rh) and the carbon rheostat (C. Rh), the ammeter reading is adjusted to a value which is slightly smaller then the calculated permissible value. Introducing also a suitable resistance in the resistance box R_b, and connecting the galvanometer by a key to the point O_2 of the two-way key, the null point is brought at the middle of the 10th wire of the potentiometer. The current in the voltameter circuit is finally adjusted to the desired highest value for which the ammeter calibration is needed.

(v) The voltameter circuit is opened and the dummy cathode is removed and the clean weighed cathode is placed in its position. The plate is inserted in the voltameter. The anode and the cathode plates must remain parallel. The key K_2 is closed and simultaneously a stopwatch is started.

(vi) Closing the key K_1 the jockey is set at the null point corresponding to the potential lead P_2 of the standard low resistance. The constancy of the current flowing in the voltameter circuit is checked by observing the fixed null point where the jockey is set at small intervals. Any deflection of the galvanometer indicates a change of current in the voltameter circuit. The deflection is brought to zero by adjusting the carbon rheostat (C. Rh) in the voltameter circuit. The current is passed for at least half an hour and then the current and the stopwatch are stopped simultaneously. The potentiometer circuit current is also stopped.

(vii) The cathode plate is removed very carefully and washed in a slow stream of water. Then it is dried completely and weighed again as before. Let the weight be m_1 gm. Thus the amount of copper ion deposited on the cathode in the given time (t) is $W = (m_2 - m_1)$; gm Employing Eq. (9.6a), the value of the true current (i) is obtained if E.C.E. (Z) of copper is known. The ammeter reading was noted during the flow of current which gave the value of the observed current (i'). Thus the correction ($i - i'$) for the observed current can be determined.

(viii) The experiment is repeated for at least two more decreasing values of the specified ammeter current and the method is repeated as in previous operations. The initial weight of the cathode in the second set of observation will be the same as the final weight of the first set and similarly for the third set. The cathode plate need not be rubbed each time for different sets and the initial weights need not be taken.

(ix) Finally for each value of the observed ammeter currents, the true currents are calculated and the corresponding corrections are determined. Graphs as shown in (Figs. 9.7 and 9.8) are drawn as before which give the calibration curves.

(x) If the two-way key is not available, the point P_1 of the standard low resistance may be directly connected to the end A of the potentiometer and the point P_2 directly to the galvanometer G.

Results :

(A) Preliminary observations :

Table 1

Approx. volume of the voltameter (cc)	Mass of the solute required (gm)	Total immersed area of the cathode surface (sq. cm)	Maximum permissible current in the voltameter (ampere)	e.m.f. of the battery E_1		
				Before Expt. (volt)	After Expt. (volt)	Mean (volt)

(B) Determination of ammeter corrections :

Table 2 (See next page)

(C) Checking constancy of the voltameter current :

Table 3 (See next page)

Value of the standard low resistance (r) = ... ohm

Calculations :

From Eq. (9.6a), $i = \dfrac{W}{Z \times t}$

From Table 2, the value of copper ion deposited (W gm) in a given time (t sec) is obtained. Knowing the value of Z for copper (= 0.000329 gm/coulomb), the value of the true current (i) is calculated. The corresponding ammeter reading gives the observed current (i').

Drawing of Graphs :

As in Expt. No. 9.5, two graphs are drawn; one with the observed ammeter current (i') against the true calculated current (i) and the other with the observed current (i') against the corrections ($i - i'$). The nature of the graphs will be as shown in (Figs. 9.7 and 9.8). These are known as the calibration curves.

Precautions :

(i) For getting a good deposition, the solution should be freshly prepared and requisite amount of H_2SO_4 and ethyl alcohol, as stated, should be added.

(ii) The current should nevr exceed the maximum permissible limit.

(iii) The copper plate should be thoroughly cleaned, washed and dried carefully to get a good deposit.

(iv) Weighing of the cathode must be done very carefully using riders or up to 1 mg.

(v) The current in the voltameter circuit must be kept constant during deposition. This should be carefully checked at regular intervals by noting the constancy of the null point. If necessary, the carbon rheostat should be constantly operated to keep the current constant.

(vi) The time of deposition should preferably be at least for 30 min for each set. In any case, this time should never be less than 20 min for a set of deposition. Less the time, the less will be the amount of deposition, involving thereby greater source of error in the experiment.

Error Calculations :

From Eq. (9.6a), $i = \dfrac{W}{Z.t}$.

Proportional error in the measurement of i is given by

$$\frac{\delta i}{i} = \frac{\delta W}{W} + \frac{\delta Z}{Z} + \frac{\delta t}{t}.$$

(B) Determination of ammeter corrections :

Table 2

No. of Obs.	Ammeter reading (i' A)	Mass of the cathode		Mass of copper ion deposited (W gm)	Time of deposition (t sec)	Calculated current (i A)	Corrections $(i - i')$ A
		Before deposition	After deposition				
1.	(i'_1)	... gm + ... gm + ... mg + ... mg = m_1 gm	... gm + ... gm + ... mg + ... mg = m_2 gm	$W_1 = (m_2 - m_1)$ gm	$30 \times 60 = \ldots$ sec	$i_1 = \frac{W_1}{Z.t}$ = ...	...
2.	(i'_2)	m_2 gm	m_3 gm	W_2	"	$i_2 = \frac{W_2}{Z.t}$	...
3.	etc.	etc.	...	...	...	...	...
4.	etc.	etc.	...	...	...	...	...
5.	etc.	etc.	...	...	...	...	...

(C) Checking constancy of the voltameter current :

Table 3

Value of the standard low resistance (r) = ... ohm

No. of Obs.	Value of R_b in the pot. circuit (ohm)	Time interval for null point observations (min)	Null point against standard resistance for					
			P_1			P_2		
			Wire No.	Scale reading (cm)	Total reading (cm)	Wire No.	Scale reading (cm)	Total reading (cm)
1.	......	0—5 5—10 ...... 25—30	1	... ...	... ...	10	... ...	... ...
2.	......	0—5 5—10 ...... 25—30						
etc.	......	etc.						
5.	......	......						

But since the value of Z is assumed to be constant, hence,

$$\frac{\delta i}{i} = \frac{\delta W}{W} + \frac{\delta t}{t}.$$

Thus, the measurement of the weights of the cathode and the time by the stopwatch, contribute to the proportional errors. To minimise these errors, the weights are measured by a sensitive balance using a rider and large time intervals are measured using a stopwatch reading up to $\frac{1}{5}$th or $\frac{1}{10}$th of the second. The maximum error, however, in this experiment creeps in from the determination of the cathode weights. Hence, very special precautions and care should be taken in that regard. The percentage error is thus $\frac{\delta i}{i} \times 100 = \ldots\ldots\%$.

Questions and Answers

(On Expt Nos. 9.5 and 9.6)

Q.1. What is meant by 'calibration of an ammeter'?

A. : It means the checking up the accuracy of the readings of the ammeter.

Q.2. Why is it necessary to calibrate the ammeter?

A. : If the ammeter readings are inaccurate, then by calibrating it, one can determine the correct value of the current corresponding to any reading of the ammeter.

Q.3. Is the calibration curve a smooth one, if not, why?

A. : It is not a smooth curve—rather a point to point curve, since it is not necessary that the correction to be applied may vary in a regular way.

Q.4. When does a calibration curve become parallel to the X-axis?

A. : A calibration curve becomes parallel to the X-axis when the same correction has to be applied to every reading. This happens only when the ammeter has no error other than zero error.

Q.5. How are 'error' and 'correction' related?

A. : Error is minus times the correction.

Q.6. Can the ammeter be calibrated with the help of another accurate ammeter?

A. : Yes, but not as precisely as done by a potentiometer.

Q.7. Does the arrangement measure current directly?

A. : No, from the measurement of potential drop, current is measured applying Ohm's law.

Q.8. What are the functions of the potential leads?

A. : See the construction of the 'standard resistance'.

Q.9. What type of galvanometer is required for potentiometer work?

A. : A voltage sensitive galvanometer is required.

Q.10. What are 'Coulomb', 'Faraday', 'Ampere' and E.C.E.?

A. : See any textbook.

Q.11. What is meant by electrolytic dissociation?

A. : See 'Questions' under Expt. No. 8.8.

Q.12. How does copper deposit on the cathode?

A. : From $CuSO_4$ solution, due to flow of current, copper ions are deposited on the cathode, while the sulphate ions attack the copper on the anode forming copper sulphate back and maintains the strength of the solution constant.

Q.13. Is the gain in the weight of the cathode equal to the loss in the weight of the anode?

A. : Pure copper from the anode forms copper sulphate while the impurities on the anode fall down. Thus loss in weight of the anode cannot be equal to the gain in weight of the cathode.

Q.14. What happens if the current in the voltameter circuit is reversed by a commutator?

A. : By this, anode and cathode will change and thus there will be no deposition.

Q.15. What type of cell is preferred in the voltameter circuit?

A. : Since constant current has to be supplied for a fairly large time, hence storage batteries are to be used there.

Q.16. What happens if electrodes other than those of copper are used?

A. : In that case, copper deposition on the cathode will take place from the solution and the strength of the latter will gradually decrease.

Q.17. Of the two methods of calibration of ammeter, which is superior and why?

A. : The experiment procedure in the voltameter method is laborious while the potential drop method is quicker and more accurate than the other method.

Q.18. What is the difference between a carbon rheostat and an ordinary rheostat?

A. : In an ordinary rheostat, the variation of resistance takes place by intermittent values, while in a carbon rheostat, the variation is continuous by increasing or decreasing the air-gap in between the carbon plates.

Q.19. Why carbon plates are used in this type of rheostat and no other metal like aluminium, copper, etc.?

A. : Since carbon is a very good conductor of electricity and also cheap and readily available.

9.7 To draw the e.m.f.-Temperature curve for the given thermocouple and hence to determine the thermoelectric power at a given temperature

Theory :

When one junction of a thermocouple is kept at a constant temperature (normally an ice bath) t_1 and the other junction is heated to a higher temperature t_2, an e.m.f. (e) is developed in the couple. The direction of the e.m.f. depends on the pair of metals used and the temperature difference between the junctions. The e.m.f. so developed can be measured by balancing it against a potential drop on a calibrated potentiometer.

In (Fig. 9.10), if E be the e.m.f. in volts of the driver cell in the potentiometer circuit, R_b is the resistance in series with it and R_p is the resistance of the potentiometer having wires of length L cm, then the potential drop (ρ) per cm of the potentiometer wire is

$$\rho = \frac{E.R_p}{(R_b + R_p).L} \text{ volt} \quad \text{... (9.7a)}$$

If l is the length of the potentiometer wire against which the thermo-e.m.f. (e) is balanced, then

$$e = l.\rho = \frac{E.R_p.l}{(R_b + R_p).L} \text{ volt} \quad \text{... (9.7b)}$$

By measuring the thermo-e.m.f. (e) for different temperatures of the hot junction, employing Eq. (9.7*b*), a curve may be plotted with the temperature of the hot junction as abscissa and thermo-e.m.f. as ordinate. This curve is known as the *calibration curve* (Fig. 9.11).

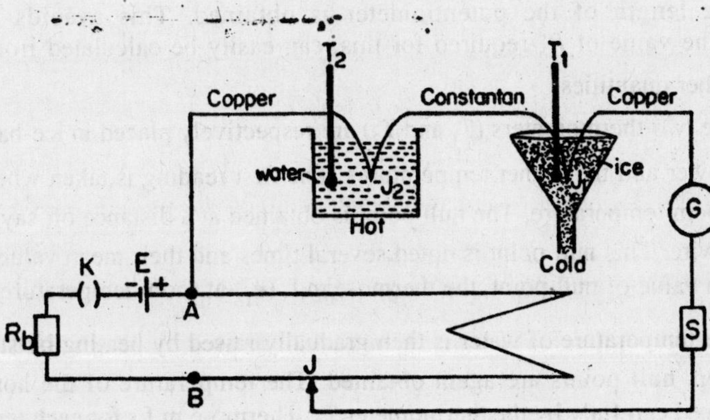

Fig. 9.10 : Thermocouple circuit

To find the thermo-electric power $\frac{de}{dt}$ for the given pair of metals at a given temperature (T°C), a tangent is drawn to the curve at the point corresponding to T°C of the hot junction. An ordinate is drawn on the X-axis from a point whose value is T°C. The value of the ordinate divided by that of the intercept made by the tangent on the X-axis, gives the value of $\frac{de}{dt}$ at T°C. Its value will be microvolts/°C if the thermo-e.m.f. is measured in micro-volts.

Apparatus :

Potentionmeter, accumulator, key, resistance boxes, thermocouple (Copper constantan or iron-copper), ice-bath, water bath, two glass tubes, galvanometer and thermometer $\left(\frac{1}{10}\right)$°C.

The connections of the circuit are made as shown in (Fig. 9.10). To construct the couple, two pieces of copper wire and one piece of constantan wire, each of about 1 metre long, are taken. The two ends of the constantan wire, after being cleaned, are joined to the one end of each of the copper wire and properly soldered. The two glass tubes are introduced in the copper wires to ensure that the metals touch at the junctions only.

Method :

(i) The thermo-couple is constructed and the resistance (R_p) of the potentiometer is measured by a P.O. Box. The e.m.f. of the driving cell (E) is measured by a voltmeter before and after the experiment and its mean value noted.

(ii) After making the connections of the circuit as shown, the junction (J_1) of the couple is placed in a bath of melting ice. The other junction (J_2) is placed in a large beaker containing water. A sensitive thermometer $\left(\frac{1}{10}\right)$°C is immersed in the beaker with its bulb placed very close to the junction. The junctions are so placed that they get equal coverage of ice and heat from all sides.

(iii) The value of R_b is to be so adjusted that a drop of potential of 10 micro-volts per centimetre length of the potentiometer is obtained. This avaoids unnecessary calculations. The value of R_b required for this, can easily be calculated from Eq. (9.7a) knowing the other quantities.

(iv) The two thermometers (T_1 and T_2) are respectively placed in ice-bath and water to record the lower and the higher temperatures. The first reading is taken when the water-bath is at the room temperature. The null point is obtained at a distance of, say l cm, on the potentiometer wire. This null point is noted several times and their mean value is obtained. From this mean value of null point, the thermo-e.m.f. (e_1) at room temperature is obtained.

(v) The temperature of water is then gradually raised by heating by steps of 10°C and at each step, null points are again obtained. The temperature of the hot junction is always determined carefully by the thermometer T_2. Thermo-e.m.f.s for each temperature of the hot junction, are obtained employing Eq. (9.7b). Care should be taken to see that the cold junction (J_1) remains always at the ice-temperature (0°C) and this is ensured from the thermometer (T_1).

(vi) At least five different readings are taken at five different temperatures and the thermo-e.m.f. against temperature graph (Fig. 9.11) is drawn from the results.

(vii) The thermoelectric power $P = \frac{de}{dt}$ at a given temperature (T°C) can be determined from the (e-t) graph.

(viii) The whole experiment may be repeated with a different value of ρ (= 5 microvolts per cm) of the potentiometer wire. To do this, the value of R_b to be employed, has to be freshly calculated as before.

(ix) From these two sets the mean value of $P = \frac{de}{dt}$ at the given temperature is calculated.

Drawing of Graphs :

On the graph paper, the temperature of the hot junction (t_2°C) is plotted along the X-axis and the corresponding thermo-e.m.f.s along the Y-axis. The temperature of the cold junction has to be mentioned on the graph. The curve will pass through a point on the X-axis corresponding to the temperature of the cold junction. As the neutral temperature of the couple is far away from the highest temperature employed normally for the hot junction and as the temperature range is small, the curve would be almost a straight line. On the desired point, where $\frac{de}{dt}$ is to be determined, a tangent is drawn on the curve (it will be the line itself) and the ordinate PQ and the abscissa QR are drawn. As the curve is a straight line, the thermo-electric power at T°C is given by

$$P = \frac{de}{dt} = \tan \phi = \frac{PQ}{QR} = \ldots\ldots \text{ volt per °C.}$$

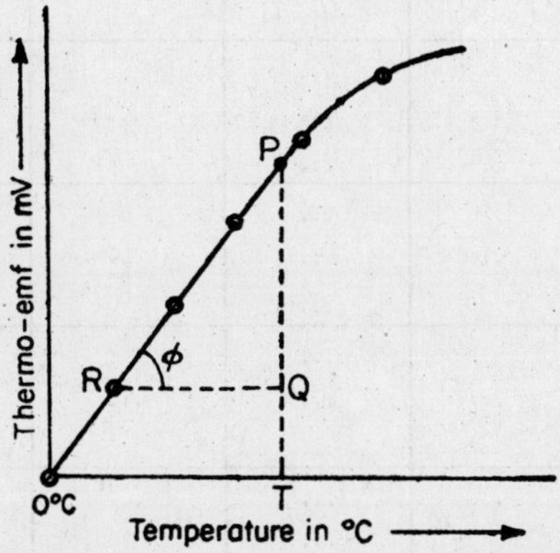

Fig. 9.11 : Temp.-thermo e.m.f. curve

Results :

e.m.f. of the battery (before expt.) = ... volts

e.m.f. of the battery (after expt.) = ... volts

Temperature of the cold junction = 0 °C

(A) Determination of the resistance of the potentiometer :

Table 1

[Make a table of P. O. Box method as shown in Table 1 of Expt. No. 8.5].

(B) Calculation of the value of R_b to make a potential drop of 10 microvolts per cm of the wire :

Normally the potentiometer has a length of wire equal to 1000 cm. For a drop of 10 μV/cm the total drop on the potentiometer wire = 10^4μV = 10^{-2} volt. If E is the e.m.f. of the battery used, then the remaining voltage (E – 0.01) volt has to be dropped over the series resistance R_b employed. Thus, if R_b is the value of the series resistance to be inserted, then

$$\frac{R_b}{(E-0.01)} = \frac{R_p}{0.01}$$

$$\therefore \quad R_b = R_p \times \frac{(E-0.01)}{0.01} \text{ ohm.}$$

The arrangement makes the potentiometer directly calibrated and from the position of null point, the value of the thermo-e.m.f. can be directly calculated.

(C) Termperature-Null point record :

Table 2

No. of Obs.	Temperatures of		Null points				Thermo-e.m.f. (*e*) developed in millivolts
	Cold junction (J_1)°C	Hot junction (J_2)°C	On wire No.	At scale reading (cm)	Mean scale reading (cm)	Total length (*l* cm)	
1.	0 °	t_1°	...	... ...	...	...	...... (e_1)
2.	0 °	t_2°	...	... ...	...	...	...... (e_2)
3.	0 °	t_3°	...	... ...	...	...	...... (e_3)
4.	0 °	t_4°	...	... ...	...	...	...... (e_4)
...	...	...	...	... ...	...	...	......
7.	0 °	t_7°	...	... ...	...	...	...... (e_7)
8.	0 °	t_8°	...	... ...	...	...	...... (e_8)

(D) Temperature-thermo e.m.f. data :

Table 3

Temp. of hot junction (*t*) in °C→	...	...	...	...	...	...	...
Thermo-e.m.f. developed in millivolts (*e*)→	...	...	...	...	...	...	...

(E) Determination of thermoelectric power $\frac{de}{dt}$ from graph :

Table 4

Temp. (*T*) in °C	Thermo e.m.f. (*de*) in millivolt (= *PQ*)	Value of *de* in microvolt	Value of temperature (*dt*) in °C (= *QR*)	Value of $\frac{de}{dt}$ at *T*°C (μV/°C)
...	...	...	...	...

Calculation :

From Eq. (9.7b), $e = \dfrac{E.R_p.l}{(R_p + R_b).L}$.

Substituting the values of the quantities on the R.H.S. of the equation, the value of the thermo-e.m.f. is obtained as *e* = ... millivolt.

For two different values of ρ (10 μV/cm and 5 μV/cm), two sets of readings are separately noted and the thermo-e.m.f.s are calculated separately. The thermoelectric power is, however, determined finally from the mean value of the two sets of readings. The above settings need not require any separate calculations as the potentiometer becomes direct-reading for e.m.f.

Precautions :

(i) Care must be taken to see that the cold junction always remains at 0 °C, the temperature of ice-bath. From time to time, ice pack should be introduced inside the bath.

(ii) It must be checked that similar poles of the driver cell and the thermocouple are connected together. For Cu-Constantan, copper is *positive* at the hot junction.

(iii) Water in the hot junction beaker must be slowly heated and stirred althrough.

(iv) Temperature difference between the junctions should not be very large, otherwise the theory will not be valid.

(v) The e.m.f. of the battery must remain constant during the experiment.

(vi) The hot junction should never be placed in boiling water. If steam temperature is required, a hypsometer should be used.

Error calculations :

From the equation $e = \dfrac{E.R_p.l}{(R_b + R_p).L}$, since E, R_p and L are constants, the proportional error in the value of e is given as

$$\frac{\delta e}{e} = \frac{\delta l}{l} + \frac{\delta R_b}{(R_b + R_p)}.$$

Since the values of l and $(R_b + R_p)$ are large, the proportional error in e is not large, for the errors in the measurement of l and R_b are very small.

[Questions and Answers at the end of the next experiment.]

9.8 Determination of the melting point of a solid by means of a thermocouple

Theory :

If one junction of a thermocouple be kept at a lower temperature while the other function at a higher temperature, then the relation between the thermo-e.m.f. (E) developed and the temperature difference (t) between the junctions are given by

$$E = at + bt^2$$

where a and b are constants, the values of which depend on the range of temperature and the metals used in the couple. If one junction (cold) be kept at ice-bath (0°C), then the relation reduces to

$$E = at + bt^2 \qquad \text{... (9.8a)}$$

where t is the actual temperature of the hot junction.

The constants a and b may be determined by keeping one junction at 0 °C and the other junction successively at two known temperatures t_1°C and t_2°C and measuring the corresponding thermo-e.m.f.s. E_1 and E_2.

$$\left.\begin{aligned} \text{Hence, } E_1 &= a.t_1 + b.t_1^2 \\ \text{and } E_2 &= a.t_2 + b.t_2^2 \end{aligned}\right\} \qquad \ldots (9.8b)$$

From this pair of equations,

$$a = \frac{E_1 t_2^2 - E_2 t_1^2}{t_1 t_2 (t_2 - t_1)} \qquad \ldots (9.8c)$$

and $$b = \frac{E_1 t_2 - E_2 t_1}{t_1 t_2 (t_1 - t_2)}. \qquad \ldots (9.8d)$$

Thus the values of a and b can be determined. Now, if the cold junction is placed in ice (0°C) and the hot junction within a bath of unknown temperature (T°C), then by measuring the corresponding e.m.f. (E') produced, the value of the unknown temperature (T°C) of the bath can be determined from Eq. (9.8a).

Apparatus :

Same as in Expt. No. 9.7. A hard glass test-tube of wide bore with the experimental substance, has to be taken.

Connection of the circuit :

[Same as in Fig. 9.10 in Expt. No. 9.7]

Method :

(i) The operations from (i) to (iv) under Method of Expt. No. 9.7 are followed. The null points are recorded at intervals of 2 minutes till it is constant, when the water-bath is at the room temperature.

(ii) The water-bath is replaced by a hypsometer, keeping the junction just above the liquid surface. When steam starts flowing, null points are recorded at intervals of two minutes till they are found constant. The hypsometer is removed and the barometer readings are recorded.

(iii) Calculating the values of E_1 and E_2 from Eq. (9.8b), knowing the values of t_1°C (room temperature) and t_2°C (steam temperature), the values of a and b are obtained from Eqs. (9.8c) and (9.8d).

(iv) A quantity of the solid, whose melting point is to be determined, is taken in a thin hard glass test-tube. The solid is melted first by heating and the hot junction of the couple is inserted in it in such a way that it is covered equally by the molten liquid on all sides. The test-tube is introduced in a water-bath so that the molten liquid freezes with the junction.

(v) The water-bath is gradually heated very slowly. With the rise of temperature, the null points are noted which vary continuously. The null points are recorded at intervals of one minute and the first sign of melting is noted very carefully. The recording of null point is continued until the whole of the solid melts. The null point during the melting away of the solid will remain fairly constant. This null point is carefully noted. Heating is still continued till melting is over and the null point again starts changing with time.

(vi) The burner is removed and the molten liquid is allowed to freeze spontaneously. Null points are again recorded during freezing. It will be found again that null points remain constant with time and this indicates the freezing of the solid. When the freezing is complete, the null point starts changing again. The constant null point during freezing is carefully noted.

(vii) This process of melting the solid and freezing the molten liquids is repeated several times alternately. For a pure crystalline solid, the two values of the null points are generally the same.

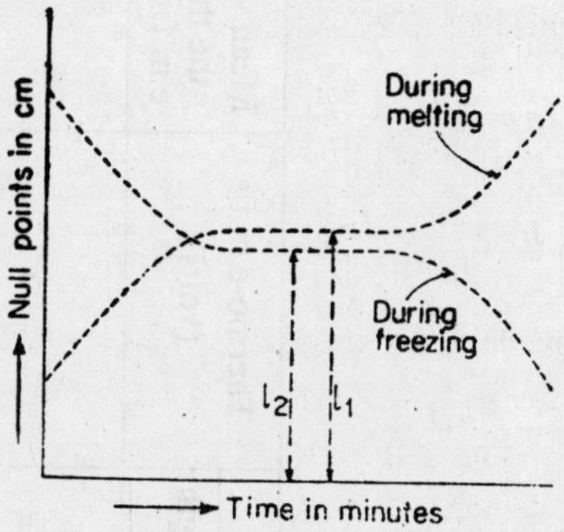

Fig. 9.12 : Time-Null point curve

(viii) Two curves are drawn by plotting time in minutes along X-axis and corresponding null points along Y-axis (Fig. 9.12). The nature of the curves will be as shown. Both the curves show a horizontal part when the melting or freezing occurs. The mean (l) of l_1 and l_2 will give the actual null point, which is employed to find the e.m.f. (E') corresponding to the temperature (T°C) of the melting point of the given solid. The value of E' is obtained from Eq. (9.7b). The values of a and b being already determined, the desired value of the melting point (T°C) of the solid is obtained employing Eq. (9.8a).

Results :

e.m.f. of the driving cell (before expt.) = ... volts

e.m.f. of the driving cell (after expt.) = ... volts

(A) Determination of the resistance of the potentiometer :

Table 1

[Make a table of P.O. Box method as shown in Table 1 of Expt. No. 8·5]

(B) Calculation of series resistance required to produce a drop of 10μV/cm of the potentiometer wire :

[Calculate in the same way as shown in (B) under Results of Expt. No. 9.7]

(C) Temperature-Null points record for two known temperatures :

Table 2 (See next page)

Barometric height = ... cm of Hg

Temperature of cold junction = 0 °C

Potentiometer resistance (R_p) = ... ohm

(C) Temperature—Null points record for two known temperatures :

Table 2

Barometric height = ... cm of Hg

Temperature of cold junction = 0°C

Potentiometer resistance (R_p) = ... ohm

No. of Obs.	Temp. of hot junction (°C)	Resistance in the potentiometer circuit (ohm)	Null points				Thermo-e.m.f.s (volts)	Mean value of the thermo e.m.f. (volts)
			On wire no.	At scale reading (cm)	Mean scale reading (cm)	Total length (l cm)		
1.	t_1 °C (room temp.)	......	...	... ...	...	...		
2.	"	......	...	... ...	...	...	...	...... (E_1)
3.	"	......	...	... ...				
1. 2. 3.	t_2 °C (Boiling point of water).							...... (E_2)

(D) Time-null point record during melting and freezing :

Table 3

Resistance in the potentiometer circuit = ...ohm

Time in minutes	Null Points					
	During melting			During freezing		
	Wire No.	Scale reading (cm)	Total length (cm)	Wire No.	Scale reading (cm)	Total length (cm)
0.	...	...	...	...	...	...
1.	...	...	...	...	...	...
2.	...	...	...	...	...	...
3.						
4.						
5.						
...						
...						
...						
etc.			... ... } Constant			... ... } Constant
etc.			... ...			... ...

(E) Evaluation of the constants *a* and *b* :

Table 4

Temperatures of		Values of thermo-e.m.f's		Values of	
Cold junction (°C)	Hot junction (°C)	E_1 (volt)	E_2 (volt)	a (volt/°C)	b (volt/°C)
0°C	$t_1°$	...		...	...
	$t_2°$		...		

(F) Determination of the melting point of the solid :

Table 5

Room Temp (t_1°C)	Boiling point of water t_2° C	e.m.f. at room temperature (E_1) (volt)	e.m.f. at t_2°C (E_2) (volt)	Value of a (volt/°C)	Value of b (volt/°C)	e.m.f. at unknown temp. (T_2°C) (E') (volt)	M.P. of solid (°C)	
							From calculation	From graph
...	...	...	...	...	...	...	...	...
...	...	...	...	...	...	...	...	...

Calculations :

From Eqs. (9.8a) and (9.8c),

$$a = \frac{E_1 t_2^2 - E_2 t_1^2}{t_1 t_2 (t_2 - t_1)} = \text{........ volt/°C}$$

$$b = \frac{E_1 t_2 - E_2 t_1}{t_1 t_2 (t_1 - t_2)} = \text{........ volt/°C}$$

For unknown temperature T°C, we have

$$E' = aT + bT_2$$

$$\text{or,} \quad T = \frac{-a \pm \sqrt{a^2 + 4bE'}}{2b} = \text{........°C}$$

Precautions :

(i) Same precautions as discussed in Expt. No. 9.7 are to be followed.

(ii) Care should be taken and guarded against spurious null points obtained, if any.

(iii) Supercooling may occur during freezing, for which necessary precautions are to be taken.

Error calculations :

From the expressions of a and b given in Eqs. (9.8c) and (9.8d), it can be shown that the proportional errors in their measurements are given by

$$\frac{\delta a}{a} = \frac{\frac{\delta l_1}{t_1^2} + \frac{\delta l_2}{t_2^2}}{\frac{l_1}{t_1^2} - \frac{l_2}{t_2^2}}$$

where l_1 and l_2 are the null points obtained on potentiometer corresponding to t_1°C and t_2°C.

$$\text{Similarly,} \quad \frac{\delta b}{b} = \frac{\frac{\delta l_1}{t_1^2} + \frac{\delta l_2}{t_2^2}}{\frac{l_2}{t_2} - \frac{l_1}{t_1}}$$

Again, for the e.m.f. E' for unknown temperature T°C, error in T is given by

$$\frac{\delta T}{T} = \frac{\frac{\delta E'}{E'}.E' + \frac{\delta a}{a}.aT + \frac{\delta b}{b}.bT^2}{T(a + 2bT)}$$

Hence, the percentage error is $\frac{\delta T}{T} \times 100 =$ .... %.

[The value of $\frac{\delta E'}{E'} = \frac{\delta \rho}{\rho} + \frac{\delta l}{l}$]

Questions and Answers

(On Expt. Nos. 9.7 and 9.8)

Q.1. How does the e.m.f. in a thermocouple arise? What are Seebeck, Peltier and Thomson effects?

A. : See any textbook on Electricity.

Q.2. What is the difference between Joule effect and Peltier effect?

A. : In Joule effect, the heat generated is proportional to the square of the current and thus it is independent of the direction of the current, but in Peltier effect, the heating or cooling of a junction is proportional to the current and hence on reversing the current direction, a heated junction will be cooled.

Q.3. What is meant by 'neutral temperature' of a couple?

A. : It is the temperature of the hot junction (cold junction being at 0°C) at which the thermo-e.m.f. in the couple is maximum.

Q.4. What is meant be 'temperature of inversion'?

A. : As the temperature of the hot junction increases, the value of the e.m.f. generated also increases and it becomes maximum at the neutral temperature. If the temperature is further increased, the e.m.f. then begins to decrease and finally it becomes zero at a particular temperature characteristic of the couple. This temperature where the e.m.f. becomes zero, is known as the inversion temperature.

Q.5. What is the order of magnitude of the thermo-e.m.f. generated?

A. : For a copper-constantan couple, it is about 41 μV/°C.

Q.6. Why is the hot junction at a higher potential here?

A. : In case of Seebeck effect, copper is higher than constantan.

Q.7. Why is the cold junction kept at 0°C?

A. : Because this junction has to be kept at a constant temperature during the experiment.

Q.8. Does the temperature of the cold junction affect the variation of e.m.f. in any way?

A. : Yes, the inversion temperature is lowered if the cold junction is at a higher temperature than zero.

Q.9. What are the laws of intermediate metals and inermediate temperature?

A. : See any textbook.

Q.10. Is the temperature-e.m.f. relation linear?

A. : No, $e = at + bt^2$ gives a quadratic relation but for a small range of temperature, it is assumed to be fairly linear.

Q.11. Why does the null point remain steady during melting or freezing?

A. : During melting or freezing, because of latent heat phenomenon, the temperature remains constant and hence steady balance point is obtained for a definite value of thermo-e.m.f.

Q.12. What is thermo-electric power and how does it vary with temperature?

A. : The thermo-electric power is $\frac{dE}{dT}$ and it varies linearly with temperature. The slope may be positive or negative.

Q.13. Can the neutral temperature of the couple be determined from the graph?

A. : Since $\frac{dE}{dT} = 0$ at neutral temperature, by differentiating the relation $E = at + bt^2$, one gets $a + 2bt_n = 0$ from which neutral temperature t_n can be found out knowing the values of a and b.

9.9 Determination of Joule's equivalent of heat (*J*) by Callendar and Barnes' method

Theory :

If a steady electric current passes through the heating coil (*H*) of Callendar and Barnes' apparatus and if a steady flow of water is maintained through the tube at a constant inflow temperature, then the temperatures of all the parts of the apparatus become steady. Under such steady conditions, the amount of electrical energy supplied in a given time is utilised in heating the water passing through the tube during the same time while some amount of heat is lost to the surroundings by radiation, etc. during the same time.

If a current of C_1 ampere flows through the coil, V_1 is the potential difference across it, θ_1 and θ_2 be the temperatures of the inlet and the outlet water and m_1 be the rate of flow (gm/sec) of water through the tube, then at the steady state,

$$\frac{V_1C_1}{J} = m_1S(\theta_2 - \theta_1) + h, \quad \text{... (9.9a)}$$

where J = Joule's equivalent

S = mean specific heat of water between θ_1°C and θ_2°C

h = amount of heat radiated per sec from the tube surface under given condition.

If the values of C_1 and V_1 be changed to some other values C_2 and V_2 and the flow rate be regulated to m_2 gm/sec to make the same difference of temperature as in the previous case $(\theta_2 - \theta_1)$, then

$$\frac{V_2C_2}{J} = m_2S(\theta_2 - \theta_1) + h. \quad \text{... (9.9b)}$$

[The value of h remains the same since $(\theta_2 - \theta_1)$ remains unchanged.]

Subtracting, Eq. (9.9a) from Eq. (9.9b), we get

$$\frac{V_2C_2 - V_1C_1}{J} = S(m_2 - m_1)(\theta_2 - \theta_1)$$

$$\text{or,}\quad J = \frac{V_2C_2 - V_1C_1}{S(m_2 - m_1)(\theta_2 - \theta_1)} \text{ joules/cal} \quad \text{... (9.9c)}$$

Knowing all the quantities on the R.H.S., the value of J can be calculated.

If the temperature difference in the two sets be not exactly the same, then assuming that the value of h does not appreciably change for this small inexactness of temperature difference $(\theta_2 - \theta_1)$, it may be shown that in that case,

$$J = \frac{V_2C_2 - V_1C_1}{m_2t_2 - m_1t_1}, \quad \text{... (9.9d)}$$

where t_1 and t_2 are the values of $(\theta_2 - \theta_1)$ in the first and the second set of the experiment respectively.

In the arrangement as shown in (Fig. 9.13), C_1 and C_2 are measured from the drop of potential across a standard resistance r connected in series with the heating coil.

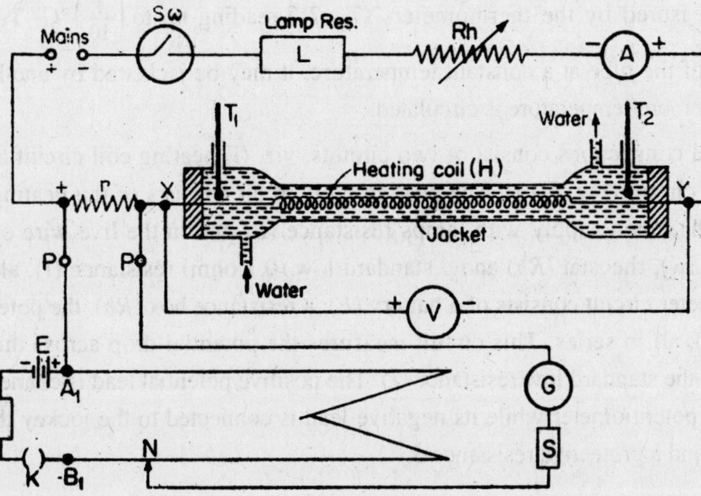

Fig. 9.13 : Callendar-Barnes' arrangement

If R_p —potentiometer resistance

R_b—series resistance with the potentiometer

E = e.m.f. in volts of the battery B

L —total length of the potentiometer wire

ρ—drop of potential per unit length of the potentiometer

l = balancing length for the potential drop across r,

then $$\rho = \frac{E}{R_b + R_p} \cdot \frac{R_p}{L} \qquad \text{... (9.9e)}$$

and the current through r and also through the heating coil is

$$C = \frac{\rho l}{r} = \frac{E.R_p.l}{r(R_b + R_p).L}. \qquad \text{... (9.9f)}$$

The potential drop V_1 and V_2 across the heating coil are measured directly by means of the voltmeter (V) connected across the heating coil (H).

Apparatus :

Callendar-Barnes' apparatus, D.C. Power Supply or Mains, Standard Resistance (about 0.1 ohm), Ammeter, Voltmeter, Sensitive Thermometers [reading $\left(\frac{1}{10}\right)$°C], Potentiometer, Battery, Resistance Boxes, Key, Galvanometer, Water-head (constant).

Callendar and Barnes' apparatus consists of a horizontal glass tube with a heating coil (H) fitted coaxially in it. The heating element is a spiral or a strip of nichrome or manganin, usually of 12-15 ohms resistance. The element is connected to the screws at the ends where electrical connections are made. At the two ends there are arrangements for inflow and outflow of water supplied from a steady water-head and the temperatures of these waters are measured by the thermometers (T_1, T_2) reading up to $\left(\frac{1}{10}\right)$°C. To keep the surroundings of the tube at a constant temperature, it may be jacketted by another tube in which water at room temperature is circulated.

The actual connections consist of two circuits, viz. (i) heating coil circuit and (ii) the potentiometer circuit (Fig. 9.13). The heating coil circuit consists of the heating coil (H), an ammeter (A), main supply with lamps resistance (L) put on the live wire of the main and a switch (Sw), rheostat (Rh) and a standard low (0.1 ohm) resistance (r), all in series. The potentiometer circuit consists of a battery (E), a resistance box (Rb), the potentiometer, a plug key (K), all in series. This circuit measures the potential drop across the potential leads (P, P) of the standard low resistance (r). The positive potential lead is connected to the point A of the potentiometer while its negative lead is connected to the jockey through the galvanometer and a protective resistance S.

Method :

(i) At first the live wire and the positive terminal of the mains are ascertained. The circuit connections are shown in (Fig. 9.13) if the live end is negative. Otherwise, if the live end is positive of the mains, then connections to the mains as well as to the voltmeter, ammeter and the cell E are to be interchanged. The technique of testing the live end (positive or negative) of the mains has to be mastered. The position of switch (Sw) has to be placed always on the live wire.

(ii) After ascertaining the live wire and the +ve terminal of the mains, water is allowed to flow through the central tube at a large rate. This will remove all air bubbles sticking on the inner surface of the glass tube. Without passing any current through the heating coil and maintaining the same flow, readings of the thermometers are taken at intervals of five minutes till the readings are found constant for three successive intervals. If t_1 and t_2 are the constant temperatures of the incoming and the outgoing water respectively, then the error (δ) of the two thermometers is $(t_1 - t_2)$°C which should be added algebraically to the final temperature difference of outflowing and inflowing water, as indicated by the thermometers T_2 and T_1 respectively.

(iii) Next the resistance of the potentiometer (R_p) is measured by a P.O. Box in the usual manner and the e.m.f. of the battery (E) is measured before and after the experiment, and the mean value is taken.

(iv) All the connections are made as shown in Fig. 9.13 and by adjusting the lamp resistance and the rheostat (Rh), the value of the current flowing through the heading coil is made very nearly equal to 1 ampere.

(v) The flow of water through the tube is then adjusted by controlling the inlet and the outlet tubes attached with screws. The constant pressure tank supplying water must have a constant pressure head during water flow through the tube. The thermometer readings are noted after an interval of 5 minutes until the difference of temperatures shown by T_2 and T_1 remains steady between 5 °C or 6 °C. This steady difference of temperatures must remain at least for 10 minutes. If the temperature difference is higher or lower than this desired value, it should be brought to the prescribed value by varying the rate of flow of the water through the tube.

(vi) At this steady state, water is collected in a large measuring cylinder for a known time and from these, the rate of flow (m) of water can be found out.

(vii) During the steady flow of water (as indicated by the steady difference of temperatures by T_2 and T_1) the potentiometer circuit is closed and null points on the potentiometer corresponding to the potential leads P_2 and P_1 are found. The value of R_b is so adjusted that the null point corresponding to P_2 is obtained preferably on the 10th wire. This null point is determined several times until it remains steady. The mean value of this null point is employed to calculate the current flowing through the heating coil. By changing the value of R_b only and without disturbing anything in the heating circuit, null point corresponding to the potential lead P_2 is obtained on the 9th wire of the potentiometer. This gives another set of observation for the same current flowing through the heating circuit. From these two sets, the mean value of the current (C) flowing through the heating coil is calculated.

(viii) The p.d. across the heating coil is obtained from the direct reading of the voltmeter (V).

(ix) For the next set of experiment, the value of the current in the ammeter is changed (say, to about 1.2 A) and the rate of flow of water is very carefully regulated and changed so that the *same difference* of temperature is recorded by the thermometers as in the first set. To get the value of the new rate of flow, the clue is to use the relation $\frac{C_1^2}{C_2^2} = \frac{m_1}{m_2}$ (if heat losses are ignored), where m_1, m_2 and C_1, C_2 are the corresponding values of the rates of flow and currents respectively in the two sets of experiment.

(x) In this way, another third set of reading is taken for another set of current value and rate of flow, keeping the same difference of temperatures in all the three sets.

(xi) Finally, employing Eq. (9.9c), the value of J is obtained—one value from Sets I and II, one from Sets II and III and one from Sets I and III. The mean of these three values of J so obtained will give finally the value of J. If the difference of temperatures are not *exactly* the same in the three sets of experiment, then Eq. (9.9d) may be employed to get the values of J from each pair of sets and the mean value therefrom.

Results :

(A) Comparison of the two thermometer readings :

(With no current but water flowing)

Table 1

Time in minutes	Temp. of incoming water (t_1°C)	Temp. of outgoing water (t_2°C)	Difference of temp. in °C	Error to be algebraically added $\delta = (t_1 - t_2)$°C
0	...	...		
2	...	...		
4	...	...		
6	...	...		
...	...	...		
...	...	...		
...	...	...		
etc.	... } Steady	... } Steady	...	...
etc.	...	...		

(B) Determination of potentiometer resistance (Rp) :

Table 2

[Make a table of P.O. Box expt. as in Table 1 of Expt. No. 8.5].

(C) Readings during flow of water :

Table 3 (See next page)

(D) Determination of heating current by potentiometer :

Table 4

(i) e.m.f. of the battery in closed circit :

Before experiment = ... volt

After experiment = ... volt

Mean e.m.f. = ... volt

(ii) Value of the standard resistance (r) = ... ohm

(iii) Potentiometer resistance (R_p) from Table 2 = ... ohm

(iv) Total length of the potentiometer wires (L) = ... cm

(v) Room temperature = ... °C

(see next, but one page after Table 3)

(C) Readings during flow of water :

Table 3

No. of Obs.	Ammeter reading (ampere)	Voltmeter reading (volt)	Time in min	Temp. of incoming water (θ_1 °C)	Temp. of outgoing water (θ_2 °C)	Observed difference of temp. $(\theta_2 - \theta_1)$°C	Corrected difference of temp. (°C) $(\theta_2 - \theta_1) + \delta$	Mass of water collected (M gm)	Time (t) of collection (sec)	Rate of flow ($m = M/t$) (gm/sec)
I.	... (C_1)	... (V_1)	0	...	...					
			5	...	...					
			10	...	...					
			15	...	...					
			20	... Steady	... Steady	...	...	... (M_1)	... (t_1)	... (m_1)
			25	...	...					
II.	... (C_2)	... (V_2)	0	...	...					
			5	...	...					
			10	...	...					
			15	...	...					
			20	...	...	...	...	... (M_2)	... (t_2)	... (m_2)
			25	...	...					
III.	... (C_3)	... (V_3)	...	...	...	...				
			...	...	...	...				
			...	...	...	...				
			...	...	...	...				
			...	...	...	...	...	...	...	...
			...	...	...	...				

Table 4

No. of Obs.	Value of R_p (ohms)	Null point obtained				Value of the current (C ampere)	Mean value of the current (C ampere)
		On wire No.	Scale reading (cm)	Mean scale reading (cm)	Total length (l cm)		
I	... ...	10	... ...	...	...	...	...... (C_1)
	...	9	... ...	...	...	...	
II	...	10	... ...	...	...	...	...... (C_2)
	...	9	... ...	...	...	...	
III	...	10	... ...	...	...	...	...... (C_3)
	...	9	... ...	...	...	...	

(E) Evaluation of *J* :

Table 5

Specific heat of water = ... cal/gm/°C

No. of Obs.	Current through the coil (C ampere)	P.D. across the coil (V volts)	Rate of water flow (m gm/sec)	Corrected temperature difference $(\theta_2 - \theta_1) + \delta$ in °C	Value of J (joules/cal)	Mean value of J (joules/cal)
I	...... (C_1)	...... (V_1)	...... (m_1)	......	...... (from Set I and II)	
II	...... (C_2)	...... (V_2)	...... (m_2)	......	...... (from Set II and III)	......
III	...... (C_3)	...... (V_3)	...... (m_3)	......	...... (from Set I and III)	

Calculations :

When the temperature difference is constant for all the three observations, then from Eq. (9.9c),

$$J = \frac{V_2C_2 - V_1C_1}{S(m_2 - m_1)(\theta_2 - \theta_1)} = \text{........ joules/cal}$$

When the temperature difference slightly varies, then from Eq. (9.9d),

$$J = \frac{V_2C_2 - V_1C_1}{m_2t_2 - m_1t_1} = \text{........ joules/cal}$$

Precautions :

(i) The current in the heating coil circuit must be kept constant whether it is drawn from the mains or from storage batteries or from power pack.

(ii) It is desirable to have the rate of flow of water between 55cc to 65cc per minute. This can be achieved by adjusting the height of the pressure head.

(iii) The apparatus gives best result for a current of approximately 2 amperes, if available. The resistance of the heating coil should preferably be of the order of 15-20 ohms.

(iv) The difference of temperature between the incoming and outgoing water should not exceed 6 °C-8 °C, otherwise Newton's law of cooling will not hold good.

(v) The thermometers used should be accurate and be able to record $\frac{1}{10}$th of a °C.

(vi) During water collection, no part of the apparatus should be displaced or touched.

(vii) If possible, the mean temperature of the water should be that of the room to avoid errors due to radiation.

(viii) The loss of heat due to radiation is given by $m(\theta - \theta_0) \times 0.05$, where θ and $\theta°$ are the mean temperature of the water at entrance and exit and room temperature round the apparatus respectively and m is the rate of flow per sec. If the radiation loss is to be accounted for, the same can be done from the above relation—adding the value of $0.05(\theta - \theta_0)$ to the difference of temperature recorded by T_1 and T_2.

Error Calculation :

If the heat loss due to radiation is neglected, then from Eq. (9.9a), $J = \frac{V_1C_1}{m_1S(\theta_2 - \theta_1)} = y$ (say). Since the errors in the measurements of V, C and m are not very large, the chief source of error lies in the measurement of the temperature difference or the actual temperatures by the thermometers. If the temperature difference $(\theta_2 - \theta_1)$ be denoted by θ, then

$$\frac{\delta J}{J} = \frac{\delta y}{y} = \frac{\delta\theta}{\theta}.$$

If the thermometers used, can read up to $\frac{1}{10}$th of a °C, then the error in the measurement of J is maximum up to 2% and if the thermometer can read up to $\frac{1}{20}$th of a °C, then the error is only 1%. Thus, the more precise and accurate are the thermometers used, the less are the errors. In fact, when the radiation correction is taken into account, the error in measuring J is about $\sqrt{2} \times \delta y$ which is greater than δy. Thus the value of J obtained when radiation loss is ignored, is better than when the correction is taken account for. Actually, in this method, the radiation correction becomes meaningless unless the thermometers can read up to $\frac{1}{20}$th of 1°C.

Questions and Answers

Q.1. What is meant by Joule's equivalent?

A. : It is the amount of work done to produce unit quantity of heat.

Q.2. What sort of work is being done in this experiment?

A. : In the metallic coil, there are free electrons. When a voltage is applied across its ends, the electrons flow. During their motion, they pass through the intra-atomic spaces when they experience a resistance. To overcome this resistance, some work is being done which appears in the form of heat.

Q.3. Why is the heating coil taken in the spiral form?

A. : Greater surface area is exposed to water and it keeps water stirred.

Q.4. Why should a constant level tank be used?

A. : Otherwise, steady flow of water and steady difference of temperatures would not be obtained.

Q.5. To get accuracy, which quantity should be measured very accurately?

A. : See 'Error Calculations'.

Q.6. Which is more important—'radiation correction' or 'accurate temperature measurement'?

A. : See 'Error Calculations'.

Q.7. How can the value of 'specific heat' of water be determined from this experiment?

A. : Knowing the value of J, the value of S can be found out from Eq. (9.9a).

Q.8. Can this method be employed to study the variation of sp. heat of water with temperature?

A. : Yes, the original experiment was designed to study that variation.

Q.9. How does sp. heat of water vary with temperature?

A. : According to this original experiment, specific heat of water is minimum at about 37 °C while it increases both for lower and higher temperatures.

Q.10. Can this method be employed to determine the specific heat of mercury?

A. : Yes, in that case mercury will be allowed to flow in place of water and the heating coil may be dispensed with, as mercury itself is a good conductor of heat.

Q.11. Why is the temperature difference not allowed to be more than 6 °C to 8 °C?

A. : In that case, Newton's law of cooling will not hold good.

Q.12. What is Newton's law of cooling?

A. : See any textbook on Heat.

Q.13. Can the experiment be performed if a voltmeter is not provided?

A. : Then the P.D. across the heating coil is measured on the potentiometer in terms of a standard cell connected appropriately in the circuit.

Q.14. What are the laws regarding generation of heat in a conductor?

A. : See 'Joule's laws' from any textbook.

Q.15. What are the advantages of this method over others?

A. : The advantage are :

(i) Observations are made at steady state, hence measurements can be made very accurately.

(ii) Thermal capacity of the tube does not come into the calculation,

(iii) Radiation correction is eliminated by taking different sets of experiment.

(iv) No external stirring is required since it is a continuous flow method.

9.10 Determination of the capacitance of a condenser by ballistic galvanometer

Theory :

When a condenser of capacitance C is charged to a potential V, the charge acquired by the condenser is CV. If this change is made to flow through a ballistic galvanometer, the latter will give a deflection d on the scale given by

$$CV = K.d\left(1+\frac{\lambda}{2}\right), \qquad \text{... (9.10a)}$$

where λ is the logarithmic decrement appropriate to the particular galvanometer circuit, and K is a constant depending upon the galvanometer and the scale distance from the galvanometer.

If a standard capacitor of known capacitance (C') is used for calibration, then the same may be charged to the same potential V as before, giving a deflection (d') in the galvanometer under the same condition. Hence, for this case,

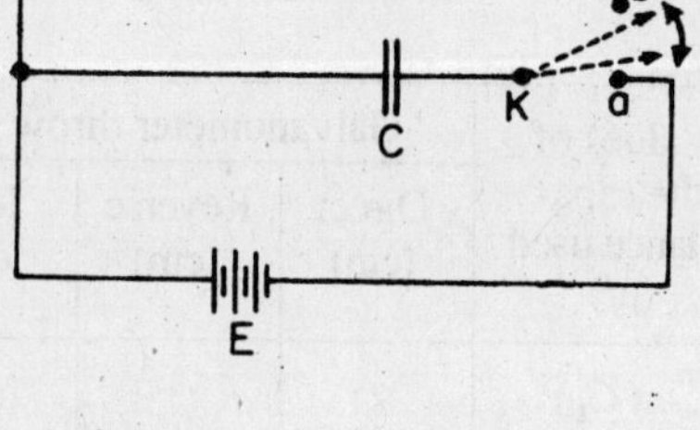

Fig. 9.14 : Determination of capacitance of a condenser

$$C'V = K.d'\left(1+\frac{\lambda}{2}\right). \qquad \text{... (9.10b)}$$

Since the circuits are open when λ is measured, hence the value of λ will be the same in both the cases.

Hence, from Eqs. (9.10a) and (9.10b), we have

$$\frac{C}{C'} = \frac{d}{d'} \text{ or, } C = C'.\frac{d}{d'}. \qquad \text{... (9.10c)}$$

Thus, knowing the value of C' and measuring d and d', the value of C can be found out.

Apparatus :

Ballistic galvanometer, battery, charge-discharge key, standard capacitor, tapping key, unknown capacitor.

For construction and principle of action of a ballistic galvanometer, **Art. 7.4(b)** may be seen.

Method :

(i) The connection of the circuit is made as shown in Fig. 9.14. One end of the condenser is connected to the central point of the charge-discharge key K' (See Art. 7.3) while one terminal from the battery and one terminal of the galvanometer are connected to the points a and b of the key. On pressing the key, the condenser is connected to the battery through Ka and it is thus charged while on releasing it, the battery is disconnected and the galvanometer is connected to the condenser through b, discharging thereby the latter.

(ii) The standard capacitor is generally taken a decade condenser box, from which different values of capacitance (C') may be introduced in the circuit. For each value of the capacitance introduced in the circuit from this decade box, the corresponding deflection (d') in the galvanometer is recorded.

(iii) The known capacitor is then replaced by the unknown capacitor (C) and by charging and discharging it, the corresponding throw in the galvanometer (d) is noted.

(iv) Employing Eq. (9-10 c), for each value of C' and d', the value of C is calculated from which, the mean value of C is determined.

(v) Otherwise, a graph may be plotted with the valus of C' plotted along X-axis and corresponding deflections (d'') and Y-axis. The graph will be a straight line passing through the origin. On this graph, the value of abscissa corresponding to the known value d, will give he value of C, the capacitance of the unknown condenser.

(vi) The experiment may be repeated for three unknown condensers taken and finding their values in series and in parallel. The experimental values of the series and the parallel connection may be compared with the theoretically expected values.

Results :

No. of Obs.	Values of the capacitance used (μf)	Galvanometer throw (d')			Value of $\frac{C'}{d'}$	Mean value of $\frac{C'}{d'}$	Value of $C = \frac{C'}{d'} \times d$ (μf)
		Direct (cm)	Reverse (cm)	Mean (cm)			
1.	...... (C_1')	...	...	...	...	...	
2.	...... (C_2')	...	...	...	...		
3.	...... (C_3')	...	...	...	...		...
etc.	etc.	...	...	...	...		.....
etc.	etc.	...	...	...	...		
7.	Unknown (C)	...	...	...	...		

Calculations :

From Eq. (9.10c), $C = C'.\frac{d}{d'} = \ldots\ldots \mu f.$

Precautions :

(i) Capacitors tend to have natural leakage. Dust and moisture increase the rate of leakage. Thus the time interval between the charging and discharging of the capacitors should be as small as possible.

(ii) Since the galvanometers (ballistic) have normally large time periods and small damping in open circuit, they thus take long time to have the oscillations dying out. They thus may be brought to rest position by using a tapping key (T.K.) across it and short-circuiting it just when the spot of light passes through the zero position.

(iii) If necessary, a resistance in series with the battery may be connected.

Error Calculation :

Since the experiment involves measurements of a set of deflections and knowledge of the values of standard capacitors, there are very little sources of error except in the exactness of the prescribed value of the standard condensers. Further, the values of d and d′ should be made neither too small, nor too large (preferably about 15-20 cm) to get more or less accurate value of C. The error in C will be

$$\frac{\delta C}{C} = \frac{\delta C'}{C'} + \frac{\delta d}{d} + \frac{\delta d'}{d'}$$

and percentage error in the measurement of the value of C is

$$\frac{\delta C}{C} \times 100\,\% = \ldots\ldots\ldots\ \%.$$

Questions and Answers

[General constructions and properties of a condenser and of the ballistic galvanometer must be known.]

Q.1. During discharge, how is it ensured that the discharge takes place through the galvanometer only and not across the air path between the condenser plates?

A. : The resistance of the air gap being very high, the charge will follow the path of less resistance through the galvanometer.

Q.2. What is the order of the resistance of the air gap?

A. : It is of the order of a few megohms.

Q.3. What is the 'time-constant' of the condenser circuit?

A. : The time constant is CR where R is the resistance in the circuit. It is equal to the time taken by the condenser to acquire nearly two-third of the total charge.

Q.4. Why is the time of charge short?

A. : If a long time is taken, electricity flows continuously and the capacity of the condenser becomes infinite for electric absorption.

Q.5. Can a Leclanche cell be used in this experiment?

A. : No, since it cannot give and maintain a constant e.m.f.

Q.6. Why should the throws be small?

A. : If they are large, the ratio of the deflections will not be equal to the ratio of the throws.

Q.7. Can any other instrument be used in place of the ballistic galvanometer?

A. : Yes, a quadrant electrometer may be used.

Q.8. Is there any other superior method for comparison of capacitances?

A. : Yes, it is De Sautry's method which is a null method and hence errors due to measurement of deflections do not occur.

9.11 Investigation of Capacitance in an a.c. circuit

A capacitor when connected in an a.c. circuit behaves in different manners and gives rise to various properties, some of which are studied here.

(A) Study of the current/voltage relationships and determinations of the value of the capacitance.

Theory :

In the a.c. circuit shown in Fig. 9.15, if *I* is the r.m.s. current and *V* is the r.m.s. voltage recorded respectively by the a.c. ammeter (*A*) and an a.c. voltmeter (*V*), then if a graph is plotted with *V* along *Y*-axis and *I* along *X*-axis, the graph will be a straight line passing through the origin. The straight line graph through the origin verifies that the r.m.s. current in the circuit is proportional directly to the r.m.s. potential difference across the capacitor, i.e., Ohm's law also applies.

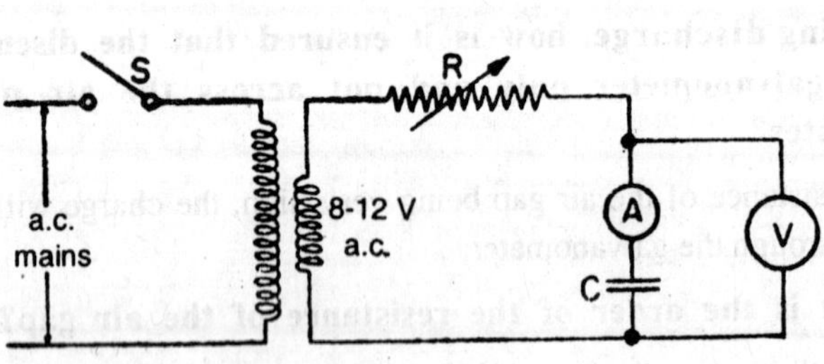

Fig. 9.15 : Behaviour of capacitor in a.c. circuit

From two convenient and wellseparated points *P* and *Q* on the graph (Fig. 9.16), the slope $\frac{PN}{QN}$ is obtained. This is the average numerical value of the ratio $\frac{V}{I}$ for the capacitor in the circuit and is the magnitude of the capacitive reactance X_c of the capacitor.

Now, since $X_C = \frac{1}{2\pi fC}$, where *f* is the frequency of the current (50 Hz) and *C* is the capacitance, the value of *C* can be found from

$$C = \frac{1}{2\pi f X_c}. \qquad \text{... (9.11a)}$$

Apparatus :

Source of low-voltage (8-12V a.c., i.e., a step-down transformer from mains), a.c. ammeter (0-0.1A), high resistance a.c. voltmeter (0-10V), non-inductive variable resistance (0-1000 ohms) and capacitor of capacitance about 5-10 μF.

Method :

(i) The circuit connections are made as shown in (Fig. 9.15). All the component connections are to be completed before inserting the mains plug or closing the mains cicuit switch (*S*).

(ii) The variable resistance *R* is set to its maximum value and the current is switched on.

(iii) Readings of the ammeter and voltmeter are noted and by varying the value of *R*, different sets are recorded, using as full as possible of the ammeter and the voltmeter scales.

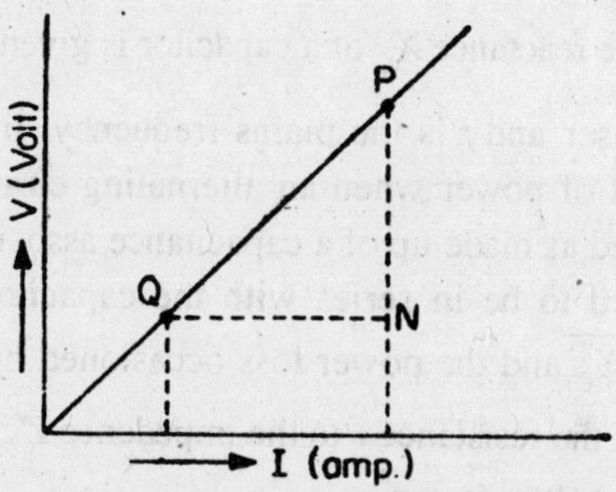

Fig. 9.16 : Current-voltage curve

(iv) From the data obtained, a graph is plotted with *I* versus *V* as shown in (Fig. 9.16).

(v) It is to be noted from the circuit diagram that owing to the high reactance of the capacitance and the low resistance of the ammeter, the readings of both ammeter and voltmeter are more reliable than if the voltmeter were placed across the capacitance only.

Results :

Frequency of the applied a.c. (f) = ... Hz

Table 1

No. of Obs.	r.m.s. value of the current (I amp)	r.m.s. value of the voltage (V volt)	Capacitive reactance of the capacitor (X_C ohm). from values	from graph	Mean value of X_C (ohm)	Value of C in μF
1.	...	...	...			
2.	...	...	...			
3.	...	...	...			
4.	...	...	...			
				...	...	...
5.	...	...	...			
6.						
7.						
8.	...	...	...			

For drawing of graph and calculations therefrom, see the second paragraph of the theory.

Precautions :

(i) The r.m.s. values of the current and the voltages should be noted.

(ii) Care should be taken in handling the a.c. mains.

(iii) Before closing the mains switch S, the circuit should be checked properly.

(B) Measurement of the reactance and loss factor in a *C-R* circuit

Theory :

The reactance X_C of a capacitor is given by $\frac{1}{2\pi f}C$, where C is the capacitance of the condenser and f is the mains frequency. In practice, every capacitor dissipates a certain amount of power when an alternating current passes through it. It may, accordingly, be regarded as made up of a capacitance associated with a small resistance in it which may be assumed to be in series with the capacitor. This arrangement has an impedence Z' of $\sqrt{r^2+X_C^2}$ and the power loss occasioned by r is expressed as a loss factor F which is the ratio of the resistance r to the impedence Z'.

Thus loss factor

$$F=\frac{r}{Z'}=\frac{r}{\sqrt{r^2+X_C^2}}=\frac{r}{\sqrt{r^2+\left(\frac{1}{2\pi fC}\right)^2}} \qquad \text{... (9.11b)}$$

Unless the loss factor is very bad, F may be taken as

$$\frac{r}{X_c}=r.2\pi fC \qquad \text{... (9.11c)}$$

The loss factor increases with the frequency of the applied a.c. source and in evaluating it, the associated a.c. frequency must accompany it.

Circuit Connection :

The connection of the circuit is done as shown in [Fig. 9.17 (a)]. The source should be a low voltage a.c. source (about 20 volts at mains frequency). The capacitor C is of capacitance about 2μF and R is a non-inductive resistor of about 1000 ohm and an a.c. voltmeter (V) of range 0-20 volts is used.

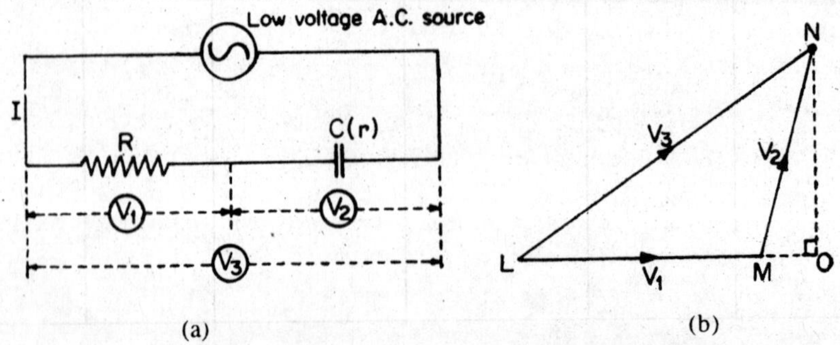

Fig. 9.17 : Reactance and loss factor of a capacitor

Method :

(i) After making the circuit connections as shown, the voltmeter is placed in turn across the standard non-inductive resistor (R), the unknown capacitor (C), and the total circuit. The corresponding voltage drops V_1, V_2 and V_3 are recorded.

(ii) From these records, the voltage vector triangle *LMN* is drawn as shown in [Fig. 9.17 (b)]. From *N*, a perpendicular is dropped to meet *LM* contained at *O*.

(iii) From this vector triangle, the loss factor is determined as shown in Calculations.

Table 2

Frequency (f) of the low voltage a.c. source = ... Hz

No. of Obs.	Value of R (ohm)	Value of C (µF)	Voltage (r.m.s.) across			Value of r (ohm)	Value of X_c (ohm)	Value of F
			R (V_1 volt)	C (V_2 volt)	Total circuit (V_3 volt)			

Calculations :

If I is the current in the circuit, then $V_1 = R.I = LM$ and if r is the loss-resistance in the capacitor C, then in-phase voltage across C due to this resistance is $r.I = MO$ (from the vector diagram). Hence from these two expressions,

$$\frac{r}{R} = \frac{MO}{LM}\ ; \text{ or, } r = R.\frac{MO}{LM} = \ldots\ldots \text{ ohm}$$

Actually NO represents the out of phase voltage due to the reactance X_c of the capacitor.

Thus, $NO = I.X_c = \dfrac{I}{2\pi fC}$

and $C = \dfrac{1}{2\pi fR}\cdot\dfrac{LM}{NO} = \ldots\ldots$ farad

Now, from Eq. (9.11c) given in Theory, the loss-factor

$$F = r.2\pi f.C$$

$$= R.\frac{MO}{LM}.2\pi f.\frac{1}{2\pi fR}\cdot\frac{LM}{NO}$$

$$= \frac{MO}{NO} = \ldots\ldots \text{ at } f = \ldots\ldots \text{ Hz}$$

Hence, from the vector diagram, measuring *MO* and *NO*, the value of the loss factor (*F*) at the given mains frequency (*f*) can be calculated.

It may be noted that for a mica capacitor the loss factor *F* is very small, which is, therefore, used as standard. But with paraffin as dielectric, *F* is much greater.

(C) Variation of the reactance with frequency of the a.c. source

Theory :

The reactance (X_c) of a capacitor is given by

$$X_c = \frac{1}{2\pi f C}, \qquad \ldots (9.11d)$$

where *C* is the capacitance and *f* is the frequency of the a.c. source. It is evident from the equation that for a given capacitor, its reactance varies inversely with the frequency of the source. Hence, if the reactances are determined for different frequency values, and if a graph is plotted with X_c against *f*, it will be a rectangular hyperbola while the graph of $\frac{1}{X_c}$ against *f* will be a straight line (Fig. 9.18). From the latter graph, the value of the capacitance can also be determined. The value of X_c of the capacitor is determined from the the measurement of the r.m.s. values of the current and the voltage as discussed in Expt. No. 9.11(A) previously.

Apparatus :

The same as in Expt. 9.11(A) and shown in (Fig. 9.15) with the only change that the mains transformer is replaced by a signal generator capable of producing variable frequencies. The circuit connections are made similarly as shown in (Fig. 9.15).

Method :

(i) After making the connections as in (Fig. 9.15) with the signal generator in place of the mains transformer, a low frequency value of *f* (say, 500 Hz) is selected. The output voltage of the signal generator is adjusted until conveniently high readings of I and V are obtained in the ammeter and the voltmeter scales. The respective r.m.s. values are recorded.

(ii) The frequency is next increased to some other higher value and again the output voltage of the signal generator is adjusted until convenient values are obtained on the ammeter and voltmeter scales.

(iii) The process is repeated for several other values of the frequency until the fullest use is being made of the scales on the various measuring instruments

(iv) Finally, two graphs are drawn ($X_c - f$) and $\left(\frac{1}{X_c - f}\right)$ as shown in (Fig. 9.18), from which the value of *C* can also be found out.

Results :

Table 3

No. of Obs.	Frequency (f Hz)	Output r.m.s. voltage (V volt)	Current (r.m.s.) (I amp)	Reactance ($X_c = \frac{V}{I}$) (ohm)	Value of $\frac{1}{X_c}$ (ohm^{-1})
1.	...	...	...	...	...
2.	...	...	...	...	...
3.	...	...	...	...	...
4.	...	...	...	...	...
5.	...	...	...	...	...
6.	...	...	...	...	...
7.	...	...	...	...	...

Graph drawings :

Two graphs are drawn

(a) With X_c ohms along Y-axis and the corresponding frequencies (f Hz) along X-axis.

(b) With $\frac{1}{X_c}$ along Y-axis and the corresponding frequencies (f) along X-axis.

The graphs are shown in (Fig. 9.18). From the second graph (which is a staight line graph), the slope $\left(\frac{a}{b}\right)$ is measured. Since the reactance, $X_c = \frac{1}{2\pi f} C$, this graph will have slope equal to $2\pi C$ provided the capacitor has negligible losses.

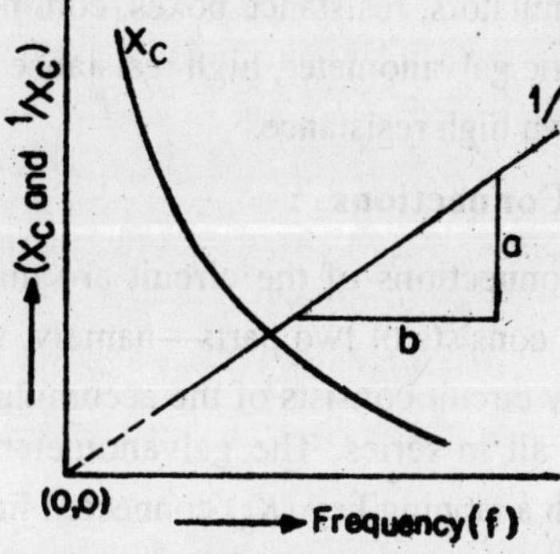

Fig. 9.18 : Frequency-reactance curve

or, $C = \frac{1}{2\pi}\left(\frac{a}{b}\right)$ = ... farad = ... μF.

This value so obtained can be compared with that marked on the capacitor.

These graphs show how the reactance of the capacitor varies with the frequency of the a.c. source.

9.12 Measurement of the value of a high resistance by the method of leakage of charge

Theory :

If a condenser of capacitance C farad be charged by a charge q_0 coulombs and then is allowed to be discharged through a very high resistance R ohms for t seconds, then the charge q that remains on the condenser is given by

$$q = q_0 e^{-t/CR}$$

or, $$\log_e \frac{q}{q_0} = -\frac{t}{CR}$$

$$\text{or, } R = -\frac{t}{C.\log_e \frac{q}{q_0}} = \frac{t}{2.303C\log_{10}\frac{q_0}{q}} \text{ ohms} \quad \dots (9.12a)$$

Now, if θ and θ_0 be the first throws of the light spot of a ballistic galvanometer when q and q_0 coulombs of electricity on the condenser are respectively discharged through the ballistic galvanometer instantaneously, then $\frac{q}{q_0} = \frac{\theta}{\theta_0}$. Thus, Eq. (9.12a) reduces to

$$R = \frac{t}{2.303C\log_{10}\frac{\theta_0}{\theta}} \text{ ohms} \quad \dots (9.12b)$$

This relation eanables to measure the value of R. If a graph be plotted with various times (t) of leakage of charge along X-axis and the corresponding values of $\log_{10}\frac{\theta_0}{\theta}$ along Y-axis, then from the slope of the graph and the known value of C, the value of R can be calculated.

Apparatus :

Accumulators, resistance boxes, commutator, charge-discharge key, decade condenser box, ballistic galvanometer, high resistance box, tapping key; plug type key, stopwtach and the unknown high resistance.

Circuit Connections :

The connections of the circuit are made as shown in (Fig. 9.19) where the circuit essentially consists of two parts—namely, the battery circuit and the galvanometer circuit. The battery circuit consists of the accumulator (E), the commutator (C), the resistance box R_1 and R_2 all in series. The galvanometer circuit consists of the ballistic galvanometer (B.G.) with a tapping key (K_3) connected in parallel across it. One end of the galvanometer is connected to the charge-discharge key (K_2) and its other end to one terminal of the condenser (C). The other end of the condenser is connected to the rocker of the CD key and the ends of R_2 are connected as shown in the figure. The unknown high resistance (R) is connected in parallel across the condenser through a plug type key (K_1).

Method :

(i) The values of the resistance R_1 and R_2 are properly adjusted according to the voltage of the battery E, so that after charging the condenser C with the p.d. across R_2, the galvanometer's light spot will have its first throw within the scale of the lamp and scale arrangement, when the condenser is discharged through the galvanometer. The central mark of the scale should be taken as the zero reference point.

(ii) Keeping the values of R_1 and R_2 fixed, the condenser is then charged with the charge q_0 by the charging key and during this, the key K_1 must remain open. After charging, the condenser is quickly discharged through the galvanometer by properly setting the discharging key (K_2). The first throw (θ_0) of the spot of light for this discharge is noted. This operation is repeated several times both for direct and reverse currents and the mean value of the throw (θ_0) is obtained.

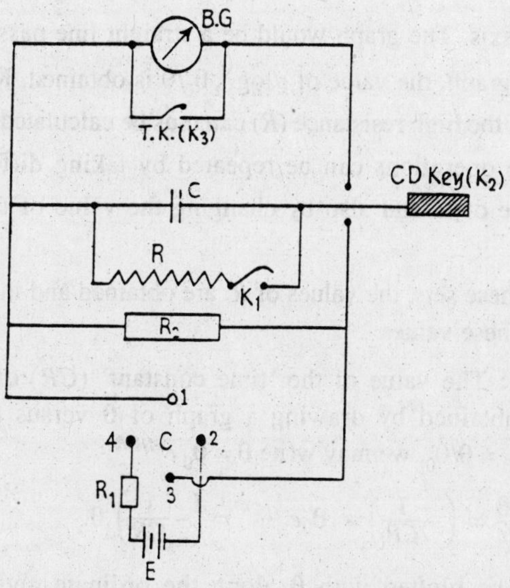

Fig. 9.19 : High resistance by leakage

(iii) Again the condenser is charged with q_0 as before and then the charge-discharge key (K_2) is placed in its neutral position. The key K_1 in the high resistance circuit is now closed and simultaneously with it, a stopwatch is started. The charge on the condenser now leaks out through the high resistance (R). After a certain time (t) which is of the order of 5 or 6 seconds after the closing of the key K_1, some charges leak out through R and a residual charge q remains on the condenser.

(iv) Next the key K_1 is opened and simultaneously by placing the discharging key properly, the remaining charge (q) on the condenser is discharged through the galvanometer. The first throw (θ) of the spot of light is noted. The operation is repeated several times both for direct and reverse currents and the mean value of the throw (θ) is noted.

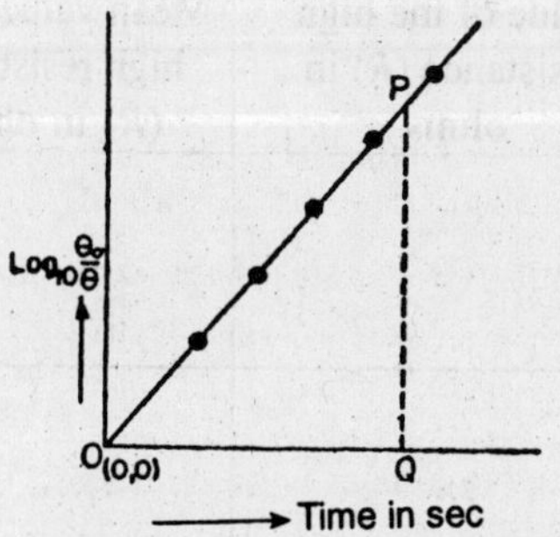

Fig. 9.20 : ($\log_{10}$ —t) curve

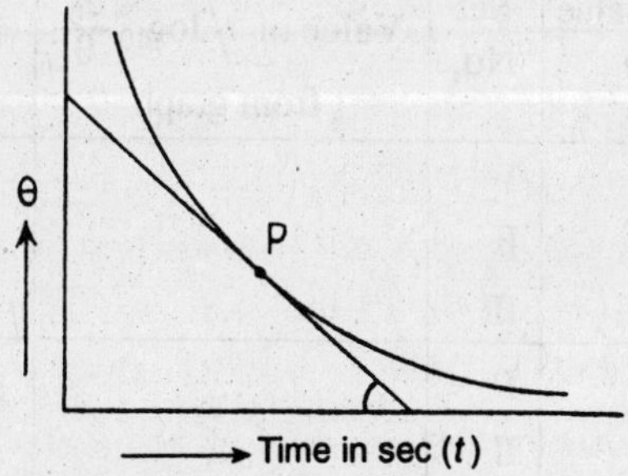

Fig. 9.20(a) : (θ—t) curve

(v) Keeping the values of R_1 and R_2 same as before, the operation (iii) is repeated for leakage of charge on C for three or more different values of t (cf. 8, 10, 12 sec etc.) and in each case the throws θ are noted, from which the values of $\log_{10}\theta_0/\theta$ are calculated.

(vi) A graph is then drawn (Fig. 9.20) with time (t) being plotted along the X-axis and $\log_{10}\frac{\theta_0}{\theta}$ along Y-axis. The graph would be a straight line passing through the origin. From the slope of the graph, the value of $t/\log_{10}\theta_0/\theta$ is obtained. Knowing the value of C employed, the value of the high resistance (R) can thus be calculated from Eq. (9.12b).

(vii) The entire operations can be repeated by taking different values of R_2 but having the same value of C and also by changing the value of C (i.e., by taking other capacitance values).

(viii) From all these sets, the values of R are obtained and the final mean value of R is thus obtained from these values.

Time Constant : The value of the 'time constant' (CR) of the circuit at a given value of θ may be obtained by drawing a graph of θ versus t. From the equation, $q = q_0 e^{-t/CR}$, since $q/q_0 = \theta/\theta_0$, we may write $\theta = \theta_0 e^{-t/CR}$.

Differentiating, $$\frac{d\theta}{dt} = \left(-\frac{1}{CR}\right) \times \theta_0 e^{-t/CR} = \left(-\frac{1}{CR}\right).\theta. \qquad \text{... (9.12c)}$$

Thus if a graph be plotted with θ along the ordinate and t along abscissa, an exponential curve will be obtained. The slope at any point of this curve $\left(\frac{d\theta}{dt}\right)$ will give the value of $\left(\frac{1}{CR}\right)$ at that particular value of θ, as evident from Eq. (9.12c). This slope is obviously negative [Fig. 9.20 (A)]. Putting the value of θ where the slope is measured, the value of $\frac{d\theta}{dt}$ at that point is obtained which gives the value of the *time-constant* (CR) at that point.

(A) Determination of the first throws (θ_0 and θ) :

Table 1 (See next page)

(B) Evaluation of the high resistance (R) :

Table 2

Condenser value (C farad)	Set No.	Value of $t/\log_{10}\frac{\theta_0}{\theta}$ from graph	Value of the high resistance (R) in ohms	Mean value of the high resistance (R) in ohms
	I	...	...	
C_1	II	...	...	
	III	...	...	
	I	...	...	
C_2	II	...	...	....
	III	...	...	
	I	...	...	
C_3	II	...	...	
	III	...	...	

(A) Determination of the first throws (θ_0 and θ) :

Table 1

Set No.	Value of C (F)	Values of R_1 (ohm)	Values of R_2 (ohm)	No. of Obs.	Values of first throw (θ_0) with charge q_0: Direct current (cm)	Reverse current (cm)	Mean (cm)	Grand mean of q_0 in cm	Time (t) of leakage (sec)	Value of first throw θ with charge q: Direct current (cm)	Reverse current (cm)	Mean (cm)	Grand mean of θ in cm	Value of $\log_{10} \frac{\theta_0}{\theta}$
I	C_1	R_1	R_2	1	...	...	...	......	...	...	...	...	...	......
				2	...	...	...			...	...	...		
				etc.								...	...	
II	C_1	R'_1	R'_2	1	...	...	...	......	...	...	...	...	...	...
				2	...	...	...		...	...	...	...	...	...
				etc.										
III	C_1	R''_1	R''_2	1										...
				2										...
				etc.										
I	C_2	R_1	R_2	etc.										...
II	C_2	R'_1	R'_2	etc.										...
III	C_2	R''_1	R''_2	etc.										...

Similar sets for other values of the condenser

Graph drawing :

As stated earlier, the graph is drawn with t along the X-axis and $\log_{10}\frac{\theta}{\theta_0}$ on the Y-axis. The graph would be a straight line passing through the origin (0,0). On the graph, any arbitrary point P is taken from which an ordinate PQ is drawn. From the ratio of $\frac{OQ}{PQ}$, the value of $t\Big/\log_{10}\frac{\theta_0}{\theta}$ can be obtained (Fig. 9.20).

Another graph of $(q - t)$ is drawn as shown in [Fig. 9.20 (a)] and the slope at a point P (for a particular value of θ) is obtained. The slope gives the value of $\frac{1}{CR}$, where CR = time constant of the circuit. From this slope, the value of 'time-constant' at P is obtained.

Precautions :

(i) The capacitor, the keys and the unknown resistance must all be well insulated. The condenser should not be touched by hand after it is being charged.

(ii) Closing of the key K_1 and starting of the stopwatch for counting of time, must be precisely done simultaneously.

(iii) While opening the key K_1 after leakage of charge, it should be done simultaneously with placing of the rocker of the discharging key, so that the residue charge can pass through the B.G.

(iv) Proper value of *CDR* must be placed in series with the galvanometer to make it critically damped.

(v) The time-constant of the circuit is equal to $C.R$. Thus if C is of the order 2 μF and R is of the order of 10 megohms, then the charge on the condenser will fall to one-third of its maximum value in a time of 20 sec. Thus this method is suitable for the measurement of resistances of the order of 10 megohms and above.

(vi) The e.m.f. of the battery should remain constant during the experiment.

Error Calculations :

Since from the Eq. (9.12b), the value of R depends on the time measurement and the throw measurement, the accuracy of the result depends on how precisely these quantities can be measured. Since the throws are fairly large and can be measured precisely up to 1 mm on the scale, little errors are contributed by their measurements. Since the time interval (t) is very small (a few seconds), hence the accuracy in the measurement of R entirely depends on how accurately the time can be measured. Accordingly, a stopwatch reading up to $\frac{1}{10}$th of a second, should be used. Moreover, the value of C should be precisely known. For this, standard condenser or decade condenser box should be used and not the commercial varieties of condensers. Since the measurable value of the high resistance is of the order of 10 megohms or more, a tolerance of 5%-10% error in the measured value is permissible in this method.

Questions and Answers

Q.1. What is meant by a high resistance?

A. : A resistance of the order of one megohm or more is called a high resistance.

Q.2. Can this high resistance be measured by a P.O. Box or metre bridge or by similar method?

A. : No, because those methods are based on Wheatstone's principle where the bridge is sensitive only when the resistance of the four arms are of the same order.

Q.3. Why the logarithmic decrement is not measured in this case?

A. : Since, the ratio $\frac{\theta_0}{\theta}$ is obtained under the same condition of the galvanometer circuit, the factor $\left(1+\frac{\lambda}{2}\right)$ cancels while taking the ratio.

Q.4. What is meant by time-constant of the circuit?

A. : It is the product of the capacitance and the resistance. [See (V) of Precautions].

Q.5. What cares must be taken in the experiment?

A. : See 'Precautions'.

Q.6. What order of capacitance is preferred in this experiment?

A. : It is of the order of 1-5μF.

Q.7. Should the possible leakge of charge through the dielectric of the condenser be accounted for?

A. : Yes, because the dielectric may not be perfectly insulated and some charge may leak out through the condenser even without connecting to R. This is known as self-leakage resistance (S) of the condenser.

Q.8. How is that resistance taken care of in the calculation?

A. : The condenser may be allowed to leak through itself for different values of time and to discharge through the galvanometer, as is done in the experiment. Similarly, plotting the graph as done in case of leakage through the high resistance, the value of the self-leakage resistance (S) is calculated. Then the corrected value of the unknown high resistance (R') will be given by $\frac{1}{R'}=\frac{1}{R}-\frac{1}{S}$, since the condenser and the high resistance are connected in parallel. Here R represents the value of the high resistance as calculated from the original experiment. S is also known as the insulation resistance of the capacitor.

[Apart from these, questions on condenser and ballistic galvanometer are to be prepared.]

9.13 Determination of the mutual inductance of two coils by a ballistic galvanometer

Theory :

If a current of i ampere flowing through the primary coil (P) of the circuit shown in (Fig. 9.21) be suddenly broken, then a charge of amount $\frac{Mi}{R}$ will flow through the secondary coil (S) and the Ballistic galvanometer (B.G.), if the points 3 and 4 of the commutator is closed by a plug, where M is the mutual inductance between P and S, and R is the total resistance in the galvanometer circuit. This includes the resistance of the coil S, R_2 and that of the galvanometer (G).

Hence,

$$\frac{Mi}{R} = \frac{T}{2\pi}\cdot\frac{c}{nAH}\cdot\theta\left(1+\frac{\lambda}{2}\right), \quad \text{... (9.13a)}$$

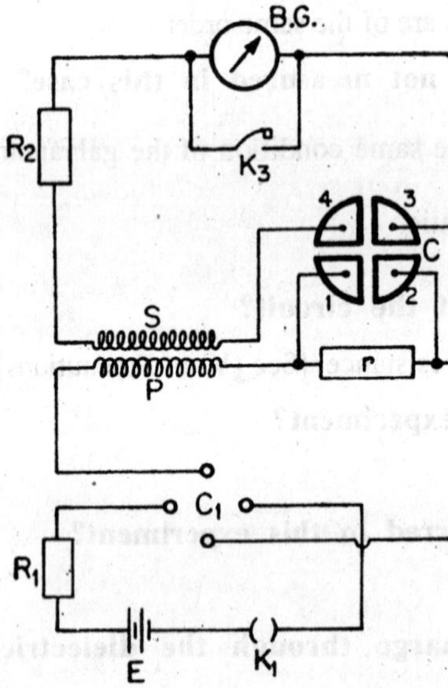

Fig. 9.21 : Determination of mutual inductant

where T = time period of the galvanometer in the open circuit

n = number of turns of the galvanometer coil

c = torque per unit twist of the galvanometer suspension

H = the magnetic field in which the coil moves

A = mean area of each turn of the coil

θ = the observed maximum throw of the galvanometer

λ = logarithmic decrement under the experimental condition.

To calibrate the galvanometer, which means the determination of the value of $\frac{c}{nAH}$, the plug connecting 3 and 4 of the commutator (C) is removed, and 1 and 4 are connected by a plug and 2 and 3 are connected by another plug. This gives a steady deflection to the galvanometer because of a steady potential drop $i.r$ across the low resistance r. The current passing through the galvanometer is $\frac{i.r}{R}$, (since the conditions are unaltered) and the steady deflection θ' of the B.G. is given by

$$\frac{ir}{R} = \frac{c}{nAH}\cdot\theta'. \quad \text{... (9.13b)}$$

$$\therefore \quad \frac{c}{nAH} = \frac{ir}{R\theta'}. \quad \text{... (9.13c)}$$

Substituting this value in Eq. (9.13a), we get

$$\frac{Mi}{R} = \frac{T}{2\pi}\cdot\frac{ir}{R\theta'}\cdot\theta\left(1+\frac{\lambda}{2}\right).$$

$$\text{or,} \quad M = \frac{rT}{2\pi}\cdot\frac{\theta}{\theta'}\cdot\left(1+\frac{\lambda}{2}\right). \quad \text{... (9.13d)}$$

If the angles of deflection are not very large, then $\frac{\theta}{\theta'} = \frac{d}{d'}$, where d is the maximum ballistic throw corresponding to θ and d' is the steady deflection of the light spot corresponding to θ'.

Hence Eq. (9.13d) reduces to

$$M = \frac{rT}{2\pi}\cdot\frac{d}{d'}\left(1+\frac{\lambda}{2}\right). \quad \text{... (9.13e)}$$

Knowing the value of r (in ohms), and measuring T (in sec) and d, d', the value of M can be calculated.

The value of the logarithmic decrement (λ) is obtained from the relation,

$$\lambda = \frac{2.303}{n-1}(\log_{10}\theta_1 - \log_{10}\theta_n), \quad \text{... (9.13f)}$$

where θ_1 and θ_n are the amplitudes of the first and the nth oscillation taken successive to right and left. The oscillations are noted in an open circuit under the given condition.

Apparatus :

Mutual inductance coil (variable and dial reading type), standard low resistance, two commutators, resistance boxes, keys, battery, ballistic galvanometer with damping key.

Connection of the circuit

The circuit connections are made as shown in (Fig. 9.21). The low resistance (r) may be a box having 0.01 ohm to 0.1 ohm. The damping key (K_3) must be connected across the galvanometer and the value in R_2 must exceed the external critical damping resistance of the galvanometer.

Method :

(i) To get a satisfactory result in this experiment, the values of r in ohms and of M in henries must be of the same order. But since the value of M is not initially known, the appropriate value of r may be determined by trial.

(ii) To do this, the primary current is adjusted to have a value of about 0.1 ampere taking $r = 0.1$ ohm. The p.d. thus formed across r is applied to the galvanometer circuit by inserting plugs between 1, 4 and 2, 3. The value of R_2 is adjusted until the deflection (steady) is nearly full scale. Even if by making R_2 nearly zero, the full scale deflection is not reached, then the value of primary current is to be increased.

(iii) The plugs are then removed and the points 3 and 4 are connected by a plug. The primary current is suddenly broken and the throw of the galvanometer is observed. If this throw is much smaller than the steady deflection, then the value of r is reduced till the two deflections are of the same order of magnitude. This value of r is kept unchanged for the rest of the experiment. The value of R_2 is also kept constant.

(iv) The mutual inductance coil is set for a particular value and for a particular value of r, the values of d and d' are noted for the ballistic throw and the steady deflection. The observations are made both for the direct and the reverse currents. The observations may also be repeated for different values of d (by changing R_1 and R_2) and of d' (by changing R_2 only). Changing the value of the primary current, different sets are observed.

(v) The time period (T) of oscillations of the moving part of the galvanometer is measured in its open circuit. The time for 30 complete oscillations are noted thrice and from them the mean value of the time period (T) is obtained.

(vi) The value of the logarithmic decrement (λ) is also measured by observing the successive amplitudes of the oscillations of the noving part of the galvanometer and employing Eq. (9.13f).

(vii) When the mutual inductance coil is available in the variable form and if the dial readings (in degrees) are to be calibrated, then the above methods are followed for a particular value of the dial reading (ϕ). Changing the dial readings (ϕ) at a step of 30°, the values of M in millihenries are calculated for each dial reading. Finally a graph is drawn by plotting the dial readings (ϕ) along X-axis and the corresponding value of M in millihenries along Y-axis. This curve will be as shown in (Fig. 9.22) which is known as the calibration curve. From this curve, the value of M for any deial readings can be found out.

Results :

e.m.f. of the battery before experiment = ... volt

e.m.f. of the battery after experiment = ... volt

Mean e.m.f. (E) of the battery = ... volt

(A) To find the ballistic throw (d) and steady deflection (d') :

See next page.

(B) Determination of the time period (T) of oscillations of the galvanometer in the open circuit :

Table 2

No. of Obs.	Time for 30 complete oscillations	Mean time (t)	Time period (T) in seconds
1.	...... min ...... sec		
2.	...... min ...... sec	min sec	
3.	...... min ...... sec	= .... sec	
4.	...... min ...... sec		

(C) Determination of logarithmic decrement (λ) in the open circuit :

Table 3

No. of Obs.	Displacement towards		$\lambda = \frac{2.303}{n-1}(\log_{10}\theta_1 - \log_{10}\theta_n)$
	Left (cm)	Right (cm)	
1.	...... (θ_1)		
2.		...... (θ_2)	
3.	...... (θ_3)		
4.		...... (θ_4)	....
5.	......		
6.		......	
7.	......		
8.		...... (θ_n)	

[The central line of the scale is taken as zero point.]

(D) Evaluation of M for different dial readings :

Table 4

No. of Obs.	Dial readings (ϕ degrees)	Value of r in ohms	Value of d/d'	Value of M in henry	Value of M in millihenry
1.	ϕ_1	......	......	......	......
2.	ϕ_2	......	......	......	......
etc.	etc.	etc.	etc.	etc.	etc.
7.	ϕ_7	......	......	......	......

(A) To find the ballistic throw (d) and steady deflection (d') :

Table 1

Value of external critical damping resistance = ……… ohm Value of the low resistance (r) = ……… ohm

Dial reading in degree (ϕ)	No. of Obs.	Current	Value of		Ballistic throw (d)		Value of R_2 changed, if necessary	Steady deflections (d')	
			R_1 (ohm)	R_2 (ohm)	Throws in cm	Mean (cm)		in cm	Mean (cm)
…… (ϕ_1)	1.	Direct	… …	… …	… …	……	… …	… …	……
		Reverse	… …	… …	… …		… …	… …	
	2.	Direct	… …	… …	… …	……	… …	… …	……
		Reverse	… …	… …	… …		… …	… …	
…… (ϕ_2)	1.								……
	2.								……
…… (ϕ_3)	1.								…… ……
	2.								……
etc.	…	……	…	……	…	……	……	……	……
…… (ϕ_7)									

(E) Drawing of graph :

To draw the graph (calibration curve) of M against ϕ, the latter is plotted along X-axis while the former along Y-axis. The graph will be as shown in (Fig. 9.22). The origin is taken as (0, 0). If the instrument is already calibrated in millihenries, then the readings can be checked from the calculated values. The graph is then to be drawn with the dial reading of M along X-axis and corresponding error at that reading of M along X-axis and corresponding error at that reading along Y-axis. The successive points when joined, the curve will be the calibration curve. If M' be the dial reading and M be the corresponding obtained value, then $(M—M')$ is the error for M', which should be algebraically added to M' to get the correct value.

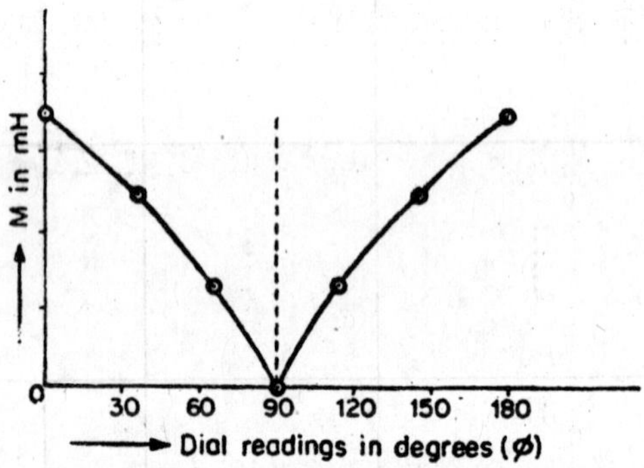

Fig. 9.22 : (M-ϕ) curve

Calculations :

From Eq. (9.13e),

$$M = \frac{rT}{2\pi}\cdot\frac{d}{d'}\left(1+\frac{\lambda}{2}\right),$$

the value of M can be calculated when the values of T, d, d' and λ are determined and the value of r is known.

Hence M = ... henry = ... millihenry

Precautions :

(i) The values of R_1 and R_2 are to be so adjusted that the ballistic throw and the steady deflection, both remain within the scale and near about 12 to 16 cm of the scale readings.

(ii) The values of R_2 in the galvanometer circuit should be made higher than the external critical damping resistance of the galvanometer, otherwise the galvanometer would be dead-beat.

(iii) During determination of d and d' for a particular set, the value of the primary current must be kept constant.

(iv) The ballistic throw should be noted during break of the primary circuit and not during closing.

Error Calculations :

From Eq. (9.13e), the proportional error in the measurement of the value of *M* is given by

$$\frac{\delta M}{M} = \frac{\delta r}{r} + \frac{\delta T}{T} + \frac{\delta d}{d} + \frac{\delta d'}{d'} + \frac{\delta(1+\lambda/2)}{1+\lambda/2}.$$

As *r* is kept constant for observing *d* and *d'*, $\delta r = 0$; the value of λ being very small, $\delta\lambda = 0$. If a stopwatch reading up to $\frac{1}{10}$th of a second be used and if the number of oscillations for determination of *T* is taken large, the little error is contributed in the measurement of time-period. Further, in measuring the values of *d* and *d'*, if the scale can read up to 1 mm and the readings are fairly large (about 12 to 16 cm on the scale), then δd and $\delta d'$ are also very small. However, the error can finally be calculated from the equation.

$$\frac{\delta M}{M} = \frac{\delta T}{T} + \frac{\delta d}{d} + \frac{\delta d'}{d'}.$$

$$\therefore \text{ percentage error} = \frac{\delta M}{M} \times 100 \,\ldots\ldots\, \%.$$

Questions and Answers

Q.1. What is meant by mutual inductance between two coils? On what factors does its value depend?

A. : For the first part, see any textbook and the value of *M* depends on the angle of setting between the two coils and on the number of turns on the primary and the secondary coils.

Q.2. What are the values of *M* maximum and minimum?

A. : When the axes of the two coils are parallel, the value of *M* is maximum while they are perpendicular, *M* is zero.

Q.3. Can the value of *M* between two coils be equal to the self-inductance of either?

A. : If two identical coils are so wound that they almost coincide with each other everywhere, then the value of *M* between them will be equal to the self-inductance of either.

Q.4. If an iron core is introduced in this primary, does the value of *M* change?

A. : Yes, the valuè of *M* will increase due to the linkage of greater magnetic flux by the secondary.

Q.5. Which is greater—the rate of growth or the rate of decay of the primary current?

A. : Rate of decay is greater due to smaller value of the time constant of the primary during break. That is why the throw is measured during break.

Q.6. Why are the values of *d* and *d'* made nearly equal?

A. : To avoid the error due to non-uninformity of the flux-density of the permanent magnet of the galvanometer.

9.14 Measurement of self-inductance of a coil by Anderson's method

Theory :

A method devised by Anderson on the basis of Wheatstone's bridge principle, given a convenient way of finding the value of an inductance (self) in terms of a capacitance and resistance.

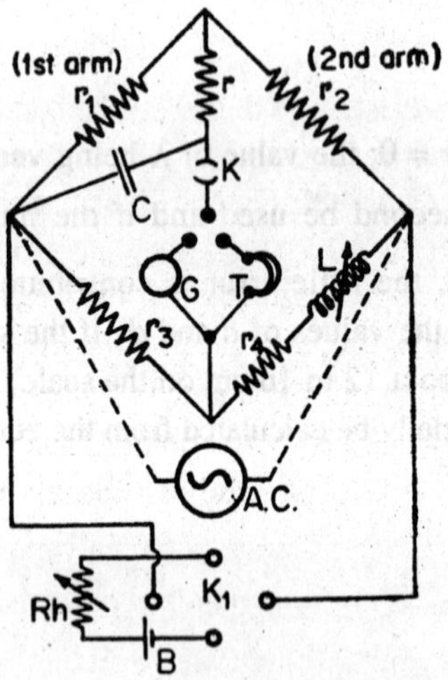

Fig. 9.23 : Anderson's bridge

The bridge, as shown in (Fig. 9.23), is first balanced for steady currents using a battery between *MN* and a galvanometer *G* between *QP*. When the battery circuit is first closed by the commutator *O* and then the galvanometer by the key *K*, the balance is obtained by adjusting the resistance in the box inserted in the 4th arm (r_4). The balance for steady current has no effect by the presence of the condenser *C* and any resistance *r* in the galvanometer circuit. Keeping $r = 0$, we have for balanced condition

$$\frac{r_1}{r_2} = \frac{r_3}{r_4}, \qquad \text{... (9.14a)}$$

where r_4 denotes the total resistance of the arm *PN*, including that of inductor.

The network is next balanced for varying currents. This may be done either by using the galvanometer and an intermittent current (direct) or an alternating current from a buzzer and a telephone receiver (*T*). If the galvanometer be used as a detector with intermittent direct current, then an additional tapping key has to be inserted in the battery circuit and *r* is so adjusted for the particular value of *C*, that on making or breaking the circuit, no deflection of the galvanometer is produced. Thus the bridge is balanced both for direct steady and alternating currents. When this condition holds, it may be shown, by the application of Kirchhoff's laws to the circuits *MQRPM* and *MQNPM* that the self-inductance *L* of the coil is given by

$$L = C\{r_2 r_3 + r(r_3 + r_4)\} \times 10^{-3} \text{ millihenry}, \qquad \text{... (9.14b)}$$

where *C* is the value of the capacitance in microfarad.

Apparatus :

Condenser (*C*) having capacitance of 0.1 μf or more, P.O. Box, galvanometer (*G*); commutator (*O*), two non-inductive resistance boxes (*r* and r_4, the former being a high resistance box and the latter containing fractional resistances), storage cell, rheostat, inductance coil (*L*), a.c. source, telephone (*T*), keys.

Connections of the circuit :

The scheme of connections is shown in (Fig. 9.23). The Wheatstone net r_1, r_2, r_3 and r_4 (r_4 being the total resistance of the arm *NP*) is as shown. The condenser *C* is connected from M to a variable point R in the resistor circuit r (QR). The storage cell B is connected to the points M and N through a commutator O and a rheostat Rh. The galvanometer G is a dead-beat type of resistance of about 100 ohms with a short-circuiting key (not shown). The galvanometer circuit can be closed or opened by another key K.

Method :

(i) After properly connecting the circuit, the resistances in the 1st, 2nd and the 3rd arms (QM, QN and MP) are each made equal to 100 ohms. The battery circuit is closed and by inserting a suitable resistance in the fourth arm (PN) from the resistance box (r_4), the galvanometer circuit is closed and the balance condition is seen. By altering the resistance in r_4 suitably, the galvanometer deflections are made in opposite directions for a change of 1 ohm in the box of r_4. Then proper fractional resistances are inserted in r_4 to get the exact balance point. If R_1 is the total resistance applied from the box and if R_i is the resistance of the inductance coil, then in the balanced condition $r_4 = R_1 + R_i = 100$ ohms. Thus the value of $R_i = (100 - R_1)$ ohms. The bridge here is most sensitive since the resistances in all the four arms are 100 ohms which is equal to the approximate resistance of the galvanometer.

(ii) To balance the bridge for direct intermittent current or alternating current, the values of the resistances of all the four arms, already set in operation (i) above, are kept unchanged. Then the galvanometer key (K) is first closed and then the main circuit commutator is closed to observe any kick in the galvanometer. The resistance (r) is gradually changed and each time the kick in the galvanometer is observed by closing and opening the commutator. Finally, for a certain value of the resistance r, the galvanometer will show no kick on closing or opening the commutator keys. Thus the bridge is balanced for direct intermittent current. Knowing the value of r_2, r_3, r_4, r and C, the value of L can be calculated from Eq. (9.14b).

(iii) If the balance is observed with an a.c. source, then the galvanometer is replaced by a telephone and the balance condition is obtained when there will be minimum or zero sound in the telephone for a particular value of r. A key has to be then added in addition to the commutator in series and by this key the closing and opening of the main current can be operated.

(iv) If direct intermittent current is used, then with the help of the commutator (O), the battery current should be reversed and operation (i) has to be repeated for the reverse current as well.

(v) For different dial readings of the variable inductance coil (L), the operations (i), (ii), (iv) are repeated (when intermittent d.c. used) or, (i) and (iii) are repeated (when a.c. used) and the value of self-inductances for each dial reading is calculated individually.

(vi) If time permits, the experiment may be repeated with a different condenser value.

(vii) If a calibration curve of the given variable self-inductance coil is required, then the dial readings are plotted along X-axis of a graph and the corresponding value of L along Y-axis. This curve would be the calibration curve and the value of the self-inductance of the coil for any of its dial reading can be obtained.

Results :

(A) Balancing the bridge with a steady current :

Table 1

No. of Obs.	Dial reading	Resistance applied in ohms in					Galvanometer deflections	Total resistance in 4th arm (ohm) ($r_4 = R_1 + Ri$)	Inductance resistance (R_i ohm) = ($r_4 - R_1$)
		1st arm (*QM*)	2nd arm (*QN*)	3rd arm (*MP*)	Box (R_1) in 4th arm (*PN*)	Box (*r*)			
		100	100	100	0.0	0	left		
		"	"	"	5000.0	0	right		
		"	"	"	1000.0	0	right		
1.	...	"	"	"	......	0	......		
		"	"	"	48.0	0	left		
		"	"	"	49.0	0	right		
		"	"	"	48.6	0	null point	100	51·4
					......				
					......				
2.	...	......	......	......				......	......
					......				
					......				
etc.	...	......	......	......	etc.	......	etc.	......	etc.
etc.	etc.	......	......	......	etc.	......	etc.	......	etc.

[The numerical data given are exemplary and when direct intermittent current is used in the above table, provisions are to be made for observations for steady reverse current.]

(B) Balancing the bridge with a.c. or intermittent d.c. :

Value of C = ... μ*F*.

Table 2

Current	No. of Obs.	Dial reading	Resistance applied in ohms in					Sound intensity in telephone/galvano-meter kick	Total resistance in 4th arm (ohm)	Value of L in millihenry
			1st arm	2nd arm	3rd arm	Box (R_1) in 4th arm	Box (r)			
a.c. or int. d.c.			100	100	100	48.6	......	......		
			"	"	"	"	......	......		
	1.	...	"	"	"	"	......	......		
			"	"	"	"	(r)	Null point	100	......
							......			
							......			
"	2.	...	......	......	......	......	......	......	......	....
							......			
							......			
"	etc.	etc.	......	......	......	etc.	etc.	......	......	etc.
"	etc.	etc.	......	......	......	etc.	etc.	......	......	etc.

[The figures are hypothetical and if intermittent d.c. is applied, then observations are also to be made for reverse current.]

Results :

(A) Balancing the bridge with a steady current :

Table 1

[The numerical data given are exemplary and when direct intermittent current is used in the above table, provisions are to be made for observations for steady reversed current.]

(B) Balancing the bridge with a.c. or intermittent d.c. :

Value of C = ………… μF.

[The figures are hypothetical and if intermittent d.c. is applied, then observations are also to be made for reversed current.]

[See previous Page (Table 1 & Table 2)]

Calculations :

From Eq. (9.14b), knowing the values of r_2, r_3, r_4 and r and the value of the capacitance (C), the value of self-inductance (L) of the coil for any of its dial reading can be calculated.

$$\therefore\ L = C\left\{r_2 r_3 + r\left(r_3 + r_4\right)\right\} \times 10^{-3} \text{ millihenry,}$$

where the value of C is in micro-farad.

Drawing of curve :

The graph of the value of L against the dial reading (ϕ) is plotted as discussed in (vii) under Method. This graph is the calibration curve.

Precautions :

(i) The frequency of the applied a.c. source must be between 300 Hz and 500 Hz. If this is too low, then a little induction effect is observed.

(ii) The bridge must be made sensitive so that the balanced condition of the bridge changes even for a small change in the value of r, when a.c. or intermittent d.c. is used. To make the bridge sensitive, the resistances in the four arms of the bridge are made equal and almost equal to the galvanometer resistance.

(iii) All the connecting wires must be kept straight without winding.

Questions and Answers

Q.1. What is meant by self-inductance of a coil and what will be its unit?

A. : See any textbook. Its unit is henry or millihenry.

Q.2. On what factors does its value depend?

A. : Its value depends on the number of turns of the coil per unit length, the diameter of the coil and the permeability of the medium inside the coil.

Q.3. What is the difference between self- and mutual-inductances?

A. : The self-inductance is the effect of inductance on the coil due to its own while the mutual-inductance is the effect of one coil on another when flux due to one is associated with the other.

Q.4. The connecting wires in this experiment are kept straight—why?

A. : To make the self-inductance of the wires minimum.

Q.5. What is the effect of self-inductance in a circuit and when is it effective?

A. : The effect is to oppose the starting or stopping of current in the circuit when an e.m.f. is set up which acts in the opposite or in the same direction of the current, when the current starts flowing or stops respectively. The effect is only present when the growth or decay of current takes place and not during the steady flow.

Q.6. What is meant by non-inductive winding?

A. : If a wire is doubled on itself and a current is allowed to flow through the wire, the magnetic field produced by the current in one wire is neutralised by that produced by the reverse flow of current in another part of the wire placed on the first. Thus the self-inductance of the doubled wire becomes nil and this particular type of winding is known as the non-inductive winding.

Q.7. Why during balancing for steady current, first the battery key and then the galvanometer key are closed, while for finding balance point with a.c. or intermittent d.c. the converse is followed?

A. : Since in the first case, there must not be any effect of self-induction, the circuit is closed first and the galvanometer key is closed in a steady state when there will be no self-induction. But in the second case, since the effect of self-induction needs to be introduced, hence by keeping the galvanometer circuit closed, the current is made to flow, for which the induction effect comes into play.

9.15 Measurement of the strength of a magnetic field by a ballistic galvanometer

Theory :

If a small circular coil (search coil) of mean area A, containing N number of turns of wire be placed with its plane at right angles to the magnetic field of strength H, then if the coil is suddenly withdrawn from the field, a quantity of charge (q) can be made to flow through the galvanometer circuit in which the coil is placed. If this charge while flowing through the galvanometer, produces a first linear displacement d of the light spot on the scale, then

$$q = \frac{NAH}{R \times 10^9} \text{ coulomb} = kd\left(1 + \frac{\lambda}{2}\right), \quad \ldots (9.15a)$$

where R = total resistance in the galvanometer circuit

K = characteristic constant of the galvanometer depending also on scale distance from the mirror.

λ = logarithmic decrement.

Again, if a current I' ampere while flowing through the primary coil (P_1P_2) of a standard solenoid having a mean cross-sectional area A' and N_1 number of turns of wire per cm length of the coil, be suddenly reversed, then a charge q' will flow in the galvanometer circuit in which a secondary (S) of standard solenoid, having N_2 number of turns, is also placed. If this charge q' produces a linear displacement d' of the spot of light on the scale set at the same position as before, then

$$q' = \frac{0.8\pi I' \ N_1 N_2 A'}{R \times 10^8} \text{coulomb} = Kd'\left(1+\frac{\lambda}{2}\right). \qquad \ldots (9.15b)$$

From Eqs. (9.15a) and (9.15b),

$$\frac{NAH}{0.8\pi I' N_1 N_2 A'} = \frac{d}{d'}$$

$$\text{or, } H = \frac{0.8\pi N_1 N_2 I'}{N} \cdot \frac{A'}{A} \cdot \frac{d}{d'}. \qquad \ldots (9.15c)$$

If D and D' represent the mean diameters of the search coil and the primary (P_1P_2) respectively, then

$$\frac{A'}{A} = \frac{D'^2}{D^2}.$$

Thus Eq. (9.15c) reduces to

$$\left.\begin{aligned} H &= 0.8\pi I'\left(\frac{N_1 N_2}{N}\right)\left(\frac{D'^2}{D^2}\right)\frac{d}{d'} \\ &= 0.8\frac{N_1 N_2}{N}\cdot\left(\frac{D'}{D}\right)^2\left(\frac{I'}{d'}\right)d \end{aligned}\right\}. \qquad \ldots (9.15d)$$

Measuring the various defections of d' for different values of I', the mean value of I'/d' can be obtained graphically and substituting this value in Eq. (9.15d) and knowing the other quantities, the value of H can be calculated.

The length of air-gap in between the pole-pieces (l) can be changed and the variation of the field strength (H) with air-gap (l) can be studied from a graph drawn. Also by changing the magnetising current (I) of the electromagnet, the variation of the field strength (H) with the magnetising current (I) can be studied from a graph drawn.

Apparatus :

Solenoid, electromagnet with variable pole-pieces, ballistic galvanometer, search coil, ammeter, two-way key, Pohl's commutator, battery, resistance boxes, Plug type keys.

Connections of the circuit :

The schematic diagram shown in (Fig. 9.24) shows the plan of the connection. The value of R_1 should exceed the external CDR of the galvanometer. The box R should be of variable type having high current bearing capacity.

Method :

(i) The length and the total number of turns of the primary coil are known from which the value of N_1 is obtained. The other constants like N_2, N, D', etc. are obtained. The value of the mean diameter (D) of the search coil is obtained by measuring the external and the internal diameter of the coil and taking the mean value of the same.

(ii) The search coil is suspended in the gap in between the pole-pieces (X, Y) of the electromagnet and its plane is kept at right angles to the magnetic field to be produced by the E.M. The coil should be hung symmetrically within the gap and its size being small in comparison to the area of the pole-pieces, the field H is supposed to remain uniform all over its area.

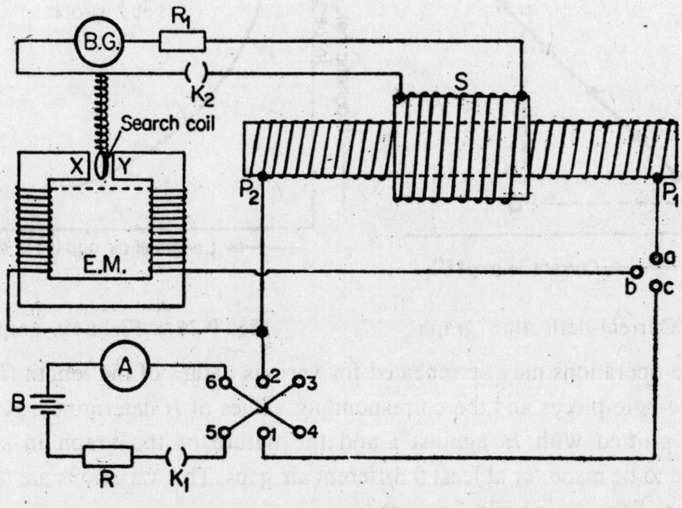

Fig. 9.24 : Measurement of magnetic field

(iii) The value of R_1 in series with the B.G. is made high to exceed the required C.D.R. of the galvanometer and the light spot is fixed at the central zero line of the scale, keeping the key K_2 open.

(iv) In the two-way key, the points b and c are connected and the electromagnet is put in the battery circuit. By properly setting the value in R, the key K_1 is closed when a current of I ampere flows in the battery circuit (obtainable from the ammeter) producing the desired magnetic field in the air-gap between XY.

(v) By closing the key K_2 in the galvanometer circuit, the search coil is *suddenly* withdrawn from the air-gap and the throw (d) of the spot of light is noted, which should preferably be between 12 to 16 cm on the scale. The throws are noted for both the direct and reverse currents in the E.M. coil and the mean value of d is noted from several readings. The length (l) of the air-gap between the pole-pieces is measured accurately several times and the mean value is obtained.

(vi) To determine the value of I'/d', the key K' is kept open and the points a and c of the two-way key is connected. The key K_1 is closed and for a lower value of I' in the primary coil, the key K_2 is now closed. Keeping the same resistance in the galvanometer circuit as before (when d was observed), the primary current (I') is reversed from +ve to –ve and again vice versa. In each case, the throw of the galvanometer is noted. From these throws, the mean value of d' is obtained and hence the value of I'/d' is found out for a particular value of the air-gap (l).

(vii) With this value of I'/d' for a particular value of air-gap distance (l), the value of H can be calculated from Eq. (9.15d) and from observations made under (v).

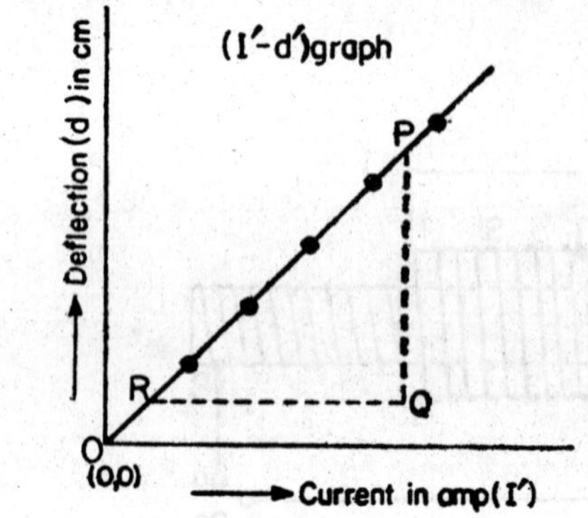

Fig. 9.25 : Current-deflection graph

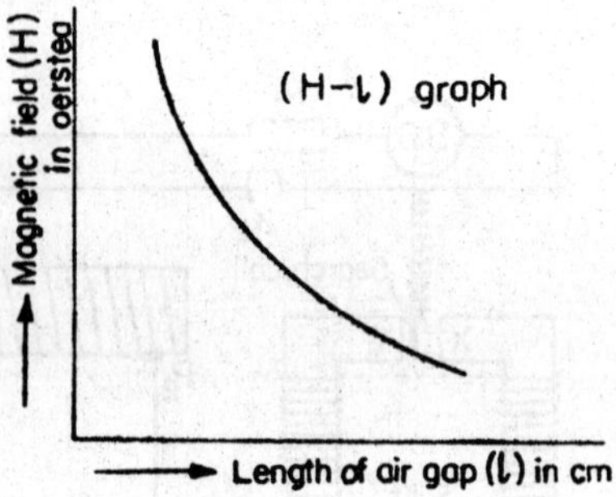

Fig. 9.26 : Field-air graph

(viii) The operations may be repeated for various values of the length (l) of the air-gap between the pole-pieces and the corresponding values of H determined. A graph (Fig. 9.26) may be plotted with H against l and the nature of the graph in shown. The observations are to be made for at least 5 different air-gaps. This variations are to be studied for a given value of the magnetising current I.

(ix) The observations made under (vi) may also be represented graphically. By increasing the value of I' by steps of about 0.5 amp., five different sets of readings may be taken for a particular value of l . The variations of I' against d' are shown in a graph (Fig. 9.25).

(x) Finally, by changing the value of the magnetising current (I), the strength of the magnetic field can be varied. The value of I is changed by steps of 0.3 ampere and for each value, the strength (H) of the magnetic field is obtained. Plotting I against H, a graph as shown in (Fig. 9.27), is obtained. This is known as the (I-H) curve.

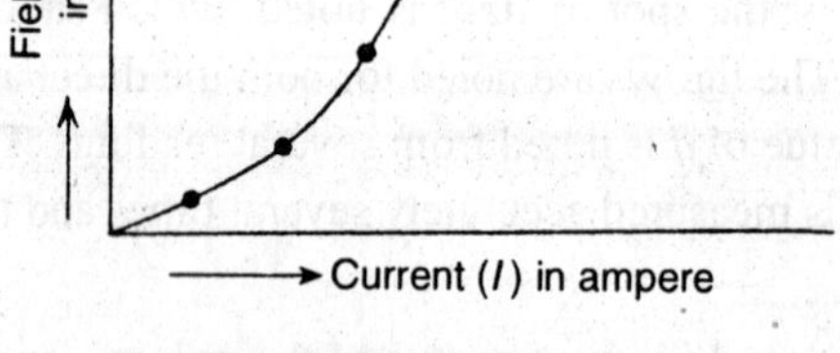

Fig. 9.27 : (I-H) curve

It must be remembered that before changing the magnetising current (I) to a new value, each time the electromagnet must be brought to the steady cyclic state by repeatedly reversing (at least 8 to 10 times) the current. This ensures demagnetisation of the specimen before taking readings for a new set.

Results :

(A) Measuring constants of the apparatus :

Table 1 (See next page)

Results :

(A) Measuring constants of the apparatus :

Table 1

Constants of the search coil				Constants of the standard solenoid				
Internal diameter (D_1 cm)	External diameter (D_2 cm)	Mean diameter ($D = \frac{D_1 + D_2}{2}$ cm)	No. of turns (N)	Mean diameter of primary coil (D' cm)	Total No. of turns (n)	Total length of the primary (L cm)	No. of turns per cm ($N_1 = \frac{n}{L}$)	Total No. of turns of the secondary (N_2)
......	......	......	......	......	......	......	......	......

(B) Observation of throws (d) due to withdrawal of the search coil :

Table 2

Fixed resistance (R_1) in the galvanometer circuit = ……… ohms

Fixed current (I) in the magnetising circuit = ……… amperes

Set No.	Length of air-gap (l cm)	No. of Obs.	Throw of light spot				Value of (I'/d') from graph (from Table 3)	Value of H in oersted
			Direct current (cm)	Reverse current (cm)	Mean (cm)	Grand mean (d cm)		
I	…	1.	…	…	…	……	……	……
		2.	…	…	…			
		3.	…	…	…			
II	…	1.	…	…	…	……	……	……
		2.	…	…	…			
		3.	…	…	…			
etc.	etc.		etc.	etc.	etc.	etc.	……	etc.
V	…	1.	…	…	…	……	……	……
		2.	…	…	…			
		3.	…	…	…			

(C) Finding the values of d corresponding to I' :

Table 3

Fixed distance between the pole-pieces (l) = ……… cm

No. of Obs.	Current changes from I' to zero (ampere)	Corresponding throw (d'_1) in cm	Current changes from zero to $-I'$ (ampere)	Corresponding throw (d'_2) in cm	Mean throw in cm $d' = \frac{d'_1 + d'_2}{2}$	Value of $\frac{I'}{d'}$ from graph in ampere/cm
1.	0.5 to zero	……	Zero to –0.5	……	……	
2.	1.0 to zero	……	Zero to –1.0	……	……	
3.	1.5 to zero	……	Zero to 1.5	……	……	………
4.	2.0 to zero	……	Zero to 2.0	……	……	
……	………	……	……	……	……	
7.	3.5 to zero	……	Zero to 3.5	……	……	

(B) Observation of throws (d) due to withdral of the search coil :

Table 2 (See previous page)

Fixed resistance (R_1) in the galvanometer circuit = ... ohms

Fixed current (I) in the magnetising circuit = ... amperes

(C) Finding the values of *d* corresponding to *I′* :

Table 3 (See. previous page)

Fixed distance between the pole-pieces (l) = ... cm

(D) Magnetic fields (*H*) for different values of the magnetising current (*I*) :

Table 4

No. of Obs.	Length of the air-gap (ℓ cm)	Magnetising current (I ampere)	Deflection of light spot				Value of H in oesrted
			Direct current (cm)	Reverse current (cm)	Mean (cm)	Grand mean (d cm)	
			......	......	......		
1.		0.3				......	......
			......	......	......		
			......	......	......		
2.		0.6				......	......
			......	......	......		
			......	......	......		
3.	......	0.9				......	......
			......	......	......		
etc.		etc.	etc.	etc.	etc.	etc.	etc.
			......	......	......		
8.		2.4	......	......	......		

Graph drawings :

Three graphs are drawn—

(a) Primary current (I') along X-axis and corresponding deflections (d') along Y-axis. The graph (Fig. 9.25) is a straight line passing through the origin (0,0).

(b) Magnetic field strength (H) along Y-axis and different values of air-gap distance (l) along X-axis. The graph is as shown in (Fig. 9.26).

(c) Magnetising current (I) along X-axis and the corresponding field strength along Y-axis. The graph is also a straight line as shown in (Fig. 9.27).

Precautions :

(i) The search coil should be withdrawn from the gap very rapidly and the plane of the coil must be placed at right angles to the field.

(ii) The coil area must be small compared to the area of the pole-pieces so that it may remain in an uniform field.

(iii) The search coil should be suspended by twined flexible wires, so that they may not embrace any lines of force.

(iv) It is better to observe the throw by withdrawing the search coil rather than by breaking the circuit; since in the latter case, as there may be always a residual magnetism in the iron when the current in the electromagnet is withdrawn.

(v) Although H increases linearly with I, care should be taken not to reach the saturation value. Of course, within the range employed normally in the experiment, the relationship between I and H remains linear and safely far away from the saturation.

(vi) Demagnetisation of the specimen must be done by repeated reversal of the primary current each time before changing the value of I.

Questions and Answers

Q.1. Is it necessary that the value of R_1 in the galvanometer circuit should remain constant?

A. : Yes, thus the effect of resistance present in the galvanometer circuit is eliminated from the final value of H.

Q.2. Can the value of H be measured by observing the throw of the galvanometer when the magnetising current is suddenly reversed?

A. : Due to greater time-constant of the E.M. circuit, the rate of change of flux is low and ballistic condition of the galvanometer is not satisfied.

Q.3. Can the value of H be measured by any other method?

A. : Resistance of bismuth spiral varies with the strength of the magnetic field. Hence, from the variation of resistance of a bismuth spiral placed in a magnetic field, the latter can be measured.

Q.4. On what factors does the strength of the magnetic field depend?

A. : It can be increased by diminishing the air-gap between the pole-pieces by increasing the ampere-turns, by increasing the permeabilility of the magnetic material.

Q.5. Under given conditions how is it possible to get a greater amount of throw?

A. : Instead of withdrawing the search coil from the air-gap, if it can be rotated through 180° in the air-gap itself, then the amount of throw becomes much larger.

[Apart from these questions, informations about magnetic saturation, magneto-striction, magnetic induction should be known. In general, magnetic properties of materials should be well acquainted with.]

9.16 Investigations of an inductance in an a.c. circuit

An inductance coil connected in an a.c. circuit gives rise to various properties and behaves in different manners, some of which are studied below.

(A) Study of the current/voltage relationship and determination of the value of the inductance

Theory :

In the a.c. circuit shown in (Fig. 9.28), if I is the r.m.s. current and V is the r.m.s. voltage recorded respective by the a.c. ammeter (A) amd an a.c. voltmeter (V), then if a graph be plotted with V along Y-axis and I along X-axis, the graph will be a straight line passing through the origin. This graph, therefore, verifies that the r.m.s. current in the circuit is directly proportional to the r.m.s. voltage across the inductance (L). Therefore, Ohm's law also applies in this case.

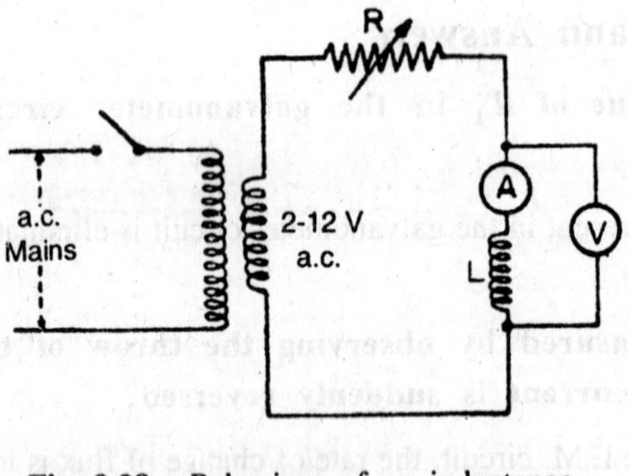

Fig. 9.28 : Behaviour of an inductance in a.c. circuit

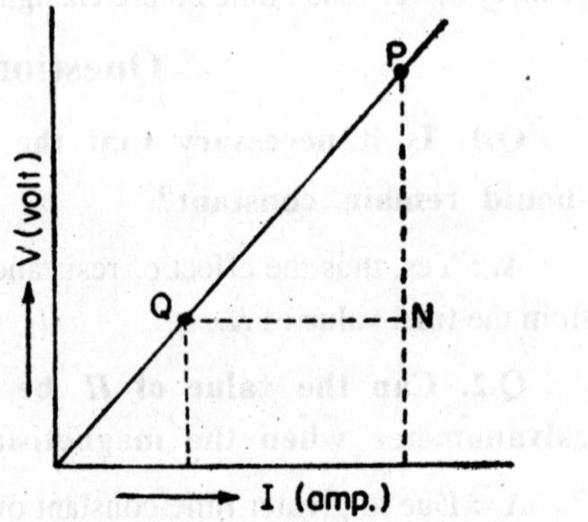

Fig. 9.29 : Current-voltage curve

From two convenient and well separated points P and Q on the graph (Fig. 9.29), the slope $\frac{PN}{QN}$ is obtained. This is the average numerical value of the ratio $\frac{V}{I}$ for the inductor in the circuit and is the magnitude of the inductive reactance (X_L) of the coil.

Now, since $X_L = 2\pi fL$, where f is the frequency of the current (50 Hz) and L is the self-inductance of the coil, usually called the inductance, the value of L can be found from

$$L = \frac{X_L}{2\pi f}. \qquad \text{... (9.16a)}$$

Apparatus :

Source of low voltage (8-12V) a.c., Ammeter (a.c.), high resistance a.c. voltmeter (0-10V), variable resistance 0-1000 ohms, inductance coil (about 100 mH), key or switch.

Method :

Follow the same steps from (i) to (v) given under Method in Experiment No. 9.11(A).

Results :

Frequency of the applied a.c. (f) = ... Hz

For drawing of graph and calulations therefrom, see the second paragraph of the theory.

If the coil possesses a resistance R_1 in addition to its inductance L, the ratio $\frac{V}{I}$ is the impedance Z of the coil, where $Z = \sqrt{(R_1^2 + X_L^2)}$. If R_1 is negligibly small, then $Z = X_L$.

Table 1

No. of Obs.	R.M.S. value of the current (*I* ampere)	R.M.S.value of the voltage (*V* volt)	Inductive reactance of the coil (X_L ohm) from		Mean value of X_L (ohm)	Value of *L* in mH
			Values	Graph		
1.	...	...	...			
2.	...	...	...			
3.						
4.				...	......	......
	...					
...						
8.	...		...	...		

Precautions :

Same as in Experiment No. 9.11 (A).

(B) Measurement of the reactance of an inductive coil in an L-R circuit, using a voltmeter

Theroy :

In the circuit shown in [Fig. 9.30(a)], if V_1, V_2 and V_3 be the voltage drops across the non-inductive resistor (R_1), the unknown inductor coil (L) and the total circuit respectively and if f is the mains frequency and R be the unknown resistance, then for these observations, the voltage vector triangle [Fig. 9.30(b)] can be constructed. From the point C of the triangle, a perpendicular is dropped to meet AB continued at D.

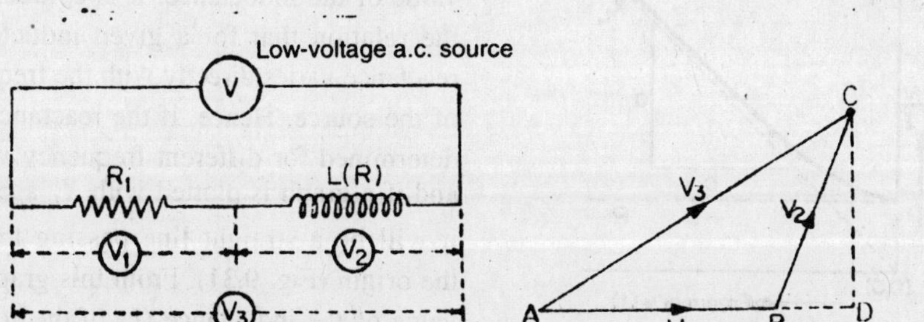

(a) (b)

Fig. 9.30 : Reactance by voltmeter

If I is the circuit current, then

$$V_1 = R_1 I = AB \qquad \text{... (9.16b)}$$

and if R is the resistance of the unknown coil, then the inphase voltage across the coil due to this resistance is equal to

$$RI = BC. \qquad \text{... (9.16c)}$$

Hence, from (9.16b) and (9.16c),

$$R = \frac{BD}{AB}.R_1 = \text{........ ohm} \qquad \text{... (9.16d)}$$

In the vector triangle, CD represents the out of phase voltage due to the reactance X_L of the coil.

Thus, $CD = X_L I = 2\pi f L.I$

$$\text{or, } L = \frac{CD}{2\pi f.I} = \frac{CD}{2\pi f}\cdot\frac{R}{BD} = \frac{CD}{AB}\cdot\frac{R_1}{2\pi f}, \text{ from (9.16c) and (9.16d)}$$

$$\text{or, } L = \frac{R_1}{2\pi f}\cdot\frac{CD}{AB}. \qquad \text{... (9.16e)}$$

Thus knowing R_1 and f and measuring the values of CD and AB from the vector triangle, the value of the inductance of the coil can be determined.

Method :

Same as under Method given in Experiment No. 9.11(B).

Results :

Table 2 [See next page]

Frequency (x) of the a.c. source = ... Hz

(C) Variation of the reactance of an inductor with the frequency of the a.c. source

Theory :

The reactance (X_L) of an inductor is given by $X_L = 2\pi f L$, where f is the frequency of the a.c. source and L is the value of the inductance. It is evident from the relation that for a given inductor, its reactance varies directly with the frequency of the source. Hence, if the reactances are determined for different frequency values, and if a graph is plotted with X_L against f, it will be a straight line passing through the origin (Fig. 9.31). From this graph, the value of the inductance (L) of the coil can also be determined.

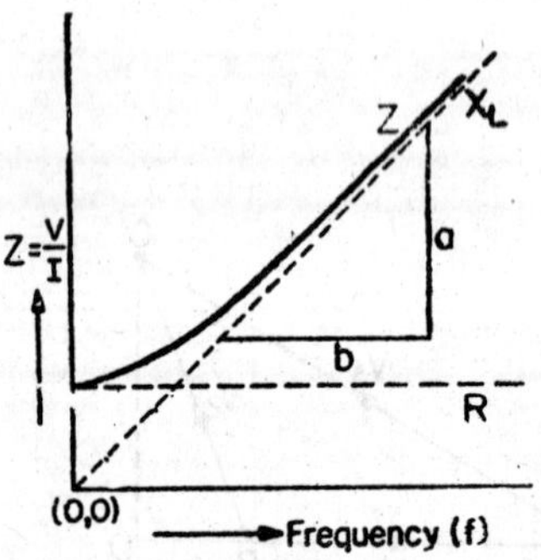

Fig. 9.31 : Frequency-reactance curve

At high frequencies, the value of the impedance (Z) may be expected to approach the reactance value as the reactance (X_L) becomes large compared to the coil resistance (R). So, if an asymptote, passing through the origin, is drawn to the plotted curve (Z against frequency) as shown, then from the value of the slope $\frac{a}{b}$ of the line, the value of L will be obtained.

Hence,

$$L = \frac{X_L}{2\pi f} = \frac{1}{2\pi}\left(\frac{a}{b}\right) = \text{...... henry.} \qquad \text{... (9.16f)}$$

Results :

Table 2

Frequency (x) of the a.c. source .... Hz

No. of Obs.	Value of R_1 (ohm)	Voltage (r.m.s.) across			From vector triangle values of			Value of coil resistance (R ohm)	Value of inductance (L in mH)
		R_1 (V_1 volt)	L (V_2 volt)	Total circuit (V_3 volt)	BD	AB.	CD		

Calculations :

See Theory :

[The experiment may be repeated with an iron rod being placed inside the coil and the effects of this introduction on the values of L and R, if any, may be studied.]

Apparatus :

The same as in **Expt. 9.16 (A)** with the mains transformer being replaced by a signal generator capable of producing variable frequencies.

Method :

The operations from (i) to (iii) given under **Method of Experiment No. 9.11 (C)** are to be followed in this case also.

Results :

Table 3

No. of Obs.	Frequency (f Hz)	Output r.m.s. voltage (V volt)	Current (r.m.s.) (I amp)	Impedence $(Z = \frac{V}{I})$ (ohm)	Reactance (X_L) from the graph (ohm)	Value of inductance (L Henry)

Graph drawings :

If the coil possesses a resistance (R) in addition to the inductance L, the ratio $\frac{V}{I}$ gives the value of the impedence Z of the coil where $Z = \sqrt{(R^2 + X_L{}^2)}$. A graph is ploted with frequency (f) along X-axis and impedence Z along Y-axis. The graph will be non-linear as shown in (Fig. 9.31). Along the straight portion of the curve (for high values of f), an asymptote is drawn through the origin. The slope $\left(\frac{a}{b}\right)$ of this straight line gives the value $\left(\frac{X_L}{f}\right)$ and from the relation given in Eq. (9.16f), the value of L can be found out. The graph shows the variation of the impedence (Z) and the reactance (X_L) of the coil with the mains frequency.

[Note, how from this graph, the value of the resistance (R) of the inductor coil can also be determined.]

(D) To draw the resonance curve of a series *L-C-R* circuit and hence to determine the *Q*-factor of the circuit

Theory

In an LCR circuit, the frequency ω at resonance is given by $\omega l = \frac{1}{\omega C}$, ... (9.16g)

where L is the inductance in henry applied in the circuit and C is the capacitance used in farads. To obtain the resonance in the LCR circuit, the frequency ω of the applied e.m.f. ($E = E_0 \sin \omega t$) should be varied. This can be varying the capacitance C and by noting the value $C = C_m$, where the voltage drop across the resistance is maximum, which gives the resonance position. If the resonance is at $C = C_m$ then from (9.16g),

$$\omega = \frac{1}{\sqrt{LC}} = \frac{1}{\sqrt{LC_m}}. \qquad \text{... (9.16h)}$$

For other values of capacitance, when the voltage drop is not maximum,

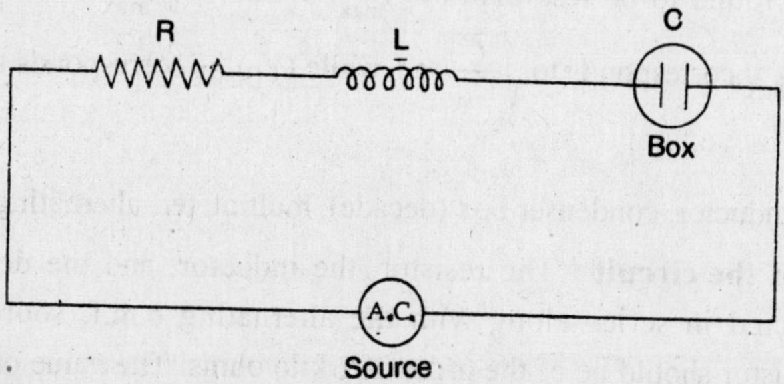

Fig. 9.32 : *L-C-R* Series Resonance circuit

we have $\omega_0 = \frac{1}{\sqrt{LC}}$.

$$\therefore \frac{\omega_0}{\omega} = \sqrt{\frac{C_m}{C}}\ ;\ \text{or,}\ \frac{\omega}{\omega_0} = \sqrt{\frac{C}{C_m}} = 1, \text{ when } C = C_m. \qquad \text{... (9.16i)}$$

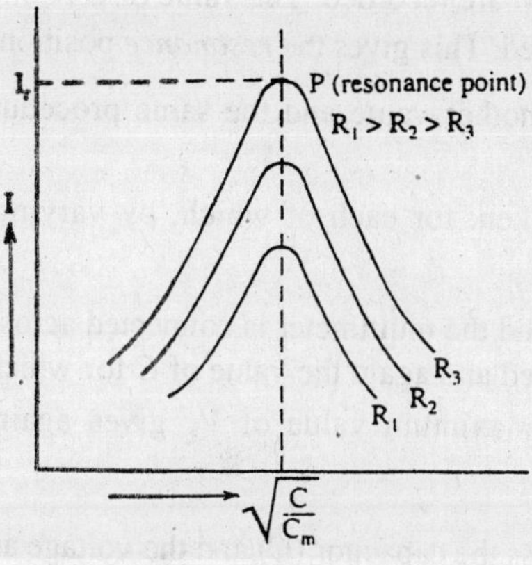

Fig. 9.33 : Resonance curve

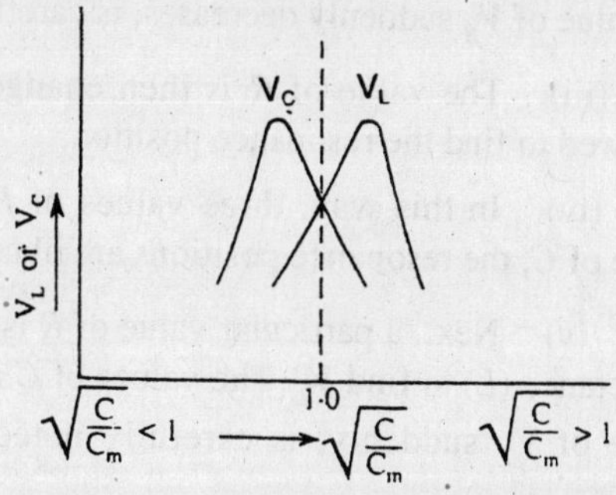

Fig. 9.34 : $\left(\frac{V_L}{V_C} - \sqrt{\frac{C}{C_m}}\right)$ curve

The current I through the resistance (R) is given by $I_R = \frac{V_R}{R}$, where R is the resistance employed in the LCR circuit.

Thus if a graph be plotted with $\sqrt{\frac{C}{C_m}}$ against I, it will be as shown in Fig. 9.33. The point P corresponding to $\sqrt{\frac{C}{C_m}} = 1$ denotes the maximum voltage drop across the resistor and shows the resonance position.

The voltage across the inductance V_L lags the applied voltage while the voltage across the capacitance leads the applied voltage. Hence, if graphs be plotted between V_L and V_C

versus $\sqrt{\frac{C}{C_m}}$, they will be as shown in (Fig. 9.34). The point of intersection of the two curves corresponds to $\sqrt{\frac{C}{C_m}} = 1$ showing the position of maximum voltage across the resistor. $(V_C)_{max}$ is found to be less than $(V_R)_{max}$. While $(V_L)_{max}$ is found greater than $(V_R)_{max}$. Hence $(V_C)_{max}$ corresponds to $\sqrt{\frac{C}{C_m}} < 1$ while $(V_L)_{max}$ corresponds to $\sqrt{\frac{C}{C_m}} > 1$.

Apparatus :

A resistor, an inductor, condenser box (decade), multimeter, alternating e.m.f. source.

Connection of the circuit : The resistor, the inductor, and the decade condenser box are all connected in series along with the alternating e.m.f. source as shown in (Fig. 9.32). The resistor should be of the order of 1 kilo ohms. The value of the resistor has to be changed (like 1 KΩ, 2 KΩ, 3 KΩ, etc.). The decade capacitors should be of the order of 0.01 μF to 5 μF.

Procedure :

(i) The circuit connection is made as shown in (Fig. 9.32). The resistor may be a box having different resistances of the order of kilo-ohm.

(ii) A multimeter is connected across R to measure the voltage V_R and changing the values of the condenser, V_R for different values of C are recorded. The value of C for which the value of V_R suddenly decreases, is carefully noted. This gives the *resonance* position.

(iii) The value of R is then changed to another value and the same procedure is followed to find the resonance position.

(iv) In this way, three values of R are taken, for each of which, by varying the value of C, the resonance positions are obtained.

(v) Next, a particular value of R is fixed and the multimeter is connected across the inductance (L) to find V_L. The values of C are varied and again the value of C for which the value of V_L suddenly, is carefully noted. The maximum value of V_L gives again the resonance postion.

(vi) The multimeter is then connected across the capacitor (C) and the voltage across the capacitor (V_C) is noted. The value of C is changed to different values and the maximum value of V_C is noted.

(vii) Next the value of C is determined corresponding to maximum value of V_R from three sets of observations (for 1 KΩ, 2 KΩ, 3 KΩ). The value of $\sqrt{\frac{C}{C_m}}$ is calculated for each value of C seperately and also the corresponding value of the current using the relation $I = \frac{V_R}{R}$.

(viii) Graphs between I_R and $\sqrt{\frac{C}{C_m}}$ for various values of R (1 KΩ, 2 KΩ, 3 KΩ) are plotted. The shape of the graph will be as shown in (Fig. 9.33).

(ix) Graphs between V_L versus $\sqrt{\frac{C}{C_m}}$ and V_C versus $\sqrt{\frac{C}{C_m}}$ are plotted on the same graph paper. The value corresponding to the point of intersection of the two curves (of $\sqrt{\frac{C}{C_m}}$), is noted.

Results :

(A) Determination of C_m :

Table 1

Serial No.	Capacitance C in μf	Resistance					
		R = 1 KΩ		R = 2 KΩ		R = 3 KΩ	
		V_R (volts)	I_R (mA)	V_R (volts)	I_R (mA)	V_R (volts)	I_R (mA)
1.	0.1	–	–	–	–	–	–
2.	0.2	–	–	–	–	–	–
3.	0.3	–	–	–	–	–	–
4.	–	–	–	–	–	–	–
5.	–	–	–	–	–	–	–
–	–	–	–	–	–	–	–
–	–	–	–	–	–	–	–
–	–	–	–	–	–	–	–

From the above readings, the value of C_m = ... μf.

(B) Determination of capacity corresponding to $(V_L)_{max}$ and $(V_C)_{max}$.

Table 2

Serial No.	Voltage across inductance (V_1)		Voltage across capacitor (V_C)	
	Capacitance in μf	V_L (volts)	Capacitance in μf	V_C (volts)
1.	0.1	–	–	–
2.	0.2	–	–	–
3.	0.3	–	–	–
4.	–	–	–	–
–	–	–	–	–
–	–	–	–	–

Capacitance corresponding to $(V_L)_{max}$ = ... μf.

Capacitance corresponding to $(V_C)_{max}$ = ... μf.

(C) Determination of $\sqrt{\frac{C}{C_m}}$

Table 3

C in μf	0.1	0.2	0.3	...	...	...	...	...	...
$\sqrt{\frac{C}{C_m}}$	...	...	...	...	...	...	...	...	...

Graphs :

The graphs plotted between $\sqrt{\dfrac{C}{C_m}}$ and I_R shows the variation of I_R with capacitance. The resonance position for all three values of resistances is found to be the same.

The point of intersection of the graph V_L verses $\sqrt{\dfrac{C}{C_m}}$ and V_C versus $\sqrt{\dfrac{C}{C_m}}$ is near about $\sqrt{\dfrac{C}{C_m}} = 1$ which verifies the proposed theory.

Determination of the *Q*-Factor

From Eq. (9.16a), the value of $\omega = \dfrac{1}{\sqrt{LC}}$ and if f_r denotes the actual resonant frequency of the circuit, then $f_r = \dfrac{\omega}{2\pi} = \dfrac{1}{2\pi}\dfrac{1}{\sqrt{LC}}$.

$$\text{or, } f_r = \frac{1}{2\pi}\cdot\frac{1}{\sqrt{LC}}, \quad \text{... (9.16j)}$$

where L is in henry and C in farad. For L = 1 mH and C = 1 μF, the value of f_r comes out to be nearly 5 kHz. The circuit is connected with an audio or radio frequency oscillator (*A.F./R.F*) when the frequency applied may be varied.

As in the previous *L-C-R* reasonance circuit experiment, the voltage/current across the inductor/resistor/capacitor ($V_L/V_R/V_C$) are measured for different applied frequencies. A graph is then plotted with V_R/I_R against the frequency (Fig. 9.35) for each of the inductor/resistor/capacitor.

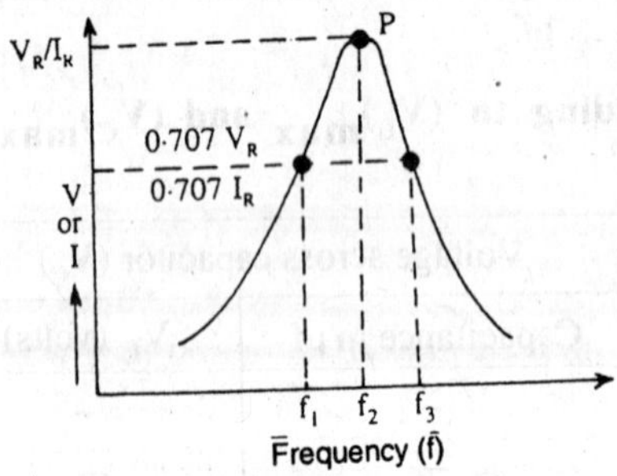

Fig. 9.35 : Q-factor

The *Q*-factor of the circuit, which determines the sharpness of resonance is given by

$$Q = \frac{f_r}{(f_2 - f_1)}, \quad \text{... (9.16k)}$$

where f_r is the resonant frequency and f_1, f_2 are the lower and higher frequencies on resource curve corresponding to current 0.707 I_r, where I_r is the current at resonant frequency.

The values of Q of individual part in series circuit are given below :

$Q_L = \dfrac{\omega L}{R} = \dfrac{i\omega L}{iR} = \dfrac{V_L}{V_R}$ (Inductive resistance is very small compared to the non-inductive resistance).

$$\text{and } Q_C = \frac{1}{\omega CR} = \frac{V_C}{V_R} \quad \text{... (9.16}\ell\text{)}$$

The voltage magnification at resonance in case of series resonance circuit is given by $V_m = \dfrac{V_{Lr}}{E} = \dfrac{V_{Cr}}{E}$ where $V_{Lr} = V_{Cr}$ is the voltage across the induction or capacitor at resonance and E is the applied *EMF*.

From the $\frac{V_R}{I_R}$ versus frequency (f) curves for each component (*L.R.C.*), the *Q*-factors for each component can be calculated using relation (9.16i) and measuring f_r, f_1, f_2.

Precautions :

(i) Multimeter should be used in proper range for measurement of voltages across each component.

(ii) Proper values of $E_0 \sin \omega t$ should be chosen to get resonance position.

(iii) Proper values of *L*, *C* and *R* must be chosen.

(iv) Thermo-milliammeter may be used in circuits to measure high frequency currents, while ordinary A.C. milliammeter may be used for measuring low frequency current.

(v) A *V.T.V.M.* may be used to measure the voltages across *L*, *C* or *R*. The same may be used to measure the output of the source.

9.17 Determination of the value of Planck's constant by using a photoelectric cell

Theory :

Einstein's equation for the maximum energy of a photoelectron emitted when a monochromatic light of frequency ν falls on a metal surface is

$$E = h\nu - w_0, \qquad \text{... (9.17a)}$$

where h = Planck's constant

w_0 = Work function of the metal.

When a monochromatic light is incident on the cathode surface of a photoelectric cell, the former liberates photoelectrons. These electrons can be detected as the photoelectric current in the external closed circuit connected through a galvanometer. If now in the external circuit a potential difference is applied between the anode and the cathode with the reversed sign, then the photocurrent drops to the zero value or a particular value of the potential applied. This external potential is then known as the *Stopping potential* (V_0). If electrons of charge e are just stopped by this potential (V_0), then

$E = eV_0 = h\nu - w_0$, from Eq. (9.17a)

The work function (w_0) is equal to $h\nu_0$, where ν_0 is the thresold frequency.

Hence,

$$eV_0 = h\nu - h\nu_0$$

or,
$$V_0 = \frac{h}{e}\nu - \frac{h}{e}\nu_0$$

$$= \frac{h}{e}(\nu - \nu_0). \qquad \text{... (9.17b)}$$

If c is the velocity of light, then $\nu = \frac{c}{\lambda}$ and $\nu_0 = \frac{c}{\lambda_0}$. Eq. (9.17b) thus reduces to

$$V_0 = \frac{hc}{e}\left(\frac{1}{\lambda} - \frac{1}{\lambda_0}\right). \qquad \text{... (9.17c)}$$

So, if a graph is plotted with the cut-off voltage (V_0) along Y-axis and ν along X-axis, then the graph will be a straight line. The slope of the curve will give the value of $\frac{h}{e}$ from which the value of h can be evaluated, substituting the values of e (Fig. 9.37).

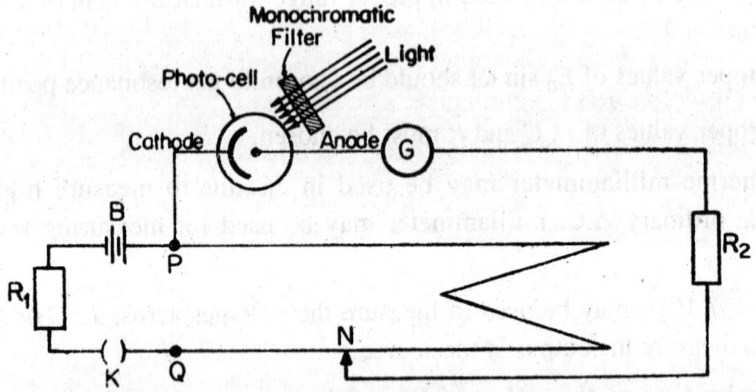

Fig. 9.36 : Planck's constant determination

The stopping potential (V_0) is measured by a potentiometer by balancing the potential on a length l cm of the wire having a potential drop of ρ per unit length. The potentiometer circuit and the photocell would have to be duly connected with their polarities, as shown in (Fig. 9.36).

Thus when balanced,

$$V_0 = \rho l \qquad \text{...(9.17d)}$$

Apparatus :

Photocell, monochromatic filters, potentiometer, storage cells, key, resistance boxes, galvanometer, source of light.

Connections of the circuit :

The connections are made as shown in (Fig. 9.36). The positive polarity of the driving cell (B) and the cathode of the photocell are connected to the same point (P) of the potentiometer.

Method :

(i) The apparatus is properly set, and setting the source of light (an incandescent lamp) close to the photocell, the monochromatic filter is set in front of the source. The potentiometer resistance is also determined.

(ii) After closing the potentiometer circuit, null point is obtained on the wire and the balancing length is noted (l).

(iii) Without changing the value of the resistance R_1 in the potentiometer circuit, similar readings are taken for other colour filters.

(iv) The entire operations may be repeated for another particular value of R_1 which gives another set of readings.

(v) From both the sets, the values of the stopping potentials for each colour of light are obtained and graphs are plotted, as discussed, for the stopping potentials against the frequency or reciprocal of wavelengths of the colours.

(vi) From the graph, the slope is calculated and therefrom the value of h obtained employing Eq. (9.17c).

Results :

e.m.f. of the battery (before experiment) = ... volt

e.m.f. of the battery (after experiment) = ... volt

(A) Determination of the potentiometer resistance :

Table 1

[Make a table of P.O. Box Expt. as given in Table 1 of Expt. No. 8.5]

(B) Determination of the value of cut-off voltage :

Table 2

Resistance of the potentiometer wire (R_p) = ... ohm

Resistance in the potentiometer circuit (R_1) = ... ohm

Length of the potentiometer wire (L) = ... cm

e.m.f. of the battery used (E) = ... volt

No. of Obs.	Colour of filter used	Null points obtained				Value of ρ volt/cm	Value of potential (V_0 volt = ρl)
		On Wire No.	On Scale reading (cm)	Mean (cm)	Total reading (l cm)		
1.		......	......				
2.	Violet	......	......				
3.		......	......		......	......	......
4.		......	......				
...	Blue			......	......	......	......
...	Green			......	......	......	......
...	Yellow			......	......	......	......
...	Red			......	......	......	......

(C) Data for drawing of graph :

Table 3

Velocity of light (c) = 3×10^{10} cm/sec

No. of Obs.	Colour of filter	Wavelength (λ cm)	Value of frequency (ν cm)	Stopping potentials (V volt)	Slope of the curve (XY/YZ)	Value of e (coulomb)	Value of h (erg-sec)
1.	Violet	......	......	......			
2.	Blue	......	......	......			
3.	Green	......	......	......			
4.	Yellow	......	......	......	......	......	......
5.	Red	......	......	......			

Drawing of Graph :

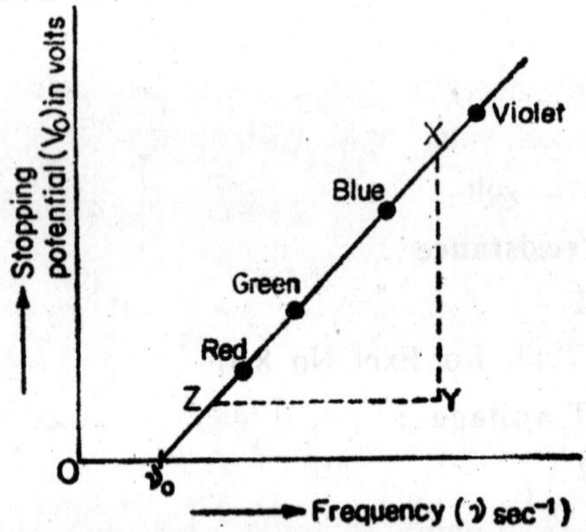

Fig. 9.37 : Frequency-potential curve

A graph is plotted with stopping potential (V_0) along Y-axis and frequency (ν) of radiation along X-axis. The graph will be a straight line as shown in (Fig. 9.37). The slope of the curve is obtained from the ratio of XY/YZ. This is equal to $\frac{h}{e}$. From the value of the slope and putting the value of e, the value of Planck's constant (h) can be obtained.

Instead of (V_0-ν) curve, $\left(V_0 - \frac{1}{\lambda}\right)$ curve can also be plotted when the slope of the curve will give the value of $\frac{hc}{e}$, where c is the velocity of light. The straight line cuts the X-axis at a point which corresponds to the thresold frequency (ν_0).

Calculation :

From Eq. (9.17b),

$$V_0 = \frac{h}{e}(\nu - \nu_0).$$

From the straight line curve of V_0 against ν, the slope

$$\frac{XY}{YZ} = \frac{h}{e} \quad \text{or, } h = e.\frac{XY}{YZ} = \text{.........} \text{ erg-sec}$$

For determination of the stopping potential, if l is the balancing length on the potentiometer wire, then from Eq. (9.17d), $V_0 = \rho l$, where ρ is the potential drop per unit length of the potentiometer wire.

Now, if R_p is the potentiometer resistance and R_1 is the resistance in the potentiometer circuit, then

$$\rho = \frac{E.R_p}{(R_p + R_1)L} \text{ volt/cm}$$

$$\therefore V_0 = \rho.l = \frac{E.R_p.l}{L(R_p + R_1)} = \text{......} \text{ volt}$$

Precautions :

(i) The value of h so obtained is likely to have the correct order but to differ from the accepted value. Samples of different types of photocells give differing values since the emissive surface is neither a pure metal, nor is it free from adsorbed gas.

(ii) The filters used should be as far as possible monochromatic and the accuracy of the result depends very much on this monochromatic nature of the filter.

(iii) The exact null point should be carefully determined and checked for different values.

(iv) The cathode and the positive polarity of the driving cell should be properly connected as shown in the circuit diagram.

(v) The main source of error in this experiment depends on the monochromatic nature of the filter used; on the property of the photocell and properties of its metal surfaces; and on the accurate determination of the null point on the potentiometer for obtaining the cut-off voltage.

Questions and Answers

Q.1. What is Planck's constant and what is its value?

A. : See any textbook.

Q.2. Does it depend on any factor?

A. : No, it is a universal physical constant.

Q.3. Is this method accurate for determination of *h*?

A. : No, because of several inaccuracies in the method, it would be wrong to give too much weight to evidence based on this experiment, but it may reasonably be regarded as an illustration of the principles involved in the measurement of Planck's constant.

Q.4. What will happen if the anode is connected to the point *P* of the potentiometer without changing any other parameter in the circuit?

A. : Because of this, no null point will be obtained on the potentiometer.

Q.5. Is the value of V_0 so obtained the correct and absolute stopping potential?

A. : No, the calculation has neglected the p.d. that will exist between cathode and anode, since they are different metals. However, this p.d. should be the same throughout the experiment and may be expected to alter the values of V_0 by a constant amount without affecting the slope of the graph.

9.18 To draw the (*B-H*) curve (Hysterisis) of iron in the form of an anchor ring

Theory :

If a current of C_1 ampere flowing through the primary winding P_2 of an anchor ring (Fig. 9.38) A.R., produces a magnetic field H_1, then assuming that the field is uniform throughout the cross-section of the ring, it may be written as

$$H_1 = 0.4\pi n_1 C_1 = m_1 C_1, \qquad \text{... (9.18a)}$$

where $m_1 = 0.4\pi n_1$, n_1 being the number of turns per cm of the primary.

The secondary S_2 having N_2 number of turns wound closely on the primary, will embrance a flux

$$\phi_1 = B_1 \alpha_1 N_1, \qquad \text{... (9.18b)}$$

where B_1 is the magnetic induction or the flux density wihin the anchor ring due to the field H_1, and α_1 represents the mean cross sectional area of the iron of the ring. If D_1 be the mean diameter of the iron of the ring, then the mean cross-sectional area $\alpha_1 = \frac{\pi D_1^2}{4}$. ...(9.18c)

If H_2 represents the magnetising field in the anchor ring due to current C_2 amperes through the primary winding P_2, then $H_2 = m_1 C_2$. ... (9.18d)

The corresponding flux embraced by the secondary S_2 is

$$\phi_2 = \beta_2 \alpha_1 N_2, \quad \text{... (9.18e)}$$

where B_2 is the magnetic induction due to H_2. The change in current from C_1 to C_2 in the primary P_2 thus produces a change of flux from ϕ_1 to ϕ_2 through the secondary S_2. This change of flux causes a flow of charge Q_1 to flow in the secondary circuit containing the ballistic galvanometer, which eventually causes a first throw θ_1 radian.

$$\text{Thus, } Q_1 = \frac{\phi_1 - \phi_2}{R' \times 10^8} = \frac{(B_1 - B_2)\alpha_1 N_2}{R' \times 10^8} = K\,\theta_1\left(1 + \frac{\lambda}{2}\right), \quad \text{... (9.18f)}$$

where R' is the total resistance in the galvanometer circuit $(R + R_g)$, K is the constant of the ballistic galvanometer and λ is the logarithmic decrement.

To eliminate K and λ, an auxiliary experiment is performed where a current of I ampere is reversed through the primary P_1 of a standard solenoid B. If the mean cross-sectional area of this primary P_1 be α_2, the number of turns per cm in it be n_3 and the total number of turns in the secondary (S_1) be N_4, then on reversal of I ampere of current through the primary (P_1) will cause a flow of charge Q_2 in the galvanometer circuit containing the secondary S_1 of the solenoid. If the first throw of the galvanometer under this condition be θ_2 radian, then it may be written as

$$Q_2 = \frac{0.8\pi n_3 \alpha_2 N_4}{R' \times 10^8} I = K\,\theta_2\left(1 + \frac{\lambda}{2}\right). \quad \text{... (9.18g)}$$

Taking ratio of Eqs. (9.18f) and (9.18g), we have

$$\frac{(B_1 - B_2)\alpha_1 N_2}{0.8\pi n_3 \alpha_2 N_4 I} = \frac{\theta_1}{\theta_2}$$

$$\text{or,} \quad (B_1 - B_2) = \frac{0.8\pi n_3 N_4}{N_2} \times \left(\frac{\alpha_2}{\alpha_1}\right) \times \left(\frac{I}{\theta_2}\right)\theta_1.$$

But since $\frac{\theta_1}{\theta_2} = \frac{d_1}{d_2}$, where d_1 and d_2 are the displacements of the galvanometer light spot on the scale, hence

$$(B_1 - B_2) = \frac{0.8\pi n_3 N_4}{N_2} \times \left(\frac{\alpha_2}{\alpha_1}\right) \times \left(\frac{I}{d_2}\right) d_1 \quad \text{... (9.18h)}$$

$$\text{or,} \quad (B_1 - B_2) = m \times d_1 \quad \text{... (9.18i)}$$

$$\text{where } m = \frac{0 \cdot 8\pi n_3 N_4}{N_2} \times \left(\frac{\alpha_2}{\alpha_1}\right)\left(\frac{I}{d_2}\right).$$

But if D_1 and D_2 be the mean diameters of the iron of the ring and the primary of standard solenoid respectively, then $\frac{\alpha_2}{\alpha_1} = \frac{D_2^2}{D_1^2}$ and hence

$$m = \frac{0 \cdot 8\pi n_3 N_4}{N_2} \times \left(\frac{D_2}{D_1}\right)^2 \times \left(\frac{I}{d_2}\right) \quad \ldots(9.18j)$$

When the current in the primary in the anchor ring is changed from $+C_1$ to $-C_1$, we get $B_2 = -B_1$. Thus from Eq. (9.18i), we get

$$2B_1 = md_1 \text{ or, } B_1 = \frac{m}{2}. \quad \ldots(9.18k)$$

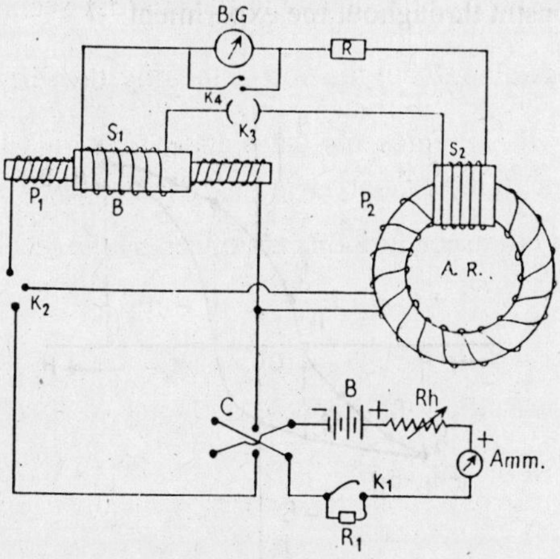

Fig. 9.38 : Circuit diagram (B-H) curve

Knowing B_1, we can get B_2 from Eq. (9.18i) and its corresponding H_2 from Eq. (9.18d). Then a graph is to be plotted with H along X-axis and its corresponding B along Y-axis. In this way, one-half of the (B-H) loop will be obtained. The other half of the loop is then drawn from symmerty of the loop.

Apparatus :

Anchor ring, standard solenoid, ballistic galvnometer, resistance boxes (R and R_1), plug type key (K_3), two-way key (K_2), Pohl's commutator (C), tapping keys (K_1 and K_4), rheostat (Rh), ammeter (range 0-5 amp), battery.

Connection of the circuit :

The schematic diagram shown in (Fig. 9.34) shows the plan of the connection. The rheostat (Rh) shall be of high-current-bearing capacity. The ammeter must have a range of 0-5 amp with steps of 0.1 ampere preferably.

Method :

(i) The constants of the apparatus are measured first. These are namely D_1, n_1 and N_2 of the anchor ring (A.R.) and D_2, n_3 and N_4 of the standard solenoid (B). Often the values of n_1 and n_3 are obtained from the measure of the total number of turns (N) and mean diameter (D). Then $n_1 = \frac{N_1}{\pi D}$ and $n_3 = \frac{N_3}{L}$, where N_1 and N_3 are the total number of turns and D and L are the mean diameter and the length of the primary of the anchor ring and the standard solenoid respectively.

(ii) Keeping the key K_3 open and K_1 closed, the two-way key K_2 is so set that the primary P_2 is connected to the commutator. The value of the rheostat is so adjusted that the maximum permissible current (about 4.0 A) passes through P_2. The iron of the anchor ring is brought to steady cyclic state by rapidly reversing the maximum current through P_2 for several times by the help of the commutator and finally the rocker is kept to the left. Next the key K_3 is closed and the spot of light of the galvanometer is brought to zero position on the scale. The current through P_2 is reversed and the first flow of the galvanometer is noted. By adjusting the resistance R, this throw is made almost full-scale. The value of R

must be kept larger than the CDR of the galvanometer. This value of R must be kept constnt throughout the experiment.

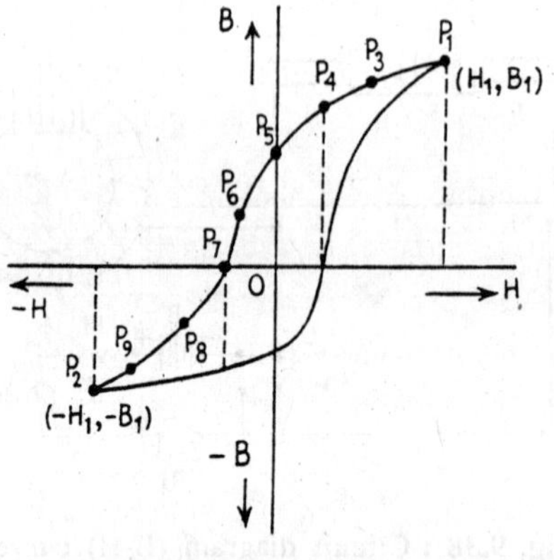

Fig. 9.39 : Hysterisis curve

(iii) From the ammeter, the maximum current is noted. This current is reversed $\frac{3}{4}$ times and each time the throw of the galvanometer is noted. The mean throw (d_1) is noted. Reversal of current and noting of throw must be done when the rocker is turned from left to right only. The values of H_1 and B_1 are thus obtained from Eqs. (9.18a) and (9.18k) respectively. The points P_1 and P_2 for values of (H_1, B_1) and ($-H_1$, $-B_1$) are plotted on the graph (Fig. 9.39).

(iv) Next the keys K_1 and K_3 are opened and suitable resistances in R_1 are inserted. The key K_1 is closed and the rocker is turned rapidly from left to right several times to bring the iron of the anchor ring into steady cyclic state as defined by P_1. Finally, the rocker is kept to the left.

(v) The key K_3 is closed and the light spot is brought to zero. Key K_1 is opened and the ballistic throw (d_1) is noted. The current C_2 is noted from the ammeter. Key K_1 is closed and the iron is brought to steady state. K_1 is opened and again the ballistic throw (d_1) is noted when the rocker is on the left. The mean value of these two throws is found when the current is reduced by opening K_1 suddenly. The value of H_2 is calculated from Eq. (9.18d) for the reduced current C_2 and value of B_2 from Eq. (9.18i) since B_1 is already determined. This gives another point P_4 (H_2, B_2) on the (B -H) loop (Fig. 9.39).

(vi) The operations (iv) and (v) are repeated with the rocker on the left, by increasing the value of R_1 in suitable steps until R_1 is infinite. Thus the points P_1, P_3, P_4, P_5 and the path thereof of the (B -H) loop are obtained.

(vii) Next key K_1 is opened and a large resistance is inserted in R_1. K_1 is closed and keeping K_3 open, the rocker is moved from left to right several times and finally kept on the left. This process not only reverses the maximum current but also reduces the reversed current. The first throw d_1 is noted and the mean value of d_1 is taken from two or three throws. Before noting the throw, the iron must be brought into the steady cyclic state. Keeping the rocker on the right, the reversed current is noted and H_2 calculated. The value of d_1 is substituted in Eq. (9.18i) and the value of B_2 is obtained from the measured value of B_1. This gives the point P_6 on the curve.

(viii) The previous step and operation is repeated by decreasing the value of R_1 in suitable stetps until it is very small or practically zero. This gives the points P_7, P_8 and P_9 on the curve.

Thus one-half of the (*B-H*) loop is obtained. The lower half is drawn from symmetry. The points for the bottom half are obtained from the corresponding points on the top half by simply reversing the signs of *B* and *H*. For example, (*H*, *B*) should be plotted as (–*H*, –*B*); or (–*H*, *B*) as (*H*, –*B*) and so on.

(ix) To find the value of $\left(\frac{I}{d_2}\right)$, the key K_3 is kept open and connection in K_2 is made between the primary (P_1) of the standard solenoid (*B*) and that of *A.R.* By closing the key K_2, the current *I* in primary P_1 of the standard solenoid *B* is made as low at 0.5 ampere by adjusting the rheostat (*Rh*). The key K_3 is closed and the rocker is moved from left to right and also from right to left. In both the cases the galvanometer throws are noted and the mean value of the throws (d_2) is found. The process is repeated serveral times by increasing the value of *I* in suitable steps of 0.5 ampere. A graph is then plotted with *I* along *X*-axis and the corresponding values of d_2 along *Y*-axis. The graph will be a straight line passing through the origin. The mean value of $\left(\frac{I}{d_2}\right)$ is obtained from this graph and this value is utilised to find the value of *m* from Eq. (9.18j).

Results :

(A) Constants of the Apparatus :

For the Standard Solenoid B

Table 1

Length of the primary (*L*) in cm	Total number of primary (P_1) turms (N_3)	No. of turns per cm of the primary (P_1) $n_3 = N_3/L$	No. of turns (N_4) on the secondary (S_1)	Mean diameter (D_2) of the primary (P_1) in cm
....	....	....	....	
....	....	....	....	....

For the Anchor Ring (A.R.)

Table 2

Mean diameter (D_1) of the iron core in cm	Total No. of turns (N_1) in the primary (P_2)	Diameter of the anchor-ring (*D*) in cm	No. of turns per cm (n_1) of the primary $n_1 = \frac{N_1}{\pi D}$	No. of turns (N_2) in the secondary	$m_1 = 0.4\pi n_1$ (Oe/ampere)
......	......	......	......	......	......

(B) Readings for (I – d_2) curve :

Table 3

[Numerical figures are given as examples only]

Current (I) in ampere	Ballistic throw (d_2) in cm when current from		Mean d_2 in cm
	$+I$ to $-I$	$-I$ to $+I$	
0.5	1.3	1.3	1.3
1.0	2.4	2.4	2.4
1.5	3.5	3.5	3.5
etc.	etc.	etc.	etc.
4.0	...	...	...

(C) Determination of the value of *m* :

Table 4

Value of n_3 from Table 1 (per cm)	Value of N_4 from Table 1	Value of N_2 from Table 2	Value of D_2 from Table 1 (cm)	Value of D_1 from Table 2 (cm)	Value (mean) of I/d_2 from (I-d) curve (ampere/cm)	$m = \frac{0.8\pi n_3 N_4}{N} \times \left(\frac{D_2}{D_1}\right)^2 \times \left(\frac{I}{d_2}\right)$ (gauss/cm)
...	...	...	...	...	...	.....

(D) Reading for (B-H) curve :

Table 5

(i) Initial values of B and H :

[Numerical figures are just for example]

Initial current (C_1) ampere	Final current ($-C_1$) ampere	Ballistic throw (d_1) cm	Mean value of d_1 (cm)	Value of $\frac{d_1}{2}$ (cm)	Initial Field $H_1 = m_1 C_1$ (Oe)	Initial Induction $\beta_1 = \frac{m}{2} d_1$ (gauss)
	–4.0	15.8				
4.0			15.7	7.85	.....	.....
	–4.0	15.6				

(ii) Final value of B and H :

Table 6

Final current (C_2) ampere	Mean value of C_2 (ampere)	Ballistic throw (d_1) in cm	Mean value of d_1 (cm)	Final field $H_2 = m_1 C_2$ (Oe)	Change of induction $\Delta B = B_1 - B_2 = md_1$ (gauss)	Final induction $B_2 = B_1 - \Delta B$ (gauss)
–4.0		15.8				
	–4.0		15.7	......	......	......
–4.0		15.6				
+3.5				......	......	......
	+ 3.5	0.6	0.6	......	......	......
+3.5		0.6				
......						
			......	......	......	......
......						
0		4.0				
	0		4.0		......	......
0		4.0				
–0.5						
	–0.5	4.8				
			4.8	......	......	......
–0.5		4.8				
......	......					
			......	......	......	......
......	......					
–3.5		15.0				
	–3.5		15.0	......	......	......
–3.5		15.0				

Graph Drawings :

Two graphs are to be drawn. In the first one, I is drawn along X-axis and the corresponding values of d_2 are plotted along Y-axis. The graph would be a straight line passing through the origin. From this graph, the mean value of $\left(\frac{I}{d_2}\right)$ should be determined by taking the co-ordinates of a point on the straight line.

Another graph (B-H curve) has to be drawn. Taking a suitable origin, two co-ordinate axes are drawn where abscissa represents $+W$ and $-H$ while the ordinate represents $+B$ and $-B$ (Fig. 9.39). Finding (B_1, H_1) and $(-B_1, -H_1)$, the two points P_1 and P_2 are plotted. Then by finding other values of B and H, other points are plotted. In this way, one-half of the cycle would be obtained. The other half of the cycle can be drawn by symmetry or by plotting the same points with their signs reversed. This curve is known as the Hysterisis Curve and the enclosed loop (Fig. 9.39) is know as the Hysterisis Loop. The significance of this loop is that its area is proportional to the loss of energy during one cyclic change of the magnetising field (H). This lost energy is covered to heat which makes the material hot. The greater the area of the loop, the greater will be the heat generation.

Precautions :

(i) Current for magnetising the iron should be taken from a fresh battery. If taken from d.c. mains, the current will fluctuate.

(ii) After each operation, the maximum current for which the cycle is to be drawn must be reversed several times (15 or 20 times) to bring the iron to steady state condition.

(iii) Before commencement of the experiment, the residual magnetism of iron should be destroyed.

(iv) The value of R should not be smaller than the external CDR of the galvanometer.

(v) Galvanometer circuit is closed only when a throw is to be observed.

(vi) If the (B-H) loop is required with the saturation value of B, then the magnetising current corresponding to the saturation value of B has to be determined first.

(vii) The value of R in the galvanometer circuit must not be changed during the experiment, once it is adjusted to give full scale deflection of the light spot with the maximum magnetising current.

Error Calculation :

The proportional error in H is obtained from Eq. (9.18a)

as $\frac{\delta H_1}{H_1} = \frac{\delta C}{C_1}$ since m_1 is constant.

Here δC is the smallest division of the ammeter.

Similarly, the proportional error in H is given by $\frac{\delta H_2}{H_2} = \frac{\delta C}{C_2}$.

To get the proportional error in B_1 we have from Eq. (9.18k), $B_1 = \frac{m}{2} d_1 = K.\frac{I}{d_2}.\frac{d_1}{2}$, where K is constant.

Thus the proportional error in B_1 is

$\frac{\delta B_1}{B_1} = \frac{\delta I}{I} + \frac{\delta d}{d_2} + \frac{\delta d}{d_1}$, where δd represents the value of the smallest division of the scale on which the spot of light moves.

Since $\Delta B = B_1 - B_2 = md_1 = K.\frac{I}{d_2}.d_1$, hence the proportional error in ΔB is

$$\frac{\delta(\Delta B)}{\Delta B} = \frac{\delta I}{I} + \frac{\delta d}{d_2} + \frac{\delta d}{d_1}.$$

The proportional error in B_2 is thus

$$\frac{\delta B_2}{B_2} = \frac{\delta(B_1 - \Delta B)}{B_2} = \frac{\delta B_1 + \delta(\Delta B)}{B_2}.$$

Knowing the values of δB, $\delta(\Delta B)$, $\frac{\delta B_2}{B_2}$ may thus be caluclated.

$$\text{The percentage error} = \frac{\delta B_2}{B_2} \times 100 = \ldots\ldots\%.$$

Questions and Answers

Q.1. Define the terms Magnetic field; Magnetic induction; Intensity of magnetisation; Retentivity; Coercivity; Permeability; Susceptibility and Hysterisis.

A. : See any textbook on Magnetism.

Q.2. What is Hysterisis loop and what is its significance?

A. : See the second paragraph in 'Graph Drawing'.

Q.3. What is the utility of finding the hysterisis loop?

A. : In the construction of armatures of motors and dynamos, the materials are so selected that their hysterisis loops are very small, so that the heat generated may be very small.

Q.4. Which has a smaller loop—steel or soft iron?

A. : Soft iron has a smaller loop.

Q.5. Why is the material demagnetised at the very outset?

A. : As the nature of the (B-H) loop depends very much on the past history of the specimen, hence the demagnetisation is essential.

Q.6. Why is the specimen taken in the form of a ring and not as a rod?

A. : As in the form of a ring, there is no free poles and as such no demagnetising effect.

Q.7. Why before each operation, the maximum current is reversed several times?

A. : This is to bring the iron in the steady cyclic condition.

Q.8. What type of material should be used in the construction of a permanent magnet?

A. : A material with high retentivity and high coercivity. The example is alnico.

Q.9. What are the nature of values of permeability for diamagnetic, paramagnetic and a ferromagnetic substance?

A. : In c.g.s. e.m.u., the value of μ is slightly less than 1 for diamagnetic, slightly larger than 1 for paramagnetic and much larger than 1 for ferromagnetic substances.

Q.10. Is the phenomenon of hysterisis observed for all types of substances?

A. : No, it is exhibited by ferromagnetic substances and not by dia- or paramagnetic substances.

9.19 Determination of the constants (calibration) of a Ballistic Galvanometer

Theory :

When an amount of charge q passes through a ballistic galvanometer producing a transient current, then

$$q = \frac{CT}{2n\pi AH}\,\theta = K\theta, \qquad \ldots (9.19a)$$

where the symbols have the significance as given under Art. 7.4(b). The quantity $\frac{CT}{2n\pi AH}$ is known as the constant of the ballistic galvanometer and denoted by K. This constant of the galvanometer may be determined by different methods. Determination of this constant is also sometimes referred to as 'calibration' of the ballistic galvanometer. If the quantity of charge q be expressed in coulombs, then $K\left(=\frac{CT}{2n\pi AH}\right)$ will be expressed in coulombs per radian. Taking damping into consideration and introducing logarithmic decrement, Eq. (9.19a) reduces to

$$q = \frac{CT}{2n\pi AH}\,\theta'\left(1+\frac{\lambda}{2}\right). \qquad \ldots (9.19b)$$

Three important methods for determination of the constant of the galvanometer are discussed. The methods are :

(A) Direct method

(B) By using a condenser of known capacity

(C) By using a standard solenoid.

(A) Direct method :

Apparatus : Ballistic galvanometer, two resistance boxes, shunt, battery, commutator, tapping key.

The circuit diagram for this method is shown in [Fig. 9.40] where a storage battery (E), a resistance box (R_1), a shunt (S) and a commutator (K) are connected in series. The ballistic galvanometer ($B.G.$) and another resistance box (R_2) are together connected in parallel to the shunt (S).

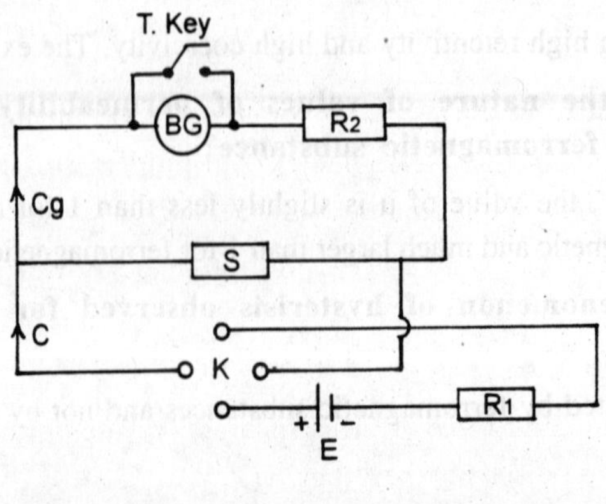

Fig. 9.40 : Direct method

If a steady current C_g amperes flowing through the galvanometer under a potential difference existing across S, produces a steady deflection θ radian of the coil of the galvanometer, then

$$C_g = \frac{c}{nAh}\theta; \text{ or, } \frac{C_g}{\theta} = \frac{c}{nAH}.$$

Here, $\frac{c}{nAh}$ represents current in amperes which produces 1 radian deflection of the galvanometer coil. If d be the deflection of the spot of light in cm on a scale placed at a distance of D cm from the galvanometer mirror, then

$\theta = \frac{d}{2D}$, where from we get

$$\frac{c}{nAh} = \frac{2DC_g}{d} \text{ ampere/rad} \quad \text{... (9.19c)}$$

If E be the e.m.f. of the battery and R_1 be its series resistance, then

$$C_g = C_r \frac{S}{S+G+R_2} = \frac{E}{R_1 + \frac{S(G+R_2)}{S+G+R_2}} \times \frac{S}{S+G+R_2}.$$

$$\text{or, } C_g = \frac{ES}{R_1(S+G+R_2)+S(G+R_2)} \quad \text{... (9.19d)}$$

$$= \frac{ES/(G+R_2)}{R_1\left\{1+\frac{S}{G+R2}\right\}+S}. \quad \text{... (9.19e)}$$

Since $S << (G + R_2)$, we can neglect $\frac{S}{(G+R_2)}$ and S from the denominator. Hence, C_g becomes.

$$C_g = \frac{ES}{R_1(G+R_2)} \text{ ampere.} \quad \text{... (9.19f)}$$

From Eq. (9.19d) or from Eq. (9.19f), C_g can be calculated.

Now, $K = \frac{cT}{2n\pi AH}$, from Eq. (9.19a)

$$= \frac{T}{2\pi} \times \frac{C}{nAH} = \frac{T}{2\pi} \times \frac{2DC_g}{d}. \quad \text{... (9.19g)}$$

[from (9.19c)]

Calculating C_g either from Eq. (9.19d) or from Eq. (9.19f) and determining the period of oscillation T in seconds of the galvanometer coil in the *open circuit*, the value of the galvanometer constant K in coulombs/radian, can be obtained using Eq. (9.19g)

[Suitable Tables may be made for recording the quantities determinable for the experiment.]

(B) Capacitance method :

Apparatus : Ballistic galvanometer, resistance box, condenser of known capacitance, commutator, battery, charge-discharge key, tapping key.

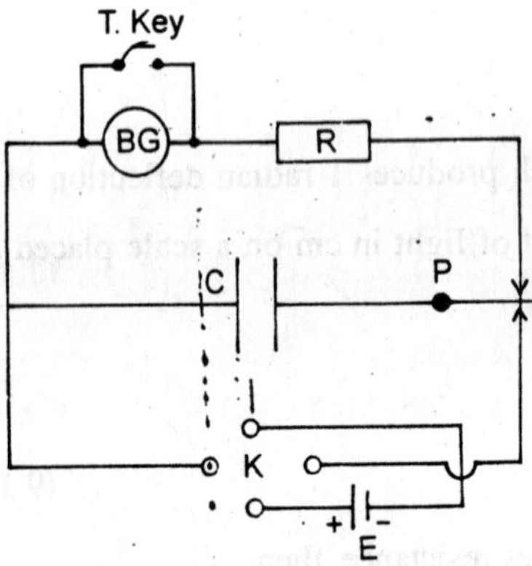

Fig. 9.41 : Capacitance method

The circuit diagram for this method is shown in (Fig. 9.41) where a storage battery (E) is connected in series with a condenser (C) of known capacitance through a commutator (K) and a charge-discharge key (a, b).

If the condenser of capacitance C farad be charged to a potential difference of V volts by a battery E, then the charge on it is $q = CV$ coulomb. If the battery is a storage type with negligible internal resistance, then the charging voltage V will be equal to the e.m.f. of the battery. Hence the charge q on the condenser is known. If this charge be very quickly sent through the ballistic galvanometer (BG) producing a first throw θ_1, then

$$q = CV = K\theta_1\left(1+\frac{\lambda}{2}\right), \quad \text{... (9.19h)}$$

where λ is the logarithmic decrement

$$\text{or,} \quad K = \frac{CV}{\theta_1\left(1+\frac{\lambda}{2}\right)} \text{ coulomb/radian} \quad \text{... (9.19i)}$$

The resistance R put in series with the galvanometer may be made zero, since when the charge is passed through the galvanometer by making the arm P of the discharge key in contact with the point (a), the condenser (whose resistance is infinity) is put in series with the galvanometer.

If the galvanometer spot of light is deflected by d cm on a scale placed at a distance of D cm from the galvanometer mirror, then $\theta = \frac{d}{2D}$. Hence, we get

$$K = \frac{CV}{\theta_1\left(1+\frac{\lambda}{2}\right)} = \frac{2CVD}{d\left(1+\frac{\lambda}{2}\right)} \text{ coulomb/radian.} \quad \text{... (9.19j)}$$

Knowing C, V, D and d, K can be obtained by determining the value of λ (logarithmic decrement) under the given condition of the experiment. It is to be noted that here λ is the open-circuit logarithmic decrement of the galvanometer.

[Suitable Tables may be made for recording the value of V,c, d. and D. The value of λ may be obtained as discussed under Art No. 7.4 (b)].

(C) Solenoidal method :

Apparatus : Ballistic galvanometer, standard solenoid (primary and secondary coils), Pohl's commutator, ammeter, resistance boxes, battery, tapping key.

The circuit diagram for this method is shown in (Fig. 9.42), where the storage battery (E), the ammeter (A), the resistance box (R_1) and the primary coil (P_1P_2) of the standard solenoid are connected in series through the Pohl's commutator (P). The secondary coil of the solenoid (S_1S_2) is connected in series with a resistance R_2 (more than the corresponding *CDR* of the *B.G.*) and the ballistic galvanometer, having a resistance, say, G.

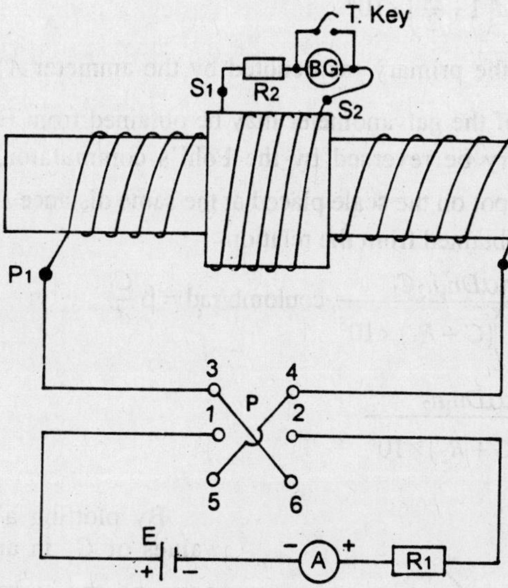

Fig. 9.42 : Solenoidal method

When a current of C_1 ampere (as indicated by the ammeter A) flows through the primary coil, the flux embraced by each turn of the secondary is αH, where α is the mean cross-sectional area of the primary coil (or the secondary coil, as the secondary coil is wound very closely over the primary coil and placed at the middle of the far-long primary coil) and H is the field inside the primary coil at its centre. Hence, the flux embraced by all the turns of the secondary coil is $N = \alpha H n_2$, where n_2 is the total number of turns of the secondary. This flux enters and leaves the secondary as and when the primary current is started or stopped. The charge passing through the ballistic galvanometer on starting or stopping the primary current is

$$q = \frac{\alpha H n_2}{(G+R_2)\times 10^9} \text{ e.m.u.} = \frac{\alpha H n_2}{(G+R_2)\times 10^8} \text{ coulomb}$$

$$\text{or,} \quad q = \frac{\alpha(0.4\pi n_1 C_1)n_2}{(G+R_2)\times 10^8} = \frac{0.4\pi\alpha n_1 n_2 C_1}{(G+R_2)\times 10^8} \text{ coulomb}$$

Since $K = \dfrac{q}{\theta_1\left(1+\dfrac{\lambda}{2}\right)}$ (Eq. 9.19h), we have

$$K = \frac{q}{\theta_1\left(1+\frac{\lambda}{2}\right)} = \frac{0\cdot 4\pi\alpha n_1 n_2 C_1}{\theta_1\left(1+\frac{\lambda}{2}\right)(G+R_2)\times 10^8} \text{ coulomb/rad} \quad \ldots (9.19k)$$

where n_1 is the no. of turns per cm^2 of the primary.

If the spot of light is deflected by d cm on a scale placed at a distance D cm from the galvanometer mirror, than $\theta_1 = \frac{d}{2D}$.

$$\therefore \quad K = \frac{0.8\pi\alpha D n_1 n_2 C_1}{(G+R_2)d\left(1+\frac{\lambda}{2}\right)\times 10^8} \text{ coulomb/rad} \qquad \ldots (9.19\ell)$$

If the current C_1 in the primary (as denoted by the ammeter A) be either started or stopped, the constant K of the galvanometer may be obtained from Eq. (9.19ℓ). But if the current C_1 in the primary be reversed by the Pohl's commutator, then by noting the deflection d' of the light spot on the scale placed at the same distance D cm from the mirror, the value of K would be obtained from the relation,

$$K = \frac{1.6\pi\alpha D n_1 n_2 C_1}{d'\left(1+\frac{\lambda}{2}\right)(G+R_2)\times 10^8} \text{ coulomb/rad} = \beta\frac{C_1}{d'} \qquad \ldots (9.19m)$$

where $$\beta = \frac{1.6\pi\alpha D n_1 n_2}{\left(1+\frac{\lambda}{2}\right)(G+R_2)\times 10^8} \qquad \ldots (9.19n)$$

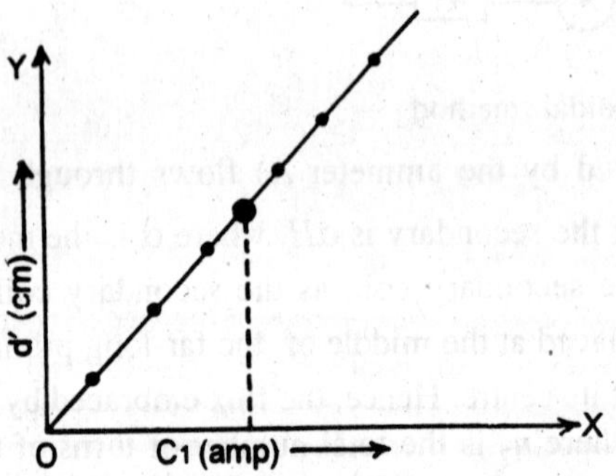

Fig. 9.43 : C_1–d' graph

By plotting a graph with various values of C_1 in ampere, along the X-axis and the corresponding throws d' along the Y-axis, a straight line is obtained. Taking a point on this straight line (Fig. 9.43), the mean value of $\frac{C_1}{d'}$ is found out. Knowing β from the constant of the apparatus, K can be found out from Eq. (9.19m). Again, here λ should be determined under the condition of the experiment and it must be open-circuited log-decrement.

[Suitable tables for noting C_1, d', R_2, etc. may be made. Data for graph of C_1 versus d' may be put in a convenient table.]

Precautions :

(i) Capacitors have the tendency to undergo natural leakage. Moisture and dust increase this tendency. Care thus should be taken on this account, In capacitance method, the time interval between charging and discharging of the capacitor, should be as small as possible.

(ii) Tapping key across the B.G. should be used to bring the galvanometer oscillation to rest quickly.

(iii) In solenoidal method, the value of R_2 must be greater that the C.D.R. of the galvanometer.

(iv) Other usual precautions in handling a ballistic galvanometer, may be taken. (Vide Expt. Nos. 9.10, 9.12, 9.13, 9.15 and 9.18).

[For Questions and Answers, see the abovenoted experiments where a Ballistic Galvanometer is involved.]

9.20 To draw the I-V characteristics of a bridge rectifier (i) without any filter and (ii) using a capacitance input filter

Theory :

A rectifier is a device which can convert a.c. powr to d.c. power. When the *p*-side of a *p-n* junction diode is made positive, then the diode can conduct but it does not practically conduct at all when its *p*-side becomes negative. So, a *p-n* diode may act as a rectifier. A rectifier using four diodes in the form of a bridge (Fig. 9.44) is known as a bridge rectifier.

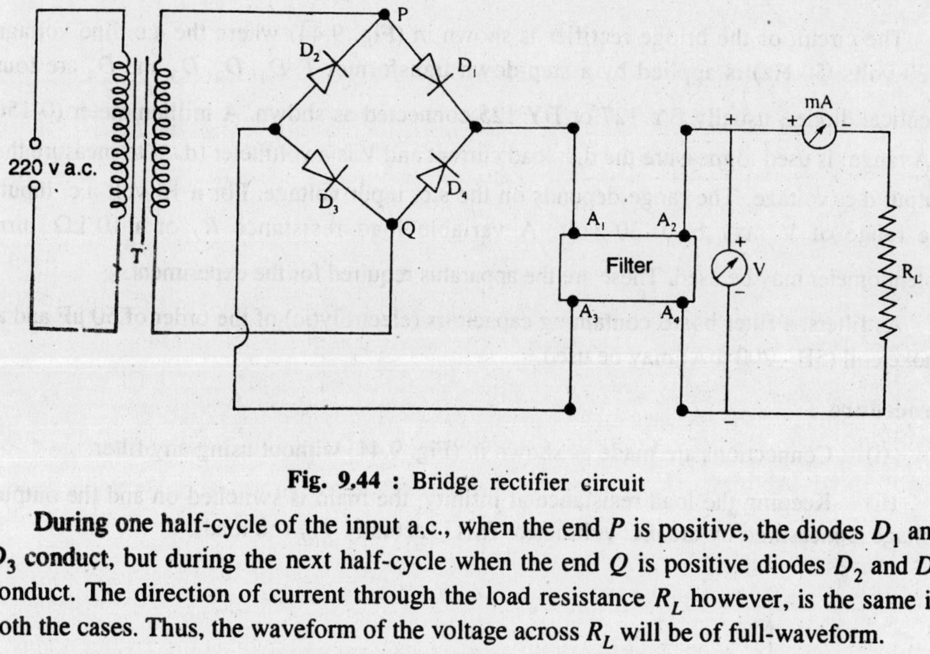

Fig. 9.44 : Bridge rectifier circuit

During one half-cycle of the input a.c., when the end P is positive, the diodes D_1 and D_3 conduct, but during the next half-cycle when the end Q is positive diodes D_2 and D_4 conduct. The direction of current through the load resistance R_L however, is the same in both the cases. Thus, the waveform of the voltage across R_L will be of full-waveform.

The variation of d.c. output voltage (V_L) as a function of d.c. load current (I_L) for a given input voltage, is known as *load regulation curve* or the *characteristic curves*. The percentage regulation (S_L) at any specific load current is given by

$$S_L = \frac{V_{NL} - V_L}{V_L} \times 100, \qquad \text{... (9.20a)}$$

where V_L — load voltage at any specific load current

and V_{NL} — no load voltage (with $I_L = 0$; $R_L = \infty$).

Output obtained from the rectifier is not a smooth d.c. but it contains also some a.c. components called *ripples*. To make the output a pure and useful d.c. source, these ripples are cut off or minimised by using components called the filters. These filters may be either a simple large shunt capacitor or a series inductor or a suitable combination of both of them. The capacitance input filter is shown in (Fig. 9.45). Such *filter* is connected between the terminals A_1, A_2, A_3 and A_4 in (Fig. 9.45) and again the characteristic (load-regulation) curves may be drawn with the filter as in case of using no such filters.

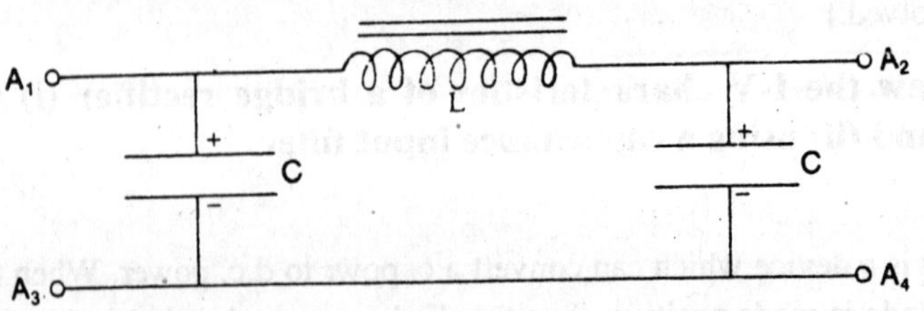

Fig. 9.45 : Filter arrangement

Circuit connection :

The circuit of the bridge rectifier is shown in (Fig. 9.44) where the a.c. line voltage (220 volts, 50 Hz) is applied by a step-down transformer T. D_1, D_2, D_3 and D_4 are four identical diodes, usually BY 127 or BY 125 connected as shown. A milliammeter (0-150 mA range) is used to measure the d.c. load current and V is a voltmeter (d.c.) to measure the output d.c. voltage. The range depends on the a.c. input voltage. For a 15 volt a.c. input, the range of V may be 0-30 volt. A variable load resistance R_L or a 10 kΩ turn potentiometer may be used. These are the apparatus required for the experiment.

As filters, a filter board containing capacitors (electrolytic) of the order of 50 μF and a choke coil (5H - 200 mA) may be used.

Procedure :

(i) Connections are made as shown in (Fig. 9.44) without using any filter.

(ii) Keeping the load resistance at infinity, the main is switched on and the output voltage is measured by the d.c. voltmeter. This is actually V_{NL} (no load).

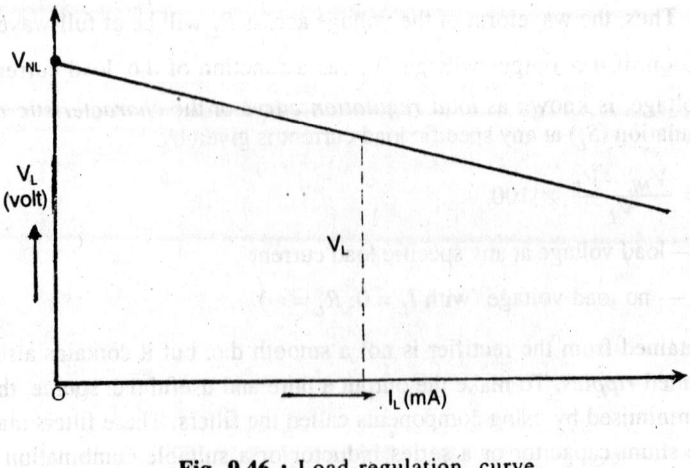

Fig. 9.46 : Load-regulation curve

(iii) Next the load resistance is set at a suitable value such that the milliammeter records a current of about 10 mA. From the voltmeter, the output voltage is measured.

(iv) The load current is then changed in steps of 10 mA by adjusting R_L and in each step the load voltage is measured from V.

(v) The characteristic curves are then drawn by plotting I_L against V_L and will be as shown in (Fig. 9.46).

(vi) Noting V_{NL} and V_L corresponding to a specific value of load current, the percentage regulation S_L may be obtained from Eq. (9.20a).

(vii) The suitable filter is then inserted in the circuit as shown and the steps (ii) to (vi) are repeated.

Results :

(a) Type of diodes use = ...

(b) Maximum permissible current through the diodes = ... A

(c) Maximum permissible reverse voltage across the diodes = ... volt

(d) Input a.c. voltage (fixed) across the secondary of the transformer = ...volt

Table 1

(A) Specifications of Meters :

Meter used	Ranges	One smallest division of the meter	Zero error, if any
d.c. milliammter (mA)	......		
d.c. voltmeter (volt)	......	......	......

(B) Data for Characteristic Curves :

Table 2

No. of Obs.	Load current (I_L) in mA	Load voltage (V_L) (volts)	
		Without filter	With filter
1.	......	......	......
2.	......		
3.	......	......	......
...			......
...	......		
...			
...			
...			
9.	......	......	......
10.	......	......	......

(C) Calculation of percentage regulation :

Table 3

Rectifier	Specific load current (I_L) mA	No load voltage (V_{NL}) in volt	Corresponding load voltage (V_L) in volt	Percentage Regulation $S_L = \frac{V_{NL} - V_L}{V_L} \times 100$
Without Filter	......	......	......	......
With Filter	......	......	......	......

(D) Drawing of graphs :

Characteristic curves for (I-V) of the bridge are drawn on a graph with I_L along X-axis and V_L along Y-axis. The curves will be as shown in (Fig. 9.46). For a specified load current, percentage voltage regulation (S_L) may be obtained from each curve (with filter and without filter) using the formula given in Eq. (9.20a).

Precautions :

(i) The diodes and the meters should be connected with proper polarities.

(ii) Value of R_L should be made infinity before switching on the mains.

(iii) Care should be taken that I_L does not exceed the maximum permissible value.

(iv) During observations, the constancy of the a.c. input voltage must be maintained.

(v) To get better regulation, S_L must be smaller.

Questions and Answers

Q.1. What is a rectifier?

A. : See Theory.

Q.2. What is the function of the filter?

A. : It eliminates the a.c. components (ripples) of the output voltage and makes to get a pure d.c. output.

Q.3. What are the advantages of a bridge rectifier?

A. : Though it is similar to a full-wave rectifier, but in this type, a transformer without a centre tap may be used; a smaller transformer can be used; suitable for high voltage applications.

Q.4. Why not either a simple capacitor or a simple inductor used as a filter? Why do we use both of them together?

A. : Neither a single capacitor (shunted) nor a service inductor in the form of a filter is capable of providing sufficiently low ripples. The capacitor filter has low ripple at low load current while inductor filter has low ripples at high load current. Thus a combination of both of them as filters can give low or less ripples at all possible load currents.

Q.5. What type of capacitance filter is preferred?

A. : Electrolytic capacitors having large capacitances.

Q.6. Rectifier converts a.c. into d.c.? Which device can do the reverse job?

A. : An oscillator may convert d.c. into a.c.

Q.7. Why is rectification needed and what are its practical application?

A. : Because of convenience, a.c. power is usually generated but for many cases, d.c. is required for practical purposes. Hence, rectification process is required for conversion of a.c. into d.c.

9.21 To calibrate a Hall probe with the help of a Ballistic Galvanometer and use it to study the variation of magnetic field of an electromagnet with (i) magnetising current and (ii) the distance between the pole-pieces

The experiment can be performed in two different ways, either by using a 3-pin Hall IC or by using a 4-terminal Hall probe. The two methods will be discussed separately.

Method-I : Using a 3-pin Hall IC

Theory :

[The first part of the Theory is the same as that of (Expt. No. 9.15) and then the following portion is to be added in the Theory.]

In a Hall probe, the Hall voltage (V_H) developed is found to be proportional to the applied magnetic field (H) for a fixed current, i.e.,

$$H = K.\ V_H, \qquad \text{... (9.21a)}$$

where K is a constant. So, if H is measured by using a search coil and a ballistic galvanometer (Expt. No. 9.15) and the corresponding Hall voltage (V_H) is measured by a voltmeter, then the value of the costant K can be obtained. Again, if the Hall voltage is measured for an unknown magnetic field, then the value of the unknown magnetic field can be determined from Eq. (9.21a)

Apparatus :

Solenoid, electromagnet with variable pole-pieces, ballistic galvanometer, search coil, two-way key, Pohl's commutator, battery, resistance boxes, plug key, one digital multimeter, three-pin Hall IC (3501U), ammeter, stabilised power supply.

Method :

[Steps (i) to (x) of (Expt. No. 9.15) are to be followed. In addition to them, the following steps are to be followed as stated.]

(xi) Next the IC Hall probe is connected as shown in (Fig. 9.47). By using a digital multimeter, the zero voltage (field) V_0 is measured. For a 3-pin Hall IC 3501U, this voltage is about 3 volts to 4 volts.

(xii) The Hall probe is then placed between the pole-pieces of the electromagnet with its flat face perpendicular to the direction of the magnetic field. The probe is rotated slightly, till the voltage developed across the probe (V_m) becomes maximum. The Hall voltage is thus ($V_m - V_0$) and it is measured.

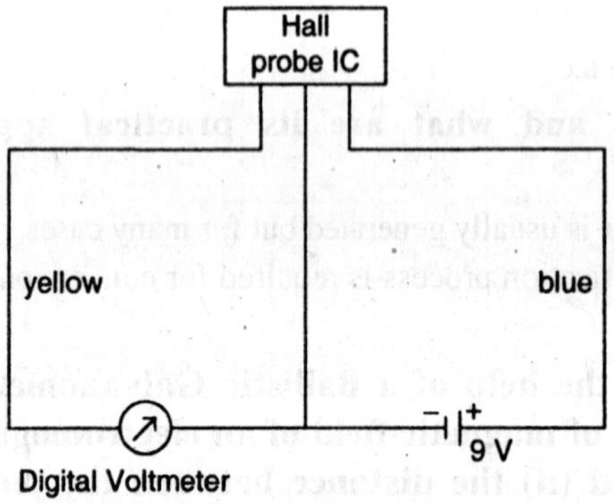

Fig. 9.47

(xiii) The Hall IC is rotated then through 180° in the magnetic field and the voltage across it (V_n) is measured when the change ($V_n - V_0$) is maximum. The Hall voltage $V_H = \frac{1}{2}(V_m - V_n)$ is calculated. It is to be noted that the Hall voltage is maximum when the probe is at perpendicular to the magnetic field.

(xiv) The experiment is repeated and each time, the value of V_H is obtained. The mean value of V_H is thus calculated.

(xv) Finally, knowing the value of H and V_H, the value of K may be had from Eq. (9.21a)

(xvi) The whole process is repeated for two other suitable values of current with the fixed air-gap between the poles. The value of K in each case is determined and hence the mean value of K is obtained.

To study the variation of magnetic field with the magnetising current (I), the air-gap between the pole-pieces is set initially at a fixed value. The electromagnet is demagnetised in the usual manner by reversing repeatedly the primary current before changing the magnetising current.

(i) The Hall probe is then inserted between the pole-pieces, and the Hall voltage (V_H) is measured as before. The magnetic field (H) is calculated by using Eq. (9.21a).

(ii) The magnetising current is increased in suitable steps up to the maximum permissible value and in each step, the Hall voltage is measured and the value of H is obtained as before.

(iii) A graph is drawn then with H versus I and it will be a graph as shown in (Fig. 9.27) [Expt No. 9.15].

(iv) To study the variation of H with the length of the air-gap (l) between the pole-pieces, the magnetising current is kept fixed at a relatively large value. The Hall voltage is measured and the value of H is calculated for at least five different values of air-gap (l).

(v) A graph is then drawn with H versus air-gap (l). It will be a graph similarly to the graph shown in (Fig. 9.26) [Expt No. 9.15]

Results :

(A) Constants of the Apparatus :

[Make Table 1 of Expt. No. 9.15]

(B) Observations of throws (*d*) with search coil :

[Make Table 2 of Expt. No. 9.15]

(C) To find mean value of (I′/d) :

[Make Table 3 of Expt. No. 9.15]

(D) Magnetic fields for different values of I :

[Make Table 4 of Expt. No. 9.15]

(E) Calibration of Hall probe :

Table 5

Zero field voltage (V_0) = ……… volt

Magnetising Current I (ampere)	Corresponding H in Oersted	Voltage across the Hall Probe (volts)		Hall voltage $V_H = \frac{1}{2}(V_m \sim V_n)$ (volt)	Mean V_H (volt)	$K = \frac{H}{V}$ (Oe/volt)	Mean K (Oe/V)
		First position (V_m)	180° rotated position (V_n)				
I_1	H_1	…	……	……	……	……	……
		……	……	……	……	……	
I_2	H_2						
		……	……	……	……	……	
I_3	H_3	……	……	……	……	……	……
		……	……	……	……	……	

(F) Magnetic fields for different magnetising current :

Table 6

K = ……… Oe/volt

Magnetising Current I (ampere)	Voltage across Hall probe (volt)		Hall Voltage $V_H = 1/2(V_m \sim V_n)$ (volt)	Mean V_H (volt)	Magnetic Field (H) $= KV_H$ (Oe)
	First position (V_m)	180° rotated position (V_n)			
0	……	……	……	……	……
	……	……	……		
I_1	……	……	……	……	……
	……	……	……		
I_2	……	……	……	……	……
……	……	……	……		
……	……	……	……	……	
I_6	……	……	……	……	……
	……	……	……		

(G) Magnetic fields for different air-gap :

Table 7

Fixed magnetising current = ... ampere

[Same Table as Table No. 6 is to be made. Only in the 1st column, "Magnetising Current I (ampere)" is to be replaced by "Length of air-gap" in cm"]

Graph drawings :

Graphs similar to graphs shown in (Figures 9.25; 9.26 and 9.27) are to be plotted [Expt. No. 9.15].

Precautions :

(i) The Hall probe should be placed with its surface perpendicular to the direction of the magnetic field, so that Hall voltage becomes maximum.

(ii) Large value of current should not be passed through Hall IC. It may produce Joule heating.

(iii) A potentiometer may also be used to measure Hall voltage instead of the digital voltmeter.

[Other precautions mentioned under Expt. No. 9.15 should be taken.]

Error Calculation :

If the Hall voltages corresponding to an unknown magnetic field (H') and the measured field (H) by the ballistic galvanometer be respectively V'_H and V_H, then

$$H' = KV'_H \text{ and } H = KV_H.$$

Combining them, we get

$$H' = H.\frac{V'_H}{V_H}$$

$$\therefore \quad \left(\frac{\delta H'}{H'}\right)_{max} = \left(\frac{\delta H}{H}\right)_{max} + \frac{\delta V_H'}{V_H'} + \frac{\delta V_H}{V_H}$$

Using Eq. (9.15d),

$$\left(\frac{\delta H'}{H}\right)_{max} = \frac{\delta d}{d} + \frac{\delta I'}{I'} + \frac{\delta d_1}{d_1} + \frac{\delta V_H'}{V_H'} + \frac{\delta V_H}{V_H}$$

Here, $\delta d = 0.1$ cm (smallest division of scale)

$\delta d_1 = 1$ smallest div. of graph.

$\delta I' = 1$ smallest div. of ammeter

$\delta V'_H = \delta V_H$ = smallest measurable voltage.

From these values, the minimum error in the measurement of H' in one set can be obtained.

Method-II : Using a 4-terminal Hall probe

Theory :

[Same as in Method-I]

Apparatus :

Electromagnet, solenoid, ballistic galvanometer, ammeter, search coil, digital multimeter, steady power supply (current 0-10 mA), one 4-terminal Hall probe (Ge crystal)

Method :

[Steps (i) to (x) of Method -I are to be followed.]

(xi) The Hall probe is then connected as shown in (Fig. 9.48). The current leads (C_1, C_2) are connected to a constant current source through a milliammeter (0-10 mA) and the potential leads (P_1, P_2) to a digital voltmeter (0-200 mV). The current through the probe is suitably adjusted by using the knob of the current source. This current must be less than the maximum permissible current through the probe. This value of the current must be kept constant all through the experiment. The zero adjustment is made on the probe screw to

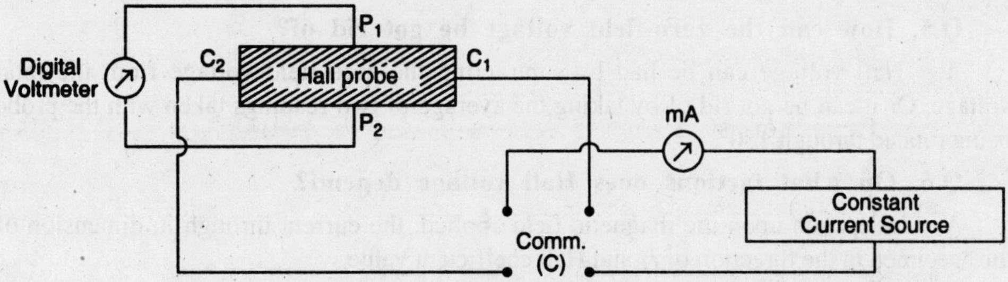

Fig. 9.48

make the zero-field voltage minimum. This minimum voltage value is recorded (V_0) from the voltmeter. This value is usually very small (< 1 mV).

(xii) The rest of the steps of Method I are followed thereafter.

Results :

[Results are to be tabulated same as in Method - I]

Only the following things are to be recorded.

(a) Fixed current through the Hall probe = ... mA

(b) Maximum allowable current through the probe = ... mA

Precautions :

[The same as in Method - I]

Other precautions for this Method are :

(i) Hall voltage measurements may be done carefully by changing the direction of the current through the probe.

(ii) Current through the Hall probe should be such that the Hall voltage developed must be appreciably measurable but care should be taken that it should never exceed the maximum permissible value prescribed.

Error Calculation :

[Same as in Method - I]

Questions and Answers

Q.1. What is Hall effect?

A. : When a current carrying conductor is placed perpendicular to a magnetic field, a potential difference is developed across the conductor and at right angles to both the directions of current flow and the magnetic field. This phenomenon is known as 'Hall Effect'.

Q.2. How does the 'Hall voltage' developed?

A. : See any textbook.

Q.3. Why one gets a voltage even at zero magnetic field in a 3-pin Hall IC?

A. : In this case, the initial voltage arises due to the ohmic voltage drop across the conductor.

Q.4. What is Hall coefficient?

A. : It is given by the potential gradient developed with applied unit magnetic field when unit current flows through the specimen.

Q.5. How can the zero-field voltage be got rid of?

A. : Hall voltage can be had by subtracting the zero-field voltage from the total voltage. Or it can be got rid of by taking the average of two readings taken with the probe being rotated through 180°.

Q.6. On what factions does Hall voltage depend?

A. : It depends upon the magnetic field applied, the current through it, dimension of the specimen in the direction of H and Hall coefficient value.

Q.7. Why one gets zero-field voltage in a 4-terminal Hall probe?

A. : It is due to the imperfect alignment of the four contact of the probe.

Q.8. For a given magnetic field, can the Hall voltage be increased?

A. : Yes, by increasing the strength of the current through the probe.

Q.9. What is the advantage of using a Hall probe over the search coil method of measuring the magnetic field?

A. : Since the Hall probe is small in size and easier to handle, it is convenient to use it and even to map a non-uniform magnetic field.

Q.10. What happens when the current direction through the probe is reversed?

A. : The sign of Hall voltage will then be reversed.

Q.11. Can the Hall coefficient be measured by this method.

A. : Yes, if the thickness t of the Hall probe is known, then from the relation R_H (Hall coeff.) $= \dfrac{V_H \times t}{I \times H}$.

Q.12. In a Hall probe, what type of materials are used?

A. : Semiconducting materials like Ge, In, As etc. metals are not used, for in them Hall voltage developed is very low.

Q.13. What are the advantages of a 3-pin Hall IC over the *A*-terminal probe?

A. : In a 3-pin IC, output voltage is relatively high and there is no problem for alignment of the terminals. It is also cheap and readily available.

Q.14. Can the Hall probe be used for any other purposes?

A. : Yes, it can be used to determine the sign, mobility and concentration of charge carriers.

Chapter 10

Electronics

10.1 Drawing the characteristic curves of a semiconductor diode and to study it as half-wave and full-wave rectifiers

Theory :

When a *p*-type and a *n*-type semiconductors are placed in contact, then this *p-n* combination is called a *semiconductor diode.* These diodes, like an electronic valve, can rectify an alternating current into a *d.c.* and hence the name (Fig. 10.1).

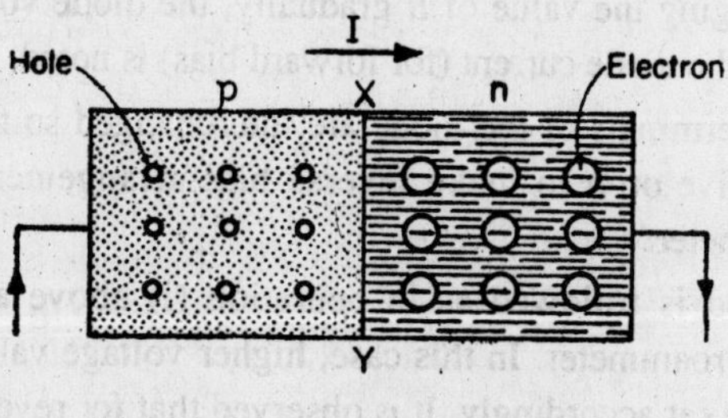

Fig. 10. 1 : Semiconductor diode

In the connections made as shown in (Fig. 10.2), positive potential will be applied on *p*-conductor and negative potential on *n*-semiconductor. This arrangement is known as the *forward bias.* Under current. When the diode terminals are interchanged (as shown by the dotted lines in the figure), then negative potential is applied on *p*, and positive on *n*. This arrangement is known as the *reverse bias.* Here, the amount of current through the diode is *very small* and hence a microammeter (μA) is used in place of the milliammeter.

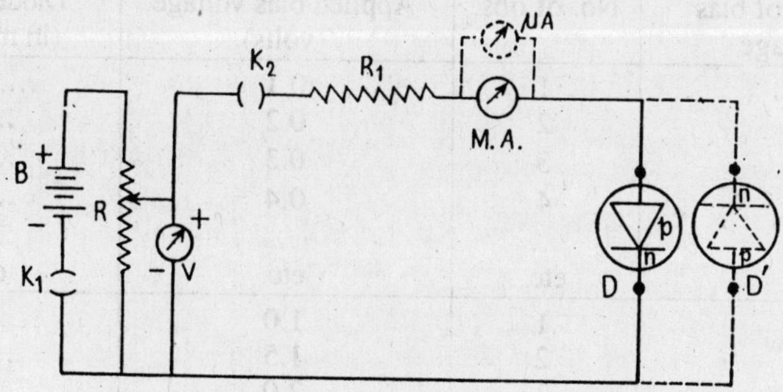

Fig. 10.2 : Characteristic of a diode determination

In the forward bias arrangement, the relation between the current flowing (I) and the diode voltage (V) is given by

$$I = I_0 \left[\exp.\left(\frac{eV}{KT}\right) - 1\right] \quad \text{.. (10.1a)}$$

where I_0 = reverse saturation current, K =Boltzmann's constant, T =absolute temperature.

The diode voltages (V), both positive and negative, are plotted along the X-axis and the diode currents (I), both forward and backward, along Y-axis. The curves obtained are known as the characteristic curves (Fig. 10.3).

Apparatus :

Semiconductor diode, milliammetre/microammeter, commutator, high impedence voltmeter (3V range), rheostat, resistance (about 100 ohm, range); key and battery.

Method :

(i) Connections are made as shown in (Fig. 10.2). The value of R is so adjusted that the voltmeter reads about 0.1 volt. Readings for forward bias current are noted from the milliammeter. By changing the value of R gradually, the diode voltage is changed step by step, and for each step, the diode current (for forward bias) is noted.

(ii) Next, the terminals of the diode are interchanged so that negative potential is applied on p and positive on n to make reverse bias arrangement. The milliammeter is replaced by a microammeter.

(iii) The method is followed as in operation (i) above and the weak current is measured from the microammeter. In this case, higher voltage values will be required and the value of R has to be set accordingly. It is observed that for reverse bias voltage value of I volt and above, the diode-current becomes saturated, and this is known as *reverse saturation current.*

(iv) In case of forward bias, the diode-current is very small for bias voltage up to 0.3 volt and for onward values of bias voltage, the current begins to increase exponentially according to Eq. (10.la)

Results :

The diode used :

Table 1

Nature of bias voltage	No. of obs.	Applied bias voltage (volts)	Diode-current (in mA/μA)
Forward	1	0.1	...... ma
	2	0.2	......
	3	0.3	......
	4	0.4	......
	...	...	......
	etc.	etc.	etc.
Reverse	1	1.0	...... μA
	2	1.5	......
	3	2.0	......
	4	...	......
	...	...	......
	...	...	......
	etc.	etc.	etc.

Drawing of graph

On the graph, forward bias voltage is plotted along OX and reverse bias voltage along OX_1. Similarly, forward bias current is plotted along OY and reverse bias current along OY_1. Current along OY is expressed in milliampere while that along OY_1 is expressed in microampere. The nature of the curves, known as the *characteristic curves* is shown in Fig. 10.3.

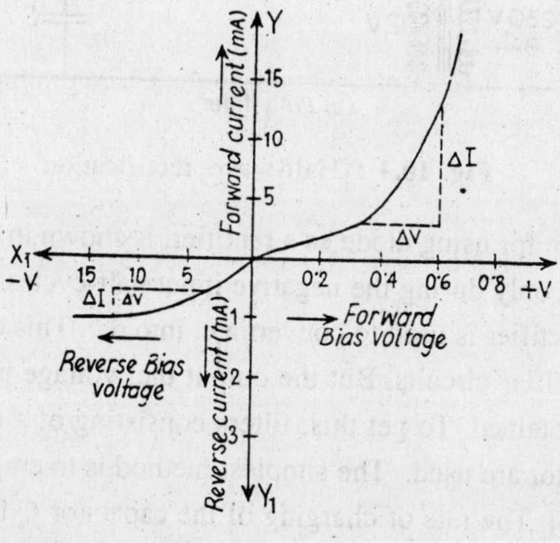

Fig. 10.3 : Characteristic curves of a diode

From this graph, the values of forward and reverse resistances of the diode can be had by measuring ΔV and ΔI for both the bias voltages, as shown in the figure.

Precautions :

(i) A rheostat R_1 must always be used in series with the diode, so that direct voltages may not be incident on the diode. This may damage the diode.

(ii) For forward bias, the voltage applied should not exceed 1.5 volt under any circumstances. This limit has to be controlled by the rheostat.

(iii) The shielding should be perfect and there should not be any air-gap in shielding. A small air-gap may have a very high resistance.

(iv) Very high voltages also should not be applied in case of reverse bias, which may also damage the diode.

(v) The static characteristic curve in the reverse bias condition has not any saturation shape because of the leakage of charge from the diode surface.

Half-wave and Full-wave Rectification

The diode conducts only when *p*-type potential is positive with respect to the *n*-type. For its negative potential, the current passing is almost zero. So, if an alternating potential is connected across the *p* and *n* side of a diode, then current will pass through it only during the positive half-cycle but not during the negative half-cycle. Thus the diode acts as a rectifier.

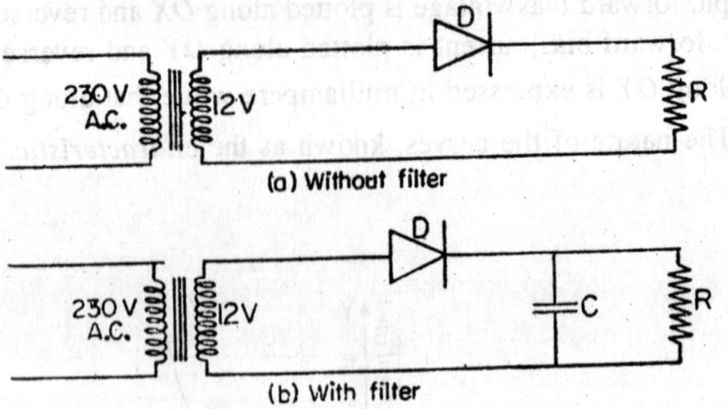

Fig. 10.4 : Half-wave rectification

The circuit diagram for using diode as a rectifier, is shown in (Fig. 10.4). This rectifier curcuit gives an output only during the negative input half-cycles. This is thus a half-wave rectifier circuit. The rectifier is used to convert a.c. into d.c. This can be done, in fact, with half and full-wave rectifier circuits. But the output d.c. voltage being pulsating in nature, steady supply is not obtained. To get this, filters consisting of a capacitor or combination of capacitor and inductor are used. The simplest method is to employ a single capacitor as shown in [Fig. 10.4(b)]. The rate of charging of the capacitor C from the rectifier must be greater than its rate of discharge through the load R for satisfactory smoothing to be obtained. Then the capacitor C will charge up to almost the maximum potential during the positive half-cycle but during the subsequent half-cycle, when the rectifier output drops to zero and then rises again, the potential across the capacitor will fall exponentially due to its discharge through the load. This variation of potential is called the *ripple voltage*. The nature of the output waveforms with filter and without filter, are shown in (Fig. 10.5).

Let the mean steady potential across C is V_c and the total variation of V_c is ΔV_c. This ΔV_c is the *ripple voltage*. The change of charge on the capacitor for a potential difference variation of ΔV_c is $C.\Delta V_c$. This equals $I_\ell.T$ where I_ℓ is the discharge current through the load and T the time of one half-cycle.

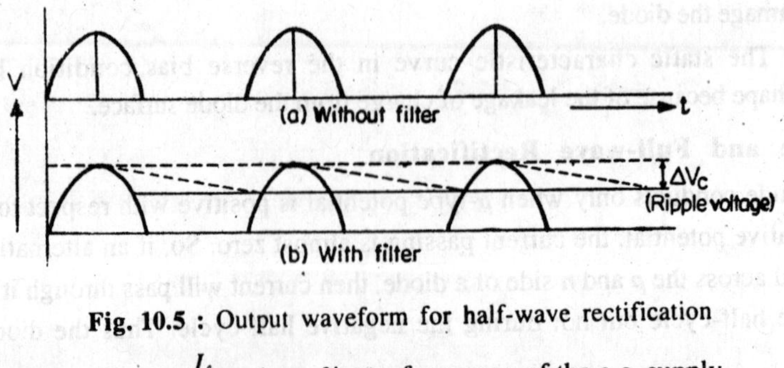

Fig. 10.5 : Output waveform for half-wave rectification

$$\therefore \quad \Delta C.\Delta V_c = I_\ell.T = \frac{I_\ell}{2f}, \text{ where } f \text{ is the frequency of the a.c. supply.}$$

But $I_\ell = \frac{V_c}{R}$

$$\therefore \quad \Delta V_c = \frac{I_\ell}{2fC} = \frac{V_c}{2fCR}$$

The percentage ripple is given by

$$\frac{\Delta V_c}{V_c} \times 100 = \frac{100}{2fCR}$$

The ripple voltage or current is measured by the ripple factor (γ) defined as

$$\gamma = \frac{\text{effective value of alternating component of voltage or current}}{\text{direct value of voltage or current}}$$

In modification of the circuit diagram shown in (Fig. 10.4), if two diodes D_1 and D_2 are connected as shown in (Fig. 10.6) and an a.c. Voltage is applied, then for the positive half-cycle the diode D_1 will be conducting, while the diode D_2 would not conduct since it is reverse biased at that time. For the negative half-cycle, the diode D_1 will be non-conducting and D_2 will conduct. So, output voltage for both positive and negative half-cycles are obtained. The nature of the output voltages are shown also in (Fig. 10.7), for without filter and with filter.

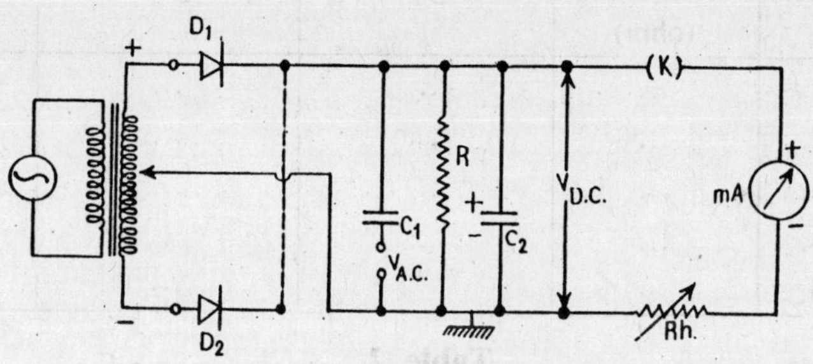

Fig. 10.6 : Full-wave rectification

When a capacitor C_2 is connected in addition to the circuit in parallel to the load resistance (Fig. 10.6), it will become charged during the conduction period and deliver the charge through the load resistance during the non-conduction period. In this way, the time during which current passes through the load, is prolonged and the ripple is considerably decreased. The nature of the output voltage in this case is shown in (Fig. 10.7).

Regulation :

The variation of d.c. output voltage as a function of the d.c. load current is know as *regulation*. The percentage regulation is given by

$$\% \text{ regulation } R = \frac{V_{NL} - V_L}{V_L} \times 100\%, \quad \text{.. (10.1b)}$$

where *NL* implies no load, *i.e.*, zero load current. Thus V_{NL} represents the output voltage when the output is on open circuit or R_L is infinite. The V_L represents the normal load voltage when load resistance is finite. For an ideal regulator, the output voltage is independent of the load and the percentage regulation is zero.

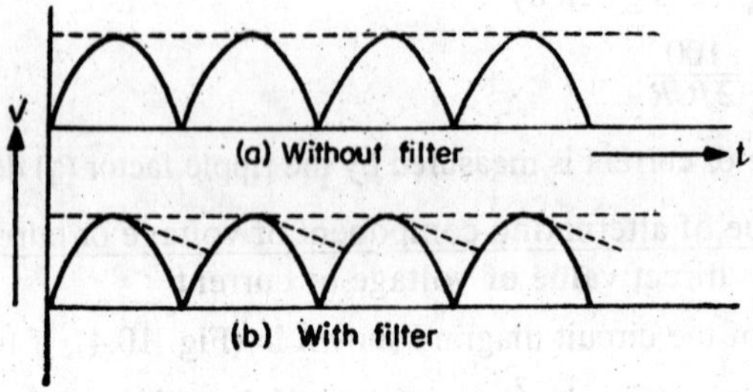

Fig. 10.7 : Output waveform for full-wave rectification

Results :

(A) For half-wave rectification :

(i) **Without filter**

Table 1

Diode used	No. of Obs.	Value of load resistance (ohm)	Input a.c. voltage (volt)	Output d.c. voltage (volt)	Frequency of a.c. supply (Hz)
	1.	...	...	...	
	2.	...	...	...	
...	3.	...	...	...	......
	4.	...	...	...	
	5.	...	...	...	

Table 2

(ii) With Filter (See next page)

(ii) With Filter

Table 2

Diode used	No. of Obs.	Value of load resistance (ohm)	Value of filter (C) (μF)	Value of a.c. voltage (volt)	Value of output d.c. voltage (volt)	Value of ripple voltage (ΔV_c volt)	Percentage ripple $\frac{\Delta V_c}{V_c} \times 100$ (volt)	Ripple factor (γ)
	1.	...	...	...	...	...	...	...
	2.	...	...	...	...	...	...	...
	3.	...	...	...	...	...	...	...
	4.	...	...	...	...	...	...	...
	5.	etc.	etc.	etc.	etc.	etc.	etc.	etc.
	6.							

Table 3

Determination of % Regulation for half-wave rectifier

	No load voltage V_{NL} in volt	Given load current (mA)	Corresponding load voltage V_L in volt	Percentage Regulation $R = \frac{V_{NL} - V_L}{V_L} \times 100\%$
Without filter	......	......	......	......
With filter	......	......	......	

(B) For full-wave rectification :

Table 4

[Same table as **Table 1,** for without filter.]

Table 5

[Same table as **Table 2,** for with filter.]

Table 6

[Same table as **Table 3**]

Discussion :

(i) The ripple in full-wave rectifier is small compared to half-wave rectifier. Due to high ripple in half-wave rectifier, an expensive smoothing filter system is required.

(ii) The ripple voltage is small when high capacitacne filter in the rectifier circuit is used.

(iii) Since the output of the rectifier circuit is low, the transfer efficiency is also low.

Questions and Answers

Q.1. What is the difference between *p*-type and *n*-type of semiconductors?

A. : See any textbook.

Q.2. Why are they called 'semiconductors'?

A. : Since there exists current during forward bias but not during the reverse bias.

Q.3. If there be no current during reverse bias, then wherefrom does the diode-current come?

A. : This is the leakage current through the diode whose value is very small and of the order of few microamperes.

Q.4. Why is a filter used in the rectifier circuit?

A. : Rectifiers are used to convert a.c. into d.c. But the output d.c. voltage being pulsating in nature, comprising of a mixture of a.c. and d.c. voltages, the a.c. part of the voltagee is allowed to go through the capacitor while the d.c. fraction of the output voltages is blocked by the capacitor. Thus, the capacitor filters the output voltage and hence it is known as a 'fillter'.

10.2 Drawing characteristic curves for a Zener diode and to study it as a voltage regulator

Theory :

Diodes which are designed with adequate power dissipation capabilities to operate in the breakdown region, are known as Zener diodes. These may be employed as voltage reference on constant voltage devices, and hence they may be used as voltage regulators.

These have unique reverse-bias current and voltage characteristics. When the diode is forward biased, it acts like a closed switch, and forward current increases with an increase of applied voltage and the former is limited by the parameters of the circuit. When the diode is reverse biased, a small current I_s, called *saturation current,* flows and which remains relatively constant despite an increase in reverse bias, until zener breakdown region in the vicinity of zener voltage V_z, is reached. Here, reverse current starts rising rapidly because of avalanche effect. Finally, zener breakdown occurs when the zenear voltage V_z is reached. The current-voltage curves of a zener diode, for forward and reverse bias, are known as its *characteristic curves.*

Zener diodes are used as voltage regulators and as voltage reference standards. The diode maintains a constant voltage across a load resistance, within required limits.

For drawing the characteristic curves, the supply voltage and resistance are so selected in the circuit that initially, the diode may be operating in the breakdown region. Plotting the diode voltages along the *X*-axis and the diode currents along the *Y*-axis of a graph, the characteristic curves, both for forward bias and reverse bias, are obtained.

Apparatus :

Regulated power supply, zener diode, milliammeters, voltmeters, switch and resistors.

Circuit connections :

The circuit connections for studying the characteristic curves of a zener diode is shown in (Fig. 10.8). The circuit is drawn for the forward-bias arrangement. For reverse-bias, keeping everything in their positions, only the diode connection has to be reversed in the circuit. *E* is a variable power supply and *S* is a switch. *R* is a resistance of the order of 500 ohms. The diode current has to be measured by a microammeter or a milliammeter as and when necessary.

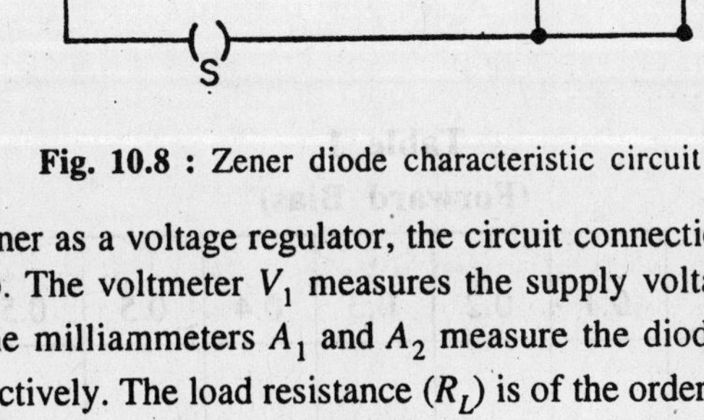

Fig. 10.8 : Zener diode characteristic circuit

For using the zener as a voltage regulator, the circuit connections have to be made as shown in (Fig. 10.9). The voltmeter V_1 measures the supply voltage while V_2 measures the load voltage. The milliammeters A_1 and A_2 measure the diode current and the total circuit currents respectively. The load resistance (R_L) is of the order of 3.3 kilo-ohms while the series resistance *R* is of the order of 500 ohms.

Method :

(i) For forward bias study, after making the connections properly, the switch *S* is kept open. *E* is set at zero volt initially. Closing the switch, the diode current I_Z is measured wih *E* set at zero, if there be any current.

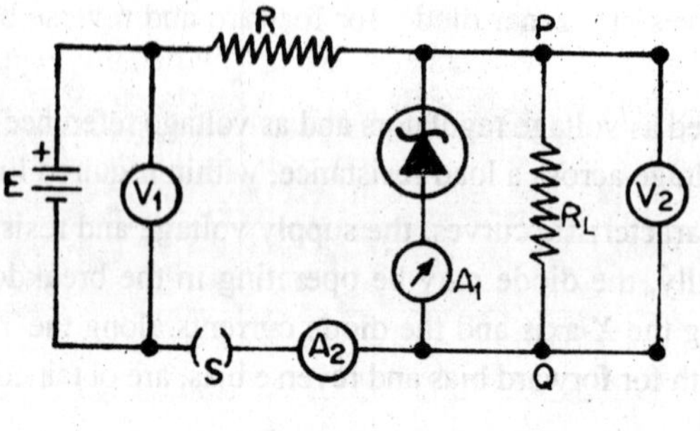

Fig. 10.9 : Voltage regulator circuit

(ii) The value of E is gradually changed and the forward current in the diode, at each level of diode voltage recorded by V, is measured. The diode voltages are changed at intervals of 0.1 V and corresponding diode currents are measured for each step and they are tabulated. From the data the forward resistance of the diode can be computed from the relation $R_F = \frac{V}{I_Z}$.

(iii) For reverse bias, keeping others in their respective positions, the diode is reversed in the circuit. Similarly, as in case of forward bias, the operations are repeated.

(iv) The initial diode voltage is kept at 2.0 volts and for change of the voltages at intervals of 0.4 volt, diode currents are measured in each case. The measurements are to be made till the zener voltage V_z is reached.

(v) Finally, graphs are plotted with diode currents (I_z) against diode voltages (V), both for forward and reverse bias. These graphs are characteristic curves as shown in (Fig. 10.10).

(vi) From the value of I_Z and V for the diode in the reverse bias arrangement, the value of the resistance (R_Z) of the diode can also be calculated.

Results :

The diode used : ....

Table 1
(Forward Bias)

Diode voltage (V volt)	0	0.1	0.2	0.3	0.4	0.5	0.55	0.60	0.65
Diode current (I mA)	...	...	...	...	...	...	...	...	...
Diode Resistance (R_Z ohm)	...	...	...	...	...	...	...	...	...

Table 2
(Reverse Bias)

No. of Obs.	Diode voltage (V volt)	Diode curent (I_Z mA)	Forward resistance (R_F ohm)	No. of Obs.	Diode voltage (V volt)	Diode current (I_Z mA)	Forward Resistance (R_F ohm)
1.	0	...	...	7.	4.0	...	...
2.	2	...	...	8.	4.4	...	...
3.	2.4	...	...	9.	4.8	...	...
4.	2.8	...	...	10.	5.2	...	...
5.	3.2	...	...	11.	...	...	...
6.	3.6	...	...	etc.	V_Z	...	...

Voltage Regulation

For using the zener diode as a voltage regulator, the circuit connections are made as in (Fig. 10.9). The switch is kept open and the output of the power supply E is set at zero volt. The milliammeters are kept on the 100 mA range.

The switch (S) is closed and slowly the supply voltage is increased until current through the diode (I_Z) measures 20 mA. Voltage are measured by V_1 and V_2. The total current is also measured (I_T) by A_2. The observations are tabulated.

The range of variation of E over which the voltage across the load as determined by V_2, remains constant within $\pm$ 0.1V of its value is determined and recorded. The variation of I_Z and I_T within this range, are recorded.

Table 3
(Voltage Regulation)

Voltage across load PQ (V_{PQ} volt)	Diode current (I_Z mA)	Total current (I_T mA)	Supply voltage (E volt)
V_{PQ}			
$V_{PQ} + 0.1$			
$V_{PQ} - 0.1$			

Drawing of Graphs

From data recorded in Table 1 and Table 2, a graph of diode current (I_Z) is plotted along Y-axis and the diode voltage (V) along X-axis. The nature of the graph will be as shown in (Fig. 10.10). An expanded graph of diode current versus voltage in the zener region is drawn. The value of the *zener voltage* (V_Z) is obtained from the graph.

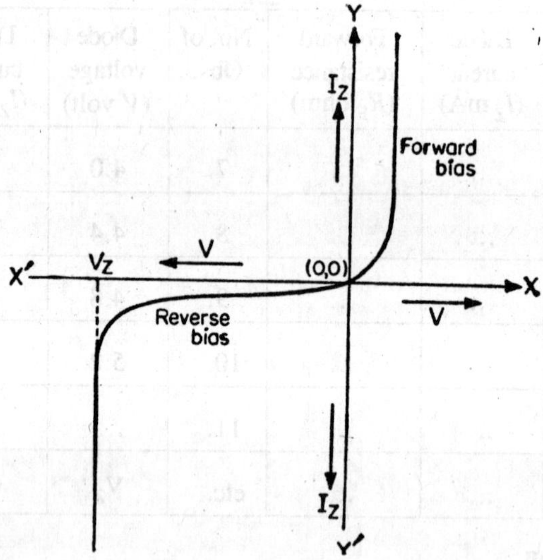

Fig. 10.10 : Characteristics curves of a zener diode

Discussions :

(i) In case of an ideal zener diode, the voltage for the knee of the reverse bias characteristic curves should be about thirteen times greater than the voltage for the knee of the forward bias characteristic curves.

(ii) In the zener breakdown region, small voltage changes cause large current changes. Obviously, there are dramatic changes in effective resistance at the *p-n* junction in this region.

(iii) Zener breakdown occurring at the zener voltage (V_Z), need not result in the destruction of the diode. As long as the current through the diode is limited by the external circuit to a level within its power-handling capabilities, the diode functions normally. Further, by reducing reverse bias below the zener voltage, the diode can be brought out of its breakdown level and restored to the saturation-current level.

(iv) The process of switching the diode between its zener and non-zener current states can be repeated again and again without damaging the diode. It should be noted, however, that there is a certain time lag, called recovery time, in switching the diode from one state to the other.

[For Questions and Answers on this experiment, see the theory, method and discussions above.]

10.3 To draw the static characteristics of a triode valve and to determine the parameters

Theory :

If the anode of a triode valve be kept at a constant positive potential with respect to the filament, and different voltages, both negative and positive, are applied at the grid, then the plate-current will vary with the grid potential. Curves representing the relation between the plate-current and the grid-potential for a fixed value of the plate-potential and also relation between the plate-current and the plate-voltage for a given value of grid-voltage, are known as the *static characteristić curves* of the valve. These curves furnish useful informations about the performance of the valve under different conditions.

By the term *parámeters* or *constants* of a valve are meant its

(i) amplification factor (μ)

(ii) plate or internal resistance (R_p)

(iii) mutual conductance (g_m).

The amplification factor (μ) of a valve is defined as the ratio of the change in plate-voltage (δV_a) required to produce a small change in plate-current to the change in grid-voltage (δV_g) required to produce the same change in the plate-current.

i.e., amplification factor $(\mu) = \left(\frac{\delta V_a}{\delta V_g}\right)_{I_a}$ .. (10.3a)

The plate-resistance (R_p) of a valve is defined as the ratio of a small change in plate-voltage to the small chage in plate current it produces, grid-voltage remaining the same.

i.e., plate-resistance $(R_p) = \left(\frac{\delta V_a}{\delta I_a}\right)_{V_g}$ .. (10.3b)

The *mutual conductance* (g_m) of a valve is defined as the ratio of the change in plate-current to the change in grid-voltage which produces it, the plate-voltage remaining constant.

i.e., mutual conductance $(g_m) = \left(\frac{\delta I_a}{\delta V_g}\right)_{V_a}$ .. (10.3c)

Plotting graphs of grid-voltages against anode-currents and anode-voltages against anode-currents, the constants of the valve can be determined.

Apparatus :

The triode valve, two voltmeters, three rheostats, ammeter, milliammeter, plug keys, battery source (24 volts), battery (6 volts), high tension mains.

Circuit connections :

The circuit connections are made as shown in (Fig. 10.11). The polarities of the high tension mains and those of the batteries should be properly and carefully connected.

Method :

(i) The anode voltage may be started from 90 volts and negative voltage is applied to the grid. The grid-voltage may be changed by step of 2 or 3 volts.

(ii) The anode voltage should be set at 90, 100 and 110 volts successively and for each setting, the anode current (I_a) is measured for different values of applied grid-voltages (V_g), starting from negative and passing to positive values through zero value. Hence, for three values of anode voltage (V_1, V_2 and V_3), three sets of readings are obtained, corresponding to which, three graphs are drawn (Fig. 10.12).

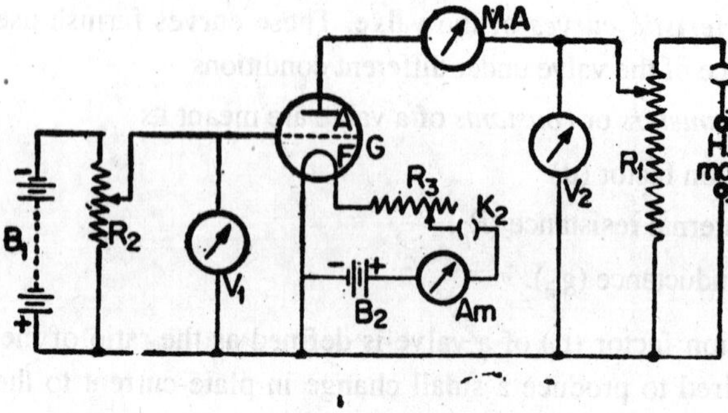

Fig. 10.11 : Circuit for drawing static characteristics

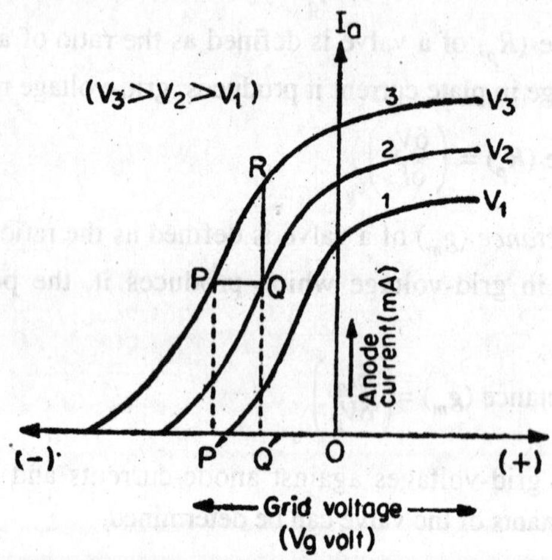

Fig. 10.12. : Static characteristics (V_g-I_a)

(iii) Next, by applying a fixed grid voltage (V_g), if different voltages are applied at the anode, then the anode current will also change. In this way, applying three different fixed values of grid-voltages (V_{g1}, V_{g2}, V_{g3}); the values of anode-voltage and the corresponding anode-curents are to be observed for each value of the applied fixed grid-voltages. The values of anode-currents corresponding to anode-voltages are plotted on a graph, corresponding to each of the fixed grid-voltages. Thus, a set of three curves (Fig. 10.13) are obtained.

(iv) From these curves, the values of μ, R_p and g_m of the valve, are determined.

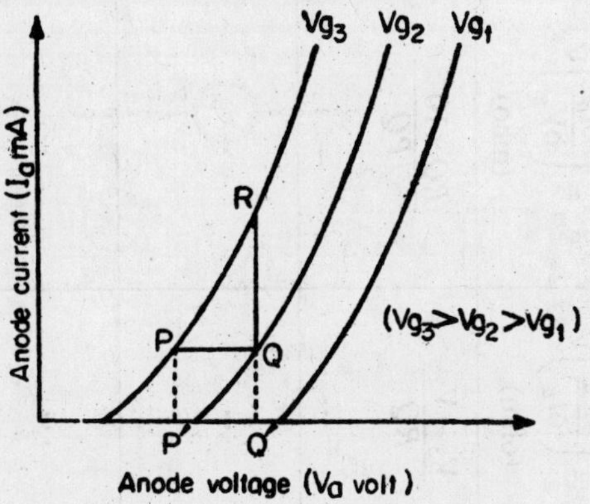

Fig. 10.13 : Static chracteristics (V_a-I_a)

Results :

Triode valve used : ......

(A) Observations for drawing (V_g-I_a) curve :

Table 1

Filament current = ... ampere

No. of Obs.	Anode voltage V_1 = ... volt		Anode voltage V_2 = ... volt		Anode voltage V_3 = ... volt	
	Grid voltage (V_g volt)	Anode current (I_a mA)	Grid voltage (V_g volt)	Anode current (I_a mA)	Grid voltage (V_g volt)	Anode current (I_a mA)
1.	...	...	...	...	...	...
2.	...	...	...	...	...	...
3.	...	...	...	...	...	...
4.	...	...	...	...	...	...
5.	...	...	...	...	...	...
6.	...	...	...	...	...	...
7.	...	...	...	...	...	...

(B) Observations for drawing (V_e–I_a) curve :

Table 2

Filament current = ... ampere

No. of Obs.	Grid voltage V_{g_1} = ... volt		Grid voltage V_{g_2} = ... volt		Grid voltage V_{g_3} = ... volt	
	Anode voltage (V_a volt)	Anode current (I_a mA)	Anode voltage (V_a volt)	Anode current (I_a mA)	Anode voltage (V_a volt)	Anode current (I_a mA)
1.	...	...	...	...	...	...
2.	...	...	...	...	...	...
3.	...	...	...	...	...	...
4.	...	...	...	...	...	...
5.	...	...	...	...	...	...
6.	...	...	...	...	...	...
7.	...	...	...	...	...	...
8.	...	...	...	...	...	...

(C) Determination of the valve constants :

[From (V_g-I_a) curve]

Table 3

From the graphs	Change in V_a (δV_a volt)	Change in V_g (δV_g volt)	Change in I_a (δI_a volt)	$\mu = \left(\frac{\delta V_a}{\delta V_g}\right)_{I_a}$	$R_p = \left(\frac{\delta V_a}{\delta I_a}\right)_{V_g}$ (ohm)	$g_m = \left(\frac{\delta I_a}{\delta V_g}\right)_{V_a}$ (mho)
3 and 2	$(V_3 - V_2)$ = ...	PQ = ...	RQ = ...	$\frac{V_3 - V_2}{PQ}$ = ...	$\frac{V_3 - V_2}{RQ}$ = ...	$\frac{RQ \times 10^{-3}}{PQ}$ = ...
2 and 1	...	...	...	...	...	...
3 and 1	...	...	...	...	...	...

[From (V_a-I_a) curve]

Table 4

From the graphs	Change in V_g (δV_g volt)	Change in I_a (δI_a ampere)	Change in V_a (δV_a volt)	$\mu = \left(\frac{\delta V_a}{\delta V_g}\right)I_a$	$R_p = \left(\frac{\delta V_a}{\delta I_g}\right)V_g$ (ohm)	$g_m = \left(\frac{\delta I_a}{\delta V_g}\right)V_a$ (mho)
3 and 2	$(V_{g3} - V_{g2})$ = …	$RQ \times 10^{-3}$ = …	PQ = …	$\frac{PQ}{(V_{g3} - V_{g2})}$ = …	$\frac{PQ}{RQ} \times 10^3$ = …	$\frac{RQ \times 10^{-3}}{V_{g3} - V_{g2}}$ = …
2 and 1	…	…	…	…	…	…
3 and 1	…	…	…	…	…	…

Hence,

Mean value of amplification factor (μ) = ……

Mean value of plate-resistance (R_p) = …… ohm

Mean value of multual conductance (g_m) = …… mho

Precautions :

(i) The filament current must remain steady during the experiment.

(ii) For calculations, the straight portions of the curves have to be taken into consideration.

(iii) During connections, the polarities of the battery and the high tension source, must be properly connected.

(iv) Grid-bias voltage should not be very high (preferably less than 16 volts).

[**Questions and Answers** on **triode valve** are given at the end of the next experiment together.]

10.4 To draw the dynamic characteristics of a triode valve and to find its voltage gain

Theory :

The characteristic curves of a triode valve show how the anode-current varies with the two independent variables, namely the plate-voltage and grid-voltage. In an amplifier circuit, these two variables are no longer independent. When the plate circuit is connected through a high resistance R_L (called the load resistance), the plate-voltage (V_a) and the anode current (I_a) are related by the equation,

$$V_a = E - R_L.I_a \qquad \text{... (10.4a)}$$

A curve showing the relationship between the anode current (I_a) and grid-voltage (V_g) under such a condition is known as the *dynamic characteristic*.

From the characteristics of a triode valve, it is seen that when its operating point is on the straight portion, a change in potential difference between the grid and the cathode produces an amplified change in the potential difference between the anode and the cathode. Thus a triode valve operating at the straight portion of the curve can be made to act as a voltage amplifier. The voltage gain across the given load resistance (R_L) is given by

$$\alpha = \frac{\delta V_a}{\delta V_g} \qquad \text{.. (10.4b)}$$

where δV_a is the change in voltage across the load due to a change of voltage δV_g across the grid. If R_p is the internal impedance of the valve, then the change of anode current (δI_a) due to a change of grid voltage δV_g, would be

$$\delta I_a = \frac{\mu}{R_L + R_p}.\delta V_g, \qquad \text{... (10.4c)}$$

where μ is the amplification factor.

The change in voltage across the load is given by

$$\delta V_a = \delta I_a.R_L = \frac{\mu.R_L}{R_L + R_p}.\delta V_g$$

$$\text{or,} \quad \alpha = \frac{\delta V_a}{\delta V_g} = \frac{\mu.R_L}{R_L + R_p} = \frac{\delta I_a}{\delta V_g}.R_L \qquad \text{... (10.4d)}$$

Also the voltage gain (α) is given by

$$\alpha = \frac{\delta V_a}{\delta V_g} = \frac{\mu . R_L}{R_L + R_p} = \frac{\mu}{1+\frac{R_p}{R_L}} \qquad \text{... (10.4e)}$$

Obviously, the value of α is smaller than μ and when $R_p << R_L$, the value of α reaches the value of μ

From the dynamic characteristic curve, the value of α can be calculated.

Circuit connections :

The determination of the dynamic characteristics can be done with the same circuit as shown in (Fig. 10.11) in Expt. No. 10.3 with the only modification that a high resistance (of the order of 10 kilo-ohm or more) has to be inserted in the anode circuit, somewhere between the anode and the milliammeter. Alternatively, the experiment can be performed with a more simplified circuit as shown in (Fig. 10.14), where the grid-bias should be a variable supply capable of employing negative voltage to the grid at a step of 1.5 volt.

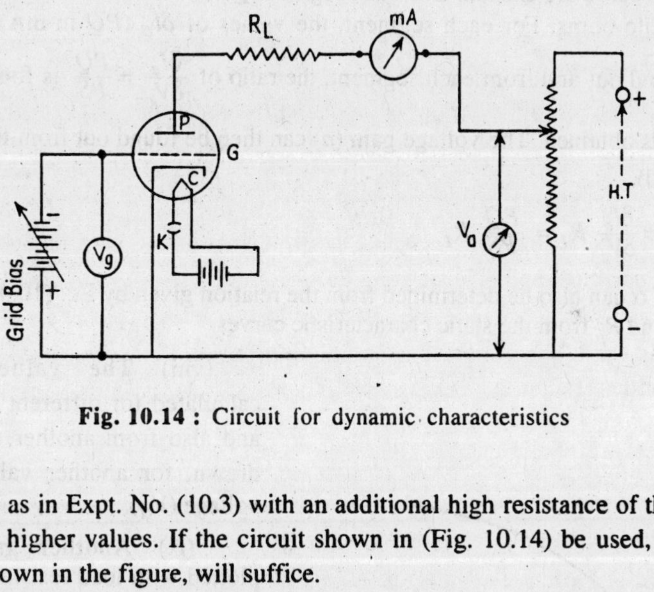

Fig. 10.14 : Circuit for dynamic characteristics

Apparatus :

The same as in Expt. No. (10.3) with an additional high resistance of the order of 10 kilo-ohms and higher values. If the circuit shown in (Fig. 10.14) be used, then only the components shown in the figure, will suffice.

Method :

(i) The circuit is set up according to the diagram. A load resistance of 10 k-ohm is placed in the anode circuit. A high tension supply, appropriate to the value, is used for the anode voltage.

(ii) The different tappings of the grid bias battery are checked by a voltmeter.

(iii) For drawing the dynamic characteristics, a grid-bias of about −15 volt is employed, and the anode current corresponding to this, is measured. The grid-bias is decreased by steps of 1.5 volt and each time, the corresponding anode current is measured.

(iv) The plate current (I_a) against grid voltage (V_g) are plotted for the given load (R_L) and for a given anode voltage (V_a) on a graph paper.

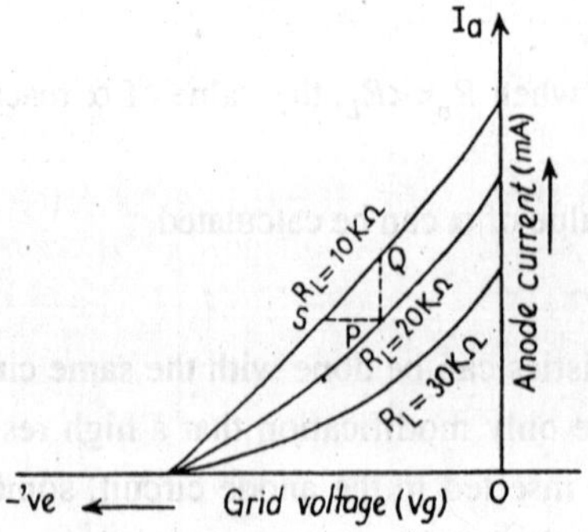

Fig. 10.15 : Dynamic characteristic curves

(v) Next the load resistance is increased to a value of 20 kilo-ohms and the operations are repeated. Again, another set is obtained for the value of load-resistance of 30 kilo-ohms.

(vi) All the curves are plotted on the same graph paper (Fig. 10.15). If necessary, the experiment may be repeated for another anode voltage also and another set of curves may be plotted with the same set of load resistances but for this changed anode voltage.

(vii) To find the voltage gain, for a given load resistance, the dynamic characteristic for that load resistance R_L is selected. Three or more segments of the straight portion of this characteristic curve are chosen. One such segment QS is shown on the curve corresponding to R_L = 10 kilo-ohms. For each segment, the values of δI_a (PQ in mA) and δV_g (PS in volt) are found out and from each segment, the ratio of $\frac{\delta I_a}{\delta V_g} = \frac{PQ}{PS}$ is found out and their mean value is obtained. The voltage gain (a) can then be found out from the relation given by Eq. (10.4d).

$$\therefore \quad \alpha = \frac{\delta I_a}{\delta V_g} R_L = \frac{PQ}{PS} . R_L$$

The value of α can also be determined from the relation given by Eq. (10.4e) by finding the values of μ and R_p from the static characteristic curves.

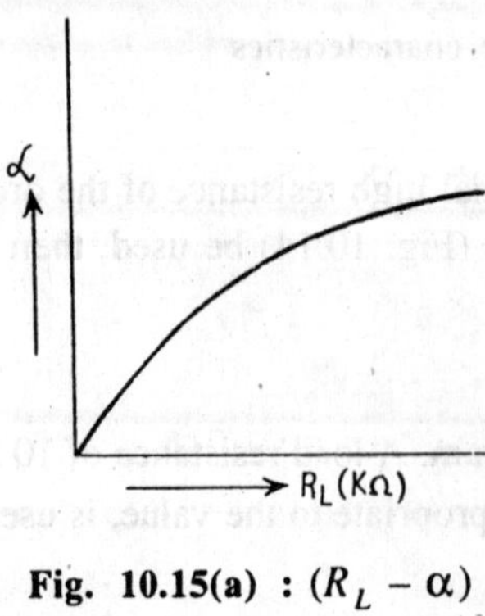

Fig. 10.15(a) : (R_L – α) curve

(viii) The values of α are calculated for different load resistances and also from another set of curves, if drawn, for another value of the plate voltage (V_a).

(ix) Another graph is to be plotted with load resistance (R_L) in kilo-ohms. against the corresponding voltage gain (α) for a given value of anode voltage (V_a). The nature of the graph is shown in [Fig. 10.15(a)].

Results :

(A) Data for drawing (V_g-I_a) curves for different values of load-resistances :

Table 1

Anode voltage (V_a) = ... volts

Filament voltage = ... volts

Filament ccurrent = ... ampere

No. of valve used = ...

No. of Obs.	Load Resistance R_L = ... kilo-ohms		Load Resistance R_L = ... kilo-ohms		Load Resistance R_L = ... kilo-ohms	
	Grid voltage (V_g volt)	Anode current (I_a mA)	Grid voltage (V_g volt)	Anode current (I_a mA)	Grid voltage (V_g volt)	Anode current (I_a mA)
1.	– 15	......	......	......	......	......
2.	– 13.5	......	......	......	......	......
3.	......	......	......	......	......	......
4.	etc.	etc.	etc.	etc.	etc.	etc.
5.						
6.						
7.						
8.						

[Similar tables are to be made for other values of plate voltages applied.]

(B) Measurement of voltage-gain :

Table 2 (See next page)

From curves corresponding to plate-voltage (V_a) = ... volts.

(C) Data for (R_L-α) curve :

Table 3

Serial No.	Values of load resistances (R_L kilo-ohms)	Value of voltage gain (α)
1.	......	......
2.	......	......
3.	......	......
4.	......	......
...		
...		
...		
etc.	etc.	etc.

Precautions :

(i) Grid bias should never be made positive

(ii) Only the straight portion of the dynamic characteristic curves should be taken to calculate the slope.

(iii) To get higher voltage gain, a value of greater amplification factor, is preferable.

(B) Measurement of voltage-gain :

Table 2

From curves corresponding to plate-voltage (V_a) = ……… volts.

Load resistance R k-ohm	Value of δI_a (= PQ) from graph (mA)	Value of δV_g (= PS) from graph (volt)	Value of $\frac{\delta I_a}{\delta V_g} = \frac{PQ}{PS}$	Mean value of $\frac{PQ}{PS}$	$\alpha = \frac{PQ}{PS} \times RL$	Calculated value of $\alpha = \frac{\mu}{1+\frac{R_p}{R_L}}$
	…	…	…			
10	…	…	…	…	…	…
	…	…	…			
20	etc.	etc.	etc.	…	…	…
etc.	etc.	etc.	etc.	…	…	…
50	etc.	etc.	etc.	…	…	…

[Similar tables for other curves for different plate voltages.]

Questions and Answers

Q.1. What are meant by triode parameters?

A. : The constants of a triode valve (cf. amplification factor, internal impedance, mutual conductance and voltage gain) are known as the parameters of the valve.

Q.2. In what respect does a triode differ from a diode?

A. : Diode has two electrodes, while a triode has three— cathode, anode and the grid.

Q.3. What is the function of the grid?

A. : It is an electrode placed between the cathode and anode of a tharmionic valve or a cathode ray tube serving to control or modulate the movements of the electrons towards the anode. A positive grid accelerates the electron motion while a negative grid retards the motion.

Q.4. What are the uses of a triode valve?

A. : It can be used as an amplifier, a rectifier and an oscillator.

Q.5. What is meant by 'space-charge'?

A. : When the filament is heated by a small voltage, the electrons emitted by the cathode do not reach the anode. They remain present in the space between the cathode and the anode. These electrons are known as the 'space-charge'. They prevent further emission of electrons from the cathode and thereby weaken the plate current.

Q.6. What is meant by the mutual characteristic curve?

A. : It is a curve showing the variation of anode current with the grid voltage, under a given anode potential.

Q.7. Can a.c. be used for applying the plate voltage?

A. : Yes, but the voltage will have to be rectified before applying it on the plate.

Q.8. Is there any relation between the constants?

A. : Yes, the amplification factor = plate resistance × mutual conductance.

Q.9. What is the filament made of?

A. : Normally, it is made of tungsten.

Q.10. What types of filament are used nowadays.

A. : Thoriated tungsten or oxide-coated filaments.

Q.11. What is the construction of the grid?

A. : It is made of a very fine nickel wire of uniform diameter and is wound on two copper leads which support the wire inside the valve. It surrounds the filament or the cathode.

Q.12. What is the construction of the plate?

A. : It is a cylinder of plated copper or nickel and surrounds the grid and the cathode.

Q.13. What is the shining deposit inside the valve and why is it given?

A. : It is the deposit of metallic barium or magnesium (called the *getter*), which absorbs the residual gases inside the valve and thus helps in maintaining the vacuum.

Q.14. What is the use of drawing the characteristic curves?

A. : Apart from drawing the constants of the valve, these curves enable us to study the suitability of a valve for a particular use.

Q.15. Is the voltage gain of a triode valve constant?

A. : No, it depends upon the load resistance employed and for a particular value of plate voltage applied.

Q.16. Why is the latter curve called the dynamic characteristic?

A. : Since, even for a fixed value of plate voltage, the curves become different for different values of the load resistances.

10.5 To draw the collector-current collector-voltage characteristics of a *p-n-p* transistor (CE mode) at different base currents

Theory :

When a semiconductor of small thickness is placed in between two other similar semiconductors and all are placed in contact, then the arrangement so formed is known as a transistor or a semiconducting triode. These transistors may be of two types—either a *p-n-p* type or a *n-p-n* type. When two *p*-type semiconductors are on either side of a n-type semiconductor (Fig. 10.16), then it is called a *p-n-p* transistor and the reverse arrangement is called a *n-p-n* transistor. In a *p-n-p* type, the left semiconductor is the emitter (*E*), the right one is the collector (*C*) and the central thin one is known as the base (*B*).

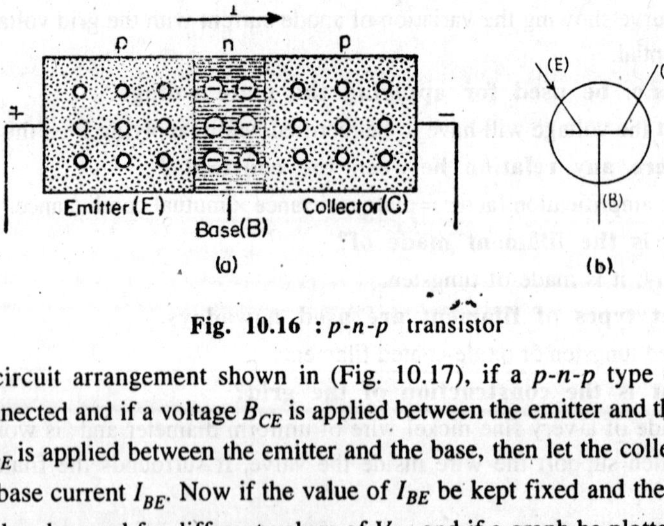

Fig. 10.16 : *p-n-p* transistor

In the circuit arrangement shown in (Fig. 10.17), if a *p-n-p* type of transistor is properly connected and if a voltage B_{CE} is applied between the emitter and the collector and a voltage V_{BE} is applied between the emitter and the base, then let the collector current be I_{CE} and the base current I_{BE}. Now if the value of I_{BE} be kept fixed and the changes in the value of I_{CE} be observed for different values of V_{CE} and if a graph be plotted for them (I_{CE} $-V_{CE}$), then this graph is called the static output characteristic curves of the said transistor. From this curve, the *incremental output impedence* of the transistor can be calculated.

In this grounded-emitter (*CE*) configuration the input signal is applied to the base and the current gain is designated by β, which is defined as the ratio of the change in collector current (δI_{CE}) effected by a change in base current (δI_{BE}) with collector voltage (V_{CE}) maintained at a constant value. Then β may be called the current amplification factor in a grounded-emitter amplifier. Another designation for β is h_{fe}.

Thus $\beta = \frac{\delta I_{CE}}{\delta I_{BE}} (V_{CE} \text{ constant}) = h_{fe}$ .. (10.5a)

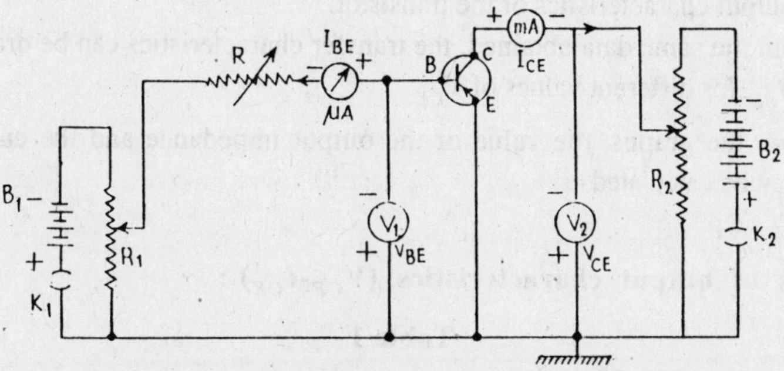

Fig. 10.17 : Circuit for transistor characteristics

For each curve, beyond the knee, the ratio of $\frac{\delta V_{CE}}{\delta I_{CE}}$ gives the output impedance for the particular value of the base current. From the graph, for a particular value of V_{CE}, the ratio $\frac{\delta I_{CE}}{\delta I_{BE}}$ gives the current amplification factor or the current gain.

Another set of characteristic curves for the transistor can be drawn. These are known as the transfer characteristics, exhibiting the variation of collector current (I_{CE}) with base current (I_{BE}), while the collector voltage is kept constant (Fig. 10.19). These curves can easily be plotted from the output characteristics, but they are generally plotted from the data obtained from a separate experiment.

Apparatus :

A *p-n-p* transistor, two variable rheostats (R_1, R_2), sliding rheostat ($R \sim 10$ k-ohm), 4-volt battery (B_1), 20-volt battery (B_2), plug type keys, d.c. voltmeters (1 volt and 20 volt ranges, with high impedance), d.c. milliammeter, d.c. microammeter.

Method :

(i) Circuit connections are made as shown in (Fig. 10.17). Here, in fact, there are two circuits—one is *B-E* circuit and the other is *C-E* circuit. The emitter (*E*) is included in both these circuits and serves as the common emitter.

(ii) Closing the key K_1, the value of R_1 is so adjusted that the value of I_{BE} is about 40 microamperes.

(iii) Now, the key K_2 is closed in the *C-E* circuit and the value of R_2 is made such that the voltage between the emitter and the collector (V_{CE}) is about 5 volts only. Under this condition, the value of I_{CE} is recorded from the milliammeter.

(iv) Keeping the value of I_{BE} constant at 40 microamperes, the value of V_{CE} is changed step by step until the value of V_{CE} reaches 20 volts. In each step, the values of I_{CE} are recorded.

(v) Next, the value of I_{BE} is changed to the next higher value of 60 microamperes and the operations are repeated as before.

(vi) In this manner, records are observed for I_{BE} values in between 80 and 100 microamperes.

(vii) Finally, graphs are plotted on the same graph paper, the variations of I_{CE} for different values of V_{CE}. Four graphs will be obtained for four different values of I_{BE}. These graphs are the output characteristics of the transistor.

(viii) From the same data obtained, the transfer characteristics can be drawn with I_{CE} plotted against I_{BE} for different values of V_{CE}.

(ix) From the graphs, the value of the output impedance and the current amplification factor can be calculated.

Results :

(A) Readings of output characteristics (V_{CE}-I_{CE}) :

Table 1

Transistor taken (*p-n-p*) = ...

No. of Obs.	I_{BE} = ... μA		I_{BE} = ... μA		I_{BE} = ... μA		I_{BE} = ... μA	
	V_{CE} (volt)	I_{CE} (mA)	V_{CE} (volt)	I_{CE} (mA)	V_{CE} (volt)	I_{CE} (mA)	V_{CE} (volt)	I_{CE} (mA)
1.	...	...	...	...	...	...	...	...
2.	...	...	...	...	...	...	...	...
3.								
4.								
5.	etc.	etc.	etc.	etc.	etc.	etc.	etc.	etc.
6.	...	...	...	...	...	...	...	...
7.	...	...	...	...	...	...	...	...
8.	...	...	...	...	...	...	...	...

(B) Readings for transfer characteristics (I_{CE}-I_{BE}) :

Table 2

No. of Obs.	V_{CE} = ...volts		V_{CE} = ...volts		V_{CE} = ... volts		V_{CE} = ...volts	
	I_{BE} (μA)	I_{CE} (*mA*)	I_{BE} (μA)	I_{CE} (*mA*)	I_{BE} (μA)	I_{CE} (*mA*)	I_{BE} (μA)	I_{CE} (mA)
1.		...	...	...	...	...	...	...
2.	...	...	...	...	...	...	...	...
3.								
4.								
5.	etc.	etc.	etc.	etc.	etc.	etc.	etc.	etc.
6.								
7.								
8.	...	...				...	...	...

Drawing of Graphs :

For *output characteristics*, values of I_{CE} are plotted in mA along *Y*-axis of the graph and the values of V_{CE} in volts, along *X*-axis. Four different curves are obtained for different values of base current (I_{BE}). The general shape of the curves are shown in (Fig. 10.18). It is evident from the graph that beyond a fairly small value of V_{CE}, the collector current depends much more on the base currents than on the collector voltage (V_{CE}). It is in this nearly saturated condition, beyond the knee of the curves, that a transistor is normally operated.

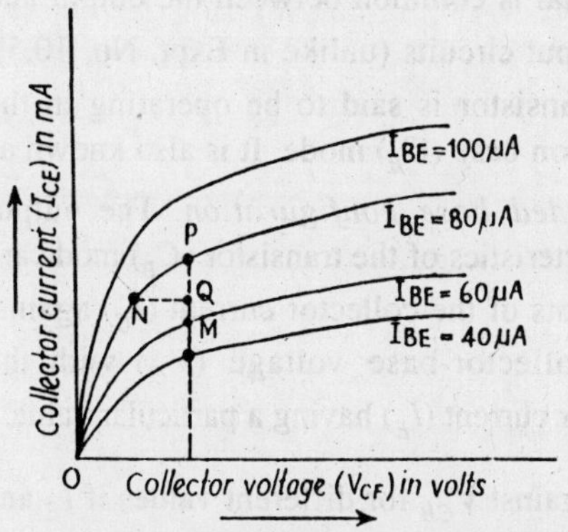

Fig. 10.18 : Output characteristics of transistor

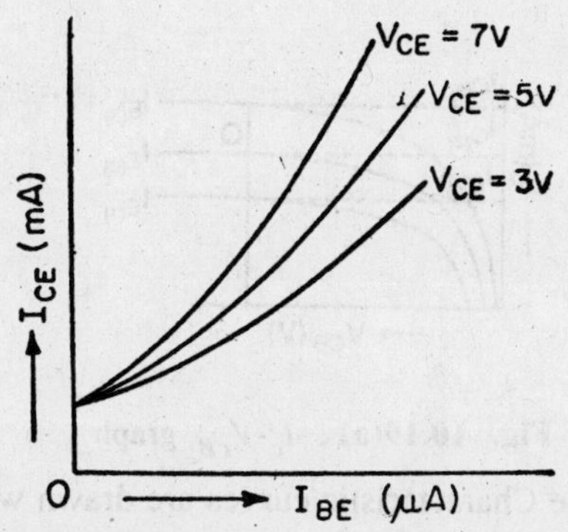

Fig. 10.19 : Transfer characteristics of transistor

The output impedence for the particular value of the base current is given by

$$\frac{\delta V_{CE}}{\delta I_{CE}} = \frac{QN \text{ volt}}{PQ \times 10^{-3} \text{ampere}} = \text{........ ohm}$$

From the curves, a particular value of V_{CE} is selected (shown by the doted line) and the ratio of change in collectors current to corresponding change in base current is calculated. This ratio gives the value of the current amplification factor or current gain (β).

$$\text{Thus } \beta = \frac{\delta I_{CE}}{\delta I_{BE}} = \frac{PM \times 10^{-3}}{20 \times 10^{-6}} = \text{.........}$$

For transfer characteristics, another set of curves are plotted with I_{CE} along *Y*-axis and I_{BE} along *X*-axis (Fig. 10.19) for different values of collector voltage (V_{CE}). The general shape of the curves are shown in the figure. From the straight portion of each curve, the slope of the curve is calculated which gives the current gain (β).

$$\text{Hence, current gain } (\beta) = \text{slope of the curve} = \frac{\delta I_{CE}}{\delta I_{BE}} = \text{.....}$$

Precautions :

(i) The value of V_{BE} should be fairly small.

(ii) While the value of V_{CE} is increased, the value of I_{BE} may fall slightly. This fall has to be adjusted by controlling the rheostat R_1.

(iii) For calculations, the straight portions of the curves have to be taken into account.

(iv) Milliammeters and microammeters should be properly connected.

(v) The voltmeters taken should preferably be of high impedance.

[Questions and Answers on transistors are given at the end of the next experiment.]

10.5(A) To draw the output characteristics of a transistor in the common base mode (CB mode) for different emitter currents

Theory :

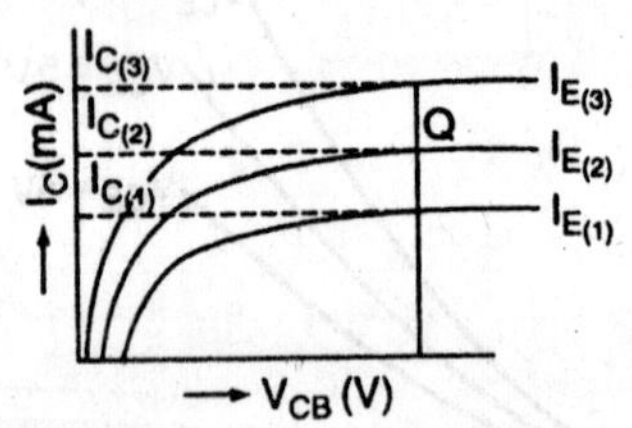

Fig. 10.19(a) : I_c-V_{cB} graph

In a transistor (*p-n-p*), if the base terminal is common between the output and the input circuits (unlike in Expt. No. 10.5), the transistor is said to be operating in the common base (C_B) mode. It is also known as *grounded base configuration*. The output characteristics of the transistor (C_B) mode are the plots of the collector current (I_C) against the collector-base voltage (V_C) with the emitter current (I_E) having a particular value.

The Characteristic curves are drawn with I_C against V_{CB} for different values if I_E and their natures are as shown in [Fig. 10.19(a)].

The a.c. current gain α and the output admittance h_{ob} for the CB mode are respectively given by

$$\left.\begin{aligned}\alpha &= \left(\frac{\delta I_C}{\delta I_E}\right)_{V_{CB}} = h_{fb} \\ &= \frac{I_{C(2)} - I_{C(1)}}{I_{E(2)} - I_{E(1)}} \\ \text{and} \quad h_{ob} & \left(\frac{\delta I_C}{\delta V_C}\right)_{I_E}\end{aligned}\right\} \quad \ldots [10.5\text{ (A) i}]$$

The value of h_{ob} is obtained by the slope of the output characteristic curve, as was obtained in Expt. No. 10.5.

Apparatus :

A *p-n-p* transistor, two variable rheostats, sliding rheostat, stabilised power supply, two d.c. milliammeter, d.c. voltmeter (0-10V).

Circuit Connections :

The circuit connections are made as shown in [Fig. 10.19(b)].

Method :

(i) Circuit connections are made as shown in [Fig. 10.19(b)]. The variable rheostats are adjusted with the power supply being put on, so that a suitable emitter current (I_E) of about (3-4)mA flows, and which is measured by the milliammeter.

(ii) The collector-base voltage (V_{CB}) is increased slowly and their values are recorded by the voltmeter (V). The increase is done in steps up to, say 10 volt. For each V_{CB}, the collector current (I_C) is noted from the milliammeter. The emitter current must remain steady and constant during the process.

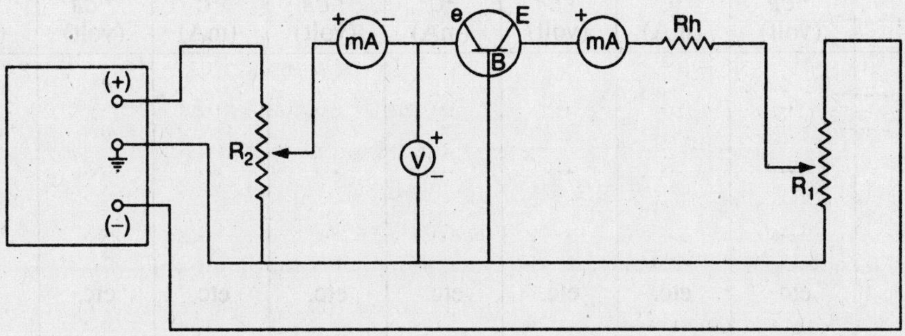

Fig. 10.19(b) : Circuit diagram

(iii) After switching off the power supply, the top terminal of R_2 is connectd to the (–)ve terminal of the power supply. Voltmeter terminal connections are also to be reversed accordingly. Negative values of V_{CB} are thus obtained.

(iv) Step (i) is repeated and value of I_E is set to the previous value.

(v) Now V_{CB} is slowly increased from zero in the negative direction in suitable steps by means of R_2 till I_C drops nearly to zero. At each step, V_{CB} is measured with the voltmeter (V) and I_C by milliammeter. I_E must remain constant during the process. Power supply is then switched off.

(vi) Then the top terminal of R_2 is disconnected and connected to the positive terminal of the power supply.

(vii) The emitter current (I_E) is then changed to a new value (say, $I_{E(2)}$) and the steps are repeated as before.

(viii) The steps are repeated for another third value of emitter current [$I_{E(3)}$]

(ix) For each value of I_E, three graphs are plotted with I_c against V_{CB}.

(x) From Eq. [10.5(A)i], the values of h_{fb} and h_{ob} are obtained from the graph by drawing tangents.

Results :

Table 1

Specifications of the meters used

Meters	Range	Value of Smallest division	Zero error, if any
Milliammeter 1	......	......	......
Milliammeter 2	......	......	......
Voltmeter	......	......	......

Transistor used = ...

Determination of output characteristics ($V_{CB} - I_C$) :

Table 2

	$I_E = 2$ mA		$I_E = 4$ mA		$I_E = 6$ mA		$I_E = 8$ mA	
	V_{CB} (volt)	I_C (mA)	V_{CB} (volt)	I_C (mA)	V_{CB} (volt)	I_C (mA)	V_{CB} (volt)	I_C (mA)
1.	...	...	...	...	...	...	...	...
2.	...	...	...	...	...	...	...	...
3.	...	...	...	...	...	...	...	...
4.	etc.	etc.	etc.	etc.	etc.	etc.	etc.	etc.
5.	etc.	etc.	etc.	etc.	etc.	etc.	etc.	etc.
6.	...	...	...	...	...	...	...	...
7.	...	...	...	...	...	...	...	...
8.	...	...	...	...	...	...	...	...

Calculations :

From the graph as shown in [Fig. 10.19(a)], the values of h_{fb} and h_{ob} are obtained after measuring the slope of the curves, as discussed earlier.

Precautions :

[Same as in Expt. No. 10.5]

[Questions and Answers are given at the end of Expt. No. 10.6 and some more questions/answers are given below.]

Q.1. What are other modes of operation of a transistor?

A. : Other than CE and CB modes, the third mode of operation is common-collector (CC) mode.

Q.2. What is meant by 'input characteristics' of a transistor?

A. : Input characteristics for a given mode of operation of a transistor consist of plotting of the input currents against the input voltages with a particular value of output voltage.

Q.3. What are the applications of a transistor?

A. : They are used in amplifiers, oscillators in analog circuits and as a switch in digital circuits.

Q.4. Which configuration gives more current gain and voltage gain (more than unity)?

A. : CE configuration gives them.

10.5(B) To draw the I-V characteristics of a solar cell

Theory :

An unbiased *p-n* junction diode, designed in such a way that it can convert sufficient percentage of light energy falling on it, directly into electrical energy, is known as a *photovoltaic cell* and a special type of photovoltaic cell having a large surface area and suitable to work with solar radiations is known as a *solar cell*. The construction and the symbol of a solar cell are shown in [Fig. 10.19(c)].

A solar cell develops an output voltage (V) when irradiated with solar radiation and the open circuit voltage is quite low and is only about 0·5 volt and a short-circuit current (I) is only a few milliamperes. But when a number of solar cells are all connected in series and parallel networks, sufficient amount of voltage and current are available. Such an array of solar cells is known as a *solar panel*

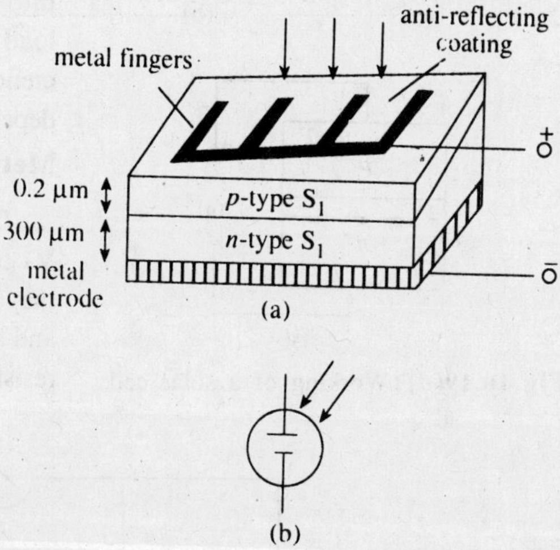

Fig 10.19(c) : Solar cell—its (a) construction and (b) symbol

Construction

A solar cell is basically an unbiased *p-n* junction diode which is usually made from silicon. [Fig. 10.19(c)] gives the constructional details of a typical solar cell and its symbol. The upper electrode of the cell is in the form of a grid with fingers which permit sunlight to pass through. The other electrode completely covers the lower surface of the cell. To prevent loss of light by reflections, an antireflecting layer is coated on the upper surface.

Working Principle

The band gap for silicon is about 1.1 eV. Thus photons of sunlight are capable of producing electron-hole pairs [Fig. 10.19(d)]. The electrons that are produced in the *p*-region near the charged depleted regions are swept away by the electric field present in the depleted region while the hole of the electron-hole pair gets deposited at the edge (c). In the same way, the holes that are produced by the absorption of photons in the *n*-region, are swept away while electrons are deposited. Thus, the net effect is that on the *p*-side of charge-depleted region, there are excess holes. This gives rise to open circuit voltage V_0. If a load is connected, then a current flows through the load on illuminating the solar cell.

A solar cell, basically being a *p-n* junction diode, is made of silicon or indium, phosphorus or gallium arsenide. It consists of a relatively thick *p*-type crystal with a thin *n*-type crystal covering with a higher rate of doping (or a thick *n*-type with a thin *p*-type).

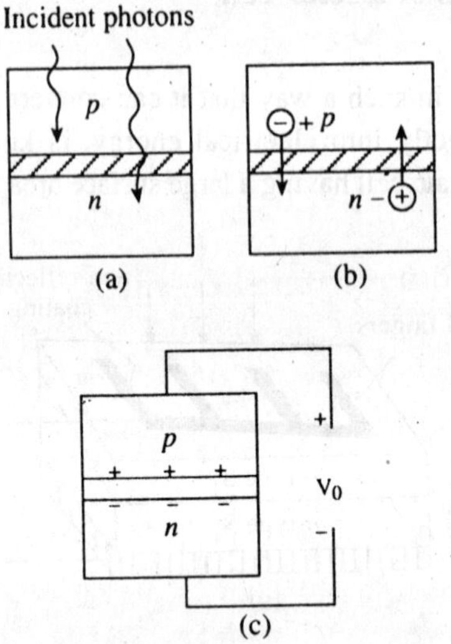

Fig. 10.19(d) : Working of a solar cell

Thickness of *n*-type crystal is so small that photons can penetrate the junction. When the photon reaches the junction, it knocks out electrons and transfers them to the conduction band. These electrons drift through the *n*-type material and then through the external circuit containing a load resistance (R). The maximum efficiency for silicon cell is about 22% and it depends on the band gap.

Method

In order to study the voltage current (I-V) characteristics, a simple ohmic circuit may be constructed with a solar cell or panel and an ammeter and a voltmeter with a load resistance.

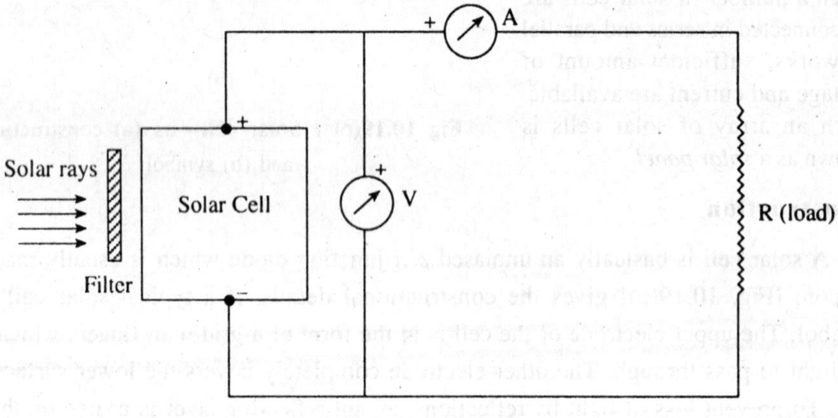

Fig. 10.19(e) : Circuit diagram

The circuit diagram may be innovated as shown in [Fig. 10.19(e)]. The intensity of the solar radiation incident on the cell may be controlled using suitable filters, and variations of incident intensity of solar radiations will make the output voltage and the current vary accordingly. Using suitbale voltmeter and the ammeter, the values of voltage (V) and the current (I) may be obtained. A graph may be drawn with V along *X*-axis and I along *Y*-axis. This graph is known as the (I-V) characteristic of the solar cell.

Discussions :

The advantages of a solar cell are that these are (a) long lasting (b) polution-free and (c) maintenance free.

If the output voltage and the current of a *single* solar cell be too small, then an a.ray of such cells in series may be taken to get sufficient voltage and current in the experiment.

Questions and Answers

Q.1. What is a solar cell and how does it work?

A. : [See any Textbook]

Q.2. How is electrical energy generated in a solar cell?

A. : See Theory and Working Principle.

Q.3. In what respect does it differ from a voltaic cell?

A. : In a voltaic cell, energy is obtained from chemical reaction between the reactants but in a solar cell, the energy is obtained from the photoelectric effect.

Q.4. What are the advantages of a solar cell?

A. : See 'Discussions'.

Q.5. What are the applications of solar cells?

A. : Solar cells are used to charge batteries for satellites and earth communication equipments. It is used to detect very low level of light. Also these are used as household gadgets.

10.6 Determination of the hybrid parameters and leakage current of a transistor

Theory :

The circuit diagram for determining the static characteristic curves of a *p-n-p* transistor in the common-emitter connection, has been shown in [Fig. 10.17]. For a transistor in the common-emitter connection, the base current I_{BE} and the collector voltage V_{CE} with respect to the emitter, are taken as *independent variables*. Since the value of V_{BE} is some function of I_{BE} and V_{CE}, and since the collector current I_{CE} is another function of I_{BE} and V_{CE}, the dependent variables can be written as

$$\left.\begin{aligned} V &= f_1(I_{BE}, V_{CE}) \\ \text{and} \quad I_{CE} &= f_2(I_{BE}, V_{CE}) \end{aligned}\right\}, \qquad \dots (10.6a)$$

where they are two different functions.

A curve representing the variations of the dependent variable V_{CE} or I_{CE} with one of the two independent variables I_{BE} or V_{CE}, is called a characteristic curve of the transistor. Thus Eqs. (10.6a) represent a set of *four* characteristic curves.

(a) $(V_{BE}\text{-}I_{BE})V_{CE}$ curves : These are the curves representing the variations of base-voltage with base-current, the collector-voltage remaining constant. These curves are called the **input characteristics.**

(b) $(V_{BE}\text{-}V_{CE})I_{BE}$ curves : These are the curves representing the variation of base-voltage with the collector-voltage, for a fixed value of the base-current. These curves are known as the **feedback characteristics** and these curves are not of any practical importance.

(c) $(I_{CE}\text{-}I_{BE})V_{CE}$ curves : These are the curves showing the variations of collector-current with the base-current for constant values of collector-voltage. These curves are known as the **transfer characteristics.**

(d) $(I_{CE}\text{-}V_{CE})I_{BE}$ curves : These are the curves which show the variations of the collector-current with the collector-voltage, for a fixed value of the base current. These curves are known as the **output characteristics.**

The ratios of the small changes of one of these variables to the small changes of another variable, are known as the hybrid or the *h*-parameters of the transistor and their definitions are given as,

(i) Input impedepe $= h_{ie} = \left(\frac{\delta V_{BE}}{\delta I_{BE}}\right)_{V_{CE}}$ .... (10.6b)

= slope of input characteristic

(ii) Reverse voltage ratio

$= h_{ic} = \left(\frac{\delta V_{BE}}{\delta V_{CE}}\right)_{I_{BE}}$ .... (10.6c)

= slope of feedback characteristic.

(iii) Forward current transfer ratio

$= h_{fc} = \left(\frac{\delta I_{CE}}{\delta I_{BE}}\right)_{V_{CE}}$ .... (10.6d)

= slope of transfer characteristic.

(iv) Output admittance

$= h_{oe} = \left(\frac{\delta I_{CE}}{\delta V_{CE}}\right)_{I_{BE}}$ .... (10.6e)

= slope of output characteristic.

= reciprocal of output impedance.

The values of these *h*-parameters of a transistor in the *CE*-mode can be determined by the slopes of the static characteristc curves. But they are true only for one or two nominal working point and apply to zero frequency. They are measured at a fixed audio frequency of about 1000 Hz by means of a special arrangement.

Leakage current :

It is found that when the base of the transistor is open-circuited and the collector is biased in the reverse direction with respect to the emitter, a small amount of collector current flows from the collector to the emitter as shown in (Fig. 10.20). This current is called the transistor's collector-to-emitter *leakage current* and it is denoted by I_{CEO}. The suffix CE denotes collector-emitter current while the suffix 0 denotes that the base current is zero. This current depends on the voltage and temperature of the collector with respect to the emitter.

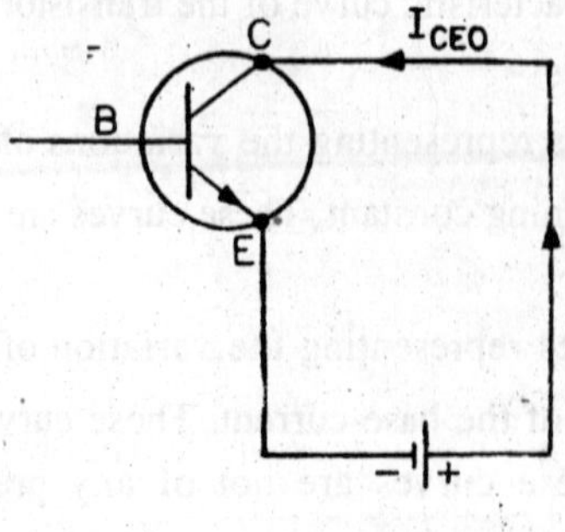

Fig. 10.20 : Transistor leakage current

Similarly, when the emitter is open-circuited and the collector-base junction is reverse-biased, a small leakage current flows through the collector-base junction and it is represented by I_{CE0}, the suffix having the similar significances as before.

If the collector is open-circuited and the emitter-base junction is reverse-biased, a small current flows through the emitter-base junction and this is known as the emitter-to-base leakage current and denoted by I_{EB0}.

From the definition of *current gain* (β) (discussed in the previous experiment), if the leakage current is considered for *CE*-mode, the part of I_E (emitter current) which reaches the collector is $I_{CE} - I_{CE0}$ and the definition of β is modified as

$$\beta = \frac{I_{CE} - I_{CE0}}{I_{BE}} \qquad \text{.... (10.6f)}$$

$$\therefore \quad I_{CE} = \beta . I_{BE} + I_{CE0} . \qquad \text{.... (10.6g)}$$

Thus, from output characteristics obtained in Expt. No. (10.5), knowing the values of I_{BE} and I_{CE}, and after finding the value of the current gain (β), the value of the leakage current (I_{CE0}) for the *CE*-mode can be determined.

Method :

(i) For determination of the hybrid parameters, it is to be noted that of the four *h*-parameters defined in the *Theory* the third and the fourth ones, given by Eqs. (10.6d) and (10.6e) are already determined in the previous experiment (Expt. No. 10.5). Of course, there the value of the output impedance (reciprocal of h_{oe}) has been found out from the graph shown in (Fig. 10.18), from which the output admittance (h_{oe}) can be calculated.

(ii) For determination of the other two *h*-parameters given by Eqs. (10.6b) and (10.6c), with the same circuit, readings are to be taken for (a) change of V_{BE} with I_{BE} for constant values of V_{CE} and (b) change of V_{BE} with V_{CE} for constant values of I_{BE}.

(iii) Two different sets of graphs are to be plotted—one for (V_{BE}-I_{BE}) for different constant values of V_{CE} and another for (V_{BE}-V_{CE}) for different constant values of I_{BE}. From the slopes of these two curves taken independently, the values of the first two *h*-parameters (h_{fe} and h_{re}) can be obtained.

Results :

Tabulate data in the same manner as shown in **Table 1** and **Table 2** in Expt. No. 10.5, for

(a) variation of V_{BE} with I_{BE} for different constant values of I_{CE}

and (b) variation of V_{BE} with V_{CE} for different constant values of I_{BE}.

Drawing of Graphs :

In addition to the graphs shown in (Figs. 10.18 and 10.19), two more graphs are to be plotted. One for V_{BE} against I_{BE} for different values of V_{CE} and another for V_{BE} against V_{CE} for different values of I_{BE}.

For determining the transistor leakage current, Eq. (10.6g) has to be applied.

Precautions :

All general precautions discussed (for transistor handling) in the previous experiment should be followed here also.

Questions and Answers

Q.1. What is the difference between *p-n-p* and *n-p-n* transistors?

A. : In the first case, *n*-type of semiconductor is placed in between two *p*-type semiconductors, while in the second case, a *p*-type semiconductor is placed in between, two *n*-type semiconductors.

Q.2. What is the difference between *p*-type and *n*-type semiconductors?

A. : The crystals from which electrons being removed by an impurity, renders it as a semiconductor, are generally *p*-type semiconductors, while the crystals which after accepting electrons from the impurities, render themselves as semiconductors, are generally *n*-type semiconductors.

Q.3. Why a non-conducting crystal becomes conducting when impurities are doped into them?

A. : Some impurities take away electrons from a crystal and thus create holes of positive charge in them. Then these crystals become semiconductors. Again, sometimes impurities donate electorns to some crystals, thus rendering them semiconductors.

Q.4. What is the advantage of using a transistor over a triode valve?

A. : There are many advantages. Firstly, transistors are very tiny and take small room in circutis. Secondly, they consume very weak current and hence their lives are prolonged and finally, since unlike in a valve, there is no cathode in a transistor. Hence, there is no need of current for heating the filament or cathode.

Q.5. The semiconductors or a transistor are made of what?

A. : Usually indium atom is doped with germanium in a *p*-type semiconductor, while in a *n*-type semiconductor, arsenic atoms are doped in germanium.

Q.6. What are the different characteristic curves of a transistor?

A. : See the Theory of Expt. Nos. 10.5 and 10.6.

[Apart from these questions, all basic properties of a transistor, their junction properties, their constructions, characteristic curves and their properties should be studied.]

10.7 Study of an Operation Amplifier (OP AMP)

An *operation amplifier* (**OP AMP**) is a high-gain, direct coupled linear amplifier whose response characteristics are externally controlled by negative feedback from the output to the input. With the development of low-cost integrated circuits, it is now possible to think of building blocks which can be assembled to form the required system. Low-cost, small size, well defined parameters and wide versatility have led to an enormous increase in the use of IC **OP AMPs**. They are widely and freely used nowadays in all electronic magazines like computers, video and audio amplifiers, oscillators, etc., and also in both digital and linear circuits. Although there are different types of **OP AMPs**, the particular one (**OP AMP 741**) is studied here and its properties are discussed.

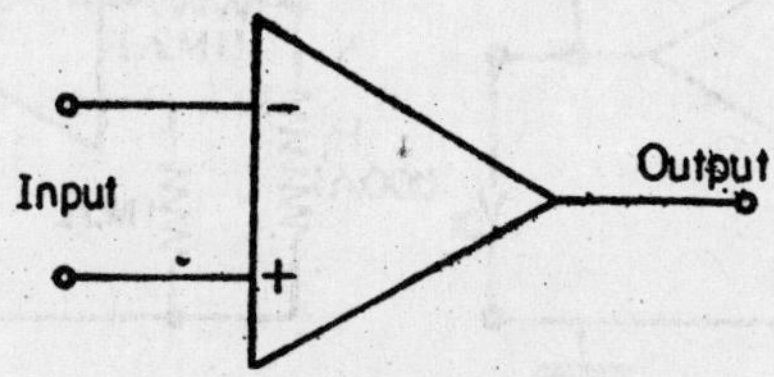

Fig. 10.21 : OP AMP

IC 741 OP AMP is a monolothic silicon integrated circuit with 19 transistors and 8 resistors, consisting of a differential input amplifier that effectively drives a gain and level-shifting stage having a complementary emitter-follower output. It provides internal phase compensation, output short-circuit protection and latch-free operation. The experiments that are discussed here with an OP AMP 741 are namely :

(A) Measurement of input offset voltage and offset current.

(B) Measurement of input bias current at inverting and non-inverting inputs.

(C) Closed-loop gain.

(D) Adjustment of offset null and use as a differential amplifier.

10.7 (A & B) Measurement of input offset voltage, offset current and input bias currents

Theory :

Input offset voltage is defined as the d.c. input voltage V_{io} that must be provided between the two input terminals of the OP AMP to get zero output voltage.

Input offest current is the difference between the separate base currents entering the input terminals of a balanced amplifier.

Input bias current is one-half the sum of the separate bias currents entering the two input terminals of a balanced amplifier.

The two transistors of the first stage of an OP AMP must draw some base current for a proper operating conditions. Normally, the OP AMs are designed in such a way that the input bias current is extremely small for which the input signal does not go loaded. Theoretically, because of the symmetry of the differential amplifier stage, the two base currents I_{B1} and I_{B2} should be the same. This implies that a resulting offset current defined as $I_{io} = (I_{B1} - I_{B2})$ would always flow and would cause a non-zero output voltage from the amplifier without any input. This can, of course, be nulled or balanced by applying a small potential, known as the *input offset* voltage V_{io} at any one of the internal offset balance terminals of the OP AMP.

To measure the input offset voltage, the connections to the OP AMP are made as shown in [Fig. 10.22(a)]. Since the source resistances are very small, contribution due to bias currents are neglected, i.e., $I_{B_1 R_1} << V_{io}$. The input offset voltage V_{io} is given by the relation,

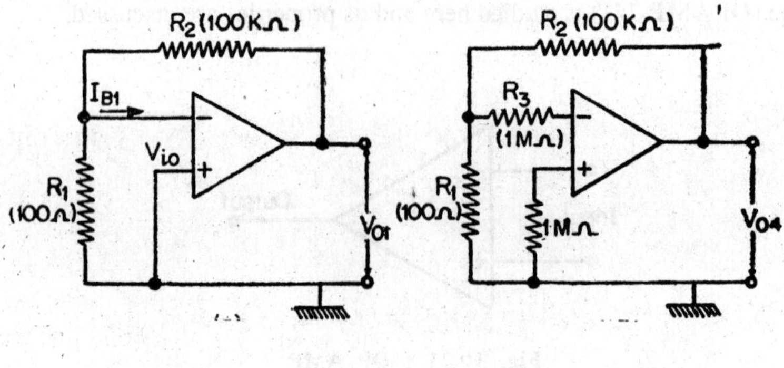

(a) (b)

Fig. 10.22 : (a) Input offset voltage and (b) current

$$V_{o1} = +\left(\frac{R_1 + R_2}{R_1}\right)V_{io}, \qquad \text{....(10.7a)}$$

where V_{01} is the output voltage.

To measure the input bias current at inverting input I_{B_I}, connections to the OP AMP are made as shown in [Fig. 10.23(a)]. Input bias currents are basically measured by forcing them to flow through large resistors (about 1 megohm) which may be by-passed to reject noise, using a capacitor of $C = 0.01\ \mu F$. From the circuit of [Fig. 10.23(a)], the output voltage V_{02} is given by

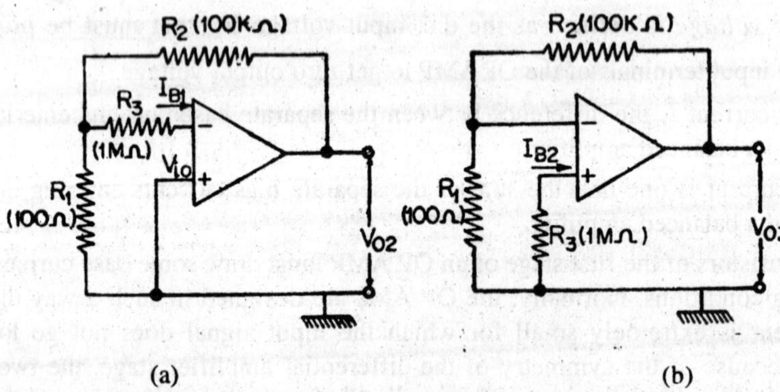

(a) (b)

Fig. 10.23 : Input bias currents (a) inverting and (b) non-inverting inputs

$$V_{02} = +\left(1 + \frac{R_2}{R_1}\right)(V_{io} + I_{B1}R_3)$$

$$\text{or,} \quad I_{B1} = \frac{V_{02} - V_{01}}{R_3\left(1 + \dfrac{R_2}{R_1}\right)}. \qquad \text{... (10.7b)}$$

To measure the input bias current at non-inverting input I_{B2}, circuit connections are made as shown in [Fig. 10.23(b)]. In this case, the output voltage V_{03} is given by

$$V_{o3} = \left(1+\frac{R_2}{R_1}\right)\left(V_{io} + I_{B2} R_3\right)$$

or, $$I_{B2} = \frac{V_{03} - V_{01}}{R_3\left(1+\frac{R_2}{R_1}\right)}. \quad \text{... (10.7c)}$$

To measure the input offset current (I_{io}) directly, connections of the OP AMP are made as shown in [Fig. 10.22(b)]. Here the output voltage V_{04} is given by

$$V_{04} = \left(1+\frac{R_2}{R_1}\right)(V_{io} + I_{io} R_3)$$

or, $$I_{io} = \frac{V_{04} - V_{01}}{R_3\left(1+\frac{R_2}{R_1}\right)}. \quad \text{... (10.7d)}$$

Results :

OP AMP 741 at room temperature = ... °C.

Table 1

Reading of the output voltage in volts		Offset parameters $R_1 = 100\,\text{ohms}, R_2 = 100\,\text{kilo-ohms},$ $R_3 = 1\,\text{megohms}.$
V_{01} = ...	Mean	V_{i0} = ...
V_{02} = ...	....	I_{B1} = ...
V_{03} = ...	....	I_{B2} = ...
V_{04} = ...	....	I_{i0} = ...

Precautions

(i) In the experiment, there may exist high frequency noise. In order to get rid of them or to reduce them, all the high-value resistors in the circuit have to be by-passed by suitably valued capacitors connected in parallel to them.

(ii) Normally there may be an initial warm up drift of V_{io} and I_B's following the application of power supply voltage. All measurements are to be taken after the initial warm up period passes on. This period varies from amplifier to amplifier. It may be as low as few minutes or even about half-an-hour.

(iii) Input offset voltage drifts over a period of time and the drift with respect to temperature is also observed. It depends very much also on the supply voltage.

10.7(C) Closed-loop gain of an OP AMP : Inverting and Non-inverting Amplifiers

The primary function of an amplifier is to amplify a signal. Generally, the more gain is derived from an amplifier, the better it is. But excessive gains can always be reduced or controlled to the required value by the external circuitry.

(i) Inverting amplifier

In an OP AMP, a basic gain block is the *inverting amplifier* circuit shown in [Fig. 10.24 (a)] The closed-loop (i.e., with feedback) gain of this amplifier is given by

$$A_{vf} = \frac{V_0}{V_i} = -\frac{R_f}{R_1}\left(1 - \frac{1}{1 + A_\beta}\right) \quad \text{... (10.7e).}$$

where A is the open-loop gain and β is the feedback fraction and is equal to $\frac{R_1}{(R_1 + R_f)}$.

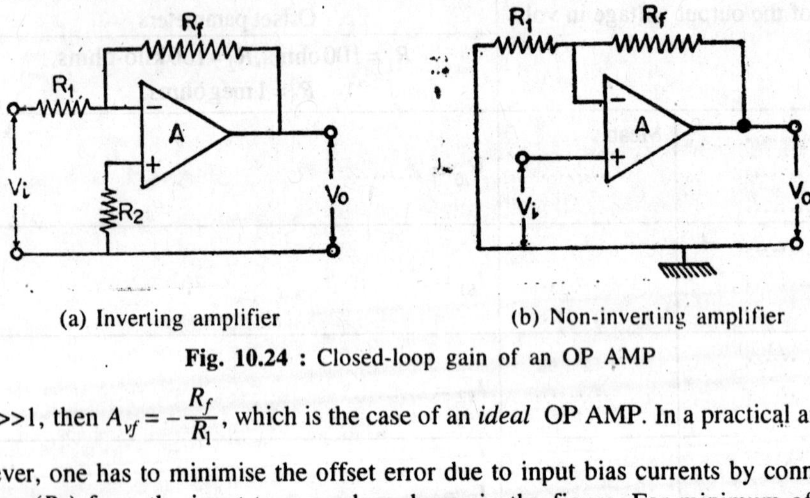

(a) Inverting amplifier (b) Non-inverting amplifier

Fig. 10.24 : Closed-loop gain of an OP AMP

If $A_\beta >> 1$, then $A_{vf} = -\frac{R_f}{R_1}$, which is the case of an *ideal* OP AMP. In a practical amplifier, however, one has to minimise the offset error due to input bias currents by connecting a resistor (R_2) from the input to ground as shown in the figure. For minimum offset, the value of R_2 is chosen as

$$R_2 = R_1 \parallel R_f.$$

(ii) Non-inverting amplifier

In case of a non-inverting amplifier as shown in [Fig. 10.24(b)], the closed-loop gain is given by

$$A'_{vf} = \frac{V_0}{V_i} = \frac{1}{\frac{R_1}{R_1 + R_f} + \frac{1}{A}} = 1 + \frac{R_f}{R_1}, \quad \text{...(10.7f)}$$

when $A \to \infty$

If again $R_1 = \infty$ and/or $R_f = 0$, then $A'_{vf} = 1$ and the amplifier then acts as a *voltage follower.*

Results :

Table 2 (See next page)

A graph may be plotted with the input voltage against the output voltage and from the graph, the value of the closed-loop can also be determined.

10.7(D) Adjustment of Offset Null and use as a Differential Amplifier

(i) **Offset Null**

The efffect of the input offset error voltages are very small no doubt and often they may be ignored. But there are certain critical cases where they cannot be ignored. For example, in the case of integration, an input voltage, however small it may be, the integral of it can eventually become large. Thus, most of the OP AMPs have provisions for measurement of V_{io} with a single potentiometer connected directly to the amplifier, as shown in [Fig. 10.25(a)].

The terminals 1 and 5 of the OP AMP are connected to A and B of the 10 kilo-ohms OFFSET NULL potentiometer. The centre of the potentiometer is already internally connected to a −12 volts terminal. The potentiometer knob is adjusted to get zero output voltage.

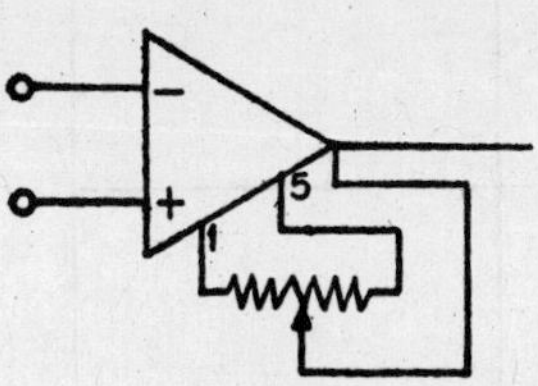

(a) Offset null

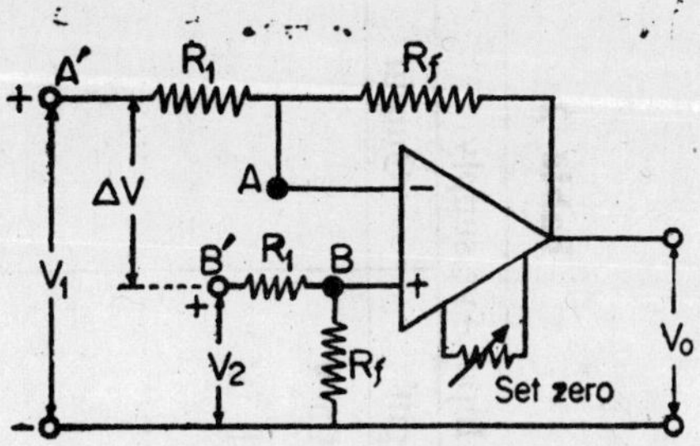

(b) Differential amplifier

Fig. 10.25 : Offset null and differential amplifier

Of course, there are many disadvantages of this internal method as under :

(i) Additional voltage drift are induced with temperature

(ii) It causes considerable variations in the gain characteristics of the amplifier.

(iii) Offset due to bias current cannot be balanced out by this technique.

Results :

Table 2

Condition of input terminal	Combination of resistances	Input +ve supply		Input –ve supply		Gain	
		Input	Output	Input	Output	Theor.	Expt.
Inverting amplifier	R_1 = ... ohm	+ 0.05 V					
		– 0.1 V					
	R_2 = ... ohm	......	......	......	......	......	......
		......	......	......	......	......	......
		......	......	......	......		
Non-inverting amplifier	... ...						
	... ...						
	$R_1 = \infty$						
	$R_f = 0$						

(ii) Differential amplifier

Often it becomes necessary to amplify the differences between two voltage levels and reject the voltage common to both. To achieve this, the OP AMP is connected as shown in [Fig. 10.25(b)]. It may be shown that ideally,

$$V_O = \frac{R_f}{R_1} \cdot \Delta V, \quad \dots (10.7g)$$

where $\Delta V = V_2 - V_1$.

The overall amplifier operates in such a way as to maintain near zero volts between A and B, with the result that the differential input inpedance for ΔV between A' and B' is $2R_1$.

Results :

Table 3

R_1 in ohms	R_f in ohms	V_2 in volts	V_1 in volts	ΔV volts	V_o in volts	
					Theoretical	Experimental
10k	100k	+ 8.1	+ 8.1	+ 0.1	1 volts	...
		+ 0.3	+ 0.1	− 0.1	− 1 volts	...
1k	100k	+ 0.1	+ 0.05	+ 0.05	+ 5 volts	...
...	...	...	...	...	...	...
etc.	etc.	etc.	etc.	etc.	etc.	etc.

Precautions :

The voltages difference ΔV should be chosen in such a way that the input offset voltage may be neglected in comparison to ΔV. Otherwise, OFFSET NULL adjustment has to be made before taking the readings.

Graph Drawing :

A graph may be drawn with ΔV $(=V_2 - V_1)$ against V_O, taking all possible combinations of positive and negative input voltages. At least five sets with different inputs (both positive and negative) are to be taken during recording observations.

Discussions :

The versatile use of an OP AMP has already been stated. To name a few of them, where OP AMP has been freely used are, *Voltage Follower, Adder, Log Amplifier, Integrator, Differential Amplifier and Voltage Comparator.*

10.7 (E) To study the performance of a Schmit Trigger using an OP AMP

Theory :

A voltage comparator circuit that develops a regenerative or positive feedback can be constructed using an OP AMP which can switch or trigger from one voltage level to another. This shows the phenomenon of hysterisis. Such constructed circuit is usually know as a 'regenerative comparator' or more commonly a 'Schmitt Trigger'. One such basic OP AMP circuit is shown in [Fig. 10.25(a)].

By the superposition theorem,

$$V_f = \frac{V_R R_2}{R_1 + R_2} + \frac{V_o R_1}{R_1 + R_2}. \quad \text{... [10.7(E)(i)]}$$

When $V_i < V_f$, the output $V_o = V_{sat}$, where V_{sat} is the positive saturation voltage of the amplifier.

$$\text{So,} \quad V_f = \frac{V_R R_2}{R_1 + R_2} + \frac{V_{sat} R_1}{R_1 + R_2} = V_1 \text{ (say)} \quad \text{... [10.7(E)(ii)]}$$

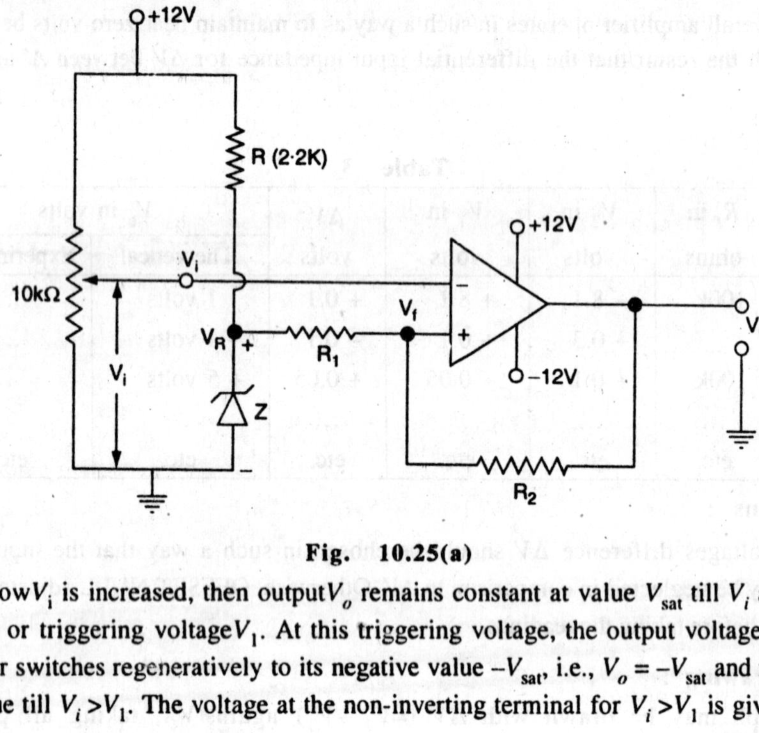

Fig. 10.25(a)

If now V_i is increased, then output V_o remains constant at value V_{sat} till V_i attains the thresold or triggering voltage V_1. At this triggering voltage, the output voltage V_o of the amplifier switches regeneratively to its negative value $-V_{sat}$, i.e., $V_o = -V_{sat}$ and remains at this value till $V_i > V_1$. The voltage at the non-inverting terminal for $V_i > V_1$ is given by

$$V_f = \frac{V_R R_2}{R_1 + R_2} - \frac{V_{sat} R_1}{R_1 + R_2} = V_2 \text{ (say)}. \quad \text{... [10.7(E)(iii)]}$$

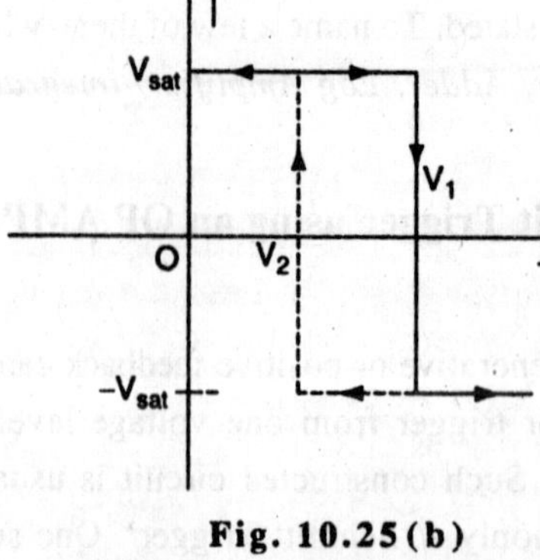

Fig. 10.25(b)

If the voltage V_i is decreased, the output remains at $-V_{sat}$ until $V_i = V_2$. At this voltage, the output regeneratively switches to $+V_{sat}$. The complete characteristics of the schmitt trigger is shown in [Fig. 10.25(b)].

The characteristics shows that the circuit exhibits hysterisis. The circuit triggers at different voltages depending on whether the signal is increasing or decreasing. The width of the hysterisis loop is given by

$$V_H = V_1 - V_2 = \frac{2R_1}{R_1 + R_2} \cdot V_{sat} \qquad \text{... [10.7(E)(iv)]}$$

Apparatus :

An OP AMP (IC741), a zeneer diode, dual power supply (± 12 v), d.c. voltmeter (0-15V), one 10 kΩ potentiometer, carbon rheostats (1kΩ, 1kΩ, 10kΩ), breadboard and wires.

Method :

(i) Circuit connections are made as shown in [Fig 10.25(a)] on a breadboard. Resistance values of R_1 and R_2 are measured by a multimeter.

(ii) The reference voltage V_R across the zener diode is obtained. The variable input voltage V_f can be obtained from the 12V supply of the OP AMP by using a 10 kΩ potentiometer as shown in the circuit diagram.

(iii) Values of R_1 and R_2 are so chosen as to make the loop gain greater than 1. For example, R_1 = 1 kΩ and R_2 = 100 kΩ or R_1 = 100 Ω and R_2 = 10 kΩ.

(iv) With chosen parameters and assumings V_{sat} = 10 Volt, the values of V_1 and V_2 are calculated using Eqs. [10.7(A)(ii)] and [10.7(E)(iii)] in **Theory.**

(v) The input V_i is varied by adjusting the potentiometer and is made slightly smaller than V_2 up to a voltage slightly greater than V_1 in small steps (0.1 V). The output voltage V_o is measured by d.c. voltmeter in each step.

(vi) Next, V_1 is decreased in similar small steps down to the starting value. Again, in each step V_i and V_o are measured.

(vii) A graph is then drawn with V_i along X-axis and V_o along Y-axis [Fig. 10.25(b)].

Results :

(A) Circuit components specifications :

OP AMP : ......

Zener : ....

Power Supply : ......

Reference Voltage (V_R) = ... volt

(B) Measurement of R_1 and R_2 :

Table 1

R_1 (ohm)		R_2 (ohm)	
As measured by multimeter	From colour code	As measured by multimeter	From colour code
.....	.....	.....	.....
.....	.....	.....	.....
.....	.....	.....	.....

(C) Data for Input-Output Voltages :

Table 2

No. of obs.	Input Voltages (V)		Output Voltages (V)	
	Increasing	Decreasing	Increasing	Decreasing
1.	......	......	......	......
2.	......	......	......	......
3.	......	......	......	......
...	......	......	......	......

(D) Hysterisis Loop-width determination :

Table 3

V_1 from graph (volt)	V_2 from graph (volt)	Hysterisis Width V_H (Volt)	
		From Graph	Theoretical
......	......	......	......
......	......	......	......

Precautions :

(i) The input voltage must be changed in steps from a low value to a high value and again from high to low values to get the hysterisis loop.

(ii) Schmitt trigger can produce square wave voltages with sharp leading and trailing edges from a slowly varying input voltage.

Questions and Answers

Q.1. What is a schmitt trigger?

A. : It is a modified bistable multivibrator which functions as a voltage comparator with hysterisis.

Q.2. What is a voltage comparator?

A. : It is nothing but a circuit which compares an input signal voltage with a reference voltage.

Q.3. Can you mention any practical application of a Schmitt trigger?

A. : It can be used to transform a slowly varying input voltage into a square wave output voltage.

Q.4. Why resistances of the order of kilo-ohm are used?

A. : For kilo-ohm resistances, current of the order of milliamperes will flow through the OP AMP. If low values of resistances are used, then the current will be of the order of ampere resulting unnecessary power drainage. Again, high resistances of the order of megohms are not used since they may be comparable with the input impedance of the OP AMP.

10.7(F) To Study the performance of an Integrator using an OP AMP

Theory :

An integrator circuit using an OP AMP is shown in [Fig. 10.25(c) and 10.25(d)], where V_o and V_1 represent the output and input voltages. Here,

$$V_o = -\frac{1}{CR}\int_0^t V_1 \, dt \quad \text{... [10.7(F)a]}$$

The output voltage is thus the integral of the input voltage multiplied by a constant factor. This is why the circuit is known as 'Integrator'.

If $V_1 = V_m \cos \omega t$, then

$$V_0 = -\frac{V_m}{\omega CR}\sin \omega t = -V_0' \sin \omega t \quad \text{... [10.7(F)b]}$$

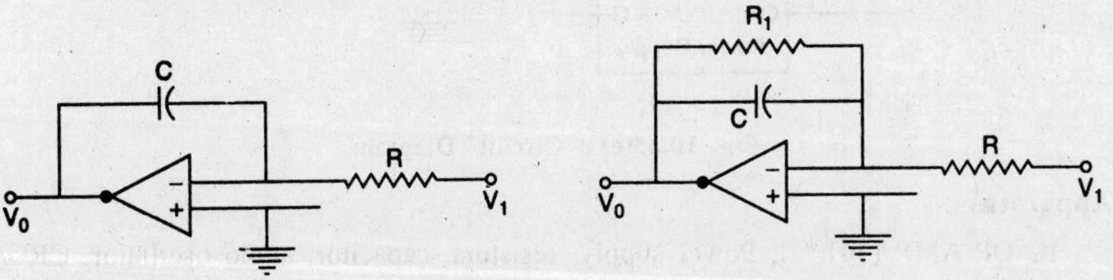

Fig. 10.25(c) and 10.25(d) : Integrator circuits

where $V_o' = \frac{V_m}{\omega RC}$ (say) $\therefore \left|\frac{V_m}{V_o'}\right| = \omega CR = 2\pi fCR$

where f = frequency of the input signal. A plot of $\left|\frac{V_m}{V_o'}\right|$ against f will be a straight line having the slope $m = 2\pi CR$ ... [10.7(F)c]

$$\text{or, } C = \frac{m}{2\pi R} \quad \text{... [10.7(F)d]}$$

If we put $m_1 = 2\pi C$, then $C = \frac{m_1}{2\pi}$ ... [10.7(F)e]

From [10.7(F) c], if taking diferent values of R, a graph is plotted with R against m [Eq. 10.7(F) c], its slope will be $2\pi C$ $(= m_1)$ wherefrom, the value of an unknown capacitance can be obtained.

Having some drawbacks in the basic circut as shown in Fig. 10.25(c), the modified circuit with a resistance connected in parallel to the capacitor, limiting the d.c. gain of the amplifier, is shown in Fig. 10.25(d). The value of this resistor must be much higher than $\frac{1}{\omega C}$. The value of R_2 shown in this circuit is approximately equal to the equivalent value of the resistances R and R_1, if those are connected in parallel.

Circuit Diagram :

The actual experimental circuit diagram is shown in [Fig. 10.25(c)]. The value of R may be of order of 10 kΩ and values of R_1 and R_2 are suitably chosen.

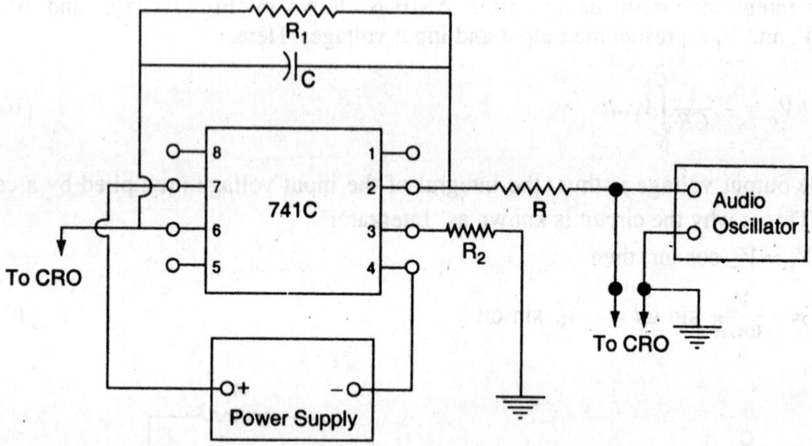

Fig. 10.25(e) : Circuit Diagram

Apparatus :

IC OP AMP (741 C), Power supply, resistors, capacitor, audio oscillator, CRO/ digital voltmeter; Breadboards,

Method :

(i) After connecting the circuit properly, power supply is switched on.

(ii) An input voltage is applied from the audio oscillator.

(iii) A minimum value of the frequency is set and waveforms of V_1 and V_0 are observed on the two channels of a dual trace CRO to check that V_0 is the integrated form of V_1. The peak input and output voltages V_m and V_0' are measured from the CRO. This measurements can also be done using an electronic voltmeter.

(iv) The value of f is increased in suitable steps and voltage measurements for each value of f, are done. Finally, the power supply and the signal are switched off.

(v) A graph is then plotted f against $\left|\frac{V_m}{V_0}\right|$ which will be a straight line having slope m = 2πCR.

(vi) Taking different values of R, the above steps are repeated and for each value of R, the values of R_2 are to be chosen accordingly.

(viii) A graph is drawn then with R against m, which is a straight line and from the slope of this line, the value of the capacitance can be obtained.

Results :

OP AMP used = ...

Power Supply voltage = ± .... volt

(A) Frquency Response of Integrator

Table 1

Audio Oscillator Type = ...

R = ... kΩ, R_1 = ... kΩ, R_3 = ... kΩ

Input signal frequency f (Hz)	Input signal voltage amplitude Vm (volt)	Output signal voltage amplitude V_o (volt)	$\left\lvert\frac{V_m}{V_o'}\right\rvert$
...	...	...	...
...	...	...	...
...	...	...	...

[Similar Tables are made for other values of R taken]

(B) m versus R data :

Table 2

Resistance value (ohm)	Change in frquency (δf) from graph. (Hz)	Change in $\left\lvert\frac{V_m}{V_o'}\right\rvert = \delta\left\lvert\frac{V_m}{V_o'}\right\rvert$ from graph	Slope (in) Hz^{-1}
...	...	...	...
...	...	...	...
...	...	...	...

A graph is plotted (m vs R) from above data

(C) Determination of Capacitance value :

Table 3

Change in R (δR) ohm (from graph)	Change in m (δm) Hz^{-1} (from graph)	Slope $m_1 = \left(\frac{\delta m}{\delta R}\right)$ (farad)	Value of the Capacitor $C = \frac{m_1}{2\pi}$ (farad)
...	...	...	...
...	...	...	...

Calculation of Percentage Error :

We have

$$\left\lvert\frac{V_m}{V_o'}\right\rvert = 2\pi fCR, \text{ or, } C = \frac{1}{2\pi fR}\left\lvert\frac{V_m}{V_o'}\right\rvert$$

Here, R is known and hence the proportional error in C will be

$$\frac{\delta C}{C} = \frac{\delta f}{f} + \frac{\delta V_m}{m} + \frac{\delta V_o'}{V_o'}$$

The errors δf, δV_m and $\delta V_o'$ correspond to the smallest divisions of the scales used to measure the quantities f, V_m and V_o' respectively. Multiplying $\frac{\delta C}{C}$ by 100, we get the percentage error.

Questions and Answers

[For Questions and Answers, it is advised to consult general discussion of an OP AMP and its various applications, from any Textbook].

10.7(G) To study the performance of a Differentiator using an OP AMP

Theory :

The basic circuit of a differentiator using an OP AMP is shown in [Fig. 10.25(f)]. If V_1 and V_0 are the input and output, then

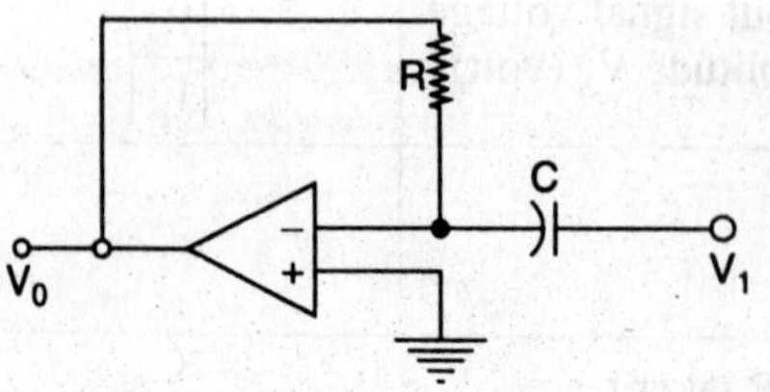

Fig. 10.25(f) : Differentiator circuit

$$V_0 = -CR\frac{dV_1}{dt} \quad \text{... [10.7(G)a]}$$

Since the output voltage is proportional to the derivative of input voltage w.r.t. time, hence the circuit is known as 'differentiator'.

If $V_1 = V_m \cos \omega t$, then

$$V_0 = \omega CRV_m \sin \omega t = V_0' \sin \omega t \quad \text{... [10.7(G)b]}$$

$\therefore \left|\frac{V_0'}{V_m}\right| = \omega CR = 2\pi fCR$, where the input signal frequency is f. Thus a plot of ratio between output and input voltages (rms values) against frequency f, will be a straight line having a slope.

$$m = 2\pi RC \quad \text{... [10.7(G)c]}$$

If R is known, from the slope of the curve, the value of an unknown capacitance may be known.

Apparatus :

IC OPAMP (741C), DC Power supply, resistors (3.3 kΩ), capacitor (0.1 μf), audio oscillator, CRO or electronic voltmeter, Breadboard.

Circuit Diagram

The actual experimental circuit connection is shown in [Fig. 10.25(g)]. The values of R_1 and R_2 may be about 3.3 kΩ and C = 0.1 μf

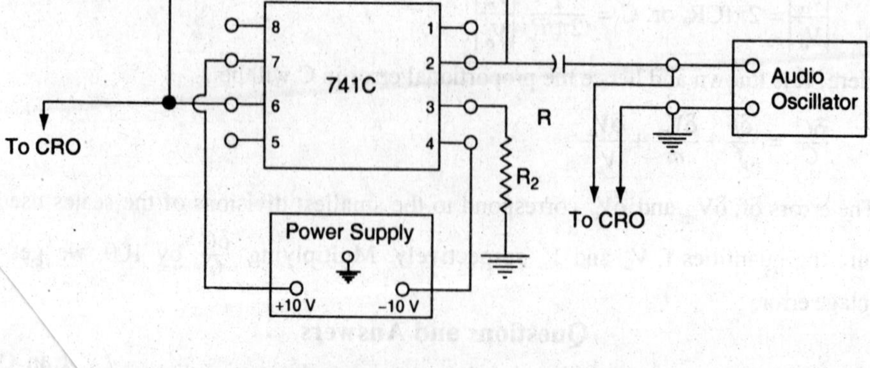

Fig. 10.25(g) : Circuit connection

Method :

(i) Circuit connections are made as shown in [Fig. 10.25(g)]. Proper values of resistors and capacitor taken and power supply is switched on.

(ii) The audio oscilletor is switched on and a suitable low input voltage of about 0.1 volt is applied.

(iii) The frequency is set at about 100 Hz and the waveforms are observed on a dual channel CRO, so that when V_1 has the form cos ωt, V_0 has the form sin ωt. which means that differentiation has been achieved.

(iv) The peak input and output voltages V_m and V_0' are measured.

(v) The frequency f is then increased by steps upto 1 kHz and for each frequency, voltage measurements are done as per (iii).

(vi) The graph is plotted with $\left|\frac{V_0}{V_m}\right|$ against f, which is a straight line having a slope m = 2πRC.

(vii) From the slope m, value of C is obtained having the knowledge of the value of R.

This value of C may be compared with the supplied value of C.

Results :

OP AMP used = ...

Power Supply voltage = ... volt

(A) Frequency Response of the Differentiator :

Table 1

Input signal Frequency f (Hz)	Input signal voltage Amplitude V_m (volt)	Output signal voltage amplitude V_0 (volt)	$\left\|\frac{V_0}{V_m}\right\|$
...	...	...	...
...	...	...	...
...	...	...	...

(B) Determination of Capacitance value (C) :

Table 2

δ(f) from graph (Hz)	$\delta\left\|\frac{V_0}{V_m}\right\|$ from graph	Slope (m) = $\frac{\delta\|V_0/V_m\|}{\delta(f)}$ (Hz^{-1})	Conclusions (whether the value of C agrees with the specified value.)
...	...	...	...

Precautions :

[Same as in Expt. No. 10.7(F)]

[For Questions/Answers, general questions and answers discussed in previous experiments, may be followed and consulted]

10.8 Construction of two-input 'OR' and 'AND' gates using diode logic and preparation of their truth tables

Theory :

OR Gate : An OR gate has two or more input signals but only one output signal. It is designed so that an output signal will occur any time, there is an input signal. In the language of digital electronics, the presence of a signal is represented as a logical 1 and the absence of a signal is represented as logical 0. The symbol for a 2-input OR gate is shown in [Fig. 10.26(a)], where *A* and *B* are the inputs and *Y* (= *A* + *B*) is the output. The circuit diagram for a 2-input OR gate for positive logic using diodes, is shown in [Fig. 10.26(b)]. This configuration is known

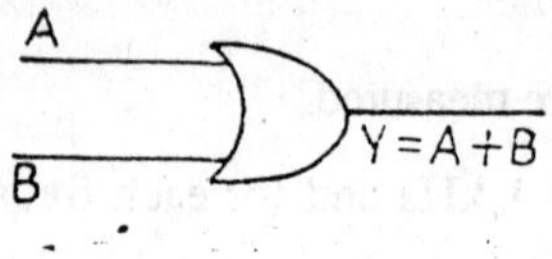

Fig. 10.26(a) : Two input OR gate symbol

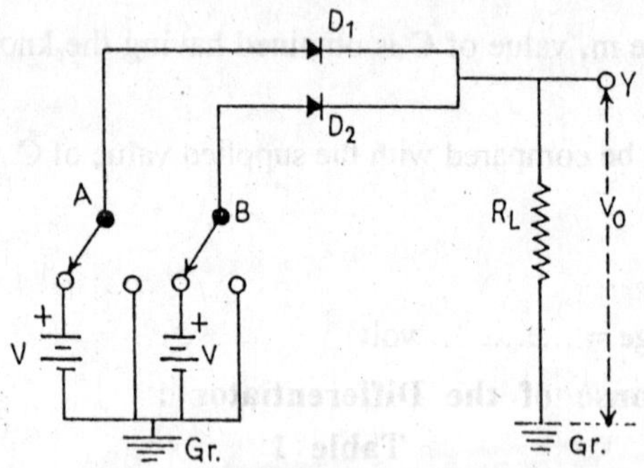

Fig. 10.26(b) : Circuit for two input OR gate

as a diode-logic (DL). Here D_1 and D_2 are two identical diodes having A and B as the two inputs to the OR gate and *Y* is the output. The output load resistance is R_L. It is possible to have either a 0 or 1 present at an input and since there are two inputs to this gate, there must be four possible combinations of inputs signals. The possibilities are :

(i) A = 0, B = 0, when output Y = 0.

(ii) A = 0, B = V volt, when Y = V volt.

(iii) A = V, B = 0, when Y = V volt.

(iv) A = V volt, B = V volt, when Y = V volt.

These are the input-output possibilities of this logic circuit which can be put in the form of a table known as the *Truth Table*. For an OR gate the said table will be as follows :

Input		Output
A	B	Y
0	0	0
0	1	1
1	0	1
1	1	1

It is possible to use any two distinct values of the input voltages and if the lower value is represented by the state 0, the higher value may be represented by the state 1.

The output voltage V_o of the gate [Fig. 10-26(b)] for the case (iv) is given by

$$V_o = \frac{V}{1+\frac{r_F}{R_L}}, \quad \ldots (10.8a)$$

where r_F is the forward resistance of each diode. If r_F is very small compared to R_L, then $V_o \simeq V$. S0, when $A = B = V$ volts, then the output voltage $V_o \simeq V$ volts. Again, for cases (ii) and (iii) when either $A = 0$ or $B = 0$, then the output voltage $V_o = \frac{V}{1+\frac{r_F}{R_L}}$ ... (10.8b)

Again, if $r_F << R_L$, then $V_o \simeq V$. Hence, when $A = 0$ and $B = V$ volts or $A = V$ volts and $B = 0$, in each case, the output voltage $V_o = V$ volts. But when both the inputs are zero as in case of (i), then no current flows through the load R_L and hence, $V_o = 0$. The digital *IC* for *OR* gate is shown in [Fig. 10.27].

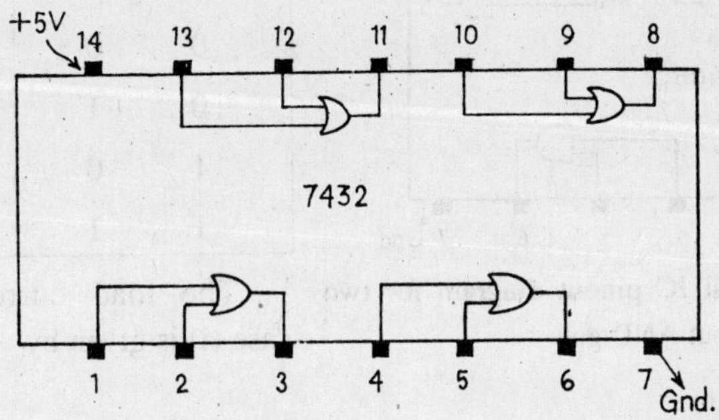

Fig. 10.27 : Digital IC for two input OR gate (Pinout diagram)

AND Gate : An AND gate is so designed that an output signal will be obtained only when there are signals at all the inputs. For a 2-input AND gate, the output assumes the state 1 if and only if both the inputs are in the state 1. The symbol for a 2-input AND gate is shown in [Fig. 10.28(a)] where *A* and *B* are the inputs and *Y* (=*AB*) is the output. The circuit diagram for a 2-input AND gate for positive logic using diodes, is shown in [Fig. 10.28(b)]. Here also, D_1 and D_2 are two diodes having *A* and *B* as the inputs and *Y* as the output. The reference voltage is V_R which is taken to be equal to *V*. The output load resistance is R_L. Here also the possibilities are :

(i) $A = 0, B = 0$, when output $Y = 0$.

(ii) $A = 0, B = V$, volt when output $Y = 0$.

(iii) $A = V$ volt, $B = 0$, when output $Y = 0$.

(iv) $A = V$ volt, $B = V$ volt, when $Y = V$ volt

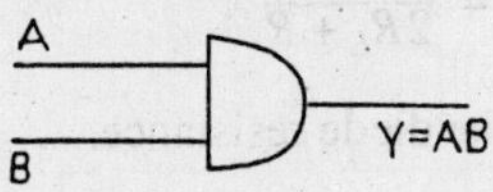

Fig. 10.28(a) : Two input AND gate symbol

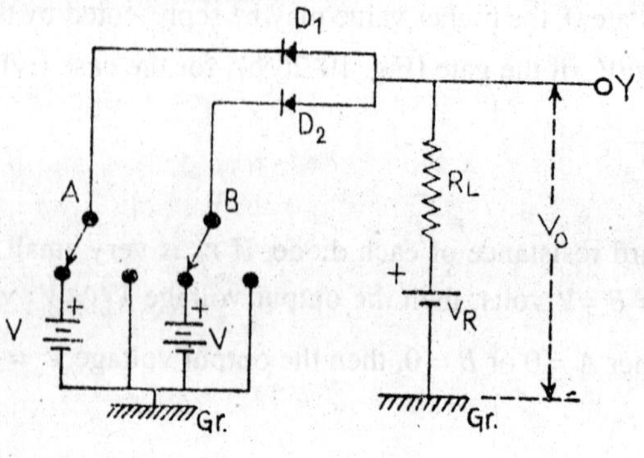

Fig. 10.28(b) : Circuit for two input AND gate with positive logic

The Truth Table of the AND gate is as follows. The digital IC pinout diagram for AND gate is shown in [Fig. 10.29].

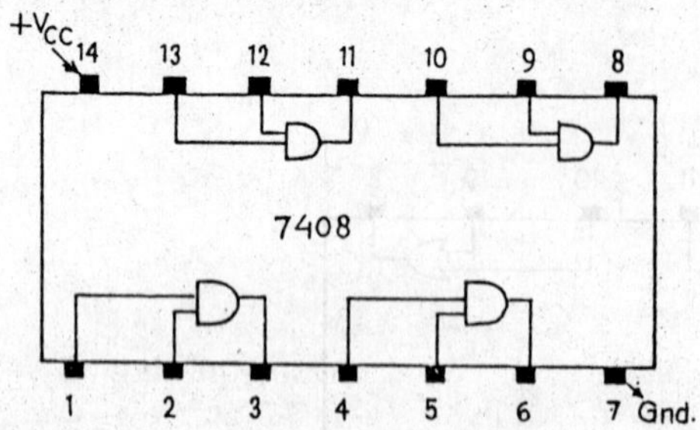

Fig. 10.29 : Digital IC pinout diagram for two input AND gate

Input		Output
A	B	Y
0	0	0
0	1	0
1	0	0
1	1	1

The load current for the first case (i) is given by

$$I = \frac{V_R}{\frac{r_F}{2} + R_L}. \qquad \text{... (10.8c)}$$

If the forward resistance (r_F) of the diode be very small compared to R_L, then $i = \frac{V_R}{R_L}$.

The output voltage (V_o) is then given by

$$V_o = V_R - iR_L = V_R - V_R = 0 \qquad \text{... (10.8}d\text{)}$$

Similarly, for case (ii) and case (iii) when one of the two inputs of the AND gate is grounded successively, then $V_o \simeq 0$, in each case. Finally, for case (iv), when each of the inputs of the AND gate is equal to V, the half of the load current is given by

$$i = \frac{V - V_R}{2R_L + R}. \qquad \text{... (10.8e)}$$

when R is the diode resistance.

Since $V_R = V$, hence $i = 0$. Thus the output voltage is given by

$$V_o = V_R + 2iR_L = V_R + 0 = V_R, \qquad \text{... (10.8}f\text{)}$$

since $i = 0$.

Apparatus : For OR Gate—Two diodes (preferably *OA* 79), two batteries each of about 5 volts, voltmeter (0-10V), one 10 kΩ 1 watt resistor and 2 two-way keys.

Method : (i) Connections for OR gate is done as shown in [Fig. 10.26(b)]. The voltage source (*V*) is a 5-volt battery each, and R_L is the load resistance (10 kΩ).

(ii) First both the inputs are grounded, i.e., both *A* and *B* are connected to the negative terminals of the batteries. Hence *A* = *B* = 0 volt. These voltages as well as the output voltage (V_O) are measured.

(iii) Next, one of the terminals, say *A*, is grounded (*A* = 0 volt) and the other (*B*) is connected to the positive terminal of the battery (*B* = 5 volts). The input and the output voltages are again measured.

(iv) The connections in the inputs are then reversed. That is *A* is made equal to 5 volts and *B* grounded. The input and the output voltages are measured.

(v) Finally, both the inputs are connected to the positive terminals of the two batteries, i.e., *A* = *B* = 5 volts each. These voltages as well as the output voltage are measured.

(vi) All these data are recorded in a tabular form as shown in Table 1.

(vii) The value of the forward resistance (r_F) of the diode is calculated employing Eq. (10.8b).

For AND Gate

Apparatus : Two diodes (preferably, *OA*79), three batteries each of about 5 volts, voltmeter (0-10V), one 10 KΩ/1-watt resistor and 2 two-way keys.

Method : (i) Connections for AND gate is done as shown in [Fig. 10.28(b)]. The value of R_L is 10 KΩ and V_R, *V* and *V* are all 5-volt batteries.

(ii) As before, both the inputs *A* and *B* are grounded (connected to the negative terminal of the batteries) and the input and the output voltages are measured.

(iii) Next, one input (say, *A*) is grounded, and the other (*B*) is connected to 5-volt battery. The input and the output voltages are measured.

(iv) The connections of the inputs are next reversed so that *B* = 0 and *A* = 5 volts. The input and the output voltages are measured.

(v) Finally, both the inputs are connected to the positive terminals of the batteries so that *A* = *B* = 5 volts. The input and the output voltages are measured.

(vi) All these data are recorded in a table as shown in Table 2.

Pinout Diagrams : To study the logic gates like OR, AND, etc. a widely used family of digital devices has been introduced, which is known as *transistor-transistor logic* (TTL). This device is inexpensive, fast and easy to use. It is an IC whose pinout diagrams are shown in [Figs. 10.27 and 10.29]. The Figure 10.27 shows four 2-input OR gates in the IC. Any two diodes can be used for a 2-input gates. After applyng +5 volts (the supply voltage for the TTL devices) to pin 14 and grounding pin 7, one or more of the OR gates can be connected. For the four diodes, the inputs are respectively pins 13, 12 (output 11), pins 10,9 (output 8), pins 1, 2 (output 3) and pins 4, 5 (output 6). similarly, for an AND gate, the pinout diagram is shown in (Fig. 10.29), where the digital IC contains four 2-input AND gates. After connecting a supply voltage of +5 volts to pin 14 and grounding pin 7, any one of the AND gates may be used. The input and the output pins for the respective diodes are symbolically the same as in (Fig. 10.27).

Results :

(A) OR Gate

Nature of diodes used : ....

IC used : ....

Table 1

Truth Table

Input		Output
A	B	Y
0	0	0
0	1	1
1	0	1
1	1	1

Table 2

Data for Truth Table

Input (V in volts)		Output V_o (in volts)
A	B	
0	0	......
0	+ 5	......
+ 5	0	......
+ 5	+ 5	......

(B) AND Gate

Nature of diodes used : .....

IC used : ....

Table 3

Truth Table

Input		Output
A	B	Y
0	0	0
0	1	1
1	0	0
1	1	1

Table 4

Data for Truth Table

Input (V in volts)		Output
A	B	(in volts) V_o
0	0	......
0	+ 5	......
+ 5	0	......
+ 5	+ 5	......

(C) Determination of the value of r_F

From Eq. (10.8b), $V_o = \dfrac{V}{1+\dfrac{r_F}{R_L}}$, the values of V and V_o as obtained from Table 2 above being substituted, [for cases (ii) or (iii)] and the value of R_L in ohm being put, the value of r_F can be obtained in ohms.

Discussions :

(i) An AND gate with a negative logic becomes an OR gate with a positive logic. It is evident from the two tables given under Theory that if the 1s of the second table be replaced by the 0s and vice versa, then a truth table is obtained which is the same truth table for an OR gate as shown in the first table. Similarly, an OR gate with a negative logic becomes an AND gate with a positive logic.

(ii) It is not essential that in OR and AND gate experiments only 5 volts and zero volts be applied to the inputs. In fact, any two values of the input voltages may be chosen remembering that they must be distinguished as being a high and a low voltage. In the case of an AND gate of course, V_R has to be replaced by a battery having voltage equal to the higher of the input voltages applied.

[For Questions, see at the end of Expt. No. 10.9]

10.8(A) Construction and Realisation of NAND/NOR Gates—their Truth Tables

NAND gates are combination logic gates in which a NOT gate follows an AND gate. The circuit symbols of a two input NAND gate is shown in [Fig 10.29(a)]. Its Truth Table is also given in Table 1.

Theory :

A NAND gate is shown in the circuit given in [Fig. 10.29(b)]. The gate is built using DTL (Diode Transistor Logic). The circuit uses only a single 5 volt supply and is very much suitable for IC uses.

If both the inputs are low in zero state, the two diodes D_1 and D_2 will be conducting. Even if one input is zero, the corresponding diode will be conducting and the voltage at P becomes the diode cut-in-voltage (about 0.6 volt). Since a voltage of $2 \times 0.6 = 1.2$ volt is required to make D_3 and D_4 conducting, these two diodes are cut-off.

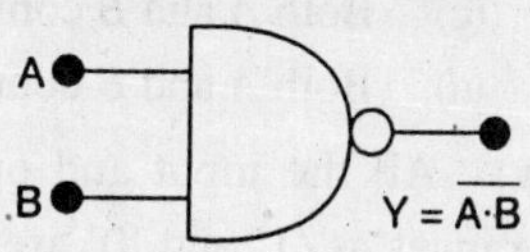

Fig. 10.29(a) : NAND gate

Morever, the transistor Q is on, it needs about $V_{BE} = 0.7$ volt. So this transistor is off and the output is high. If both the inputs are high, the input diodes D_1 and D_2 are off and the voltage at P is sufficient to drive the transistor into saturation, when the output is low. Therefore, the circuit performs the operation of a NAND gate.

Apparatus :

Stabilised d.c. power supply (5V), d.c. voltmeter (0-5V) or digital multimeter, four diodes (better IN 4147 or 4148, In 4002, etc.), transistor (SL-100 or CL-100), carbon resistors, breadboard, two switches, wires, etc.

Method :

(i) Circuit conections are made as shown in [Fig. 10.29 (b)] on a breadboard. Proper identifications of the leads of the transistors and the diodes should be made carefully. The supplied parameters of the transistor are used to calculate the values of R_1, R_2 and R_C.

(ii) The five terminal of the 5V power supply is connected to one side of the bread-board and the (–)ve terminal to another side. These are then +5V and ground line respectively. Connection for '1' state is to be made to the +5V line.

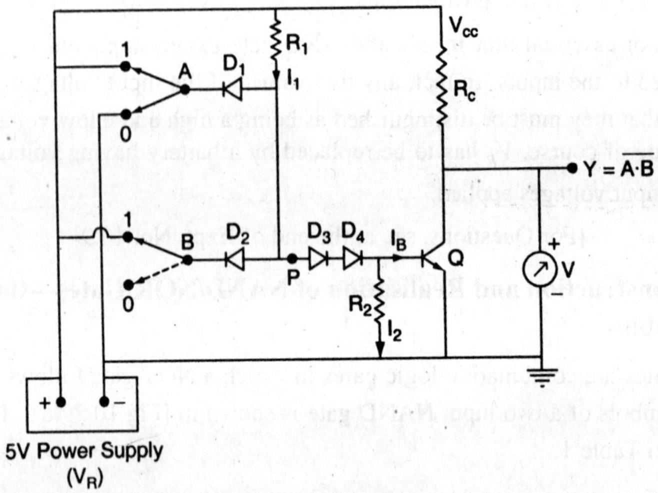

Fig. 10.29(b)

(iii) The voltages at the points *A*, *B* and *Y* are measured by a voltmeter for four different and distinct cases, namely

(a) *A* connected to '1' state and *B* to '0' state.

(b) *A* connected to '0' state and *B* to '1' state.

(c) Both *A* and *B* connected to '0' state.

(d) Both *A* and *B* connected to '1' state.

(iv) All the input and output voltages are recorded in a tabular form. Suitable voltage ranges as '1' and '0' are defined. Another table is prepared which gives the Truth Table of a NAND gate.

Results :

Choice of circuit parameters :

$R_C = 470\Omega$; $R_1 = 15K\Omega$; $R_2 = 33K\Omega$.

Diode used = ...

Cut-in voltage = ...

Transistor used = ...

Data for Input-Output voltages of NAND gate

Table 1

Input voltages (volt) at		Output voltage (volt) at
A	B	Y
0.0	0.0	4.7
5.0	0.0	4.7
0.0	5.0	4.7
5.0	5.0	0.2

[The figures given are just for illustrations.]

Truth Table for NAND gate from Table 1

Table 2

For example, define 0.0-0.8 volts as '0' level and 3.0-5.0 volts as '1' level.

Input		Output
A	*B*	*Y*
0	0	1
1	0	1
0	1	1
1	1	0

Discussions :

(i) Different resistances are so chosen that in the conducting state, currents through the transistor or the diode should not exceed the maximum rated values.

(ii) The experiment is done by using (+)'ve level logic. For (–)'ve logic, the low voltage is taken as '1' state and high voltage is taken as '0' state..

(iii) The NAND circuit should be so designed that the transistor is in saturation state over the possible temperature variation range.

(iv) Absolute values of the input and output voltages are not important. These voltages must be high or low and must lie within certain specified ranges.

Establishing NAND/NOR gates as Universal Gates

NAND and NOR gates possess the property that using only NAND or NOR gates repeatedly, one can have the logical functions of OR, AND and NOT gates. That is why NAND and NOR gates are called Universal gates. NAND is a combination gate in which a NOT gate follows an AND gate. The symbolic circuit of a two input NAND gate is shown in [Fig. 10.29(a)]. Its Truth Table is shown in Table 2.

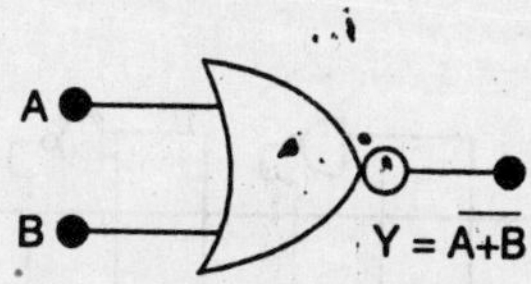

Fig.10.29(c) : NOR gate

The circuit symbol and the Truth Table of a two input NOR gate are shown in [Fig. 10.29(c)] and given in Table 3.

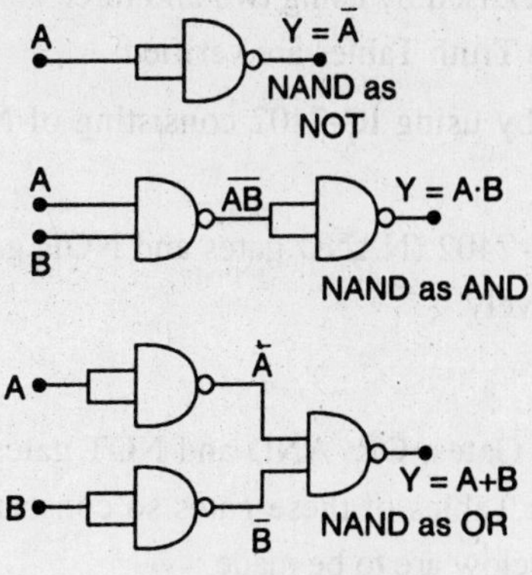

Fig. 10.29(d)

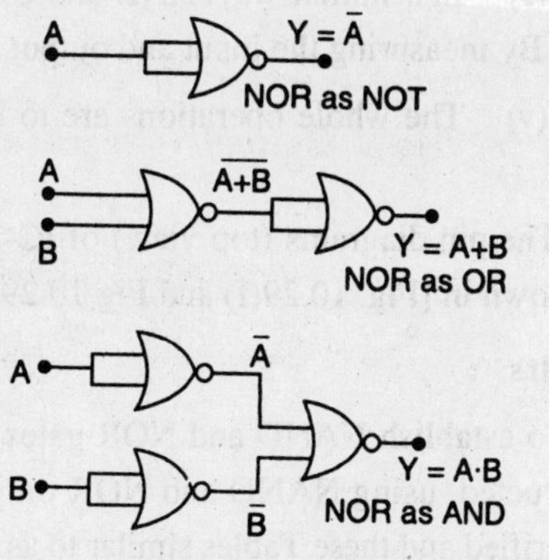

Fig. 10.29(e)

Realisation of different gates using NAND only is shown in [Fig. 10.29(d)] and realisation of different gates using NOR only is shown in [Fig. 10.29(e)] respectively.

Apparatus :

IC-7400 (for NAND gates), IC-7402 (for NOR gates), d.c. power supply (5V), d.c. voltmeter (5V), breadboard, logic switch, single str and wires, etc.

Method :

(i) At first NOT, AND and OR gates are *realised* one by one by using one or more NAND gates of IC-7400 [Fig. 10.29(d)]. As for example, to realise a NOT gate, firstly the pins (input) 1 and 2 are connected together and the common point is connected to input logic switch unit *A* or *B*. The output gate pin is connected (3) to output logic level detector *Y*. The imput terminal is connected alternately to +5V line and ground line.

(ii) The input and output voltages are measured by a voltmeter. Suitable voltage ranges are defined as '0' and '1'.

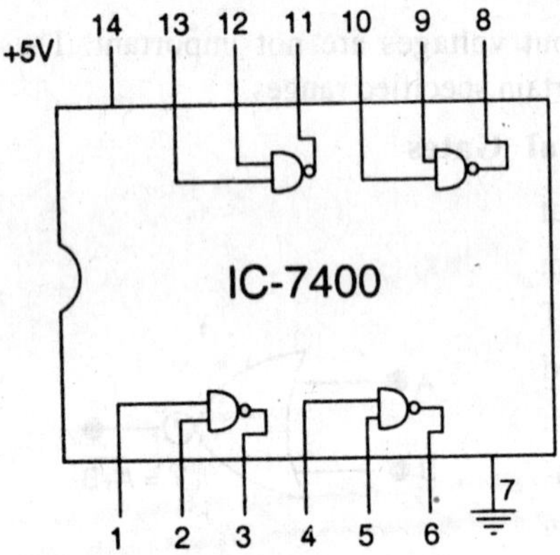

Fig. 10.29(f)

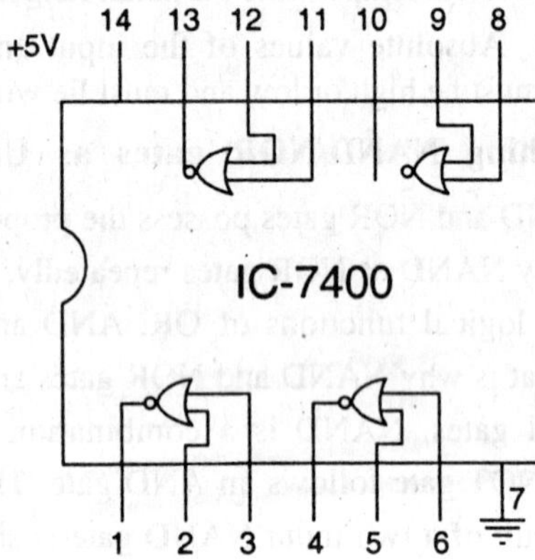

Fig. 10.29(g)

(iii) The Truth Table is obtained for a NOT gate.

(iv) In a similar way, AND and OR gates are realised by using two and three NAND gates. By measuring the input and output voltages, the Truth Tables are verified.

(v) The whole operations are to be repeated by using IC-7402 consisting of NOR gates.

The pin diagrams (top view) of IC-7400 and IC-7402 (NAND gates and NOR gates) are shown in [Fig. 10.29(f) and Fig 10.29(g)] respectively.

Results :

To establish NAND and NOR gates as Univeral Gates, OR, AND and NOT gates are constructed, using NAND and NOR only. The Truth Tables of these gates so constructed are verified and these Tables similar to as mentioned below are to be made :

For OR gates :

Prepare Table 1 and Table 2 of Expt. No. 10.8.

For AND gates :

Prepare Table 3 and Table 4 of Expt. No. 10.8.

For NOT gates :

Prepare Table 1 and Table 2 of Expt. No. 10.9.

Precautions and Discussions :

(i) In connecting +5V d.c. supplly, care should be taken in selecting the IC pins. Wrong connection may damage the IC.

(ii) NAND and NOR gates are called *Universal Building Blocks* since the basic gates like OR, AND, NOT can be constructed by using NAND and NOR only.

[Other points under 'Discussions' in Expt. Nos. 10.8 and 10.9 are to be consulted.]

For questions on this experiment the questions given after Expt. Nos. 10.8 and 10.9 may be read.]

10.9 To study the Operation of a NOT circuit

Theory : The *NOT* circuit has a single input and a single output. Its output assumes the state 1, if and *only* if, the input does not assume the state 1. So, the output is high when the input is low and vice versa. Thus a NOT circuit inverts the sense of the output with respect to the input and hence is called an *inverter*. The symbol for a NOT gate is shown in (Fig. 10.23). The *truth table* for a NOT gate will be as follows. The NOT operation is also known as *negation*.

Input	Output
0	1
1	0

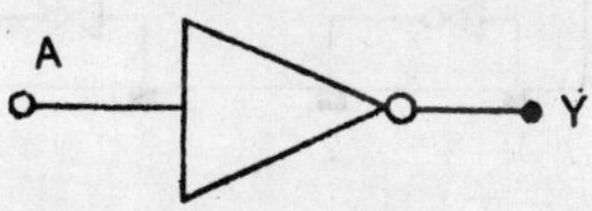

Fig. 10.30(a) : Symbol for NOT gate

Circuit Diagram : The circuit diagram for a NOT gate with positive logic is shown in [Fig. 10.30(b)]. The transistor (*Tr*) operates in the *CE*-mode. When the input voltage V_{in} is zero (connected to *P*), transistor is *off* because its emitter-base junction is reverse biased due to the polarity V_B and thus the current through R_L is almost zero and the output voltage $V_o \simeq V_{CC}$. This state of the input ($V_{in} = 0$) is known as the '0' state and the condition $V_o \simeq V_{CC}$ is referred to as the '1' state of the output. Thus, when the input is in the '0' state, the output is in state 1, as is evident from the Truth Table given above.

Again, when the input voltage $V_{in} = V$ volts (connected to *Q*), the emitter-base junction is forward biased so that the transistor is on. Under this condition, the *CE*-voltage drop V_{CE} is very low, so that the output voltage $V_0 \simeq 0$. The condition $V_{in} = V$ is known as the 1 state of the input and $V_o \simeq 0$ is known as the '0' state of the output. Hence, when the input is in state 1, the output attains the state '0', as is evident from the above Truth Table.

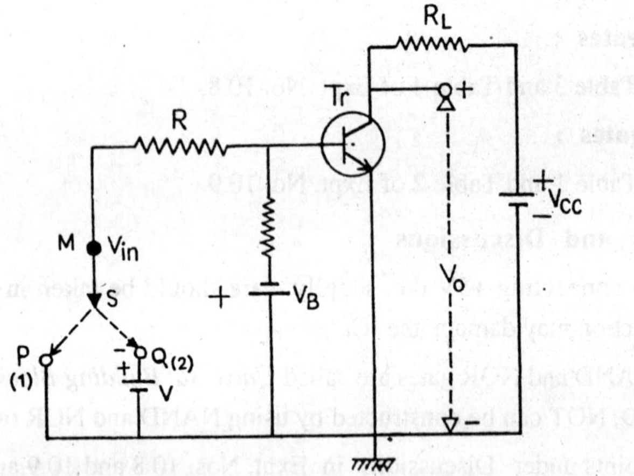

Fig. 10.30(b) : Circuit diagram of a NOT gate for positive logic

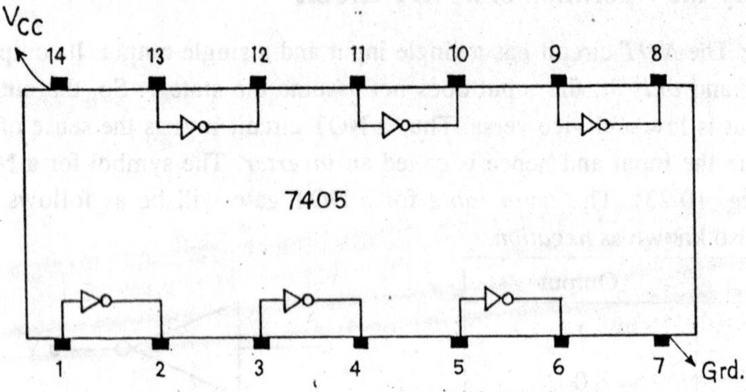

Fig. 10.31 : Pinout diagram for NOT gate

Transfer characteristics :

The value of the input voltage may be varied by using a voltage divider arrangement or a variable d.c. supply. By changing the vlaues of the input voltage in suitable steps, in each step the values of V_{in} and V_o are measured. If a graph be plotted with V_{in} as the abscissa and V_o as the ordinate, then that graph is known as the transfer characteristic of the NOT gate. The nature of the graph is shown in (Fig. 10.32). How the voltage divider arrangement can be operated is discussed under the *Method*.

Apparatus : Suitable transistor (*AC* 127, *AC* 128, *SL* 100, *SK* 100, *BC* 148, *BC* 158), voltage sources (5*V*), resistors (15 KΩ, 470 Ω, 140 KΩ), voltmeter (D.C.), switch.

Method :

(i) The circuit connections are made as shown in [Fig. 10.30(b)].

(ii) The switch (S) is connected to the point P so that V_{in} = 0 and then both V_{in} and V_o are measured by a voltmeter. (*off* position).

(iii) Next the switch (S) is connected to Q, so that $V_{in} = V = 5$ volts and again the values of V_{in} and V_o are measured by the voltmeter. (*on* postion).

(iv) Finally, from these data, the Truth Table is constructed and the operation of a NOT circuit is verified.

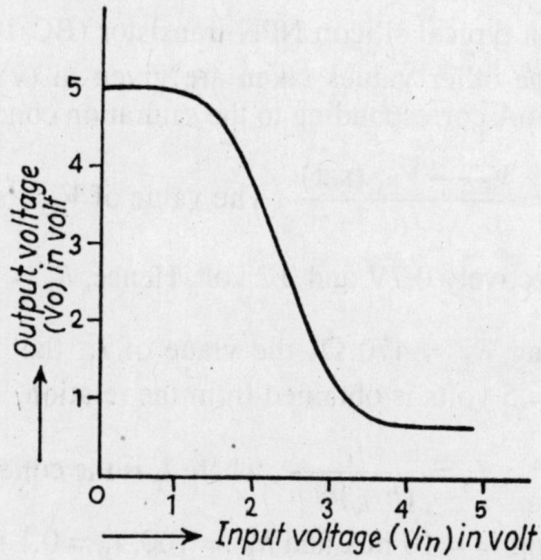

Fig. 10.32 : Transfer characteristic of the NOT gate

(v) It may be noted that typically the transistor used here is a silicon NPN transistor. The voltages V and V_{CC} are about 5 volts and the supply voltage V_B must be less than the base-emitter breakdown voltage which for a *BC* 148 transistor, is about 14 volts. Normally, V_B is taken as equal to 5 volts. The transistors used are usually any one of the following, namely AC 127, AC 128, BC 148, BC 158, SK 100 or SL 100.

(vi) For voltage divider arrangement, a suitable rheostat is connected between the terminals P and Q by removing the switch S and connecting its slider arm to terminal M in [Fig. 10.30(b)].

Results :

Transistor used : ……, Value of V_B = …… volts, Value of V_{CC} = …… volts

Table 1
Truth Table

Input	Output
0	1
1	0

Table 2
Data for Truth Table

Input (V_{in}) in volts	Output (V_o) in volts
0	……
……	……

Table 3
Data for Transfer Characteristic

V_{in} (volts)	V_o (volts)
0	……
……	……
……	……
……	……
……	……
5	……

Calculation of Circuit Parameters :

For a typical silicon NPN transistor (BC 148), the voltages V and V_{CC} are taken as 5 volts. The other values taken are given in (v) under Method. For the collector current $I_C \simeq 10$ mA corresponding to the saturation condition of the transistor, we have

$R_L = \dfrac{V_{CC} - V_{CC}(\text{sat})}{I_C}$. The value of V_{CC} (sat) for Silicon and Germanium transistors are respectively 0.7V and 0.2 volt. Hence, $R_L = \dfrac{5-0.7}{10\times10^{-3}} = 430\ \Omega$ for silicon transistor.

Using $R_L = 470\ \Omega$, the vlaue of I_B that drives the transistor to saturation when $V_{in} = V = 5$ volts is obtained from the relation,

$$I_B = \frac{I_C}{(h_{FE})min}, \text{ where } I_c \text{ is the collector current.}$$

Using $I_C = 10$ mA and $h_{FE} = 100$, $I_B = 0.1$ mA.

When the transistor is on, then the current I (through R) is given by $I = I_1 + I_B$ (where I_1 is current through R_1 and I_B is the base current [see Fig. 10.30(b)].

The currents I and I_1 are given by

$$I = \frac{V-(V_{BE})_{\text{sat}}}{R} \text{ and } I_1 = \frac{V-(V_{BE})_{\text{sat}}+V_B}{R_1},$$ where $(V_{BE})_{\text{sat}}$ is the base-emitter voltage drop under saturation condition, Hence,

$$\frac{V-(V_{BE})_{\text{sat}}}{R} = \frac{V-(V_{BE})_{\text{sat}}+V_B}{R_1} + I_B.$$

Substituting the values of $V = V_{BE} = 5$ volt, $(V_{BE})_{\text{sat}}$ and I_B, the relation between R and R_1 is obtained. Again, when the transistor is off, neglecting the reverse saturation current, the current through R (I_2) is given by $I_2 = I_3 = \dfrac{V_B}{R+R_1}$, where I_3 is the current through R_1.

The B-E voltage V_{BE} under off condition is $(V_{BE})\text{off} = I_3 \times R_1 - V_B = \dfrac{V_B R_1}{R_1 + R} - V_B$.

Taking (V_{BE}) off $= -0.5$ volt and putting $V_B = 5$ volt, one gets again relation between R and R_1. From the two relations between R and R_1 so obtained, it is possible to get the values of R_1 and R. Carbon resistors for values of R and R_1 so obtained, may be used.

Table 4

Reference data for Transistors

Type	Max V_{CB}	Max V_{CB}	Max collector current	Max Power Dissipation	Current Gain	Junction Temp	h_{FE}
AC 127 (Ge *npn*)	+ 32 V	+ 5V	+ 500 mA	300 mW	100 (I_C = 20 mA)	100 °C	100
AC 128 (Ge *pnp*)	– 32 V	– 9V	–1A	100 mW	90 (I_C = – 50 mA)	100 °C	175
BC 148 (Si *npn*)	—	+ 5V	+ 500 mA	350 mW	100 (I_C = 2 mA)	100°C	800
BC 158 (Si *pnp*)	—	– 5V	– 750 mA	120 mW	95 (I_C = – 2 mA)	90 °C	180

N.B. : Apart from the above transistors, transistors SK 100 and SL 100 may also be used in this experiment, whose reference data may be had from any standard transistor guide-book.

Verification of De Morgan's Theorem

For solving logical problems in digital computers, *Augustus De Morgan* gave some links between logic and mathematics. But *George Boole* (1854) invented a new kind of algebra, known as *Boolean Algebra,* which could be used to analyse and design computer circuits. Only three operations are employed on the variables in Boolean algebra. These operations are (i) OR addition represented by a plus (+) sign; (ii) the AND multiplication denoted by a cross (×) or a dot(.) sign and (iii) the NOT operation denoted by a bar (–) over the variable. These operations are not the same as in ordinary algebra. According to these symbols—

(a) For OR operation :

(i) when $A = 0, B = 0; Y = 0$; hence, $Y = A + B = 0 + 0 = 0$.

(ii) when $A = 0, B = 1; Y = 1$; hence, $Y = A + B = 0 + 1 = 1$.

(iii) when $A = 1, B = 0; Y = 1$; hence, $Y = A + B = 1 + 1 = 1$.

(iv) when $A = 1, B = 1; Y = 1$; hence, $Y = A + B = 1 + 1 = 1$.

(b) For AND operation :

(i) when $A = 0, B = 0; Y = 0$; hence, $Y = A.B = 0.0 = 0$.

(ii) when $A = 0, B = 1; Y = 0$; hence, $Y = A.B = 0.1 = 0$.

(iii) when $A = 1, B = 0; Y = 0$; hence, $Y = A.B = 1.0 = 0$.

(iv) when $A = 1, B = 1; Y = 1$; hence, $Y = A.B = 1.1 = 1$.

(c) For NOT operation :

(i) If $A = 0$, then $\overline{A} = 1$.

(ii) If $A = 1$, then $\overline{A} = 0$.

(iii) If $\overline{A} = 0$, then $A = 1$.

(iv) If $\overline{A} = 1$, then $A = 0$.

Hence, here $\overline{0} = 1$ and $\overline{1} = 0$.

De Morgan's Theorems : The two theorems of De Morgan are as follows :

Theorem I : The complement of the sum of variables is equal to the product of the complements of the variables.

Theorem II : The complement of the product of variables is equal to the sum of the complement of the variables.

In logical symbols $\overline{A+B} = \overline{A}.\overline{B}$ ...... (i)

and Boolean notations : and $\overline{A.B} = \overline{A} + \overline{B}$ ...... (ii)

Table for verfication of De Morgan's theorems :

Table 5

(a) To prove the first theorem : $\overline{A+B} = \overline{A}.\overline{B}$

L.H.S.				R.H.S.		
A	B	$A + B$	$Y = \overline{A+B}$	$\overline{A}$	$\overline{B}$	$Y = \overline{A}.\overline{B}$
0	0	0	1	1	1	1
0	1	1	0	1	0	0
1	0	1	0	0	1	0
1	1	1	0	0	0	0

Hence, L.H.S. = R.H.S. (*Proved*)

Table 6

(b) To prove the second theorem : $\overline{A.B} = \overline{A+B}$

L.H.S.				R.H.S.		
A	B	$A.B$	$Y = \overline{A.B}$	$\overline{A}$	$\overline{B}$	$Y = \overline{A+B}$
0	0	0	1	1	1	1
0	1	0	1	1	0	1
1	0	0	1	0	1	1
1	1	1	0	0	0	0

Hence, L.H.S. = R.H.S. (*Proved*)

Therefore, De Morgan's theorems are verified using two input OR, AND and NOT gates. This method may be extended for any number of operations and combinations, following the same principles of Boolean algebra, like determining the output (Y), using the gates circuits, of the logic $(A + B)(B + C)(C + A)$ or verifying the logic identities like (a) $ABC + A\overline{B}C + AB\overline{C}$ = A (B + C) or, (b) $AB + AC + B\overline{C} = AC + B\overline{C}$.

Discussions :

(i) The parameter h_{FE} varies with temperature. Hence, the circuit must be designed (NOT) in such a way that the transistor is in saturation over the possible range of temperature variation.

(ii) While taking data for transfer characteristics, for a particular value of V_{in}, two values of V_o may be measured—one while V_{in} is increasing and the other when V_{in} is decreasing. The mean value of V_o may thus be taken.

(iii) The pinout diagram for OR gate (Fig. 10.27) shows four diodes. The outputs of these gates may be fed as input to any successive gates as per requirements. Similarly, (Fig. 10.29) shows pinout diagrams for 2-input AND gates with four diodes. Their outputs may be utilised as inputs for any complicated logic circuits. The same is the case with the pinout diagram for the NOT gate (Fig. 10.31) where six gates have been shown.

Suggested Questions

(For Expt. No. 10.5 and 10.6)

1. What is meant by logic circuit or logic gate? What are the most common logic gates?

2. What do you mean by a digital circuit? How is its output designated? What are negative and positive logic systems?

3. An AND gate is followed by a NOT gate. with 2-inputs *A* and *B*, find the Boolean expression of the output C. (C.U. 1981, 1987)

4. How an OR gate may be constructed and AND and NOT gates?

5. What is meant by the truth table of a logic circuit?

6. Why an AND gate is also called a coincidence circuit?

7. What is Boolean algebra? Write and explain the Boolean expressions for the AND, OR and NOT gates.

8. What are meant by NAND and NOR gates?

9. Explain how all logic operations can be done by using only the NOT and OR circuits or with only NOT and AND circuits.

10. State De Morgan's theorems.

11. What is the utility and significance of the 'transfer characteristic'?

12. How the NAND and NOR gates may be constructed using OR, AND and NOT gates?

13. What are meant by the circuit parameters and how may they be calculated?

14. What are the applications of the digital circuits?

10.9(A) Verifications of Boolean expressions using IC gates

Boolean Algebra and expressions, which can be used to analyse and design computer circuits, have already been discussed in Expt. No. 10.9 under Verification of De Morgan's Theorem.

Suppose we have to verify the following three Boolean expressions :

(a) $A + \overline{A}B = A + B$

(b) $A(\overline{A} + B) = AB$

(c) $(A + B)(\overline{A} + C) = AC + \overline{A}B$.

It has already been stated that only three operations are employed on the variables in Boolean Algebra. These are :

(i) OR addition by a (+) sign

(ii) AND multiplication by a (×) or (.) dot sign.

(iii) NOT operation by a bar (‾) over the variable.

A circuit may be developed by combining AND, OR and NOT gates to verify the Boolean expressions. For examples, a circuit is shown [Fig. 10.32(a)] to verity, say, the first relation.

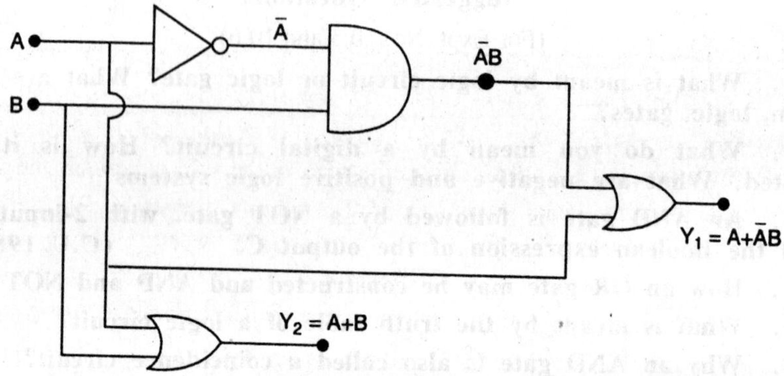

Fig. 10.32(a)

Theory :

The schematic diagram for the verification of the Boolean expression $A + \bar{A}B = A + B$ is shown in Fig. 10.32(a). The purpose of this experiment is to show only that $Y_1 = Y_2$, which obviously verifies the expression or the identity mentioned.

Apparatus :

OR, AND and NOT gates (See Expt. No. 10.8), a buffer [see Fig. 10.22(a)], power supply, wires, breadboard, One d.c. voltmeter.

Method :

(i) The gates are arranged as shown in [Fig. 10.32(a)]. The buffer may preferably be used between two successive gates, when the loading by one gate by the next one is not appreciable.

(ii) The inputs A and B may have the combinations of various high and low values and for each, the voltages corresponding to $\bar{A}$, $\bar{A}B$. and $Y_1 = A + \bar{A}B$; $Y_2 = A + B$ are all measured and tabulated.

(iii) From the data, the Truth Tables are constructed and it is shown that $Y_1 = Y_2$, which verifies the particular Boolean expression.

Results :

(A) Data in tabular form.

(B) preparation of the Truth Tables

[Tables similar to those made for verification of De Morgan's theorem under Expt. No. 10.9]

Precautions :

[Same as in Expt. No. 10.9]

Questions and Answers

(Questions on Expt. Nos. 10.8 and 10.9 and different gates, may be consulted.)

N.B. : Developments of similar circuits may be made to verify the other two Boolean expressions, namely (b) and (c) given at the beginning of this experiment.

This may be treated as exercises.

10.9(B) To design a Half-Adder circuit using basic gates and to verify its truth table

Theory :

A half-adder has basically two inputs (A and B) denoting the bits to be added and two outputs (C and D). The output D represents the sum of A and B when the sum is represented by one single digit [Fig. 10.32(b)]. But if this sum is greater then being represented by a single digit, then D gives the digit in the sum which has the same significance as when the individual digits are added. The output C generally gives the carry bit.

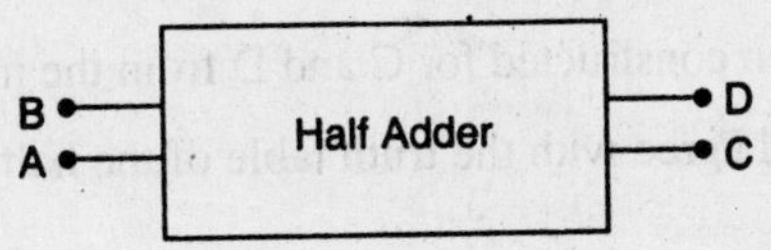

Fig. 10.32(b) : Half-adder logic circuit

Boolean expression for C and D are

$$C = AB$$

$$D = \overline{A}B + A\overline{B} = A\overline{A} + \overline{A}B + A\overline{B} + B\overline{B}$$

$$= (A+B)(\overline{A} + \overline{B})$$

$$= (A+B)\overline{AB} \text{ [since } \overline{A} + \overline{B} = \overline{AB} \text{ by De Morgan's theorem]}$$

Apparatus :

IC 7404, 7408 and 7432, power supply (5V), DC voltmeter (5V), Breadboards, wires etc.

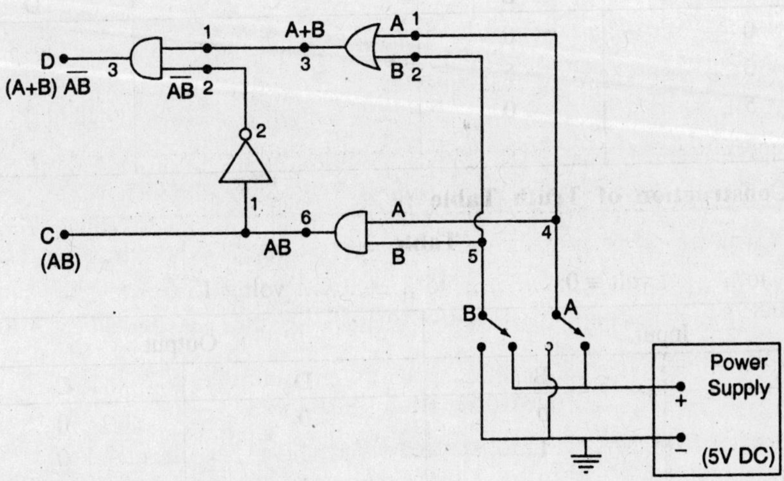

Fig. 10.32(c) : Half-adder circuit diagram

Method :

(i) All the IC's are mounted on the breadboard and it is checked that all the gates functioning properly.

(ii) The circuit connections are made as shown in [Fig. 10.32(c)]

(iii) The DC power supply is then switched on and it is set at 5 volt. This voltage is measured by the DC voltmeter.

(iv) All possible logic states of the input voltages *A* and *B* are taken by using the respective switches. For each combination of the inputs, the output voltages are measured.

(v) Proper voltage ranges are denoted by 0 and 1.

(vi) Truth tables are then constructed for C and D from the measured voltages.

(vii) This truth table will agree with the truth table of the half-adder (Table-1).

Result :

(A) Truth Table of the Half-adder :

Table 1

Input		Output	
A	B	D	C
0	0	0	0
0	1	1	0
1	0	1	0
1	1	0	1

(B) Records for Input-output voltages :

Table 2

Input voltage (volts)		Output voltage (volts)	
A	B	C	D
0	0	...	...
0	5	...	...
5	0	...	...
5	5	...	...

(C) Construction of Truth Table :

Table 3

........to .... volt ≡ 0; ....to ....volt ≡ 1

Input		Output	
A	B	D	C
0	0	0	0
0	1	1	0
1	0	1	0
1	1	0	1

Discussions :

It is to be noted that apart from the method stated above, half-adders may also be constructed by using (i) NAND gates only, (ii) NOR gates only, (iii) AND and XOR gates. Corresponding circuits are to be made accordingly.

10.9(C) To design a Full-Adder circuit using basic gates and to verify its truth table

Theory :

If two binary digits are added, then excepting the least significant bit position, there are carry-out and carry-in C_i and C_{i-1} respectively at the i-th position. A full-adder is a special device that can handle a carry-in while adding two binary digits.

Suppose A_i and B_i are the two bits to be added and C_{i-1} is the carry-in, C_i is the carrry-out, S_i is the sum. The respective inputs are A_i, B_i and C_{i-1} and the corresponding outputs are S_i and C_i.

The truth table of the full-adder will be as follows :

Inputs			Outputs	
A_i	B_i	C_{i-1}	S_i	C_i
0	0	0	0	0
0	1	0	1	0
1	0	0	1	0
1	1	0	0	1
0	0	1	1	0
0	1	1	0	1
1	0	1	0	1
1	1	1	1	1

Boolean expressions for C_i and S_i are

$$C_i = C_H + S_H C_{i-1}$$

$$S_i = S_H \overline{C}_{i-1} + \overline{S}_H C_{i-1}$$

The logic circuit, using the basic gates, for the full adder circuit is shown in [Fig. 10.32(d)]

Apparatus :

[Same as in Expt. No. 10.9(B)]

Method :

(i) After mounting the chips on the breadboard, circuit connections are made as shown in [Fig 10.32(d)].

(ii) The power source is then switched on and set at 5 volts. It is measured by a d.c. voltmeter.

(iii) All possible logic states of the input voltages A_i, B_i, and C_{i-1} are taken by using the respective switches and the output voltages C_i and S_i are measured by the voltmeter for each combination of the inputs .

(iv) Proper voltage ranges are denoted by 0 and 1.

(v) Truth tables are constructed for C_i and S_i from the measured voltages.

(vi) This truth table will agree with the truth table of the full adder given above under theory.

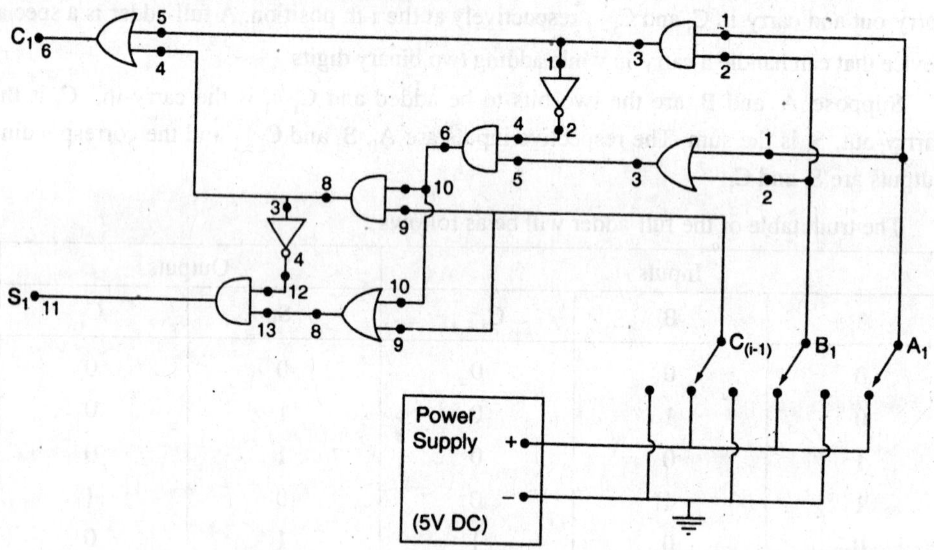

Fig. 10.32(d) : Full-adder circuit diagram

Results :

(A) Records for Input-Output voltages :

Table 1

Inputs (volts)			Outputs (volts)	
A_i	B_i	C_{i-1}	S_i	C_i
0	0	0	...	...
0	5	0	...	...
5	0	0		
5	5	0		
0	0	5	...	...
0	5	5		
5	0	5		
5	5	5	...	...

(B) Construction of Truth Table :

Table 2

...... to .... volt ≡ 0; .... to .... volt ≡ 1

Input			Outputs	
A_i	B_i	C_{i-1}	S_i	C_i
0	0	0	0	0
0	1	0	1	0
1	0	0	1	0
1	1	0	0	1
0	0	1	1	0
0	1	1	0	1
1	0	1	0	1
1	1	1	1	1

Discussions :

This truth table will agree with the full-adder truth table. Here also, full-adder may also be constructed using NOR and NAND gates only.

10.10 Construction of a regulated power supply on a breadboard using a power transistor as pass element, a second transistor as a feedback amplifier and a Zener diode as a reference voltage source and to study its operational characteristics

Theory :

A class of power supply system, which not only regulates the output, but also responds rapidly to common signals and varies the output level accordingly, is generally known as *programmable supply.* Such a basic series regulator is shown in (Fig. 10.33). The circuit achieves regulator by comparing a sample of output voltage with a reference voltage. The voltage divider network consisting of R_2 and R_3 (Fig. 10.33) provides the sample signal and across the zener diode the reference voltage is taken. The output sample is compared with V_Z by the d.c. amplifier (Amp). It produces a signal proportional to the difference between the two. The difference of error voltage is amplified and it appears across R_1. The series (pass) transistor T_1 is operated as an emitter follower. This is the control element of the system, which interprets the signal from the d.c. amplifier and adjusts its internal resistance such that a constant output voltage is maintained. The d.c. amplifier may be a single CE stage or an emitter coupled difference amplifier or a high gain operational amplifier. The choice depends on the degree of regulation and temperature stability required.

At constant temperature, the output voltage (V_o) of the system is a function of the input voltage (V_1) and the load-current (I_L). That is.

$$V_o = f(V_1, I_L) \qquad \text{... (10.10a)}$$

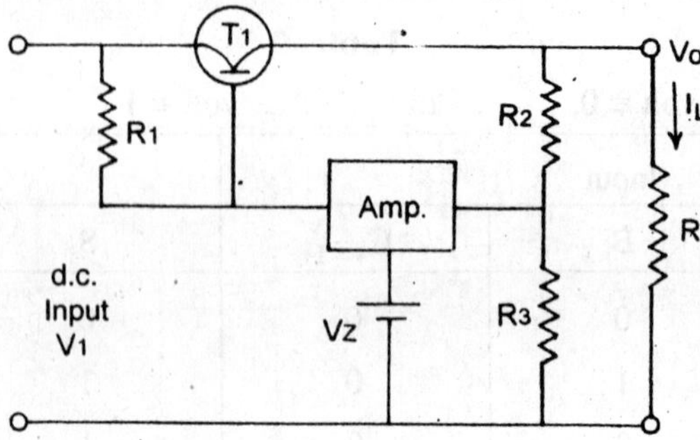

Fig. 10.33 : Basic series regulator

$$\Delta V_o = \frac{\delta V_o}{\delta V_1}\Delta V_1 + \frac{\delta V_o}{\delta I_1}\Delta I_L = S_v \Delta V_1 + R_0 \Delta I_L$$

$$= S_v v_1 + R_0 i_L, \qquad \text{... (10.10b)}$$

$$[\Delta V_1 = v_1 \text{ and } \Delta I_L = i_L.]$$

where the regulation factor $S_v = \left.\frac{\Delta V_o}{\Delta V_o}\right| I_L$ constant ... (10.10c)

and the output resistance $R_0 = \left.\frac{\Delta V_o}{\Delta I_L}\right| V_I$ constant. ... (10.10d)

A reasonably simple but effective power supply system may be analysed as shown in Fig. 10.34. Since the output resistance (R_0) of the unregulated power supply is usually small, any change in its output voltage ($\Delta V_1 = v_1$) appears across the regulator circuit as shown in the figure. Neglecting the change in the base-emitter voltage of T_1, the change in the current ($\Delta i = i$) is given by

$$i = \frac{\Delta V_1 - \Delta V_o}{R_1} = \frac{v_i - v_o}{R_1} \approx \frac{v_1}{R_1} \quad (\because \Delta V_o = v_o << v_i). \qquad \text{... (10.10e)}$$

For the output voltage to remain constant (at R_L constant), I_L and hence i_{B_1} to remain constant is

$$i = \Delta i_{C_2} = i_{C_2}. \qquad \text{... (10.10f)}$$

It can he shown that for small values of R_1, $i_{C_2} = K v_0$,

$$\text{where } K = h_{fe2}\frac{R_3}{R_2 + R_3} - \frac{1}{(R_2 + R_3) + h_{ie_2} + (1 + h_{fe_2})\, r_Z} \qquad \text{... (10.10g)}$$

(r_z– dynamic resistance of the zener)

Using the above equation, we get for

$$S_v = \frac{1}{KR_1} \qquad \text{... (10.10h)}$$

$$\text{and} \quad R_0 = \frac{r_0 + \left\{\left(\frac{R_1 + h_{ie1}}{1 + h_{fe1}}\right)\right\}}{1 + K(R_1 + r_0)} \qquad \text{... (10.10i)}$$

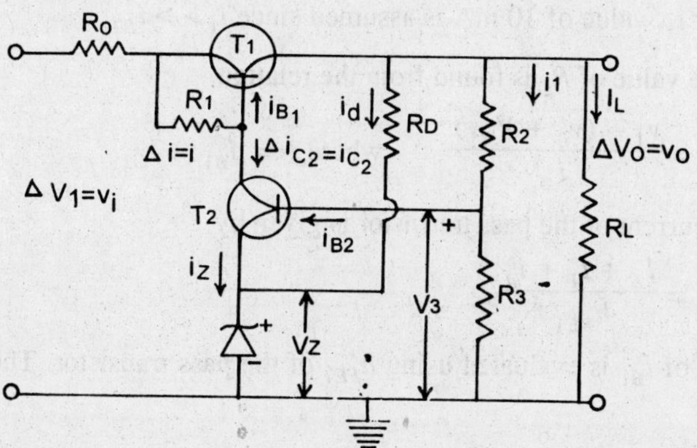

Fig. 10.34 : Voltage regulator

Apparatus :

Unregulated power supply (variable output), multimeter, transistors, zener diode, milliammeter, resistors.

The suitable choice of the different apparatus for the experiment, is suggested below :

(a) Zener diode : FZ 9.1A

(b) Pass transistor (T_1) : AC 128

(c) Transistor (T_2) : AC 125.

Method :

(i) The objective of the experiment is, say, to design a series regulated power supply to provide a nominal output of 15V at 100 mA. The unregulated power supply gives an output of 20 ± 2V and its output resistance (R_0) is 10 ohms.

(ii) A suitable silicon Zener diode with $V_z \approx \frac{v_0}{2}$ is selected. A zener F_z 9.1A may be chosen which gives a constant voltage of 9.1 volts at $i_z = 20$ mA.

(iii) $i_{c_2} = i_{e_2} = 10$ mA is chosen. AC 125 may provide this current. The values of h_{fe} h_{ie} of this transistor for i_{c_2} (10 mA) are determined.

(iv) A suitable pass transistor (AC 128) is selected. It gives a power output of 2 watt. The collector current is 500 mA. The values of h_{ie} and h_{fe} of this pass transistor, are determined.

(v) The value of R_D to be connected in series with the zener diode is obtained form the relation $R_D = \frac{(v_0 - v_z)}{i_d}$. The value of i_d is chosen as 10 mA, so that $i_z = 10 + 10 = 20$ mA.

(vi) The values of the components of the sampling network are obtained from the following relations :

$$R_2 = \frac{(v_0 - V_3)}{i_1};\ V_3 = V_2 + V_{BE2};\ R_3 \simeq \frac{V_3}{i_1}.$$

For i_1, value of 10 mA is assumed since $i_1 >> i_{B2}$.

(vii) The value of R_1 is found from the relation,

$$R_1 = \frac{v_1 - (v_0 + V_{B1})}{i}, \qquad \text{where } i = i_{B1} + i_{C2}.$$

The base current of the pass transistor is given by

$$i_{B1} = \frac{I_L + i_1 + i_d}{h_{FE1}}.$$

The value of i_{B1} is evaluated using h_{FE1} of the pass transistor. The value of $V_{BE} = 0.7$ volt.

(viii) With all the specified and recommended values of the components of the circuit, the circuit connections are made as shown in (Fig. 10.34).

(ix) Keeping the input voltage constant, R_L is varied and the output voltage is measured as a function of the load current (measured by a milliammeter connected in series with R_L). The results are represented graphically. The output impedance R_0 is measured and compared with the theoretical value as given in Eq. (10.10i).

(x) Maintaining the load at an optimum value, the input voltage is varied from 15 to 25 volts and the corresponding output voltage is measured. The results are represented graphically. The stabilisation factor is evaluated using the results and it is compared with the theoretical value given in Eq. (10.10h).

Results :

Suitable and convenient tables are to be prepared for recording the different variable quantities (output voltage, load current, output impedance, input voltage, stabilisation factor, etc.)

Precautions :

(i) Suitable values of the components are to be chosen to get the best results.

(ii) The circuit should be properly earthed.

(iii) Soldering of joints must be carefully done.

Questions

1. What is the need of voltage regulation?

2. What is programmable power supply?

3. Explain the terms : 'load regulation', 'ripple rejection'.

4. What is the action of a series regulator?

5. What are the limitations of a zener regulated supply?

6. What factors limit the maximum d.c. load current of the regulator circuit of (Fig. 10.34)?

7. What modification is necessary in the circuit of (Fig. 10.34) for providing short-circuit overload protection?—Explain.

8. What factors limit the minimum load for the regulator?

10.11 Verification of Thevenin, Norton and maximum power transform theorems using resistive Wheatstone bridge, D.C. source and D.C. meters

Theory :

Analysis of many complex electronic circuits may be facilitated by replacing all or a part of a network by an equivalent circuit which has the same voltage-current characteristics as the original. Two simple network theorems useful in arriving at the equivalent circuits, are **Thevenin's** theorem and **Norton's** theorem.

Thevenin's theorem : This theorem states that any network of resistors and batteries having two output terminals, may be reduced to one equivalent voltage source (V_{Th}) in series with one equivalent resistance (R_{Th}). The equivalent e.m.f. is just the voltage at the output terminals when the load current is zero or the open-circuit voltage. The equivalent resistance is the resistance at the output terminals when the batteries are replaced by zero resistances or by their internal resistances if they are appreciable.

Thevenin's equivalent circuit of the Wheatstone bridge is shown in [Fig. 10.35(a)]. The equivalent resistance is given in [Fig. 10.35(b)]. This equivalent resistance

$$R_{Th} = \frac{R_1 R_3}{R_1 + R_3} + \frac{R_2 R_4}{R_2 + R_4} \quad \text{.... (10.11a)}$$

The open-circuit voltage at the output terminals is simply the Potential difference between A and B.

It is given by $V_{Th} = \dfrac{VR_3}{R_1 + R_3} - \dfrac{VR_4}{R_2 + R_4}$ ... (10.11b)

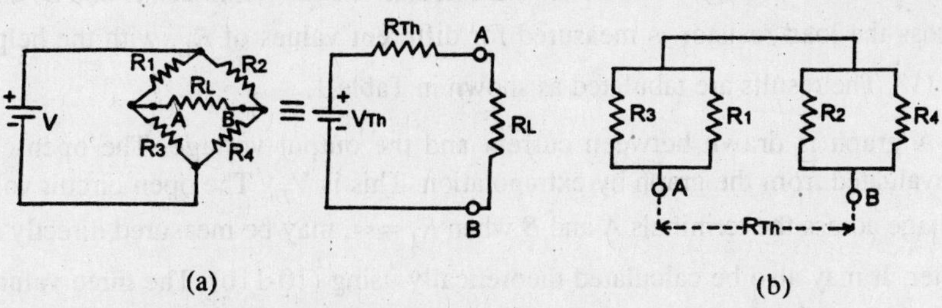

Fig. 10.35 : Wheatstone bridge and its Thevenin equivalent

Thevenin's theorem replaces a complicated circuit with a voltage source and a resistance in series.

Norton's theorem, in solving the same problem, permits replacement of the circuit by a current source I_N and resistance R_N in parallel (Fig. 10.36). The equivalent current source is the current that flows in the circuit with the output short-circuited. The equivalent resistance (R_N) is obtained in the same way as that for R_{Th}. In fact, $R_N = R_{Th}$. Then the equivalent current source $I_N = \frac{V_{Th}}{R_{Th}}$. The power in the load R_L is given by

$$P_L = \frac{V_{Th}^2}{(R_{Th} + R_L)^2} \cdot R_L. \quad \text{... (10.11c)}$$

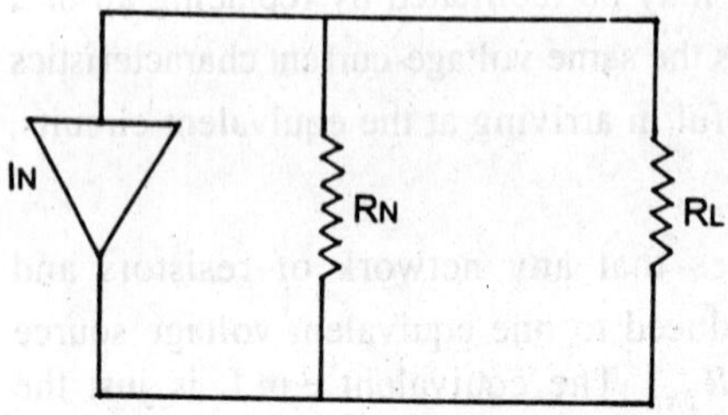

Fig. 10.36 : Norton equivalent circuit

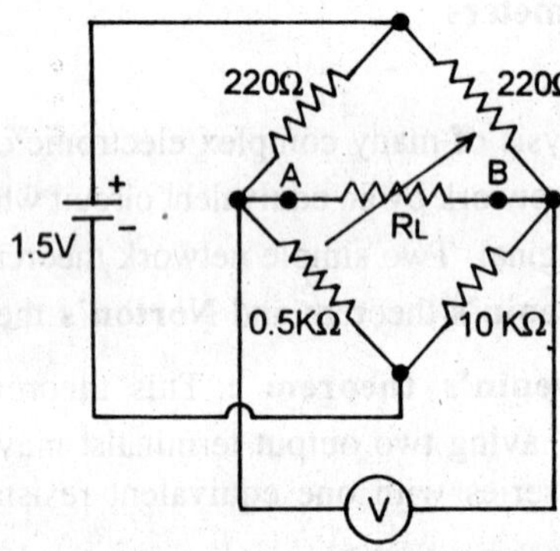

Fig. 10.37 : Equivalent circuit evaluation

It can be shown that P_L is maximum when $R_{Th} = R_L$, i.e., the source delivers maximum power when the load resistance is equal to the internal resistance of the source which is *Power Transfer Theorem*. In this experiment, both Thevenin's and Norton's theorems (equivalent of the Wheatstone bridge) are derived. The power theorem is also verified with the bridge circuit as shown (Fig. 10.37).

Apparatus :

Carbon Resistors, resistance box, multimeters, battery (1.5V).

Method :

(i) Bridge circuit is made as shown in (Fig. 10.37).

(ii) A variable resistance box R_L is connected across the terminals (*A* and *B*) and the voltage across the load resistor is measured for different values of R_L, with the help of a multimeter (*V*). The results are tabulated as shown in Table 1.

(iii) A graph is drawn between current and the output voltage. The open-circuit voltage is evaluated from the graph by extrapolation. This is V_{Th}. The open-circuit voltage, i.e., the voltage across the terminals *A* and *B* when $R_L = \infty$, may be measured directly using a multi-meter. It may also be calculated theoretically using (10-11b). The three values are then compared.

(iv) Next R_L is removed from the circuit and the short-circuit current is measured, i.e., I is measured when $R_L = 0$, with a current meter as shown in (Fig. 10.38). The short-circuit current is the equivalent current source I_N. It may be considered whether the resistance of the meter may be neglected in this experiment.

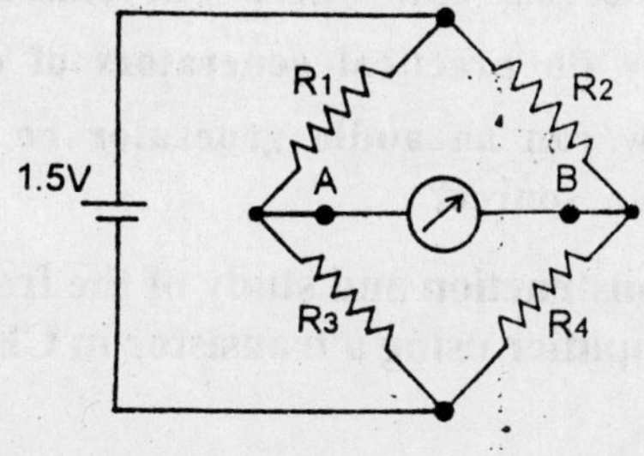

Fig. 10.38 : Norton current measurement

(v) The meter is then removed and the 1-5V battery is replaced by a short circuiting wire. The resistance at the output terminals A and B is measured in this condition, with the help of a multimeter. This is equal to $R_{Th} = R_N$.

(vi) Finally, a graph is drawn between power (P) and R_L. The power transfer theorem may thus be verified.

Results :

Table 1

(A) Thevenin Equivalent Circuit :

Serial No.	Value of R_L (ohms)	Value of V (volts)	$I = V/R_L$ (mA)	Power $P = I^2 R_L$ (watts)
1.				
2.				
3.				
4.				
...				
...				

(B) Graph Drawings :

Table 1

Serial No.	Current (mA)	Output Voltage (V)	Open circuit voltage from the graph
1.	...	...	...
2.	...	...	...
3.	...	...	...
...	...	...	...
...	...	...	...

Table 2

Serial No.	Power P (watt)	R_L (ohm)	Conclusion
1.	...	...	...
2.	...	...	...
3.	...	...	...
...	...	...	...
...	...	...	...

Questions

1. What is the essential difference between Thevenin's theorem and Norton's theorem?

2. Can you reporduce Thevenin equivalent circuit of a resistor?

3. If you are provided with two complex circuits— one involving vacuum tubes and the other involving transistors— where do you use Norton's theorem and where Thevenin's theorem? Justify your answer.

4. Why do practical generators of e.m.f. have internal impedance?

5. How can an audio generator be classified,— as a current source or as a voltage source?

10.12. Construction and study of the frequency-response of a voltage amplifier using a transistor in CE mode and to find the bandwith

Theory :

Frequency response of an amplifier is said to be the change of amplifier gain with frequency of the input source. It can be studied by constructing a circuit with a transistor, an input oscillator. From the overload characteristic study and by measuring the output voltage, the voltage gain may be obtained. By changing suitably, the frequency steps up to the highest oscillator frequency, a graph showing variation (Fig. 10.40) of frequency with voltage gain, may be plotted. The voltage gain (A_v) is given by the relation,

$$A_v = \frac{\text{output voltage } (v_o)}{\text{input voltage } (v_i)} \quad \text{... (10.12a)}$$

The bandwith (B.W.) is given by B.W. = $(f_2 - f_1)$, ... (10.12b)

where f_1 and f_2 are obtained from (Fig. 10.40).

The gain bandwidth is given by the product of $(f_2 - f_1)$ and A' as obtained from the graph (Fig. 10.40).

The circuit diagram for the experiment is shown in (Fig. 10.39).

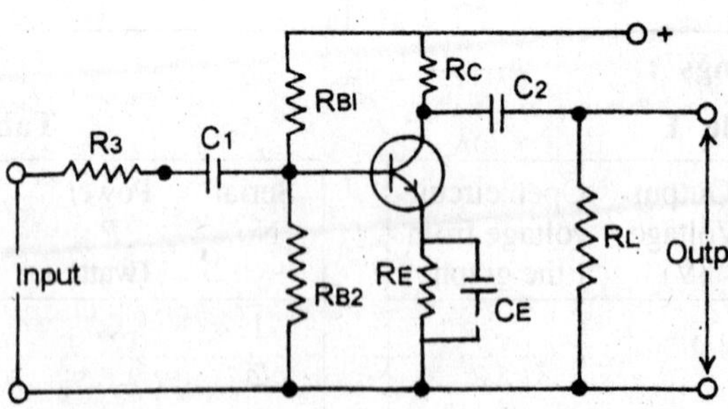

Fig. 10.39 : Study of frequency response

Apparatus :

Audio oscillator, transistor (452), capacitors, resistors.

Method :

(i) The circuit connections are made as shown in (Fig. 10.39). The prescribed suitable values of the components are as follows :

Transistor (452), $C_1 = C_2 = 1 \mu f$, $R_{B1} = 33\ k\Omega$, $R_{B2} = 5.6\ k\Omega$ $R_C = 5.6\ k\Omega$, $R_E = 1\ k\Omega$, $C_E = 10\ \mu F$, $R_L = 5.6\ k\Omega$.

Necessary modifications in these values may be necessary if situation demands.

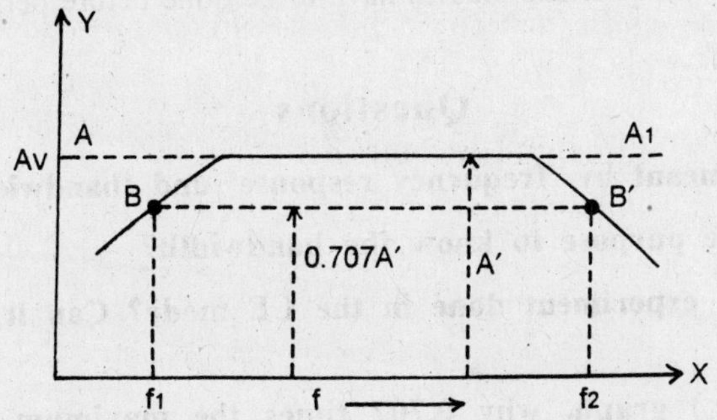

Fig. 10.40 : $(f - A_v)$ graph

(ii) The audio oscillator (input) is set at its lowest frequency (say, 30 Hz) and at the voltage level v_{im} (mV), as given by the overload characteristics study. The output voltage is measured and the voltage gain (A_v) is calculated from the relation (10.12a). The earth point of the millivoltmeter is kept at ground terminal of the amplifier while the active lead is replaced to measure v_i and v_o. However, v_i is kept constant at v_{im} for this study but it should be checked for each frequency and adjusted by the attenuator.

(iii) The experiment is repeated in suitable frequency steps up to the highest oscillator frequency and a graph is plotted (Fig. 10.40) showing variation of f against A_v.

(iv) From this graph, maximum gain level A' is noted at level AA_1, a level equal to $0.707A'$ is marked as BB' and the lower and higher cut-off frequencies (f_1 and f_2) are noted.

(v) The bandwidth (B.W.) is calculated as equal to $(f_2 - f_1)$. The point B and B' in the graph are known as *half-power points*.

Results :

Frequency Response Study

Serial No.	Fixed input voltage (V_{im}) in mV	Frequency (Hz)	Output voltage (v_o) in mV	Voltage gain (A_v)	Analytic values from graph: Max. gain level (A')	Lower cut-off frequency (f_1) Hz	Higher cut-off frequency (f_2)Hz	$(f_2 - f_1)$ Hz (Bandwidth)	Gain Bandwidth A' $(f_2 - f_1)$
1.	...	...	...	...					
2.	...	...	...	...					
3.	...	...	...	...	...	...	...	...	...
...	...	...	...	...					
...	...	...	...	...					

Precautions :

(i) Suitable components for the experiment are to be chosen.

(ii) Proper earthing has to be done.

(iii) Overhead characteristic studies have to be done before performing the actual experiment.

Questions

1. What are meant by 'frequency response' and 'bandwidth'?

2. What is the purpose to know the bandwidth?

3. Why is the experiment done in the CE mode? Can it be done in the CB mode?

4. In the (f-A_v) graph, why 0.707 times the maximum gain level (A') is taken to determine the cut-off frequencies?

5. In respect to physical significance, what is the difference between 'bandwidth' and 'gain handwidth product'?

10.13 Construction of a Wein Bridge oscillator using OP AMP and to get it stabilized. Study of the waveform using a CRO

Theory :

An IC OP AMP. may be used in designing ad constructing a Wein bridge oscillator on a breadboard. The Wein bridge oscillator is show in (Fig. 10.41), where the positive feedback is provided by the network Z_1 and Z_2. The feedback ratio is given by

$$\beta = \frac{Z_2}{Z_1 + Z_2} = \frac{R}{3R + j\left(\omega CR^2 - \frac{1}{\omega C}\right)} \quad \text{... (10.13a)}$$

At $\omega = \frac{1}{CR}$, the network introduces zero phase shift. At this frequency, the feedback ratio is $\frac{1}{3}$ which means that it introduces an attenuation of $\frac{1}{3}$. To maintain oscillations, the loop gain must be unity. This can be achieved by introducing negative feedback via R_1 and R_2. The values of R_1 and R_2 are so adjusted that $R_1 = 2R_2$. The actual Wein bridge is formed of R_1, R_2, Z_1 and Z_2.

The circuit will oscillate as long as $\beta = \frac{1}{3}$. Any departure from this condition makes the oscillator unstable. The amplitude of the oscillation may be stabilised by varying the negative feedback ratio, either by using a thermistor in place of R_1 or by connecting two diodes across R_1, or even by using a FET (Field-effect Transistor). For any reason, if the amplitude of the output voltage increases, the diodes conduct more, decreasing the effective value of R_1 and increasing the amount of negative feedback, thereby bringing down the amplitude to a suitable level.

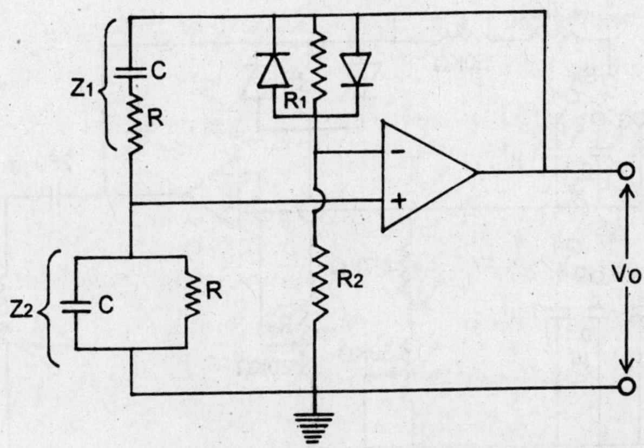

Fig. 10.41 : Wein Bridge oscillator

Apparatus :

IC 741 (OPAMP), dual power supply (15-10-15), two silicon diodes (matched), range switch, 50 kΩ linear ganged potentiometers, a 20 kΩ linear potentiometer, cathode-ray oscilloscope.

Design Procedure :

(a) *Desired ranges* : 3-30 Hz, 300–3,000 Hz, 3kHz–30 kHz.

(b) *Choice of Potentiometer* : The oscillator frequency may be varied either by varying *R* or by varying *C*. Resistance variation permits coveragence of frequency over one decade in each step. For the lowest frequency range, a capacitance of 1μF may be chosen. The maximum and the minimum values of resistance needed to cover the range are 50 kΩ and 5.3 kΩ. The standard value of 4.7 kΩ may be selected for connection in series with a 50 kΩ linear ganged potentiometer. For other ranges, the capacitors are 0.1μF, 0.01μF, and 0.001μF.

(c) The value of R_1 and R_2 are chosen at R_1 = 10 kΩ and R_2 = 20 kΩ linear potentiometer. The wiper of this potentiometer is connected to the inverting input. Variation of the potentiometer allows to start oscillations and also to control the distortion.

The entire arrangement of the circuit for multi-range oscillator is shown in (Fig. 10.42).

Method :

(i) The circuit connections are made as shown in (Fig. 10.42) with the values of the components preferably selected.

(ii) The range switch S (S_A and S_B) is kept on range 1. The waveform generated is observed on the CRO. The potentiometer $P(R_2$ is adjusted so that the oscillations are stable and the distortion is minimum.

(iii) On the 50 kΩ ganged potentiometer, a dial is fixed and the frequency of the oscillator is determined.

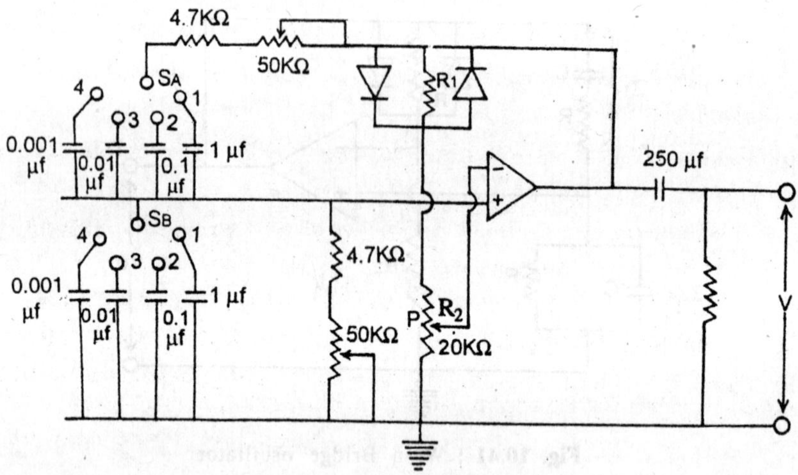

Fig. 10.42 : Four range oscillator

(iv) For different dial positions of this potentiometer, the frequencies of the oscillator are determined by forming Lissajous' figures, using a standard audio oscillator and a CRO. If a digital frequency meter be available, directly the frequencies for different dial positions can be read from it and the dial can also be calibrated.

(v) The steps are repeated for calibration in the other ranges (2,3,4).

(vi) The experiment may be tried with two closely matched diodes and also with diodes differing in characteristics. It may be interesting to note how do these affect the output waveforms.

(vii) The two diodes connected across R_1 are removed and R_1 is replaced by a thermistor. The waveform obtained with the thermistor and with the previous arrrangement, may he suitably compared.

Results :

Suitable and convenient Tables for recording the data, may be made according to the convenience of the experimenter.

Questions

1. What is meant by an 'oscillator'? What is the input to the oscillator?

2. What are the conditions for maintenance of sustained oscillations?

3. What is the feedback network in the Wein bridge oscillator?

4. What constitutes the bridge in the Wein bridge oscillator? What is the input to the bridge?

5. How is the expression given by (10.13a) obtained?

6. What are the expressions for (i) the frequency of oscillation and for (ii) maintenance of oscillations of the Wein bridge oscillator?

7. What factors determine the stability of oscillations and purity of the waveform of a Wein bridge oscillator?

8. What is the role of the OP AMP. in the Wein bridge oscillator?

9. How can a thermistor be used to stabilise oscillations of a Wein bridge oscillator?

10. Discuss how it is possible to use a FET to stabilise oscillations of the OP AMP wein bridge.

10.14 Determination of Fourier spectrum of (i) Square, (ii) Triangular and (iii) Half-sinusoidal waveform by a C.R.O.

Theory :

According to Fourier's theorem, a periodic function can be expressed by an infinite series of sine or cosine functions of harmonically related frequencies. The frequency of the sine or cosine term in the series is a harmonic of the frequency of the periodic function.

The four periodic waves namely

(i) a square-wave of amplitude A and period T,

(ii) a triangular-wave of peak value A and period T.

(iii) a half-sinusoidal wave of amplitude A and period T and

(iv) a half-sinusoidal wave (from full-wave rectifier) of amplitude A and period T

are shown in (Figs. 10.43; 10.44; 10.45 and 10.46) respectively.

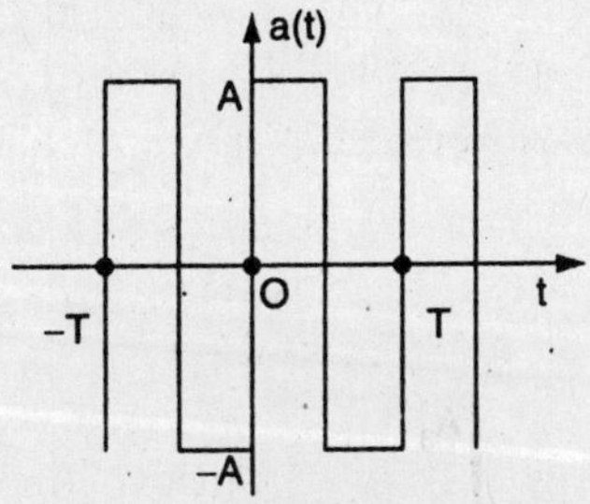

Fig.10.43

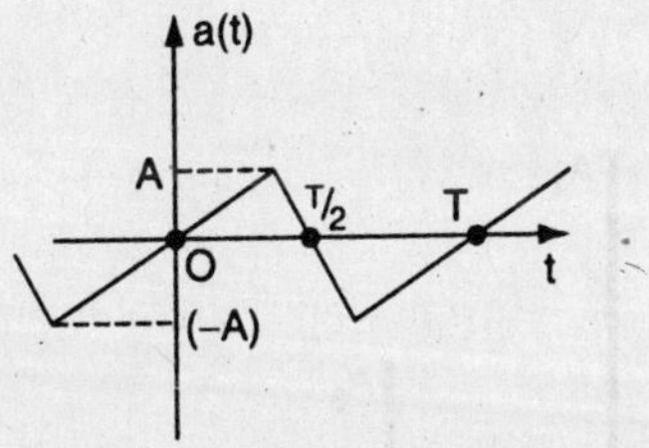

Fig.10.44

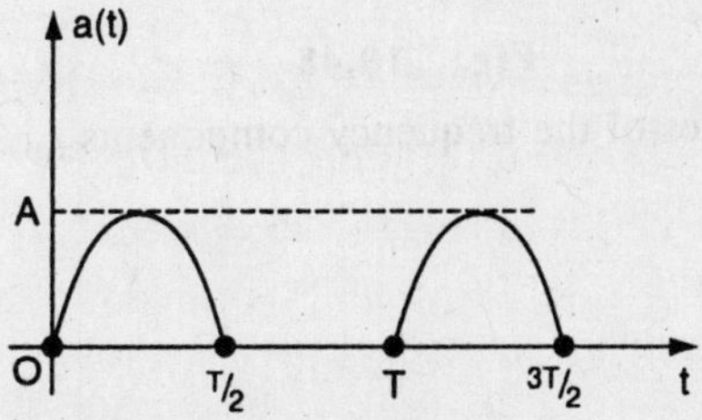

Fig.10.45

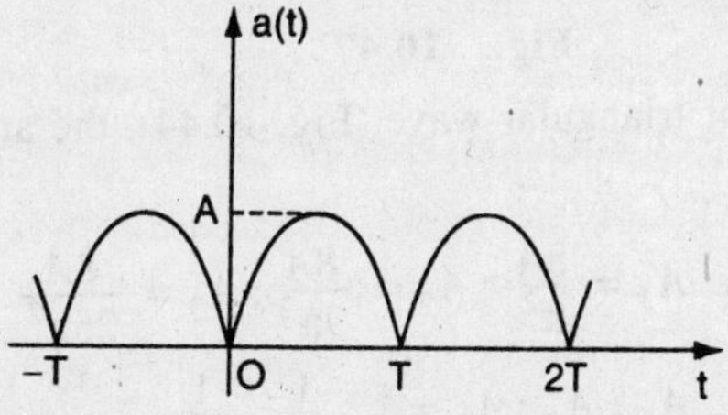

Fig.10.46

The Fourier expansions of these four periodic waves are respectively given by

(i) $a(t) = \frac{4A}{\pi} \sum_{n=1,2,\ldots}^{\infty} \frac{1}{n} \sin n\omega_0 t$

(ii) $a(t) = \frac{8A}{\pi^2}\left[\sin \omega_0 t + \frac{1}{9}\sin 3\omega_0 t + \frac{1}{25}\sin 5\omega_0 t + \ldots\right]$

(iii) $a(t) = \frac{A}{\pi} + A\left[\frac{1}{2}\sin \omega_0 t - \frac{2}{3\pi}\cos 2\omega_0 t - \frac{2}{15\pi}\cos 4\omega_0 t - \ldots\right]$

(iv) $a(t) = \frac{2A}{\pi} + \frac{4A}{\pi} \sum_{n=1,2,\ldots}^{\infty} \frac{\cos n\omega_0 t}{(1-n^2)}$,

where $\omega_0 = \frac{2\pi}{T}$ is the fundamental angular frequency.

The *amplitude spectrum* of the waveform is the plot of the amplitudes of the harmonic components of the Fourier series with frequency. The frequency components for the square wave (Fig. 10.43) are

$$f_0\left(=\frac{\omega_0}{2\pi}\right),\ 3f_0,\ 5f_0,\ 7f_0 \ldots\ldots$$

The amplitude spectrum of the square-wave is shown in (Fig. 10.47), which are $A_1 = \frac{4A}{\pi}, A_2 = \frac{4A}{3\pi}, A_3 = \frac{4A}{5\pi}$ and so on,

so that $A_1 : A_2 : A_3 = 1 : \frac{1}{3} : \frac{1}{5} \ldots\ldots$

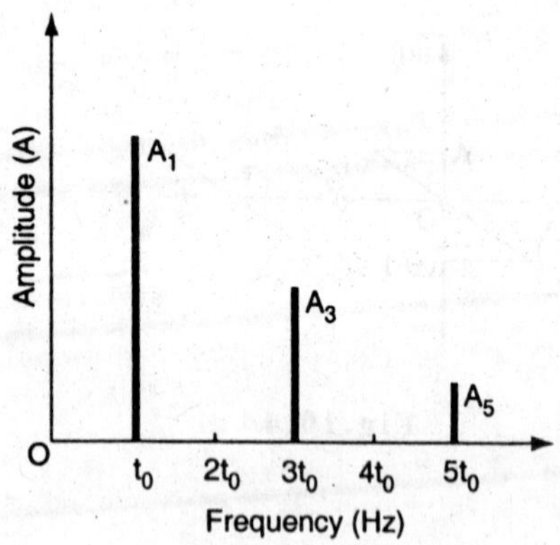

Fig. 10.47

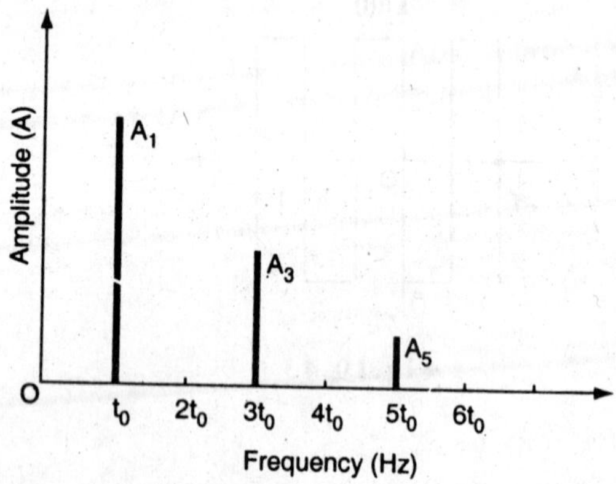

Fig. 10.48

For triangular-wave (Fig. 10.44), the amplitudes of the frequency components f_0, $3f_0$, $5f_0$ are

$$A_1 = \frac{8A}{\pi^2}\ ;\ A_2 = \frac{8A}{9\pi^2};\ A_3 = \frac{8A}{25\pi^2}.$$

$$\therefore\ A_1 : A_2 : A_3 = 1 : \frac{1}{9} : \frac{1}{25} \ldots\ldots$$

The spectrum is shown in (Fig. 10.48).

For the half-wave (sinusoiduls) of (Fig. 10.45), the component (d.c.) is $A_0 = \frac{A}{\pi}$. The amplitudes of the frequency components f_0, $2f_0$ and $4f_0$ are

$$A_1 = \frac{A}{2} \; ; A_2 = \frac{2A}{3\pi} \; ; A_4 = \frac{2A}{15\pi} \; \ldots\ldots \text{ respectively.}$$

$$\therefore \quad A_0 : A_1 : A_2 : A_4 = 1 : \frac{\pi}{2} : \frac{2}{3} : \frac{2}{15}.$$

The amplitude spectrum is shown in (Fig. 10.49). Considering only the a.c. components,

$$A_1 : A_2 : A_3 = 1 : \frac{4}{3\pi} : \frac{4}{15\pi}.$$

For the half-sinusoidal waveform of (Fig. 10.46), the d.c. component is $A_0 = \frac{2A}{\pi}$. The amplitude of the frequency components $2f_0$, $4f_0$, $6f_0$ are respectively,

$$A_2 = \frac{4A}{3\pi} \; ; A_4 = \frac{4A}{15\pi} \; ; A_6 = \frac{4A}{35\pi}$$

so that $A_0 : A_2 : A_4 : A_6 = 1 : \frac{2}{3} : \frac{2}{15} : \frac{2}{35}$.

The amplitude spectrum of the half-sinusoidal waveform is shown in (Fig. 10.50).

The first few frquencies are significant as the amplitude drops down very rapidly with increasing frequency.

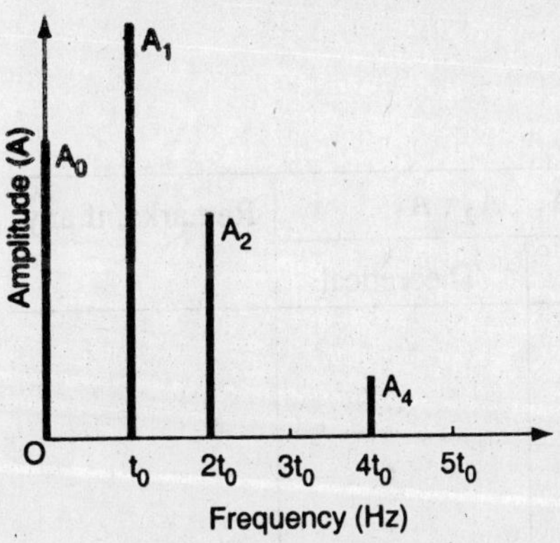

Fig. 10.49

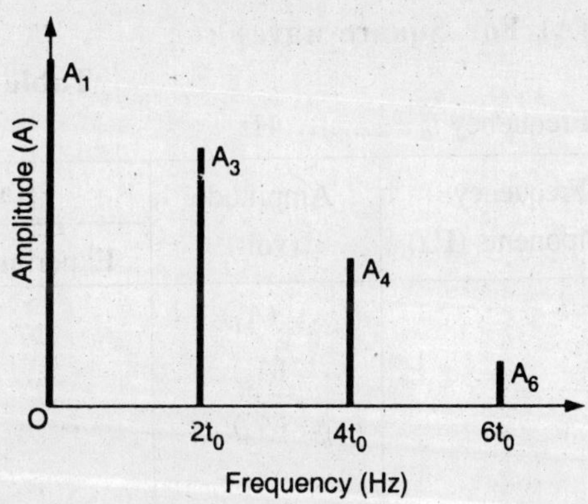

Fig. 10.50

Apparatus :

Square and triangular waves generator, half-wave and full-wave rectifiers for half sinusoidals, active band-pass filters, CRO.

Method :

(i) From square-wave generators square-waves are obtained and displayed in the CRO. A suitable frequency is chosen, like 1 kHz, for the waveform.

(ii) A band-pass filter of centre frequency 1 kHz and of narrow bandwidth is taken and set up as below. The band filter allows only the frequency component 1 kHz (f_0) to pass through and debars all other frequencies from appearing.

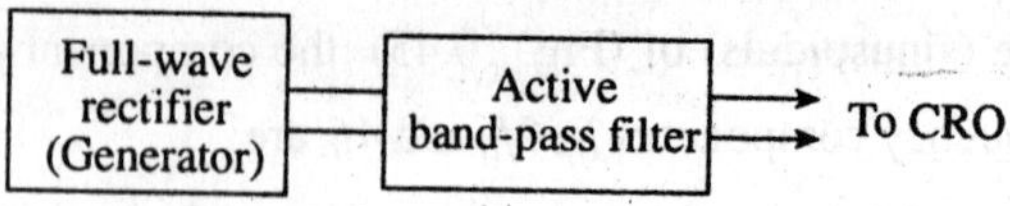

(iii) The output signal of the band-pass filter is displayed on the CRO and its amplitude is measured on CRO.

(iv) The band-pass filter is then replaced by band pass filters of centre frequencies $2f_0$, $3f_0$, $4f_0$ and $5f_0$ and the corresponding amplitudes A_2, A_3, A_4, etc. are measured on CRO. The amplitudes obtained are corrected by the gains of the active band-pass filters at the centre frequencies.

(v) The amplitude spectum is plotted as in (Fig. 10.43) and A_1, A_2, A_3 are obtained. These values are compared with the theoretical values.

(vi) The process is repeated for triangular waveform of frequency $f_0 = 1$ kHz.

(vii) Then the function generator is replaced by a half-wave rectifier giving half sinusoidal waveform of frequency $f_0 = 1$ kHz and the experiment is repeated. The d.c.. component of the Fourier expansion of half-sinusoidal wareform is not considered in this process. Only A_1, A_2 and A_4 are measured by the CRO.

(viii) The experiment is repeated by replacing the half-wave rectifier with a full-wave rectifier. The amplitudes A_2, A_4 and A_6 are measured.

Results :

(A) For Square-waves :

Table 1

Frequency f_0 = ... Hz

Frequency components (Hz)	Amplitude (volt)	Ratio $A_1 : A_2 : A_3$		Remarks, if any
		Experimental	Theoretical	
f_0 = ...	...... (A_1)	...	...	
$2f_0$ = ...	...... (A_2)	...	...	
$3f_0$ = ...	...... (A_3)	...	...	......
......	......	...	...	
......	......	...	...	

Table 2

[Similar Tables for triangular waveforms made]

Table 3

[Similar Tables for half-sinusoidal waveforms made]

Precautions :

(i) When the gains of the active band-pass filters are equal at the centre frequencies, the amplitudes need not be corrected for these gains, since the gains cancel out while the amplitude ratios are taken.

(ii) It is better to operate the CRO in the triggered mode to get the patterns on the screen as steady.

Questions and Answers

Q.1. What is Fourier's theorem?

A. : See 'Theory'

Q.2. What is the difference between an active filter and a passive filter?

A. : Filter circuits constructed with passive elements like L, C, R only is known as passive filters, while passive and active elements are used to have an active filter.

Q.3. What is a band-pass filter?

A. : It allows to pass through it, signals of all frequencies lying within a band and attenuates signals of frequencies lying outside the band.

Q.4. What is the effective or r.m.s. value of a periodic function $f(t)$?

A. : The effective value is given by

$$F_{eff} = \sqrt{\frac{1}{T}\int_0^T [f(t)]^2 \, dt}\, .$$

Q.5. Distinguish between even functions symmetry and odd function symmetry.

A. : The periodic function $f(t)$ is known to possess even function symmetry when $f(t) = f(-t)$ and to possess odd function symmetry when $f(t) = -\{f(-t)\}$.

10.15 To study the temperature dependence of reverse saturation current in a junction diode and hence to determine the band gap of a semiconductor

Theory :

The reverse saturation current I_0 in a p-n junction diode is given by

$$I_0 = A.e\left[\frac{D_p}{L_p N_D} + \frac{D_n}{L_n N_A}\right] \times n_i^2, \qquad \text{... (10.15a)}$$

where A = Junction area, e = Electronic charge

D_p = Diffusion coeff. of holes, D_n = Diffusion coeff of electrons

L_p = Diffusion lengths of holes, L_n = Diffusion lengths of electrons

N_A = Acceptor impurity density, N_D = Donor impurity density

n_i = Intrinsic carrier concentration

The value of n_i varies with temperature (T) as

$$n_i^2 \propto T^3.e^{-E_g/KT},$$

where E_g is the forbidden energy gap between valence and conduction band, K is Boltzmann constant. Moreover, the quantities D_n and D_p vary with temperature. The temperature dependence also vary with the semiconductor material type. Of course, for all practical purposes, the dependance of I_0 on temperature T can be written approximately by the following relation,

$$I_0 = \left(GT^{\alpha}\right)e^{-E_g/K\eta T}, \quad \text{... (10.15b)}$$

where G is a constant independent of temperature, α and η depending on the type of material. For example,

for Si; $\eta = 2$, $\alpha = 1.50$

for Ge; $\eta = 1$, $\alpha = 2.0$.

The temperature dependence of the first term of the Eq. (10.15b) is not so important in comparison with the latter exponential term. So, we may write

$$I_0 = G_1 \cdot e^{-E_g/K\eta T},$$

where G_1 is approximately independent of temperature.

Thus, $\log_{10} I_0 = \log_{10} G_1 - \dfrac{1}{2.30}\dfrac{E_g}{\eta K}\cdot\dfrac{1}{T}$. ... (10.15c)

Hence, if a graph is plotted with $\log_{10} I_0$ against $\frac{1}{T}$, a straight line will be obtained. The slope of this graph (m) is given by

$$m = \frac{E_g}{2.30\eta K}$$

and thus $E_g = m \times \eta K \times 2.30$. ... (10.15d)

For Ge-diodes, $\eta = 1$, $K = 8.62 \times 10^{-5}$ eV/°K and for Si-diodes, $\eta = 2$.

Thus, from the graph $\left(\log_{10} I_0 \text{ vs. } \frac{1}{T}\right)$, the value of E_g (the forbidden energy gap) can be obtained.

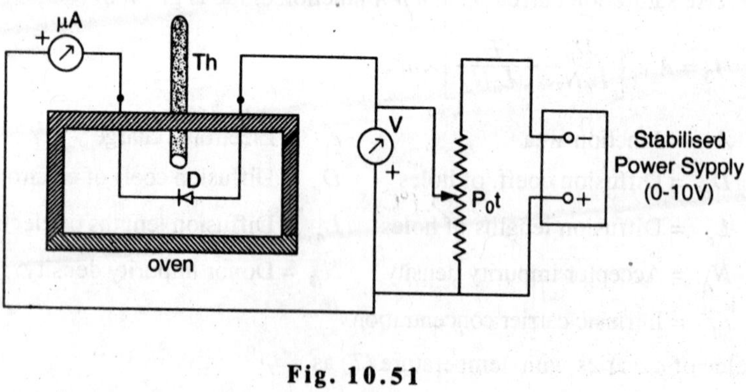

Fig. 10.51

Apparatus :

Ge and Si diodes, stabilised power supply (0-10 V), microammeter (0-100 μA), voltmeter (0-10 V), one 10 kΩ turn potentiometer, thermometer, electrically controlled thermostat.

Method :

(i) The experimental set-up is arranged as shown in (Fig. 10.51). The diode is placed inside an electrically controlled oven. It can also be placed in a water bath whose temperature may be varied as desired.

(ii) The power supply S is switched on and the potentiometer (P) is adjusted to apply sufficient reverse biasing (3 or 4 volts) and recorded by the voltmeter (V). The reverse saturation current I_0 is noted from the microammeter. The value of I_0 shuld not change with reverse biasing voltage (V). The temperature is also noted from the thermometer. This may be the room temperature.

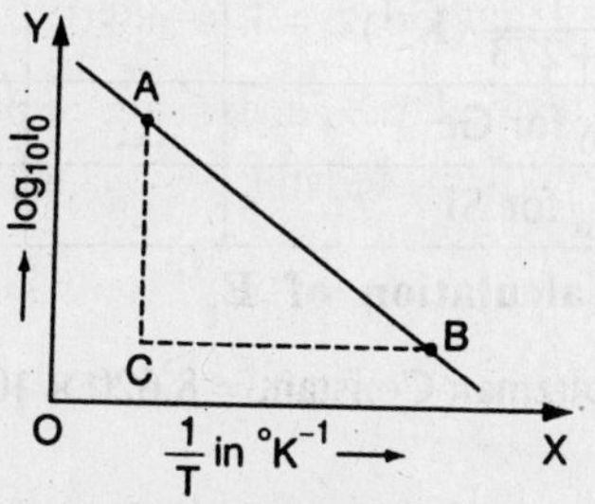

Fig. 10.52

(iii) The oven is then switched on and its temperature is adjusted to a value of about 20 °C above room temperature. Sufficient time is allowed for the current to be recorded by the microammeter. These values must attain steady ones. The readings of microammeter and the thermometer are recorded.

(iv) Keeping the applied reverse biasing fixed, the temperature of the oven in suitable steps, is increased and in each step, the values of I_0 and the temperature are noted as in (iii).

(v) Finally, a graph is plotted with $\frac{1}{T}$ in °K^{-1} along X-axis and $\log_{10} I_0$ along Y-axis. The nature of the curve will be shown in (Fig. 10.52). From the graph the value of the slope (m) is obtained and from this value of m, the value of E_g is obtained from Eq. (10.15d).

(vi) The whole process is repeated with another diode made of a different material.

Results :

(A) Reverse Saturation Current as a function of Temperature

Table 1

Diode	No. of Obs.	Temp. (°C)	Reverse Voltage (volt)	Reverse current (I_0) in μA
Ge (D_1)	1	...	...	...
	2	...	...	...
	3	...	...	...
	4	...	...	...
	...			
Si (D_2)	1	...	...	...
	2	...	...	...
	3	...	...	...
	4	...	...	...
	...			

(B) Drawing of $\log_{10} I_0$ vs. $\frac{1}{T}$ graph

Table 2

Temp. (°C)	...	...	...	...	...	...
$\frac{1}{T} = \frac{1}{t+273}$ (K^{-1})	...	...	...	...	...	...
$\log_{10} I_0$ for Ge	...	...	...	...	...	...
$\log_{10} I_0$ for Si	...	...	...	...	...	...

(C) Calculation of E_g

Boltzman Constant = 8.620×10^{-5} eV/°K

Table 3

Diode	Value of η	From graph (Fig. 10.52) in °K⁻¹		Slope $m = \frac{AC}{BC}$ (°K)	E_g in eV (= $m \times 2.3\, \eta\, K$)
		Value of AC	Value of BC		
Ge	1	...	...	...	...
Si	2	...	...	...	...

∴ E_g for Ge = ... eV.

E_g for Si = ... eV.

Precautions :

(i) The reverse bias voltage should be kept well below the reverse breakdown voltage of the diode.

(ii) Sufficient time is to be allowed after each temperature change for the reverse saturation current to attain a new steady value.

(iii) Instead of electrical oven, liquid paraffin bath may be used, taking proper steps to make good thermal contact and electrical insulation.

Questions and Answers

Q.1. What is diffusion length?

A. : It is the distance within the semiconductor at which the carrier concentration falls to $\frac{1}{e}$ times its value at the surface.

Q.2. What happens when reverse bias voltage is increased?

A. : Breakdown in the diode may occur. The current suddenly rises to a large value, which causes a damage to the diode.

Q.3. What is meant by 'reverse saturation current'?

A. : See any textbook.

10.16 To design and verify S-R flip-flop, (J-K) flip-flop and D-flip-flop circuits and to study their operations using basic gates

(A) S-R flip-flop

Theory :

The circuit shown in (Fig. 10.53) is an *S-R* flip-flop circuit. The input *S* is known as the set input and the input *R* is known as the clear or reset input. The corresponding output *Q* will be in state 1 or *set state*, when $R = 0$ and $S = 1$. But when $R = 1$ and $S = 0$, the flip-flop is set to the state 0, this is $Q = 0$. Again, when $R = S = 0$, the output *Q* can be either 0 or 1 which means that it is undetermined. Moreover, $R = S = 1$ is forbidden since in this case Q and $\overline{Q}$ (both) tend to be 1, which is not allowed.

Apparatus :

An IC 7400, d.c. voltmeter (0-5V), d.c. power supply (5V), breadbroad, wires, etc.

Method :

(i) The circuit is connected, as shown in (Fig. 10.54) on a breadboard. The respective IC pin numbers are indicated in the figure.

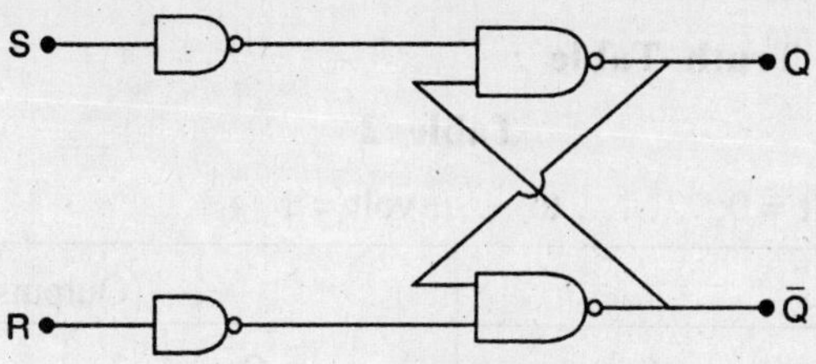

Fig. 10.53

(ii) The d.c. power supply is switched on and placed with output at 5V. The output voltage is measured by the d.c. voltage.

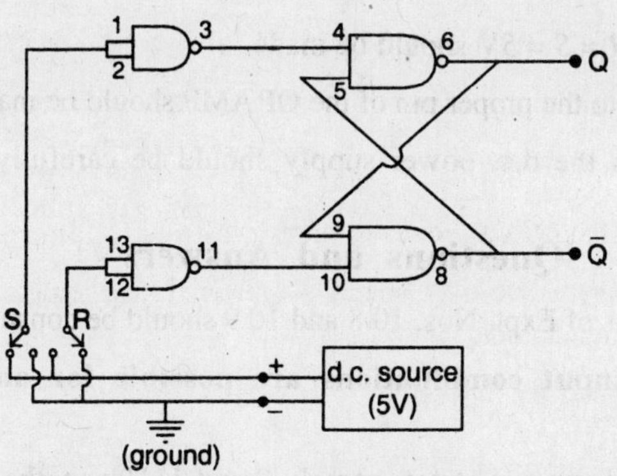

Fig. 10.54

(iii) The switches are set such that $R = 0$ volt and $S = 5$ volts. The output voltage is measured with respect to the ground at Q and $\overline{Q}$ using a d.c. voltmeter.

(iv) The switches are then changed so that $R = 5$ volts and $S = 0$ volt. The voltages are again measured at Q and $\overline{Q}$.

(v) Finally, the switches are connected to obtain $R = S = 0$ volt. The voltages at Q and $\overline{Q}$ are measured. (Never to do $R = S = 5$ volts)

(vi) Proper voltages are used to define 0 and 1.

(vii) Respective Truth Tables are then constructed from the data so obtained.

Results :

(A) Record of Input and Output voltages :

Table 1

Input voltages (volt)		Output voltages (volt)	
R	S	Q	$\overline{Q}$
0	5	...	...
5	0	...	...
0	0	...	...

(B) Construction of Truth Table :

Table 2

.... to ...... volt = 0; .... to ...... volt = 1

Inputs		Outputs	
R	S	Q	$\overline{Q}$
0	1	1	0
1	0	0	1
0	0	...	...

Precautions :

(i) At no time, $R = S = 5$V should be made.

(ii) Connections to the proper pin of the OP AMP should be made carefully.

(iii) Terminals to the d.c. power supply should be carefully checked and then connected.

Questions and Answers

[Questions of Expt. Nos. 10.8 and 10.9 should be consulted]

Q.1. How many input combinations are possible for an *n*-input binary logic gate?

A. : Each input can have two states, namely 0 and 1. Hence, the number of possible input combinations is 2^n.

Q.2. What do you mean by 'fan-in' and 'fan-out' of a logic gate?

A. : The number of gates that can be connected to the input of a logic gate, is known as 'fan-in'. The number of gates that a logic gate can drive, is known as 'fan-out'.

Q.3. Which logic is used in 7400 IC series?

A. : Transistor-Transistor Logic (TTL).

Q.4. Do you know any other IC series using TTL?

A. : The series starting with the numbr 54. These ICs are used in military purpose while the series starting with digits 74 are used in industrial purposes.

(B) J-K Flip-flop

(Fig. 10.55) shows the basic J-K flip-flop circuit. It actually consists of an *S-R* flip-flop with two NOR gates and two AND gates. The data input is applied along with Q to the input of an AND (1) and the data input along with $\overline{Q}$ to the input of the other AND (2). The Truth Table is shown below. The data inputs have four possible combinations and for each combination can be either 0 or 1.

The new value of$Y$ is found from the Truth Table of the *S-R* flip-flop. The reduced Truth Table of the J-K flip-flop is given below. In this flip-flop, the ambiguities of the *S-R* flip-flop is removed.

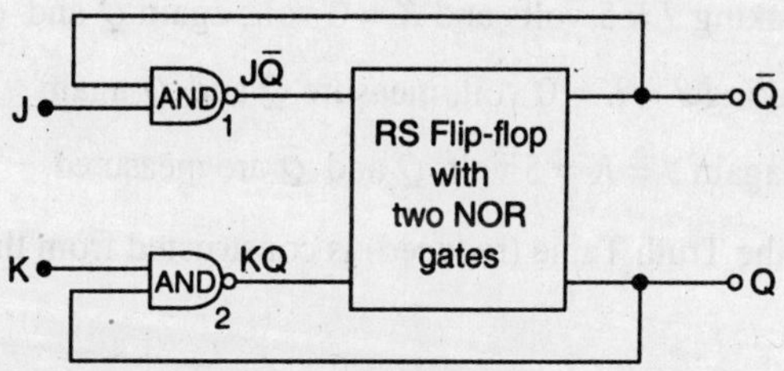

Fig. 10.55

Reducced Truth Table (J-K Flip-flop)

J	K	Q
0	0	Q_0
0	1	0
1	0	1
1	1	$\overline{Q_0}$

Apparatus :

An IC 7408 or 7409 (two inputs AND gates), stabilised power supply, breadboards, wires, etc.

Methods :

(i) The circuit is made as shown in (Fig. 10.56). The power supply is switched on and set at 5 volts.

(ii) Then Q and $\overline{Q}$ are measured making $J = 0$ volt and $K = 5$ volts.

(iii) Next again Q and $\overline{Q}$ are measured making $J = K = 0$ volt

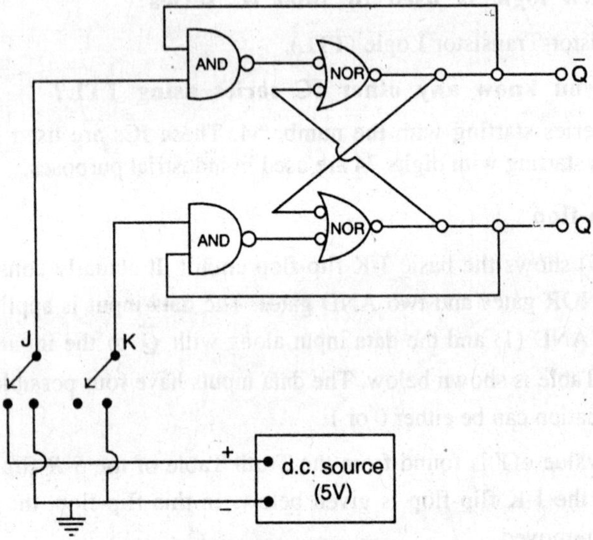

Fig. 10.56 : J-K flip-flop

(iv) Then making $J = 5$ volts and $K = 0$ volt, again Q and $\overline{Q}$ are measured.

(v) Then making $J = K = 0$ volt, measure Q and $\overline{Q}$ again

(vi) Making again $J = K = 5$ volt. Q and $\overline{Q}$ are measured.

(vii) Finally, the Truth Table (reduced) is constructed from the measured voltages.

Results :

(A) Record of Input/Output voltages :

Table 1

Inputs voltages (V)		Output voltages (V)	
J	K	Q	$\overline{Q}$
0	5	...	...
0	0	...	...
5	0	...	...
0	0	...	...
5	5	unsteady	unsteady

(B) Construction of Truth Table :

Table 2

.... to .....volt ≡ 0; ..... to ..... volt ≡ 1

Inputs		Outputs
J	K	Q
0	0	...
0	1	0
1	0	1
1	1	...

[Except for the values $J = K = 1$, results of Table - 2 are supposed to agree with that of a *J-K* flip-flop]

Precautions :

[Same as discussed in *S-R* Flip-flop case]

(C) D-type Flip-flop

(Fig. 10.57) shows a Delay (D) type flip-flop where it is obtained from the *J-K* flip-flop (Fig. 10.55) by using an inverter (NOT gate or a single input NOR gate) so that $K = \overline{J}$. Of course, a D-type latch may be obtained if a J-K flip-flop here is replaced by a S-R flip-flop. The Truth Table of the D-type flip-flop is shown below.

$D = J$	K	Q
0	1	0
1	0	1

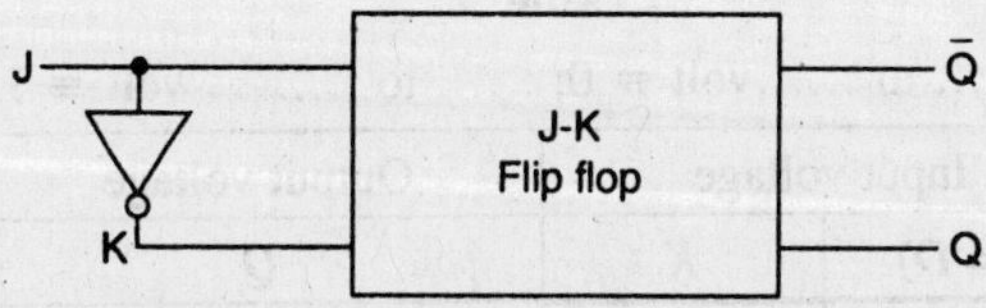

Fig. 10.57

Apparatus :

[Same as before]

Method :

(i) Circuit connections are made as shown in (Fig. 10.58). The power supply is switched on and set at 5 volts.

(ii) Two possible logic states of the input are taken by means of the switch.

(iii) Other input voltage and the output voltages Q and $\overline{Q}$ are measured by the voltmeter.

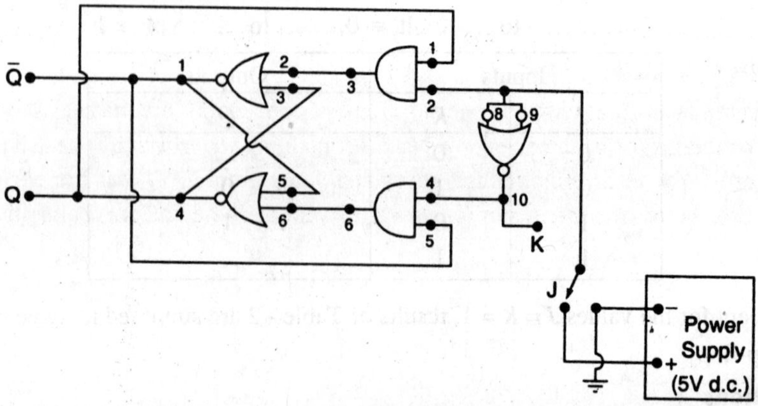

Fig. 10.58 : D-type flip-flop circuit

(iv) Finally, the Truth Table is constructed as before and it is verified with the Truth Table of the D-type flip-flop as given above.

Results :

(A) Input-Output voltage records :

Table 1

Input voltage (volt)		Output voltage (volt)	
J	K	Q	$\overline{Q}$
0	...	...	...
5	...	...	...

(B) Construction of Truth Table

Table 2

.... to ......volt ≡ 0; ...... to .... volt ≡ 1

Input voltage		Output voltage
$J\,(=D)$	K	Q
0	1	0
1	0	1

[It is to be checked that this Truth Table satisfies the Truth Table of D-type flip-flop]

Precautions :

[Same as discussed in *S-R* flip-flop case]

10.17 To design and fabricate a temperature controller and to study its Performance Characteristics

Theory :

Thermistors are special type of resistors having negative temperature coefficients and they are highly sensitive resistors to the variation of temperatures. These thus may be used for measurement of precise temperatures changes. These may be used within the temperature range of – 100 °C to + 350 °C. These are made of mixtures of metallic oxides of manganese, nickel, copper and cobalt, etc. The resistance may vary from 0.5 Ω to 100 megohms.

Calibration :

To calibrate such a sensor (thermistor), circuit connections are made as shown in (Fig. 10.59). The thermistor is immersed in a beaker containing water which is heated to different temperatures. For each temperature, the water temperature (T) and the corresponding potential drop across the thermistor (using a millivoltmeter) are noted. A calibration curve is

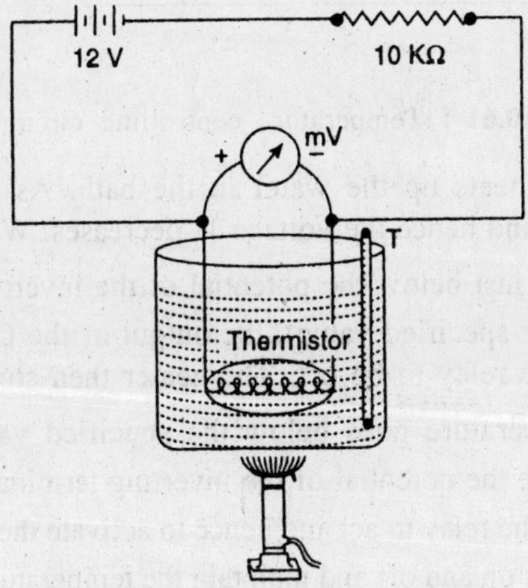

Fig. 10.59 : Calibration of thermistor

plotted with temperature (T) along X-axis and corresponding voltage drop (mV) across Y-axis (Fig. 10.60). From the graph, the p.d. across the thermistor for any temperature may be obtained.

Control of temperature bath :

For controlling the temperature bath, a circuit usually a relay may be constructed as shown in (Fig. 10.61). The thermistor along with the heater is immersed in water bath whose temperature (t °C) is to be determined.

If V_t be the p.d. across the thermistor corresponding to this temperature as obtained from the calibration curve, then the voltage at the non-inverting terminal of the OP AMP is $V_2 = \frac{R_2 + R_1}{R_1} . V_t$. The inverting terminal of

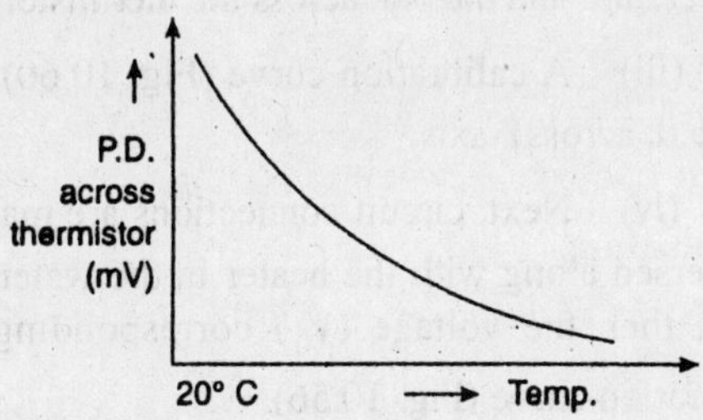

Fig. 10.60 : Calibration curve of thermistor

the OP AMP is kept at a potential below V_2 by adjusting the 47 kΩ potentiometer for which the output of the OP AMP becomes saturated to a potential at about 12 V. It causes the

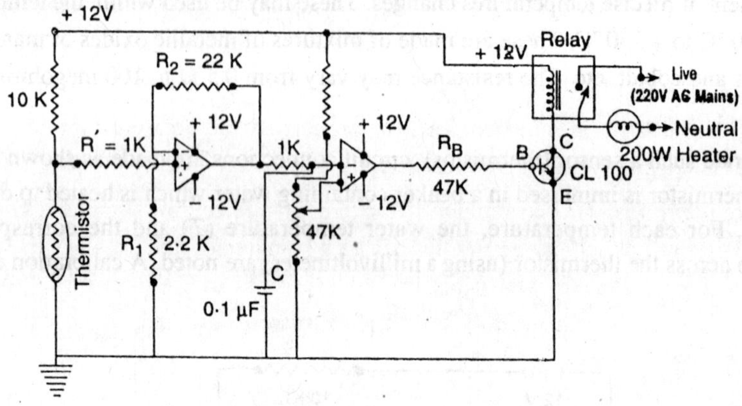

Fig. 10.61 : Temperature controlling circuit

relay to be on and thus it heats up the water in the bath. As temperature increases, resistance of the thermistor and hence the voltage V_t decreases. When the potential at the non-inverting terminal goes just below the potential of the inverting terminal (when the bath temperature goes above specified value), the output of the OP AMP switches from $^{+}V_{sat}$ to $^{-}V_{sat}$ and causes the relay to be off. The heater then stops heating water bath. Again, when the bath temperature goes below the specified value, the non-inverting terminal potential goes above the potential of the inverting terminal. The OP AMP output then switches on and causes the relay to act and hence to activate the heater. In this way, the relay will alternately switches on and off and maintain the temperature of the bath constant.

Instead of the relay, a silicon-controlled rectifier (SCR) may also be used in place of the relay.

Procedure :

(i) Circuit connections are made as shown in (Fig. 10.59). The thermistor is immersed in water bath in the beaker. The temperature of bath and the p.d. across the thermistor are noted using a thermometer and a millivoltmeter.

(ii) The bath temperature is then increased in small steps (5 °C) and at each step, the temperature and the p.d. across the thermistor are noted.

(iii) A calibration curve (Fig. 10.60) is then drawn with temperature along *X*-axis and p.d. across *Y*-axis.

(iv) Next, circuit connections are made as shown in (Fig. 10.61). The thermistor is immersed along with the heater in the water bath. If the temperature is to be kept, say, at t_1°C, then the voltage (V_1) corresponding to this temperature is obtained from the calibration curve (Fig. 10.56).

(v) The potentiometer 47 kΩ is then adjusted such that the potential at the inverting terminal of the OP AMP becomes $V_1 \times \dfrac{R_2 + R_1}{R_1}$.

(vi) The bath temperature is noted at intervals of 3 minutes till the thermometer reading remains constant for at least ten minutes. The corresponding potential difference across the thermistor is obtained from the calibration curve at this constant temperature.

(vii) This process is repeated for other specified temperatures by adjusting the potentiometer and using the calibration curve.

There are many possible ways to devise the temperature control circuit but this method is adopted to give an idea of use of an OP AMP as a comparator, the use of realy or the use of SCR as a switch. The values of the different resistors in the circuit of (Fig. 10.57) depend on the resistance of the thermistor at room temperature and the necessary trigger current and the firing potential of the SCR.

Results :

(A) Calibration of thermistor :

Table 1

No. of Obs.	Temperature of heat bath (°C)	Potential drop across the thermistor (mV)
1.	Room Temperature	...
2.	...	...
3.	...	...

(B) Time-Temperature of Heat Bath :

Table 2

Required temperature (°C)	Reference voltage of the comparator (volt)	Temp. of the heat bath (°C)	Corresponding voltage from the calibration curve (V_t) volt	Temp. of heat bath from calibration curve (°C)
t_1 °C	(i) ...	(i) ...	(i) ...	(i) ...
	(ii) ...	(ii) ...	(ii) ...	(ii) ...
	(iii) ...	(iii) ...	(iii) ...	(iii) ...
t_2 °C	(i) ...	(i) ...	(i) ...	(i) ...
	(ii) ...	(ii) ...	(ii) ...	(ii) ...
	(iii) ...	(iii) ...	(iii) ...	(iii) ...
t_3 °C	(i) ...	(i) ...	(i) ...	(i) ...
	(ii) ...	(ii) ...	(ii) ...	(ii) ...
	(iii) ...	(iii) ...	(iii) ...	(iii) ...

Precautions :

(i) The thermistor and the thermometer should be placed very close inside the water bath to get temperatues accurately.

(ii) Temperature of bath should be changed at equal steps and should be kept constant.

(iii) A 12-volt relay is preferred for convenience.

(iv) The capacitor should be of proper value so as to avoid any unwanted noise or pick-up in the circuit.

(v) Live and neutral terminals should be carefully ascertained in the beginning to avoid any shock hazards and for proper connections.

(vi) For safety purposes, use of relay is preferable to that of a SCR.

Questions and Answers

Q.1. What is a thermistor?

A. : It is a thermally sensitive resistor made of semiconductors having negative temperature coefficient. Its resistance is between few ohms to megaohms.

Q.2. What are sensors and transducers?

A. : See any textbook.

Q.3. What is the thermal time constant of a thermistor?

A. : It is the time taken for the excess temperature to drop to $\frac{1}{e}$ times the initial value. It determines, in fact, the speed of operation of a thermistor.

Q.4. What is a relay?

A. : It is an electromechanical device which can act as an electrically operated switch.

Q.5. What is a SCR?

A. : Silicon controlled rectifier is a thyristor which can switch large currents on and off.

Q.6. Which is advantageous—a relay or a SCR?

A. : A relay, which is a electromechanical system, is slow in operation but SCR being an electronic system, can act at a faster rate. But for safety purposes, relay use is preferable.

10.18 To draw the regulation characteristics of a bridge rectifier (i) without using a filter and (ii) using a filter, and to determine the ripple factors in both cases by measuring the ripple voltages with the help of an a.c. meter

Theory :

An unregulated power supply is made up of a rectifier which converts an a.c. voltage into a pulsating d.c. voltage and also of a filter network which smoothens out the rectified voltage. A bridge rectifier actually consists of four diodes arranged in the form of a bridge which acts like a full-wave rectifier.

(A) Without Filter :

When the bridge circuit does not have any filter (Fig. 10.62), then the rectifier is formed of four identical diodes (D_1, D_2, D_3, D_4), a load resistance (R_L) across which the rectified output voltage is obtained.

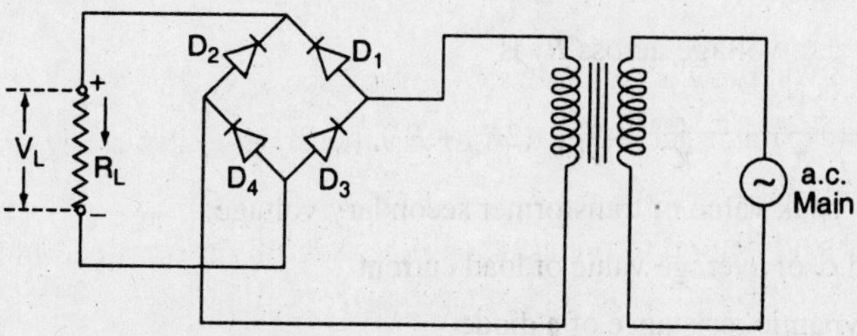

Fig. 10.62 : Bridge Circuit without filter

In one half-cycle of the input a.c. signal (when the upper end of the secondary is positive), the current passes through D_1, the load R_L and the diode D_4. The circuit is then completed through D_1 and D_4 and the current through R_L is from top to bottom. Again, for the other half of the cycle (when the lower end of the secondary is positive), the current passes through D_2, the load resistance R_L and the diode D_3. The circuit is then completed through the diodes D_2 and D_3. Again, in this half-cycle also, current through R_L flows from top to bottom, direction. Thus, during each half of the input signal, current through R_L passes in the same direction. Therefore, the rectified voltage V_L across the load R_L is though pulsating but unidirectional. So, the waveform obtained will be the same as that of a full-wave rectifier (Fig. 10.63).

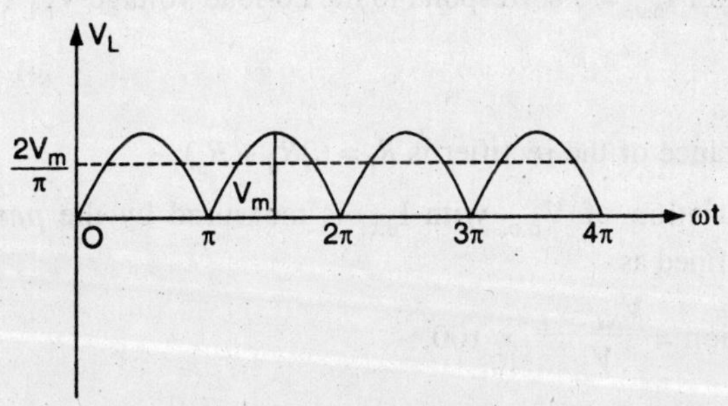

Fig. 10.63

The rectified voltage V_L contains a d.c. component having a.c. components super-imposed which are called 'ripple voltage'. Fourier analysis of V_L will give

$$V_L = \frac{2V_m}{\pi} - \frac{4V_m}{3\pi}\cos 2\omega t - \frac{4V_m}{15\pi}\cos 4\omega t - \ldots\ldots$$

where V_m — peak value of the rectifier output voltage, ω — angular frequency of the input a.c. signal.

The first term here gives the d.c. output voltage across R_L and the remaining terms denote the fluctuating a.c. components. It is evident that the lowest frequency of the a.c. component is double the input frequency, which is the output 'ripple frequency'.

Regulation Characteristics :

The output d.c. voltage across R_L is

$$V_{d.c.} = \frac{2V_m}{\pi} = \frac{2V_{peak}}{\pi} - I_{d.c.}\,(2R_d + R_s), \qquad \ldots (10.18\text{ a})$$

where V_{peak} — peak value of transformer secondary voltage

$I_{d.c.}$ — d.c. or average value of load current

R_d — dynamic resistance of a diode

R_s — resistance of transformer secondary.

$$\therefore I_{d.c.} = \frac{V_{d.c.}}{R_L} = \frac{2V_m}{\pi R_L}. \qquad \ldots (10.18\text{b})$$

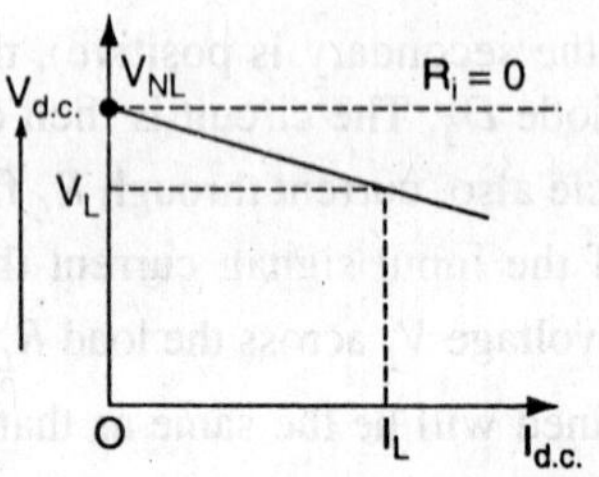

Fig. 10.64 : Regulation Characteristics

The graph showing the variation of output d.c. voltage ($V_{d.c.}$) against the load current ($I_{d.c.}$) is shown in (Fig. 10.64). This curve is known as the 'regulation characteristics' of the rectifier.

Eq. (10.18a) shows that $V_{d.c.}$ decreases with the increase of $I_{d.c.}$ having a slope of $-(2R_d + R_s)$. The intercept on the $V_{d.c.}$ axis is $\frac{2V_{peak}}{\pi}$.

The value of $V_{d.c.}$ for $I_{d.c.} = 0$ correspond to the no-load voltage V_{NL} (see graph).

$$\text{So, } V_{NL} = \frac{2V_{peak}}{\pi}. \qquad \ldots (10.18\text{c})$$

The internal resistance of the rectifier is $R_i = (2R_d + R_s)$. ... (10.18d)

The extent of variation of $V_{d.c.}$ with $I_{d.c.}$ is measured by the *percentage voltage regulation*, which is defined as

$$\%\text{ voltage regulation} = \frac{V_{NL} - V_L}{V_L} \times 100, \qquad \ldots (10.18\text{e})$$

where V_L — rated load voltage.

Better the regulation will be, less is the percentage regulation. A high value of R_i makes the regulation poor and so, an ideal source must have zero internal resistance whence $V_{NL} = V_L$ and then the percentage voltage regulation is zero.

Ripple Factor :

Ripple factor (γ) of a rectifier determines the success of it in converting an a.c. voltage to d.c. voltage. It is defind as

$$\gamma = \frac{\text{r.m.s. value of a.c. voltage}}{\text{d.c. voltage}}$$

$$= \frac{V_{rms}}{V_{d.c.}}, \qquad \ldots (10.18\text{f})$$

where V_{rms} is the r.m.s value of the ripple voltage.

If V_r is the r.m.s value of V_L, the net rectified output voltage

$$V_r^2 = V^2_{rms} + V^2_{d.c.} \quad \text{... (10.18g)}$$

$$\text{So, } \gamma = \frac{V_{rms}}{V_{d.c.}} = \frac{\sqrt{V_r^2 - V^2_{d.c.}}}{V_{d.c.}} \text{ [from Eq. (10.18g)]}$$

$$= \sqrt{\left(\frac{V_r}{V_{d.c.}}\right)^2 - 1}. \quad \text{... (10.18h)}$$

The values of V_{rms} and $V_{d.c.}$ can be measured with the help of suitable voltmeters. Thus the ripple factor (γ) can be obtained from Eq. (10.18f). The approximate theoretical value of γ is 0.482 and the experimental value obtained form Eq. (10.18f) can be compared with the theoretical value.

(B) With Filters :

In a bridge rectifier circuit, for smoothening of the ripple voltage, four types of filters may be used. Those are (i) Capacitor filter, (ii) Inductor filter, (iii) π-type LC filter and (iv) L-type LC filter. Of these, the capacitor filter only will be discussed here.

The circuit diagram of a bridge rectifier with a capacitor filter is shown in (Fig. 10.65). In one half-cycle, the capacitor is charged to almost its peak value. The diode curreent flows as a short duration pulse during the charging of the capacitor.

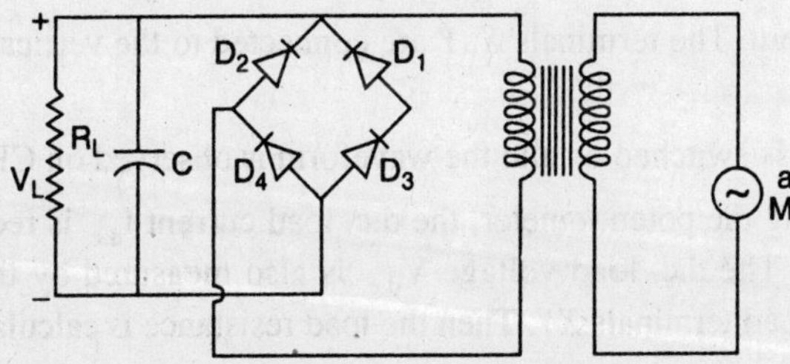

Fig. 10.65 : Capacitor Filter

The capacitor discharges through R_L when the input voltage drops below the capacitor voltage. The time constant for this discharge is CR_L. Barring the first half-cycle, the output waveform of V_L is almost a triangular waveform when $WCR_L >> 1$. The waveform is shown in (Fig. 10.66).

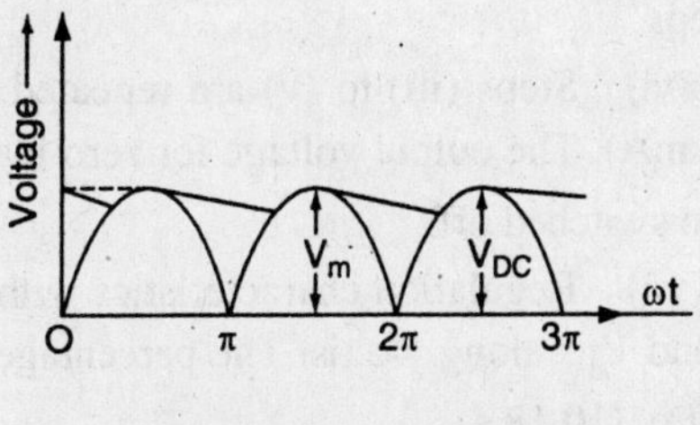

Fig. 10.66 : Waveform

The regulation characteristics or the relation between I_{DC} and V_{DC} of the rectifier with a capacitor filter, is given by

$$V_{DC} = V_M - \frac{I_{DC}}{4fC}, \quad \text{... (10.18 i)}$$

where f is the frequency of the input a.c. voltage.

Obviously, $V_{d.c.}$ decreases linearly with $I_{d.c.}$ and the no-load voltage $V_{NL} = V_m$. The percentage voltage regulation is

$$\frac{V_{NL} - V_L}{V_L} \times 100 = \frac{100}{4fCR_L} \% \quad \text{... (10.18 j)}$$

($\frac{1}{4fC}$ is the internal resistance of the rectifier)

The rms value of the ripple voltage is

$$V_{rms} = \frac{V_{d.c.}}{4\sqrt{3}\, fCR_L} \quad \text{... (10.18 k)}$$

The ripple factor γ is given by

$$\gamma = \frac{V_{rms}}{V_{d.c.}} = \frac{1}{4\sqrt{3}\, fCR_L}. \quad \text{... (10.18 } \ell\text{)}$$

The value of γ obtained experimentally may be compared with the theoretical value as given in Eq. (10.18 ℓ)

Apparatus :

Step-down transformer, bridge rectifier, choke coil, resistors, capacitor (~8μF), 10kW potentiometer, plug keys, a.c. millivoltmeter, d.c. milliameter, d.c. voltmeter, CRO.

Method :

(i) According to the circuit diagram (Fig. 10.67), the experiment is first performed *without* any filter. In this case, the key K_2 is kept open and K_1 is closed. This makes the capacitor filter circuit out. The terminals X, Y are connected to the vertical input of a CRO,. if feasible.

(ii) The circuit is switched on and the waveform is observed on CRO.

(iii) By adjusting the potentiometer, the d.c. load current $I_{d.c.}$ is recorded by the d.c. milliameter (~ 2 mA). The d.c. load voltage $V_{d.c.}$ is also measured by the d.c. voltmeter, connectted across the load terminals XY. Then the load resistance is calculated (R_L) from the relation $R_L = \frac{V_{d.c.}}{I_{d.c.}}$.

(iv) Next, removing the d.c. voltmeter, the a.c. voltmeter is connected across XY to measure the ripple voltage V_{rms} and the ripple factor (γ) is obtained from Eq. (10.18 ℓ).

(v) By adjusting the potentiometer, the values of $I_{d.c.}$ in milliampere are increased by steps.

(vi) Steps (iii) to (v) are repeated till $I_{d.c.}$ reaches the specified maximum value (~ 20 mA). The output voltage for zero load current ($R_L = \infty$) is also measured. The circuit is then switched off.

(vii) Regulation characteristics without filter is then drawn by plotting $I_{d.c.}$ along X-axis and $V_{d.c.}$ along Y-axis. The percentage regulation for a rated current is then calculated from Eq. (10.18 e).

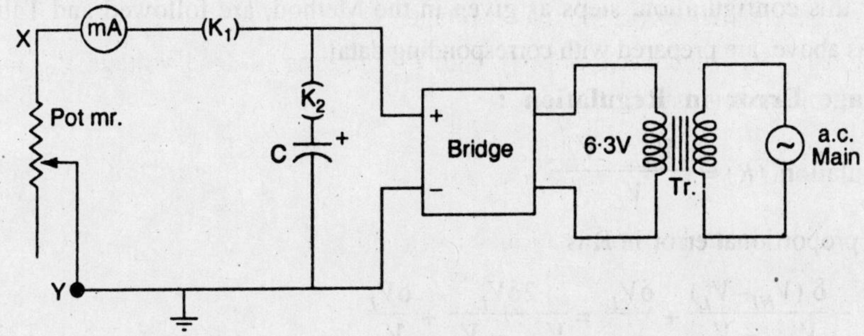

Fig. 10.67 : Experimental set-up

(viii) Ripple factor γ against R_L is plotted and the same may be compared with the theoretical value.

(ix) Now, both the keys K_1 and K_2 are closed which connects the capacitor C across the load. Now, we get the bridge rectifier with the capacitor filter. The steps (ii) to (viii) are repeated for this case.

(x) The filters (inductor, L-type LC filter, π-type LC filter, etc.) may also be used by properly connecting inductor, and capacitors in the circuit. Those are not discussed in details.

Results :

Table 1

[Diodes and Meters used and their specifications]

a.c. Line freequency = ... Hz

Voltage across transformer secondary = ... volt

Table 2

(A) Without filter :

Regulation Characteristics and Ripple Factor Data

d.c. load voltage $V_{d.c.}$ (volt)	d.c. load current $I_{d.c.}$ (mA)	Load Resitance R_L (ohm)	r.m.s. ripple voltage V_{rms} (volt)	Ripple factor (γ)	Theoritical value of γ
...	...	...	...		
...	...	...	...	...	
...	...	...	...	...	
...	...	...	...	...	

(B) Determination of percentage regulation :

Table 3

Value of V_{NL} (volt)	Specified Load currennt (mA)	Corresponding Load voltage (from graph) (volt)	Percentage Regulation $\left[\frac{V_{NL}-V_L}{V_L}\times 100\right]$
......	......	......	......

(C) With Capacitor Filter :

[For this configuration, steps as given in the Method, are followed and Table 2 and Table 3 as above, are prepared with corresponding data].

Percentage Error in Regulation :

$$\text{Regulation } (R) = \frac{V_{NL} - V_L}{V_L}$$

The proportional error in R is

$$\frac{\delta R}{R} = \frac{\delta (V_{NL} - V_L)}{V_{NL} - V_L} + \frac{\delta V_L}{V_L} = \frac{2\delta V_L}{V_{NL} - V_L} + \frac{\delta V_L}{V_L},$$

where δV_L is the Error in reading V_L, (smallest division of the voltmeter).

Precautions and Discussions :

(i) When without filter, the regulation characteristics slope gives the value of $(2R_d + R_s)$. By measuring the transformer secondary resistance (R_s), the dynamic resistance of the diode (R_d) can be obtained.

(ii) From the capacitor filter configuration, the slope of the regulation characteristics give $\frac{1}{4fC}$. Thus the slope of the curve, the value of C (capacitance) can be obtained and can be compared with the specified value.

(iii) The capacitor filter configuration gives poor voltage regulation because the ripple voltage tends to increase with increasing load current. This decreases the average voltage. Therefore, this fillter configuration is useful only when the load current is small.

Questions and Answers

Q.1. What is 'rectification'?

A. : It is a process of converting a.c. voltage or current into a d.c. component.

Q.2. What are the advantages of a full-wave rectifier over a half-wave rectifier.

A. : [See any textbook].

Q.3. What types of filter are used?

A. : Capacitor filter, inductor filter, π-type *LC* filter and *L*-type *LC* filter.

Q.4. Why is a filter used in a rectifier circuit?

A. : See Q. 4 under Expt. No. 10.1.

Q.5. Which is a better filter—capacitor or inductor?

A. : See Precautions and Discussions above.

Q.6. What is a bridge rectifier?

A. : It consists of four rectifier diode. Its function is similar to that of a full-wave rectifier.

Q.7. Of a bridge rectifier and a full-wave rectifier, which one is advantageous to use?

A. : The bridge rectifier is always preferable. [The reasons may be had from a text-book].

Q.8. Which one is a better rectifier—a semiconductor diode or a vacuum diode?

A. : A vacumm diode is a better rectifier.

Q.9. Of a silicon diode and a germanium diode, which is better for use in a rectifier and why?

A. : A silicon diode is better because its reverse saturation current is much less compared to that of a germanium diode.

Chapter 11

Appendix

TABLES OF THE PHYSICAL CONSTANTS

1. Some useful mathematical formulae :

Area of a circle = πr^2.

Volume of a sphere = $\frac{4}{3}\pi r^3$.

Circumference of a circle = $2\pi r$.

Surface area of a sphere = $4\pi r^2$.

Volume of a cylinder = $\pi r^2 l$.

Curved surface of a cylinder = $2\pi rl$.

Volume of a cone = $\frac{1}{3}\pi r^2 l$.

2. Moment of Inertia of some regular bodies :

Body	Axis of rotation	Moment of Inertia
Cylinder	Axis perpl. to length and through its centre	$M\left(\frac{l^2}{12}+\frac{r^2}{4}\right)$
Cylinder	Own axis	$\frac{1}{2}Mr^2$
Sphere	Any diameter	$\frac{2}{5}MR^2$
Rectangular lamina	Perpendicular to plane and through C.G.	$\frac{M}{12}(l^2+b^2)$

3. Conversion Table :

Mass		Length		Volume	
1 lb	= 453·6 gm	1 inch	= 2·54 cm	1 pint	= 0·568 litre
1 gm	= 15·43 grain	1 foot	= 30·48 cm	1 gallon	= 4·546 litre
1 Kgm	= 2·20 lb	1 yard	= 91·44 cm	1 litre	= 0·353 cft
1 Oz	= 28·35 gm	1 metre	= 39·37 inch	1 litre	= 1·76 pint
1 Ton	= 1016 Kgm	1 Km	= 0·621 mile	1 cft	= 28·31 litre
1 lb	= 0·4536 Kg	1 mile	= 1·61 Km	1 cubic metre	= 35·52 cft
1 grain	= 64·8 mgm	1 cm	= 0·3937 inch	1 cft	= 0·283 cub. metre
		1 metre	= 3·281 ft	1 litre	= 61·03 cub. inch

4. Densities of some common substances :

Solids (gm/cc)		Liquids (gm/cc)		Gases (gm/cc) (at N.T.P.)	
Aluminium	2·70	Alcohol	0·80	Air	0·001293
Brass	8·50	Castor oil	0·95	Argon	0·001784

Contd.

Solids (gm/cc)		Liquids (gm/cc)		Gases (gm/cc) (at N.T.P.)	
Copper	8·93	Glycerine	1·26	Hydrogen	0·000089
Ebonite	1·15	Kerosene	0·83	Helium	0·0001785
Steel	7·80	Milk	1·03	Nitrogen	0·00125
Marble	2·70	Turpentine	0·87	Oxygen	0·001429
Lead	11·37	Olive oil	0·92	Water vapour	0·000581
Ice	0·916	Chloroform	1·50	(at 100 °C)	
Cork	0·26	Mercury	13·60		
Alum	1·76	Paraffin oil	0·80		
Carbon	2·30	Benzene	0·80		
Sand	2·60	Aniline	1·02		
Sugar	1·60	Linseed oil	0·92		
Salt	2·20	Sulphuric acid	1·84		
Gold	19·32	Sea water (brine)	1·03		
Glass (Crown)	2·60	Ether	0·81		
Glass (Flint)	4·00	Carbon tetrachloride	1·59		
Silver	10·50	Carbon disulphide	1·27		
Paraffin wax	0·9				
Porcelain	2·40				
Napthalene	1·15				

5. Densities of water at different temperatures (gm/cc) :

Temp. °C	0	2	4	6	8
0	0·99987	0·99997	1·0000	0·99997	0·99988
10	0·99973	0·99953	0·99927	0·99897	0·99862
20	0·99823	0·99780	0·99732	0·99681	0·99626
30	0·99567	0·99505	0·99440	0·99371	0·99300
40	0·99220	0·99150	0·98070	0·98980	0·98900
50	0·98810	0·98720	0·98620	0·98530	0·98430
60	0·98320	0·98220	0·98110	0·98010	0·97890
70	0·97780	0·97670	0·97550	0·97430	0·97310
80	0·97180	0·97060	0·96930	0·96800	0·96670
90	0·96530	0·96400	0·96260	0·96120	0·95980

6. Elastic constants and Breaking Stresses of materials :

Substance	Young's modulus (Y) $\times 10^{11}$ dynes/cm^2	Bulk modulus (K) $\times 10^{11}$ dynes/cm^2	Rigidity modulus (n) $\times 10^{11}$ dynes/cm^2	Breaking stress (Kgm/cm^2)	Poisson's ratio
Iron	11·5	9·6	4·4	1500	0·23-0·31
Steel	19·7	17·0	7·5	11230	0·27

Substance	Young's modulus (Y) $\times 10^{11}$ dynes/cm^2	Bulk modulus (K) $\times 10^{11}$ dynes/cm^2	Rigidity modulus (n) $\times 10^{11}$ dynes/cm^2	Breaking stress (Kgm/cm^2)	Poisson's ratio
Brass	9·7	10·6	3·5	3000	0·37
Copper	12·4	13·9	4·0	4000	0·26
Aluminium	7·0	7·5	2·55	2200	0·34
Silver	7·8	10·9	2·9	3000	0·38
Manganin	12·4	12·1	4·7	……	0·33
Quartz	5·2	1·4	3·0	……	……
Phosphor Bronze	12·0	14·0	4·4	4000	0·38

7. Surface Tensions and Viscosities :

Substance	Surface tension at 20°C in dynes/cm	Viscosity coeff. at 20°C in C.G.S units
Water	72·7	0·01002
Alcohol	22·3	0·0119
Paraffin oil	25·5	0·020
Castor oil	……	9·86
Glycerine	63·0	8·30
Mercury	487·0	0·0156
Turpentine	27·3	0·0149
Olive oil	32·0	0·840
Soap soln.	20·0	……
Air	……	0·000183 (0°C)
Hydrogen	……	0·000086 (0°C)
Nitrogen	……	0·000167 (0°C)
Oxygen	……	0·000192 (0°C)
Water vapour	……	0·000087 (0°C)

8. Variation of Surface Tension and Viscosity of Water with temperature :

Temperature (°C)	Surface Tension (dynes/cm)	Viscosity Coeff. (poise)
0	75·7	0·01786
10	75·2	0·01304
20	72·7	0·01002
30	71·2	0·00798
40	69·6	0·00654
50	67·9	0·00548
60	66·2	0·00467

Contd.

Temperature (°C)	Surface Tension (dynes/cm)	Viscosity Coeff. (poise)
70	64·4	0·00405
80	62·6	0·00355
90	60·1	0·00315
100	58·8	0·00283

9. Viscosities of liquids at different temperatures :

Temp.⟶	10°C	20°C	25°C	30°C	35°C	40°C
Castor oil	24·2	9·86	6·55	4·52	3·28	2·31
Glycerine	......	8·6	5·6	3·5	2·3	.....
Turpentine Oil	00·0178	0·015	0·0133	0·0128	0·0118	0·011

10. Velocities of Sound in different media :

Medium	Velocity at 20°C (m/sec)	Medium	Velocity at 20°C (m/sec)
Brass	3500	Mercuty	1407
Copper	3700	Water	1450
Glass	5000	Air	344
Iron	5100	Hydrogen	1330
Aluminium	5000	Nitrogen	338
Steel	5130	Carbon dioxide	275

11. Pressure of Saturated Water Vapour :

Temp. (°C)	Pressure (mm of Hg)	Temp. (°C)	Pressure (mm of Hg)	Temp. (°C)	Pressure (mm of Hg)
13	11·23	23	21·02	33	37·59
14	11·98	24	22·32	34	39·75
15	12·78	25	23·69	35	42·02
16	13·62	26	25·13	36	44·40
17	14·52	27	26·65	37	46·90
18	15·46	28	28·25	38	49·51
19	16·46	29	29·94	39	52·26
20	17·51	30	31·71	40	55·13
21	18·62	31	33·57	45	71·64
22	19·79	32	35·53	50	92·30

12. Coefficients of Expansion of Solids, Liquids and Gases :

Solids (linear)	α per °C × 10^{-6}	Liquids (Cubical)	γ per °C × 10^{-5}	Gases (Cubical)	γ per °C × 10^{-4}
Aluminium	24	Alcohol	110	Air	36·7
Copper	17	Water (5°-10°)	5·4	All other gases	36·6

Solids (linear)	α per °C × 10^{-6}	Liquids (Cubical)	γ per °C × 10^{-5}	Gases (Cubical)	γ per °C × 10^{-4}
Iron	10	Water (10°-20°)	15·0	Air	36·7
Steel	11	Water (20°-40°)	30·2	All other gases	36·6
Lead	29	Glycerine	52		
Brass	18·9	Mercury	18·2		
Zinc	26	Paraffin oil	90		
Glass	9·0	Turpene oil	94		
Ice	51	Benzene	124		
Constantan	16	Ether	163		
German silver	18·5	Aniline	85		
Invar	0·9	Chloroform	126		
Nickel	12·8	Oilve oil	70		
Platinum	8·9				
Tungsten	4·4				
Quartz fibre	1·7				

13. Other Thermal Properties of Different Substances :

Substances	Melting point (°C)	Boiling point (°C)	Latent Heat (cal/gm)	Specific Heat (cal/gm/°C)	Thermal conductivity (cal/cm/sec/°C) × 10^{-3}
Aluminium	657	2450	96	0·29	503
Copper	1089	2600	49	0·091	920
Iron	1100	3000	28	0·119	114
Steel	1400	3000	49	0·112	110
Lead	327	1730	6·2	0·031	83
Brass	900	...	27	0·092	260
Zinc	419	906	24	0·091	265
Glass	1000	...	...	0·160	2·50
Wax	62	...	...	0·69	0·22
Marble	...	...	...	0·22	7·80
Ice	0	...	80	0·50	5·70
Napthalene	80	210·8	...	...	...
Water	0	100	537	1·00	1·36
Glycerine	17	290	...	0·59	0·71
Mercury	– 38·9	357	68	0·33	19·8
Paraffin oil	...	...	...	0·52	0·36
Turpene oil	– 10	159	70	0·42	...

Substances	Melting point (°C)	Boiling point (°C)	Latent Heat (cal/gm)	Specific Heat (cal/gm/°C)	Thermal conductivity (cal/cm/sec/°C) × 10^{-3}
Benzene	5·5	80·2	93	0·40	0·43
Ether	– 12	34	89	0·56	0·30
Kerosene	...	...	...	0·05	0·40
Alcohol	– 115	78	205	0·58	0·48
Air	...	...	...	0·237	0·0570
Hydrogen	...	...	...	3·42	0·3180
Oxygen	...	...	...	0·242	0·0583
Nitrogen	...	...	...	0·235	0·0581
Helium	...	...	...	1·25	0·343

13A. Thermal Conductivities of some Bad Conductors :

Substance	Value of *K* (cal/cm/sec/°C)	Substance	Value of *K* (cal/cm/sec/ °C)
Asbestos	3×10^{-4}	Paper	3×10^{-4}
Asbestos paper	6×10^{-4}	Porcelain	25×10^{-4}
Ebonite	$4{\cdot}2 \times 10^{-4}$	Rubber	$4{\cdot}5 \times 10^{-4}$
Mica	18×10^{-4}		

14. Magnetic Elements and "g" at different places :

Station	Declination	Dip	Value of *H* (oersted)	Value of '*g*' (cm/sec^2)
Allahabad	0°20' W	37°25'	0·363	978·9
Agra	0°10' E	40°40'	0·348	979·06
Ajmer	0°10' E	39°20'	0·3483	978·98
Mumbai	0°20' W	24°22'	0·376	978·6
Burdwan	...	31°0'	0·380	978·8
Calcutta	1°W'	30°30'	0·378	978·8
Cuttack	...	25°30'	0·377	978·6
Delhi	0°40' E	42°50'	0·345	979·2
Dhaka	...	32°40'	0·371	978·9
Dehradun	0°50' E	45°55'	0·331	979·1
Gauhati	...	37°0'	0·379	978·9
Gorakhpur	0°20' W	30°35'	0·357	978·9
Jadavpur	1°W	30°0'	0·378	978·8
Jaipur	0°30' E	40°30'	0·347	978·5
Jodhpur	0°0'	39°10'	0·348	978·9

Station	Declination	Dip	Value of H (oersted)	Value of 'g' (cm/sec^2)
Kharagpur	...	28°45'	0·379	978·9
Kanpur	...	38°40'	0·363	978·9
Kalyani	...	30°45'	0·377	978·8
Lucknow	0°10' W	40°0'	0·354	979·0
Chennai	0°10' W	34°40'	0·369	978·3
Meerut	0°40' E	43°30'	0·339	979·8
Patna	...	36°0'	0·360	978·9
Siliguri	...	37°20'	0·376	978·9
Santiniketan	...	32°30'	0·380	978·7
Udaipur	0°0'	35°45'	0·362	978·9
Varanasi	0°30' W	37°10'	0·364	979·0

15. Refractive Indices of different substances for λ = 5893 Å :

Substance	Ref. Index	Substance	Ref. Index	Substance	Ref. Index
Glass (Crown)	1·48-1·61	Alcohol	1·362	Canada balsam	1·53
Glass (Flint)	1·53-1·96	Paraffin oil	1·44	Potash Alum	1·456
Glass (Soda)	1·50	Glycerin	1·473	Cocoanut oil	1·45
Diamond	2·42	Kerosene	1·390	Rock Salt	1·544
Ice	1·31	Castor oil	1·472	Ether	1·354
Mica	1·56-1·60	Turpentine	1·47	Chloroform	1·449
Quartz	1·544	Olive oil	1·46	Plastic (Polysterine)	1·589
Benzene	1·504	Linseed oil	1·474	Liq. Paraffin	1·48
Water	1·33	Carbon disulphide	1·632	Sod. Chloride	1·546

16. Wavelengths of Emission Spectral Lines :

Substance	λ in Å (10^{-8} cm)	Substance	λ in Å (10^{-8} cm)	Substance	λ in Å (10^{-8} cm)
Helium	3889 *v*	Argon	4159 *v*	NaCl (in flame)	5890 *y*
	4026 *v*		4198 *v*		5896 *y*
	4471 *b*		4201 *v*		
	4713 *b*		4259 *b*		5460 *g*
	4922 *gb*		4703 *b*	Mercury Lamp	5770 *y*
	5016 *g*		6031 *r*		5790 *y*
	5876 *y*		—		6152 *o*

Substance	λ in Å (10^{-8} cm)	Substance	λ in Å (10^{-8} cm)	Substance	λ in Å (10^{-8} cm)
	6678 *r*		5765 *y*		6253 *o*
	7065 *r*	Neon	5853 *y*		—
	—		5945 *o*	Calcium	4227 *v*
Hydrogen	3790 *v*		6075 *o*	Chloride	3934 *v*
	4102 *v*		6383 *o*	(Flame and arc)	3968 *v*
	4340 *b*		6402 *o*		5589 *y*
	4861 *gb*		6507 *r*		
	6563 *r*				

17. Dispersive Powers of Standard Optical Glasses :

Name	Dispersive Power (ω)
Hard Crown	0·0166
Flint	0·0243
Extra Dense Flint	0·0330
Double EDF	0·0360

18. Specific Rotations of some substances (λ = 5893 Å) :

Solvent	Substance	Concentration (c%)	Sp. rotation in degrees
Alcohol	Camphor	45-90	+ 54·2-0·13c
Water	Cane sugar	4-28	+ 66·6-0·009c
Water	Glucose	4	– 94·4 (after 7 mins.)
Water	Dextrose	9	+ 105·2 (after 6 mins.)
Water	Pot. tartarate	8-50	+ 27·14+0·09c
Pure	Turpene oil	……	– 37

19. Specific Resistances and Temperature Coefficients of Resistances :

Substance	Sp. resistance ($\times 10^{-6}$ ohm. cm)	Temp. Coeff. of resistance ($\times 10^{-4}$/°C)	Substance	Sp. resistance ($\times 10^{-6}$ ohm. cm)	Temp. Coeff. of resistance ($\times 10^{-4}$/°C)
Aluminium	3·21	38·0	Manganin	44·5	–0·1 to 1·0
Copper	1·70	43·0	Nickel	11·8	27
Brass	6·3	10·0	Tungsten	5·5	51
Constantan	49	–0·4 to 0·1	Mercury	95·8	9·0
German Silver	33	2·5 to 6·0	Graphite	3000	– 7·0
Iron	13	62	Carbon filament	4000	negative
Steel	19	16 to 22	Phosphor bronze	5 to 10	……

Substance	Sp. resistance ($\times 10^{-6}$ ohm. cm)	Temp. Coeff. of resistance ($\times 10^{-4}$/°C)	Substance	Sp. resistance ($\times 10^{-6}$ ohm. cm)	Temp. Coeff. of resistance ($\times 10^{-4}$/°C)
Silver	1·63	40	Quartz	$1{\cdot}2 \times 10^{20}$	......
Platinum	11	37	Mica	9×10^{21}	......
Gold	2·4	40	Zinc	6·1	37
Nichrome	110	1·70	Tin	11·3	45

20. Colour Code of Resistances :

0	Black
1	Brown
2	Red
3	Orange
4	Yellow
5	Green
6	Blue
7	Violet
8	Grey
9	White

21. E.M.F. of Different Cells :

Cells	E.M.F. (volt)	Cells	E.M.F. (volt)	Cells	E.M.F. (volt)
Primary cells		*Standard cells*		*Secondary cells*	
Bunsen	1·8-1·9	Cadmium	1·01864 (at 20°C)	Acid (Pb)	2·0
Daniell	1·08	Clark	1·43	Alkali (Ni-Fe)	1·25
Leclanche's	1·46				
Grove	1·90				
Bichromate	2·0				

22. Electrochemical Equivalent of Substances :

Substance	E.C.E. (gm/coulomb)	Substance	E.C.E. (gm/coulomb)
Copper	0·0003293	Chromium	0·000180
Silver	0·0011182	Hydrogen	0·00001044
Zinc	0·000340		

23. Electrical Conductivity of Solutions :

c = Mass of anhydrous solute per 100 gm of solution.

K_{18} = Specific conductivity at 18°C in $ohm^{-1} cm^{-1}$.

η = number of gm-equivalents in 1 cc of the soluiton.

λ = equivalent conductivity $= \frac{K}{\eta}$.

Substances	c%	K_{18}	$\eta \times 10^{-4}$	$\lambda = \frac{K}{\eta}$	c%	K_{18}	$\eta \times 10^{-4}$	$\lambda = \frac{K}{\eta}$
Copper Sulphate	2·5	0·0109	3·206	34	5·0	0·0189	6·587	28·7
	10·0	0·0320	13·85	23·1	15·0	0·0421	21·92	19·2
	17·5	0·0458	26·82	17·4	20·0	0·0527	28·71	15·7
NaCl	5·0	0·0672	8·84	76·0	10·0	0·1211	18·29	66·2
	15·0	0·1642	24·8	57·8	20·0	0·1957	39·22	49·9
	25·0	0·2135	50·8	42·0	26·5	0·2156	54·17	39·8
KCl	5·0	0·0690	6·91	99·9	10·0	0·1359	14·27	95·2
	15·0	0·2020	22·1	91·5	20·0	0·2677	30·11	88·9
NaOH	2·5	0·1090	6·41	170	5·0	0·1970	13·22	149
	10·0	0·3120	27·85	112	15·0	0·3460	43·79	79
	20·0	0·3270	61·69	53	30·0	0·2020	101·0	20

24. Constants of some common Thermocouples :

Couple	Value of a (per °C)	Value of b (per °C^2)	Thermoelectric power (μV/°C)
Copper-constantan	$37·54 \times 10^{-6}$	$0·0444 \times 10^{-6}$	40
Copper-Iron	$13·4 \times 10^{-6}$	$-0·0137 \times 10^{-6}$	14

25. Variations of Thermo-e.m.f. with Temperature :

Temperature of cold junction	0°C										
Temp. of Hot junction	0	10	20	30	40	50	60	70	80	90	100
Thermo-emf(in mV) for Copper constantan	0	0·39	0·79	1·19	1·61	2·03	2·47	2·91	3·36	3·81	4·28
Thermo-emf (in mV) for Iron-constantan	0	0·52	1·05	1·58	2·12	2·66	3·20	3·75	4·30	4·85	5·40

26. Some mathematical constants and their logarithms :

Constants	Values	$\log_{10}$
π	3·1416	0·4972
π^2	9·8696	0·9943
$\frac{1}{\pi}$	0·3183	$\bar{1}$·5028
$\frac{1}{\pi^2}$	0·1013	$\bar{1}$·0057
$\sqrt{2}$	1·4142	0·1505
$\sqrt{3}$	1·7321	0·2386
$\sqrt{10}$	3·1623	0·5000
e	2·7183	0·4343
$\log_e 10$	2·3026	0·3622

FOUR-FIGURE LOGARITHMIC TABLES

LOGARITHMS

	0	1	2	3	4	5	6	7	8	9	1	2	3	4	5	6	7	8	9
10	0000	0043	0086	0128	0170						5	9	13	17	21	26	30	34	38
						0212	0253	0294	0334	0374	4	8	12	16	20	24	28	32	36
11	0414	0453	0492	0531	0569						4	8	12	16	20	23	27	31	35
						0607	0645	0682	0719	0755	4	7	11	15	18	22	26	29	33
12	0792	0828	0864	0899	0934						3	7	11	14	18	21	25	28	32
						0969	1004	1038	1072	1106	3	7	10	14	17	20	24	27	31
13	1139	1173	1206	1239	1271						3	6	10	13	16	19	23	26	29
						1303	1335	1367	1399	1430	3	6	10	13	16	19	22	25	29
14	1461	1492	1523	1553	1584						3	6	9	12	15	19	22	25	28
						1614	1644	1673	1703	1732	3	6	9	12	14	17	20	23	26
15	1761	1790	1818	1847	1875						3	6	9	11	14	17	20	23	26
						1903	1931	1959	1987	2014	3	6	8	11	14	17	19	22	25
16	2041	2068	2095	2122	2148						3	6	8	11	14	16	19	22	24
						2175	2201	2227	2253	2279	3	5	8	10	13	16	18	21	23
17	2304	2330	2355	2380	2405						3	5	8	10	13	15	18	20	23
						2430	2455	2480	2504	2529	3	5	8	10	12	15	17	20	22
18	2553	2577	2601	2625	2648						2	5	7	9	12	14	17	19	21
						2672	2695	2718	2742	2765	2	4	7	9	11	14	16	18	21
19	2788	2810	2833	2856	2878						2	4	7	9	11	13	16	18	20
						2900	2923	2945	2967	2989	2	4	6	8	11	13	15	17	19
20	3010	3032	3054	3075	3096	3118	3139	3160	3181	3201	2	4	6	8	11	13	15	17	19
21	3222	3243	3263	3284	3304	3324	3345	3365	3385	3404	2	4	6	8	10	12	14	16	18
22	3424	3444	3464	3483	3502	3522	3541	3560	3579	3598	2	4	6	8	10	12	14	15	17
23	3617	3636	3655	3674	3692	3711	3729	3747	3766	3784	2	4	6	7	9	11	13	15	17
24	3802	3820	3838	3856	3874	3892	3909	3927	3945	3962	2	4	5	7	9	11	12	14	16
25	3979	3997	4014	4031	4048	4065	4082	4099	4116	4133	2	3	5	7	9	10	12	14	15
26	4150	4165	4183	4200	4216	4232	4249	4265	4281	4298	2	3	5	7	8	10	11	13	15
27	4314	4330	4346	4362	4378	4393	4409	4425	4440	4456	2	3	5	6	8	9	11	13	14
28	4472	4487	4502	4518	4533	4548	4564	4579	4594	4609	2	3	5	6	8	9	11	12	14
29	4624	4639	4654	4669	4683	4698	4713	4728	4742	4757	1	3	4	6	7	9	10	12	13
30	4771	4786	4800	4814	4829	4843	4857	4871	4886	4900	1	3	4	6	7	9	10	11	13
31	4914	4928	4942	4955	4969	4983	4997	5011	5024	5038	1	3	4	6	7	8	10	11	12
32	5051	5065	5079	5092	5105	5119	5132	5145	5159	5172	1	3	4	5	7	8	9	11	12
33	5185	5198	5211	5224	5237	5250	5263	5276	5289	5302	1	3	4	5	6	8	9	10	12
34	5315	5328	5340	5353	5366	5378	5391	5403	5416	5428	1	3	4	5	6	8	9	10	11
35	5441	5453	5465	5478	5490	5502	5514	5527	5539	5551	1	2	4	5	6	7	9	10	11
36	5563	5575	5587	5599	5611	5623	5635	5647	5658	5670	1	2	4	5	6	7	8	10	11
37	5682	5694	5705	5717	5729	5740	5752	5763	5775	5786	1	2	3	5	6	7	8	9	10
38	5798	5809	5821	5832	5843	5855	5866	5877	5888	5899	1	2	3	5	6	7	8	9	10
39	5911	5922	5933	5944	5955	5966	5977	5988	5999	6010	1	2	3	4	5	7	8	9	10
40	6021	6031	6042	6053	6064	6075	6085	6096	6107	6117	1	2	3	4	5	6	8	9	10
41	6128	6138	6149	6160	6170	6180	6191	6201	6212	6222	1	2	3	4	5	6	7	8	9
42	6232	6243	6253	6263	6274	6284	6294	6304	6314	6325	1	2	3	4	5	6	7	8	9
43	6335	6345	6355	6365	6375	6385	6395	6405	6415	6425	1	2	3	4	5	6	7	8	9
44	6435	6444	6454	6464	6474	6484	6493	6503	6513	6522	1	2	3	4	5	6	7	8	9
45	6532	6542	6551	6561	6571	6580	6590	6599	6609	6618	1	2	3	4	5	6	7	8	9
46	6628	6637	6646	6656	6665	6675	6684	6693	6702	6712	1	2	3	4	5	6	7	7	8
47	6721	6730	6739	6749	6758	6767	6776	6785	6794	6803	1	2	3	4	5	5	6	7	8
48	6812	6821	6830	6839	6848	6857	6866	6875	6884	6893	1	2	3	4	4	5	6	7	8
49	6902	6911	6920	6928	6937	6946	6955	6964	6972	6981	1	2	3	4	4	5	6	7	8

LOGARITHMS

	0	1	2	3	4	5	6	7	8	9	1	2	3	4	5	6	7	8	9
50	6990	6998	7007	7016	7024	7033	7042	7050	7059	7067	1	2	3	3	4	5	6	7	8
51	7076	7084	7093	7101	7110	7118	7126	7135	7143	7152	1	2	3	3	4	5	6	7	8
52	7160	7168	7177	7185	7193	7202	7210	7218	7226	7235	1	2	2	3	4	5	6	7	7
53	7243	7251	7259	7267	7275	7284	7292	7300	7308	7316	1	2	2	3	4	5	6	6	7
54	7324	7332	7340	7348	7356	7364	7372	7380	7388	7396	1	2	2	3	4	5	6	6	7
55	7404	7412	7419	7427	7435	7443	7451	7459	7466	7474	1	2	2	3	4	5	5	6	7
56	7482	7490	7497	7505	7513	7520	7528	7536	7543	7551	1	2	2	3	4	5	5	6	7
57	7559	7566	7574	7582	7589	7597	7604	7612	7619	7627	1	2	2	3	4	5	5	6	7
58	7634	7642	7649	7657	7664	7672	7679	7686	7694	7701	1	1	2	3	4	4	5	6	7
59	7709	7716	7733	7731	7738	7745	7752	7760	7767	7774	1	1	2	3	4	4	5	6	7
60	7782	7789	7796	7803	7810	7818	7825	7832	7839	7846	1	1	2	3	4	4	5	6	6
61	7853	7860	7868	7875	7882	7889	7896	7903	7910	7917	1	1	2	3	4	4	5	6	6
62	7924	7931	7938	7945	7952	7959	7966	7973	7980	7987	1	1	2	3	3	4	5	6	6
63	7993	8000	8007	8014	8021	8028	8035	8041	8048	8055	1	1	2	3	3	4	5	5	6
64	8062	8069	8075	8082	8089	8096	8102	8109	8116	8122	1	1	2	3	3	4	5	5	6
65	8129	8136	8142	8149	8156	8162	8169	8176	8182	8189	1	1	2	3	3	4	5	5	6
66	8195	8202	8209	8215	8222	8228	8235	8241	8248	8254	1	1	2	3	3	4	5	5	6
67	8261	8267	8274	8280	8287	8293	8299	8306	8312	8319	1	1	2	3	3	4	5	5	6
68	8325	8331	8338	8344	8351	8357	8363	8370	8376	8382	1	1	2	3	3	4	4	5	6
69	8388	8395	8401	8407	8414	8420	8426	8432	8439	8445	1	1	2	2	3	4	4	5	6
70	8451	8457	8463	8470	8476	8482	8488	8494	8500	8506	1	1	2	2	3	4	4	5	6
71	8513	8519	8525	8531	8537	8543	8549	8555	8561	8567	1	1	2	2	3	4	4	5	5
72	8573	8579	8585	8591	8597	8603	8609	8615	8621	8627	1	1	2	2	3	4	4	5	5
73	8633	8639	8645	8651	8657	8663	8669	8675	8681	8686	1	1	2	2	3	4	4	5	5
74	8692	8698	8704	8710	8716	8722	8727	8733	8739	8745	1	1	2	2	3	4	4	5	5
75	8751	8756	8762	8768	8774	8779	8785	8791	8797	8802	1	1	2	2	3	3	4	5	5
76	8808	8814	8820	8825	8831	8837	8842	8848	8854	8859	1	1	2	2	3	3	4	5	5
77	8865	8871	8876	8882	8887	8893	8899	8904	8910	8915	1	1	2	2	3	3	4	4	5
78	8921	8927	8932	8938	8943	8949	8954	8960	8965	8971	1	1	2	2	3	3	4	4	5
79	8976	8982	8987	8993	8998	9004	9009	9015	9020	9025	1	1	2	2	3	3	4	4	5
80	9031	9036	9042	9047	9053	9058	9063	9069	9074	9079	1	1	2	2	3	3	4	4	5
81	9085	9090	9096	9101	9106	9112	9117	9122	9128	9133	1	1	2	2	3	3	4	4	5
82	9138	9143	9149	9154	9159	9165	9170	9175	9180	9196	1	1	2	2	3	3	4	4	5
83	9191	9196	9201	9206	9212	9217	9222	9227	9232	9238	1	1	2	2	3	3	4	4	5
84	9243	9248	9253	9258	9263	9269	9274	9279	9284	9289	1	1	2	2	3	3	4	4	5
85	9294	9299	9304	9309	9315	9320	9325	9330	9335	9340	1	1	2	2	3	3	4	4	5
86	9345	9350	9355	9360	9365	9370	9375	9380	9385	9390	1	1	2	2	3	3	4	4	5
87	9395	9400	9405	9410	9415	9420	9425	9430	9435	9440	0	1	1	2	2	3	3	4	4
88	9445	9450	9455	9460	9465	9469	9474	9479	9484	9489	0	1	1	2	2	3	3	4	4
89	9494	9499	9504	9509	9513	9518	9523	9528	9533	9538	0	1	1	2	2	3	3	4	4
90	9542	9547	9552	9557	9562	9566	9571	9576	9581	9586	0	1	1	2	2	3	3	4	4
91	9590	9595	9600	9605	9609	9614	9619	9624	9628	9633	0	1	1	2	2	3	3	4	4
92	9638	9643	9647	9652	9657	9661	9666	9671	9675	9680	0	1	1	2	2	3	3	4	4
93	9685	9689	9694	9699	9703	9708	9713	9717	9722	9727	0	1	1	2	2	3	3	4	4
94	9731	9736	9741	9745	9750	9754	9759	9763	9768	9773	0	1	1	2	2	3	3	4	4
95	9777	9782	9786	9791	9795	9800	9805	9809	9814	9818	0	1	1	2	2	3	3	4	4
96	9823	9827	9832	9836	9841	9845	9850	9854	9859	9863	0	1	1	2	2	3	3	4	4
97	9868	9872	9877	9881	9886	9890	9894	9899	9903	9908	0	1	1	2	2	3	3	4	4
98	9912	9917	9921	9926	9930	9934	9939	9943	9948	9952	0	1	1	2	2	3	3	4	4
99	9956	9961	9965	9969	9974	9978	9983	9987	9991	9996	0	1	1	2	2	3	3	3	4

FOUR-FIGURE ANTILOGARITHMIC TABLES

ANTILOGARITHMS

	0	1	2	3	4	5	6	7	8	9	1	2	3	4	5	6	7	8	9
.00	1000	1002	1005	1007	1009	1012	1014	1016	1019	1021	0	0	1	1	1	1	2	2	2
.01	1023	1026	1028	1030	1033	1035	1038	1040	1042	1045	0	0	1	1	1	1	2	2	2
.02	1047	1050	1052	1054	1057	1059	1062	1064	1067	1069	0	0	1	1	1	1	2	2	2
.03	1072	1074	1076	1079	1081	1084	1086	1089	1091	1094	0	0	1	1	1	1	2	2	2
.04	1096	1099	1102	1104	1107	1109	1112	1114	1117	1119	0	1	1	1	1	2	2	2	2
.05	1122	1125	1127	1130	1132	1135	1138	1140	1143	1146	0	1	1	1	1	2	2	2	2
.06	1148	1151	1153	1156	1159	1161	1164	1167	1169	1172	0	1	1	1	1	2	2	2	2
.07	1175	1178	1180	1183	1186	1189	1191	1194	1197	1199	0	1	1	1	1	2	2	2	2
.08	1202	1205	1208	1211	1213	1216	1219	1222	1225	1227	0	1	1	1	1	2	2	2	3
.09	1230	1233	1236	1239	1242	1245	1247	1250	1253	1256	0	1	1	1	1	2	2	2	3
.10	1259	1262	1265	1268	1271	1274	1276	1279	1282	1285	0	1	1	1	1	2	2	2	3
.11	1288	1291	1294	1297	1300	1303	1306	1309	1312	1315	0	1	1	1	2	2	2	2	3
.12	1318	1321	1324	1327	1330	1334	1337	1340	1343	1346	0	1	1	1	2	2	2	2	3
.13	1349	1352	1355	1358	1361	1365	1368	1371	1374	1377	0	1	1	1	2	2	2	3	3
.14	1380	1384	1387	1390	1393	1396	1400	1403	1406	1409	0	1	1	1	2	2	2	3	3
.15	1413	1416	1419	1422	1426	1429	1432	1435	1439	1442	0	1	1	1	2	2	2	3	3
.16	1445	1449	1452	1455	1459	1462	1466	1469	1472	1476	0	1	1	1	2	2	2	3	3
.17	1479	1483	1486	1489	1493	1496	1500	1503	1507	1510	0	1	1	1	2	2	2	3	3
.18	1514	1517	1521	1524	1528	1531	1535	1538	1542	1545	0	1	1	1	2	2	2	3	3
.19	1549	1552	1556	1560	1563	1567	1570	1574	1578	1581	0	1	1	1	2	2	3	3	3
.20	1585	1589	1592	1596	1600	1603	1607	1611	1614	1618	0	1	1	1	2	2	3	3	3
.21	1622	1626	1629	1633	1637	1641	1644	1648	1652	1656	0	1	1	2	2	2	3	3	3
.22	1660	1663	1667	1671	1675	1679	1683	1687	1690	1694	0	1	1	2	2	2	3	3	3
.23	1698	1702	1706	1710	1714	1718	1722	1726	1730	1734	0	1	1	2	2	2	3	3	4
.24	1738	1742	1746	1750	1754	1758	1762	1766	1770	1774	0	1	1	2	2	2	3	3	4
.25	1778	1782	1786	1791	1795	1799	1803	1807	1811	1816	0	1	1	2	2	2	3	3	4
.26	1820	1824	1828	1832	1837	1841	1845	1849	1854	1858	0	1	1	2	2	3	3	3	4
.27	1862	1866	1871	1875	1879	1884	1888	1892	1897	1901	0	1	1	2	2	3	3	3	4
.28	1905	1910	1914	1919	1923	1928	1932	1936	1941	1945	0	1	1	2	2	3	3	4	4
.29	1950	1954	1959	1963	1968	1972	1977	1982	1986	1991	0	1	1	2	2	3	3	4	4
.30	1995	2000	2004	2009	2014	2018	2023	2028	2032	2037	0	1	1	2	2	3	3	4	4
.31	2042	2046	2051	2056	2061	2065	2070	2075	2080	2084	0	1	1	2	2	3	3	4	4
.32	2089	2094	2099	2104	2109	2113	2118	2123	2128	2133	0	1	1	2	2	3	3	4	4
.33	2138	2143	2148	2153	2158	2163	2168	2173	2178	2183	0	1	1	2	2	3	3	4	4
.34	2188	2193	2198	2203	2208	2213	2218	2223	2228	2234	1	1	2	2	3	3	4	4	5
.35	2239	2244	2249	2254	2259	2265	2270	2275	2280	2286	1	1	2	2	3	3	4	4	5
.36	2291	2296	2301	2307	2312	2317	2323	2328	2333	2339	1	1	2	2	3	3	4	4	5
.37	2344	2350	2355	2360	2366	2371	2377	2382	2388	2393	1	1	2	2	3	3	4	4	5
.38	2399	2404	2410	2415	2421	2427	2432	2438	2443	2449	1	1	2	2	3	3	4	4	5
.39	2455	2460	2466	2472	2477	2483	2489	2495	2500	2506	1	1	2	2	3	3	4	5	5
.40	2512	2518	2523	2529	2535	2541	2547	2553	2559	2564	1	1	2	2	3	4	4	5	5
.41	2570	2576	2582	2588	2594	2600	2606	2612	2618	2624	1	1	2	2	3	4	4	5	5
.42	2630	2636	2642	2649	2655	2661	2667	2673	2679	2685	1	1	2	2	3	4	4	5	6
.43	2692	2698	2704	2710	2716	2723	2729	2735	2742	2748	1	1	2	3	3	4	4	5	6
.44	2754	2761	2767	2773	2780	2786	2793	2799	2805	2812	1	1	2	3	3	4	4	5	6
.45	2818	2825	2831	2838	2844	2851	2858	2864	2871	2877	1	1	2	3	3	4	5	5	6
.46	2884	2891	2897	2904	2911	2917	2924	2931	2938	2944	1	1	2	3	3	4	5	5	6
.47	2951	2958	2965	2972	2979	2985	2992	2999	3006	3013	1	1	2	3	3	4	5	5	6
.48	3020	3027	3034	3041	3048	3055	3062	3069	3076	3083	1	1	2	3	4	4	5	6	6
.49	3090	3097	3105	3112	3119	3126	3133	3141	3148	3155	1	1	2	3	4	4	5	6	6

ANTILOGARITHMS

	0	1	2	3	4	5	6	7	8	9	1	2	3	4	5	6	7	8	9
.50	3162	3170	3177	3184	3192	3199	3206	3214	3221	3228	1	1	2	3	4	4	5	6	7
.51	3236	3243	3251	3258	3266	3273	3281	3289	3296	3304	1	2	2	3	4	5	5	6	7
.52	3311	3319	3327	3334	3342	3350	3357	3365	3373	3381	1	2	2	3	4	5	5	6	7
.53	3388	3396	3404	3412	3420	3428	3436	3443	3451	3459	1	2	2	3	4	5	6	6	7
.54	3467	3475	3483	3491	3499	3508	3516	3524	3532	3540	1	2	2	3	4	5	6	6	7
.55	3548	3556	3565	3573	3581	3589	3597	3606	3614	3622	1	2	2	3	4	5	6	7	7
.56	3631	3639	3648	3656	3664	3673	3681	3690	3698	3707	1	2	3	3	4	5	6	7	8
.57	3715	3724	3733	3741	3750	3758	3767	3776	3784	3793	1	2	3	3	4	5	6	7	8
.58	3802	3811	3819	3828	3837	3846	3855	3864	3873	3882	1	2	3	4	4	5	6	7	8
.59	3890	3899	3908	3917	3926	3936	3945	3954	3963	3972	1	2	3	4	5	5	6	7	8
.60	3981	3990	3999	4009	4018	4027	4036	4046	4055	4064	1	2	3	4	5	6	6	7	8
.61	4074	4083	4093	4102	4111	4121	4130	4140	4150	4159	1	2	3	4	5	6	7	8	9
.62	4169	4178	4188	4198	4207	4217	4227	4236	4246	4256	1	2	3	4	5	6	7	8	9
.63	4266	4276	4285	4295	4305	4315	4325	4335	4345	4355	1	2	3	4	5	6	7	8	9
.64	4365	4375	4385	4395	4406	4416	4426	4436	4446	4457	1	2	3	4	5	6	7	8	9
.65	4467	4477	4487	4498	4508	4519	4529	4539	4550	4560	1	2	3	4	5	6	7	8	9
.66	4571	4581	4592	4603	4613	4624	4634	4645	4656	4667	1	2	3	4	5	6	7	9	10
.67	4677	4688	4699	4710	4721	4732	4742	4753	4764	4775	1	2	3	4	5	7	8	9	10
.68	4786	4797	4808	4819	4831	4842	4853	4864	4875	4887	1	2	3	4	6	7	8	9	10
.69	4898	4909	4920	4932	4943	4955	4966	4977	4989	5000	1	2	3	5	6	7	8	9	10
.70	5012	5023	5035	5047	5058	5070	5082	5093	5105	5117	1	2	4	5	6	7	8	9	11
.71	5129	5140	5152	5164	5176	5188	5200	5212	5224	5236	1	2	4	5	6	7	8	10	11
.72	5248	5260	5272	5284	5297	5309	5321	5333	5346	5358	1	2	4	5	6	7	8	10	11
.73	5370	5383	5395	5408	5420	5433	5445	5458	5470	5483	1	3	4	5	6	8	9	10	11
.74	5495	5508	5521	5534	5546	5559	5572	5585	5598	5610	1	3	4	5	6	8	9	10	12
.75	5623	5636	5649	5662	5675	5689	5702	5715	5728	5741	1	3	4	5	7	8	9	10	12
.76	5754	5768	5781	5794	5808	5821	5834	5848	5861	5875	1	3	4	5	7	8	9	11	12
.77	5888	5902	5916	5929	5943	5957	5970	5984	5998	6012	1	3	4	5	7	8	10	11	12
.78	6026	6039	6053	6067	6081	6095	6109	6124	6138	6152	1	3	4	6	7	8	10	11	12
.79	6166	6180	6194	6209	6223	6237	6252	6266	6281	6295	1	3	4	6	7	9	10	11	13
.80	6310	6324	6339	6353	6368	6383	6397	6412	6427	6442	1	3	4	6	7	9	10	12	13
.81	6457	6471	6486	6501	6516	6531	6546	6561	6577	6592	2	3	5	6	8	9	11	12	14
.82	6607	6622	6637	6653	6668	6683	6699	6714	6730	6745	2	3	5	6	8	9	11	12	14
.83	6761	6776	6792	6808	6823	6839	6855	6871	6887	6902	2	3	5	6	8	9	11	13	14
.84	6918	6934	6950	6966	6982	6998	7015	7031	7047	7063	2	3	5	6	8	10	11	13	15
.85	7079	7096	7112	7129	7145	7161	7178	7194	7211	7228	2	3	5	7	8	10	12	13	15
.86	7244	7261	7278	7295	7311	7328	7345	7362	7379	7396	2	3	5	7	8	10	12	13	15
.87	7413	7430	7447	7464	7482	7499	7516	7534	7551	7568	2	3	5	7	9	10	12	14	16
.88	7586	7603	7621	7638	7656	7674	7691	7709	7727	7745	2	4	5	7	9	11	12	14	16
.89	7762	7780	7798	7816	7834	7852	7870	7889	7907	7925	2	4	5	7	9	11	13	14	16
.90	7943	7962	7980	7998	8017	8035	8054	8072	8091	8110	2	4	6	7	9	11	13	15	17
.91	8128	8147	8166	8185	8204	8222	8241	8260	8279	8299	2	4	6	8	9	11	13	15	17
.92	8318	8337	8356	8375	8395	8414	8433	8453	8472	8492	2	4	6	8	10	12	14	15	17
.93	8511	8531	8551	8570	8590	8610	8630	8650	8670	8690	2	4	6	8	10	12	14	16	18
.94	8710	8730	8750	8770	8790	8810	8831	8851	8872	8892	2	4	6	8	10	12	14	16	18
.95	8913	8933	8954	8974	8995	9016	9036	9057	9078	9099	2	4	6	8	10	12	15	17	19
.96	9120	9141	9162	9183	9204	9226	9247	9268	9290	9311	2	4	6	8	11	13	15	17	19
.97	9333	9354	9376	9397	9419	9441	9462	9484	9506	9528	2	4	7	9	11	13	15	17	20
.98	9550	9572	9594	9616	9638	9661	9683	9705	9727	9750	2	4	7	9	11	13	16	18	20
.99	9772	9795	9817	9840	9863	9886	9908	9931	9954	9977	2	5	7	9	11	14	16	18	20

NATURAL SINES

Degree	0′ 0°.0	6′ 0°.1	12′ 0°.2	18′ 0°.3	24′ 0°.4	30′ 0°.5	36′ 0°.6	42′ 0°.7	48′ 0°.8	54′ 0°.9	Mean Differences 1	2	3	4	5
0	.0000	0017	0035	0052	0070	0087	0105	0122	0140	0157	3	6	9	12	15
1	.0175	0192	0209	0227	0244	0262	0279	0297	0314	0332	3	6	9	12	15
2	.0349	0366	0384	0401	0419	0436	0454	0471	0488	0506	3	6	9	12	15
3	.0523	0541	0558	0576	0593	0610	0628	0645	0663	0680	3	6	9	12	15
4	.0698	0715	0732	0750	0767	0785	0802	0819	0837	0854	3	6	9	12	15
5	.0872	0889	0906	0924	0941	0958	0976	0993	1011	1028	3	6	9	12	14
6	.1045	1063	1080	1097	1115	1132	1149	1167	1184	1201	3	6	9	12	14
7	.1219	1236	1253	1271	1288	1305	1323	1340	1357	1374	3	6	9	12	14
8	.1392	1409	1426	1444	1461	1478	1495	1513	1530	1547	3	6	9	12	14
9	.1564	1582	1599	1616	1633	1650	1668	1685	1702	1719	3	6	9	12	14
10	.1736	1754	1771	1788	1805	1822	1840	1857	1874	1891	3	6	9	12	14
11	.1908	1925	1942	1959	1977	1994	2011	2028	2045	2062	3	6	9	11	14
12	.2079	2096	2113	2130	2147	2164	2181	2198	2215	2232	3	6	9	11	14
13	.2250	2267	2284	2300	2317	2334	2351	2368	2385	2402	3	6	8	11	14
14	.2419	2436	2453	2470	2487	2504	2521	2538	2554	2571	3	6	8	11	14
15	.2588	2605	2622	2639	2656	2672	2689	2706	2723	2740	3	6	8	11	14
16	.2756	2773	2790	2807	2823	2840	2857	2874	2890	2907	3	6	8	11	14
17	.2924	2940	2957	2974	2990	3007	3024	3040	3057	3074	3	6	8	11	14
18	.3090	3107	3123	3140	3156	3173	3190	3206	3223	3239	3	6	8	11	14
19	.3256	3272	3289	3305	3322	3338	3355	3371	3387	3404	3	5	8	11	14
20	.3420	3437	3453	3469	3486	3502	3518	3535	3551	3567	3	5	8	11	14
21	.3584	3600	3616	3633	3649	3665	3681	3697	3714	3730	3	5	8	11	14
22	.3746	3762	3778	3795	3811	3827	3843	3859	3875	3891	3	5	8	11	14
23	.3907	3923	3939	3955	3971	3987	4003	4019	4035	4051	3	5	8	11	14
24	.4067	4083	4099	4115	4131	4147	4163	4179	4195	4210	3	5	8	11	13
25	.4226	4242	4258	4274	4289	4305	4321	4337	4352	4368	3	5	8	11	13
26	.4384	4399	4415	4431	4446	4462	4478	4493	4509	4524	3	5	8	10	13
27	.4540	4555	4571	4586	4602	4617	4633	4648	4664	4679	3	5	8	10	13
28	.4695	4710	4726	4741	4756	4772	4787	4802	4818	4833	3	5	8	10	13
29	.4848	4863	4879	4894	4909	4924	4939	4955	4970	4985	3	5	8	10	13
30	.5000	5015	5030	5045	5060	5075	5090	5105	5120	5135	3	5	8	10	13
31	.5150	5165	5180	5195	5210	5225	5240	5255	5270	5284	2	5	7	10	12
32	.5299	5314	5329	5344	5358	5373	5388	5402	5417	5432	2	5	7	10	12
33	.5446	5461	5476	5490	5505	5519	5534	5548	5563	5577	2	5	7	10	12
34	.5592	5606	5621	5635	5650	5664	5678	5693	5707	5721	2	5	7	10	12
35	.5736	5750	5764	5779	5793	5807	5821	5735	5850	5864	2	5	7	10	12
36	.5878	5892	5906	5920	5934	5948	5962	5976	5990	6004	2	5	7	9	12
37	.6018	6032	6046	6060	6074	6088	6101	6115	6129	6143	2	5	7	9	12
38	.6157	6170	6184	6198	6211	6225	6239	6252	6266	6280	2	5	7	9	11
39	.6293	6307	6320	6334	6347	6361	6374	6388	6401	6414	2	4	7	9	11
40	.6428	6441	6455	6468	6481	6494	6508	6521	6534	6547	2	4	7	9	11
41	.6561	6574	6587	6600	6613	6626	6639	6652	6665	6678	2	4	7	9	11
42	.6691	6704	6717	6730	6743	6756	6769	6782	6794	6807	2	4	6	9	11
43	.6820	6833	6845	6858	6871	6884	6896	6909	6921	6934	2	4	6	8	11
44	.6947	6959	6972	6984	6997	7009	7022	7034	7046	7059	2	4	6	8	10

NATURAL SINES

Degree	0′ 0°.0	6′ 0°.1	12′ 0°.2	18′ 0°.3	24′ 0°.4	30′ 0°.5	36′ 0°.6	42′ 0°.7	48′ 0°.8	54′ 0°.9	Mean Differences 1	2	3	4	5
45	.7071	7083	7096	7108	7120	7133	7145	7157	7169	7181	2	4	6	8	10
46	.7193	7206	7218	7230	7242	7254	7266	7278	7290	7302	2	4	6	8	10
47	.7314	7325	7337	7349	7361	7373	7385	7396	7408	7420	2	4	6	8	10
48	.7431	7443	7455	7466	7478	7490	7501	7513	7524	7536	2	4	6	8	10
49	.7547	7558	7570	7581	7593	7604	7615	7627	7638	7649	2	4	6	8	9
50	.7660	7672	7683	7694	7705	7716	7727	7738	7749	7760	2	4	6	7	9
51	.7771	7782	7793	7804	7815	7826	7837	7848	7859	7869	2	4	5	7	9
52	.7880	7891	7902	7912	7923	7934	7944	7955	7965	7976	2	4	5	7	9
53	.7986	7997	8007	8018	8028	8039	8049	8059	8070	8080	2	3	5	7	9
54	.8090	8100	8111	8121	8131	8141	8151	8161	8171	8181	2	3	5	7	8
55	.8192	8202	8211	8221	8231	8241	8251	8261	8271	8281	2	3	5	7	8
56	.8290	8300	8310	8320	8329	8339	8348	8358	8368	8377	2	3	5	6	8
57	.8387	8396	8406	8415	8425	8434	8443	8453	8462	8471	2	3	5	6	8
58	.8480	8490	8499	8508	8517	8526	8536	8545	8554	8563	2	3	5	6	8
59	.8572	8581	8590	8599	8607	8616	8625	8634	8643	8652	1	3	4	6	7
60	.8660	8669	8678	8686	8695	8704	8712	8721	8729	8738	1	3	4	6	7
61	.8746	8755	8763	8771	8780	8788	8796	8805	8813	8821	1	3	4	6	7
62	.8829	8838	8846	8854	8862	8870	8878	8886	8894	8902	1	3	4	5	7
63	.8910	8918	8926	8934	8942	8949	8957	8965	8973	8980	1	3	4	5	6
64	.8988	8996	9003	9011	9018	9026	9033	9041	9048	9056	1	3	4	5	6
65	.9063	9070	9078	9085	9092	9100	9107	9114	9121	9128	1	2	4	5	6
66	.9135	9143	9150	9157	9164	9171	9178	9184	9191	9198	1	2	3	5	6
67	.9205	9212	9219	9225	9232	9239	9245	9252	9259	9265	1	2	3	4	6
68	.9272	9278	9285	9291	9298	9304	9311	9317	9323	9330	1	2	3	4	5
69	.9336	9342	9348	9354	9361	9367	9373	9379	9385	9391	1	2	3	4	5
70	.9397	9403	9409	9415	9421	9426	9432	9438	9444	9449	1	2	3	4	5
71	.9455	9461	9466	9472	9478	9483	9489	9494	9500	9505	1	2	3	4	5
72	.9511	9516	9521	9527	9532	9537	9542	9548	9553	9558	1	2	3	3	4
73	.9563	9568	9573	9578	9583	9588	9593	9598	9603	9608	1	2	2	3	4
74	.9613	9617	9622	9627	9632	9636	9641	9646	9650	9655	1	2	2	3	4
75	.9659	9664	9668	9673	9677	9681	9686	9690	9694	9699	1	1	2	3	4
76	.9703	9707	9711	9715	9720	9724	9728	9732	9736	9740	1	1	2	3	3
77	.9744	9748	9751	9755	9759	9763	9767	9770	9774	9778	1	1	2	3	3
78	.9781	9785	9789	9792	9796	9799	9803	9806	9810	9813	1	1	2	2	3
79	.9816	9820	9823	9826	9829	9833	9836	9839	9842	9845	1	1	2	2	3
80	.9848	9851	9854	9857	9860	9863	9866	9869	9871	9874	0	1	1	2	2
81	.9877	9880	9882	9885	9888	9890	9893	9895	9898	9900	0	1	1	2	2
82	.9903	9905	9907	9910	9912	9914	9917	9919	9921	9923	0	1	1	2	2
83	.9925	9928	9930	9932	9934	9936	9938	9940	9942	9943	0	1	1	1	2
84	.9945	9947	9949	9951	9952	9954	9956	9957	9959	9960	0	1	1	1	2
85	.9962	9963	9965	9966	9968	9969	9971	9972	9973	9974	0	0	1	1	1
86	.9976	9977	9978	9979	9980	9981	9982	9983	9984	9985	0	0	1	1	1
87	.9986	9987	9988	9989	9990	9990	9991	9992	9993	9993	0	0	0	1	1
88	.9994	9995	9995	9996	9996	9997	9997	9997	9998	9998	0	0	0	0	0
89	.9998	9999	9999	9999	9999	1.000	1.000	1.000	1.000	1.000	0	0	0	0	0
90	1.000														

NATURAL TANGENTS

Degree	0′ 0°.0	6′ 0°.1	12′ 0°.2	18′ 0°.3	24′ 0°.4	30′ 0°.5	36′ 0°.6	42′ 0°.7	48′ 0°.8	54′ 0°.9	Mean Differences 1	2	3	4	5
0	.0000	0017	0035	0052	0070	0087	0105	0122	0140	0157	3	6	9	12	15
1	.0175	0192	0209	0227	0244	0262	0279	0297	0314	0332	3	6	9	12	15
2	.0349	0367	0384	0402	0419	0437	0454	0472	0489	0507	3	6	9	12	15
3	.0524	0542	0559	0577	0594	0612	0629	0647	0664	0682	3	6	9	12	15
4	.0699	0717	0734	0752	0769	0787	0805	0822	0840	0857	3	6	9	12	15
5	.0875	0892	0910	0928	0945	0963	0981	0998	1016	1033	3	6	9	12	15
6	.1051	1069	1086	1104	1122	1139	1157	1175	1192	1210	3	6	9	12	15
7	.1228	1246	1263	1281	1299	1317	1334	1352	1370	1388	3	6	9	12	15
8	.1405	1423	1441	1459	1477	1495	1512	1530	1548	1566	3	6	9	12	15
9	.1584	1602	1620	1638	1655	1673	1691	1709	1727	1745	3	6	9	12	15
10	.1763	1781	1799	1817	1835	1853	1871	1890	1908	1926	3	6	9	12	15
11	.1944	1962	1980	1998	2016	2035	2053	2071	2089	2107	3	6	9	12	15
12	.2126	2144	2162	2180	2199	2217	2235	2254	2272	2290	3	6	9	12	15
13	.2309	2327	2345	2364	2382	2401	2419	2438	2456	2475	3	6	9	12	15
14	.2493	2512	2530	2549	2568	2586	2605	2623	2642	2661	3	6	9	12	16
15	.2679	2698	2717	2736	2754	2773	2792	2811	2830	2849	3	6	9	13	16
16	.2867	2886	2905	2924	2943	2962	2981	3000	3019	3038	3	6	9	13	16
17	.3057	3076	3096	3115	3134	3153	3172	3191	3211	3230	3	6	10	13	16
18	.3249	3269	3288	3307	3327	3346	3365	3385	3404	3424	3	6	10	13	16
19	.3443	3463	3482	3502	3522	3541	3561	3581	3600	3620	3	7	10	13	16
20	.3640	3659	3679	3699	3719	3739	3759	3779	3799	3819	3	7	10	13	17
21	.3839	3859	3879	3899	3919	3939	3959	3979	4000	4020	3	7	10	13	17
22	.4040	4061	4081	4101	4122	4142	4163	4183	4204	4224	3	7	10	14	17
23	.4245	4265	4286	4307	4327	4348	4369	4390	4411	4431	3	7	10	14	17
24	.4452	4473	4494	4515	4536	4557	4578	4599	4621	4642	4	7	11	14	18
25	.4663	4684	4706	4727	4748	4770	4791	4813	4834	4856	4	7	11	14	18
26	.4877	4899	4921	4942	4964	4986	5008	5029	5051	5073	4	7	11	15	18
27	.5095	5117	5139	5161	5184	5206	5228	5250	5272	5295	4	7	11	15	18
28	.5317	5340	5362	5384	5407	5430	5452	5475	5498	5520	4	8	11	15	19
29	.5543	5566	5589	5612	5635	5658	5681	5704	5727	5750	4	8	12	15	19
30	.5774	5797	5820	5844	5867	5890	5914	5938	5961	5985	4	8	12	16	20
31	.6009	6032	6056	6080	6104	6128	6152	6176	6200	6224	4	8	12	16	20
32	.6249	6273	6297	6322	6346	6371	6395	6420	6445	6469	4	8	12	16	20
33	.6494	6519	6544	6569	6594	6619	6644	6669	6694	6720	4	8	13	17	21
34	.6745	6771	6796	6822	6847	6873	6899	6924	6950	6976	4	9	13	17	21
35	.7002	7028	7054	7080	7107	7133	7159	7186	7212	7239	4	9	13	18	22
36	.7265	7292	7319	7346	7373	7400	7427	7454	7481	7508	5	9	14	18	23
37	.7536	7563	7590	7618	7646	7673	7701	7729	7757	7785	5	9	14	18	23
38	.7813	7841	7869	7898	7926	7954	7983	8012	8040	8069	5	9	14	19	24
39	.8098	8127	8156	8185	8214	8243	8273	8302	8332	8361	5	10	15	20	24
40	.8391	8421	8451	8481	8511	8541	3571	8601	8632	8662	5	10	15	20	25
41	.8693	8724	8754	8785	8816	8847	8878	8910	8941	8972	5	10	16	21	26
42	.9004	9036	9067	9099	9131	9163	9195	9228	9260	9293	5	11	16	21	27
43	.9325	9358	9391	9424	9457	9490	9523	9556	9590	9623	6	11	17	22	28
44	.9657	9691	9725	9759	9793	9827	9861	9896	9930	9965	6	11	17	23	29

NATURAL TANGENTS

Degree	0′ 0°.0	6′ 0°.1	12′ 0°.2	18′ 0°.3	24′ 0°.4	30′ 0°.5	36′ 0°.6	42′ 0°.7	48′ 0°.8	54′ 0°.9	Mean Differences				
											1	2	3	4	5
45	1.0000	0035	0070	0105	0141	0176	0212	0247	0283	0319	6	12	18	24	30
46	1.0355	0392	0428	0464	0501	0538	0575	0612	0649	0686	6	12	18	25	31
47	1.0724	0761	0799	0837	0875	0913	0951	0990	1028	1067	6	13	19	25	32
48	1.1106	1145	1184	1224	1263	1303	1343	1383	1423	1463	7	13	20	27	33
49	1.1504	1544	1585	1626	1667	1708	1750	1792	1833	1875	7	14	21	28	34
50	1.1918	1960	2002	2045	2088	2131	2174	2218	2261	2305	7	14	22	29	36
51	1.2349	2393	2437	2482	2527	2572	2617	2662	2708	2753	8	15	23	30	38
52	1.2799	2846	2892	2938	2985	3032	3079	3127	3175	3222	8	16	24	31	39
53	1.3270	3319	3367	3416	3465	3514	3564	3613	3663	3713	8	16	25	33	41
54	1.3764	3814	3865	3916	3968	4019	4071	4124	4176	4229	9	17	26	34	43
55	1.4281	4335	4388	4442	4496	4550	4605	4659	4715	4770	9	18	27	36	45
56	1.4826	4882	4938	4994	5051	5108	5166	5224	5282	5340	10	19	29	38	48
57	1.5399	5458	5517	5577	5637	5697	5757	5818	5880	5941	10	20	30	40	50
58	1.6003	6066	6128	6191	6255	6319	6383	6447	6512	6577	11	21	32	43	53
59	1.6643	6709	6775	6842	6909	6977	7045	7113	7182	7251	11	23	34	45	56
60	1.7321	7391	7461	7532	7603	7675	7747	7820	7893	7966	12	24	36	48	60
61	1.8040	8115	8190	8265	8341	8418	8495	8572	8650	8728	13	26	38	51	64
62	1.8807	8887	8967	9047	9128	9210	9292	9375	9458	9542	14	27	41	55	68
63	1.9626	9711	9797	9883	9970	2.0057	2.0145	2.0233	2.0323	2.0413	15	29	44	58	73
64	2.0503	0594	0686	0778	0872	0965	1060	1155	1251	1348	16	31	47	63	78
65	2.1445	1543	1642	1742	1842	1943	2045	2148	2251	2355	17	34	51	68	85
66	2.2460	2566	2673	2781	2889	2998	3109	3220	3332	3445	18	37	55	73	92
67	2.3559	3673	3789	3906	4023	4142	4262	4383	4504	4627	20	40	60	79	99
68	2.4751	4876	5002	5129	5257	5386	5517	5649	5782	5916	22	43	65	87	108
69	2.6051	6187	6325	6464	6605	6746	6889	7034	7179	7326	24	47	71	95	119
70	2.7475	7625	7776	7929	8083	8239	8397	8556	8716	8878	26	52	78	104	131
71	2.9042	9208	9375	9544	9714	9887	3.0061	3.0237	3.0415	3.0595	29	58	87	116	145
72	3.0777	0961	1146	1334	1524	1716	1910	2106	2305	2506	32	64	96	129	161
73	3.2709	2914	3122	3332	3544	3759	3977	4197	4420	4646	36	72	108	144	180
74	3.4874	5105	5339	5576	5816	6059	6305	6554	6806	7062	41	81	122	163	204
75	3.7321	7583	7848	8118	8391	8667	8947	9232	9520	9812	46	93	139	186	232
76	4.0108	0408	0713	1022	1335	1653	1976	2303	2635	2972	53	107	160	213	267
77	4.3315	3662	4015	4374	4737	5107	5483	5864	6252	6646					
78	4.7046	7453	7867	8288	8716	9152	9594	5.0045	5.0504	5.0970			Mean differences		
79	5.1446	1929	2422	2924	3435	3955	4486	5026	5578	6140			cease to be		
80	5.6713	7297	7894	8502	9124	9758	6.0405	6.1066	6.1742	6.2432			sufficiently		
81	6.3138	3859	4596	5350	6122	6912	7720	8548	9395	7.0264			accurate.		
82	7.1154	2066	3002	3962	4947	5958	6996	8062	9158	8.0285					
83	8.1443	2636	3863	5126	6427	7769	9152	9.0579	9.2052	9.3572					
84	9.5144	9.677	9.845	10.02	10.20	10.39	10.58	10.78	10.99	11.20					
85	11.43	11.66	11.91	12.16	12.43	12.71	13.00	13.30	13.62	13.95					
86	14.30	14.67	15.06	15.46	15.89	16.35	16.83	17.34	17.89	18.46					
87	19.08	19.74	20.45	21.20	22.02	22.90	23.86	24.90	26.03	27.27					
88	28.64	30.14	31.82	33.69	35.80	38.19	40.92	44.07	47.74	52.08					
89	57.29	63.66	71.62	81.85	95.49	114.6	143.2	191.0	286.5	573.0					
90	∞														